Physics and Necessity

Rationalist Pursuits from the Cartesian Past to the Quantum Present

OLIVIER DARRIGOL

OXFORD
UNIVERSITY PRESS

Great Clarendon Street, Oxford, OX2 6DP,
United Kingdom

Oxford University Press is a department of the University of Oxford.
It furthers the University's objective of excellence in research, scholarship,
and education by publishing worldwide. Oxford is a registered trade mark of
Oxford University Press in the UK and in certain other countries

Published in the United States of America by Oxford University Press
198 Madison Avenue, New York, NY 10016, United States of America

British Library Cataloguing in Publication Data
Data available

Library of Congress Control Number: 2013953857

ISBN 978-0-19-871288-6

Printed in Great Britain by
Clays Ltd, St Ives plc

À la mémoire de mes parents,

Simone Darrigol, née Heinzelmann (1919–2011)
et Guy Darrigol (1913–2007)

PREFACE

We have found a strange foot-print on the shores of the unknown. We have devised profound theories, one after another, to account for its origin. At last, we have succeeded in reconstructing the creature that made the foot-print. And Lo! It is our own.[1](Arthur Eddington, 1920)

What really interests me is whether God could have made the world any differently; in other words, whether the demand for logical simplicity leaves any freedom at all.[2] (Albert Einstein, late 1940s, as remembered by his assistant Ernst Straus)

Our best physical theories explain an astonishing diversity of phenomena on the basis of a few general definitions and laws. The power and simplicity of the premises invite rational justification: is it possible to arrive at such theories by reasoning only, without consulting experiment? Since the early modern period, a significant number of philosophers and physicists have indulged in this kind of rationalism. Some of them have constructed new theories; others have justified older theories in this manner. They may have claimed to be using reason only, or they may have conceded their reliance on broad empirical assumptions. At any rate, they believed they had proven the rational necessity of some physical theory. Were they the victims of excessive intellectual self-confidence? Or did they truly prove the inevitability of some of our physical laws and theories?

When historians of physics encounter rationalist arguments, they tend to regard them as short-lived illusions implying unconscious assumptions, cultural biases, and even errors of reasoning. At best they recognize that these arguments function as useful motors of research. Today's physicists have a more ambiguous attitude. They may profess empiricism and yet secretly flirt with rationalism. Although they commonly assert that experience is the only criterion of truth, some of them engage in a quest for "natural axioms" from which the most fundamental theories would follow. Philosophers of science have a divided opinion. Most of them are empiricists of some kind and they are eager, as Ernst Mach was in the late nineteenth century, to denounce flaws in allegedly rational deductions. Yet some of them emphasize the role of a priori considerations in physical theory, often in modified forms of Immanuel Kant's transcendental idealism.[3]

As a historian of physics, I have come across a large variety of rational necessity arguments. Some of them, for instance Jean le Rond d'Alembert's deduction of the principles of dynamics or Hermann Helmholtz's deduction of the locally Euclidean character of space,

---

[1]Eddington 1920, p. 182.

[2]Einstein, in Straus 1956, p. 72. The complete citation ends with the parenthetical remark: "Note, however, that this is much less the case than one would have thought fifty years ago."

[3]In defense of the physicists' hybridization of empiricism and rationalism, see Scheibe 1994. On the recent revival of transcendentalism among philosophers, see, e.g., Bitbol, Kerszberg, and Petitot 2009.

are truly ingenious and they highly impressed me. It seemed to me that these arguments could resist later criticism and beautifully captured the necessity of some basic laws, although the implied kind of necessity was not purely rational. This experience motivated the more systematic inquiry presented in this book.

The least we can demand of a rational deduction is that it should provide a small number of simple hypotheses or axioms from which all the laws of the theory can be derived. This in itself is not a minor achievement, because a theory is rarely given to us in a clean deductive manner. Claims of necessity, however, imply more than a restructuration of the theory: they include arguments for the naturalness or inevitability of the premises. The character of these arguments varies. In a rough order of decreasing rationalism, the premises may be theological, metaphysical, transcendental, conceptual, or empirical. At one end of this spectrum, the theory is derived in a purely a priori manner and it is regarded as immune to experimental refutation. At the other end, the theory is deduced from a small number of hypotheses whose truth can be directly established by observation or by experience.

In conformity with the historian's pessimism, rational deductions rarely achieve as much as their authors claim: even when they do not lead to false theories, they are flawed in various manners. The theological, metaphysical, and transcendental arguments, no matter how popular they were in the past, now appear to rest on unwarranted or unconscious assumptions. They may, nonetheless, have revealed important structural properties of the derived theory. For instance, even though Pierre Louis Moreau de Maupertuis introduced the principle of least action for theological reasons hard to maintain nowadays, no one would deny the ability of this principle to simplify the deductive apparatus of classical mechanics. The more moderate arguments, those purporting to be purely conceptual or even a bit empirical, often imply gaps in reasoning and most of them have been subsequently criticized or refuted. However, in a few felicitous cases the rational deduction can be mended to provide a genuinely convincing proof of necessity. These cases fall under the category of what I call *comprehensibility arguments*, in homage to Helmholtz's idea that nature, if it is at all comprehensible, must a priori comply with certain theoretical structures.[4]

Comprehensibility arguments share with Kant's transcendental deductions the basic reliance on a priori conditions of experience. They answer the question: how should the world be in order that we should be able to make sense of it? Yet, comprehensibility criteria depart from Kant's transcendental apparatus in several ways. They are more diverse as they may involve measurability, causality, homogeneity, correspondence, reduction, decoupling of scales, and more. They are always tentative.[5] They have not always been accepted: most of them were foreign to Aristotelian natural philosophy. They are not forever imposed on us: they may someday fail when a wider domain of experience is explored. For this reason, they do not carry any ontological weight: the entities assumed in the comprehensibility requirements need not be the most fundamental components of the theoretical representation of phenomena.

---

[4]On Helmholtz's *Begreiflichkeit der Natur*, cf. Krüger 1994, Darrigol 1994, Jurkowitz 2010.

[5]According to Gregor Schiemann, the tentative character of the Helmholtzian comprehensibility of nature is a symptom of a broader, époque-making hypothesizing of truth in science (Schiemann 1997, 2009).

This multiplicity and adaptability of comprehensibility arguments makes them similar to the relativized varieties of transcendentalism proposed by the Marburg school of neo-Kantianism, by its member Ernst Cassirer, and by a few living philosophers of physics. There are important differences, however. Whereas these philosophers try to identify the constitutive principles of physics by mostly philosophical means, I try to find them by improving on the most successful necessity arguments found in the history of physics. A second difference is in the way the constitutive principles are chosen at a given stage of the history of physics. Whereas in relativized transcendentalism this choice is either free or determined by contingent cultural conditions, comprehensibility conditions derive from the success of very broad empirical assumptions in a given domain of experience. A last difference concerns the purpose of the constitutive principles. In relativized transcendentalism, these principles never completely determine a physical theory; they only define a frame that needs to be filled with empirical data and laws. In contrast, some comprehensibility arguments purport to completely determine a theory.[6]

The comprehensibility conditions discussed in this book indeed have far-reaching consequences when applied to idealized physical systems. It will be seen that some of our most successful theories or theoretical frames, including Euclidean geometry, classical mechanics, pseudo-Riemannian spacetime, Maxwell's electromagnetic theory, Einstein's gravitation theory, and quantum mechanics, can be justified in this manner. To be true, these deductions are not purely rational: they rely on empirically refutable assumptions of comprehensibility, as well as on empirical background knowledge inspiring the definition of the ideal systems to which these assumptions are applied. They are not completely rigorous, and there is always a remnant of subjectivity in judging the premises of comprehensibility. They nonetheless make it very difficult to imagine, in given domains of experience, theories different from or not equivalent to those already known to be successful.

This is not to say that natural philosophers, had they been cleverer, would have discovered all our best theories without consulting experiment. The comprehensibility criteria on which the best rational deductions are based usually come to us after the fact, when we seek deeper justifications of theories we have earlier reached in a more groping, experimentally guided manner. Time may elapse before the full legitimacy of these criteria is recognized. For instance, it is only in the course of the eighteenth century that the measurability of physical phenomena began to be generally assumed; and the analysis of concrete phenomena into simpler, homogenous, idealized processes is a conquest of Galilean physics.

If necessity arguments do no more than justify already known theories and if the premises of these arguments are doomed to be someday refuted, do they still serve some useful purpose? Worse: is not there a risk that such arguments would block the progress of physics? To the first question it can be answered that the refutation of the comprehensibility premises of a good necessity argument is only partial. The comprehensibility criteria that lead to classical mechanics did not become futile when the failure of this theory to

---

[6]I return to the comparison between Helmholtzian comprehensibility and relativized transcendentalism in Chapter 9, pp. 343–5.

describe atomic motion was recognized. Rather, it became clear that these criteria applied only to a limited domain of experience. In brief, the comprehensibility of nature is a regional notion. This being recognized, comprehensibility arguments seriously improve our understanding of the structure and contents of the theory they imply. For instance, the truth of Pythagoras' theorem in Euclidean geometry becomes palpable when we learn from Helmholtz that Euclid's axioms express the possibility of measuring space through freely mobile rigid bodies. Or the Hilbert-space structure of quantum mechanics becomes far less artificial when it is derived from natural assumptions about the statistical correlations between discrete measurements.

As for the danger of fossilizing physics, it should not exist if the regional character of comprehensibility assumptions is recognized from the start. Comprehensibility arguments may even help to design alternatives to an empirically endangered theory by relaxing one of the comprehensibility assumptions that served to justify it and by requiring the new theory to contain its predecessor as an approximation. It may still be thought that necessity arguments rigidify a theory by privileging one of its possible forms, whereas the multiplicity of forms or formulations is known to be cognitively and heuristically important.[7] This criticism does not apply to the comprehensibility arguments, because these arguments, being essentially based on empirical accessibility, only imply the necessity of the empirical predictions of the deduced theory. They do not imply the necessity of the form of the theory that most directly derives from them. Although they admittedly favor this form, they also help us recognize the contingency of past forms and free us to conceive newer forms. For instance, d'Alembert's rational derivation of the laws of mechanics was a good cure against the pre-existing ontology of forces and opened the possibility of multiple formulations of mechanics. Or Helmholtz's discussion of the foundations of geometry, originally meant to corroborate Euclidean geometry, served to open the spectrum of conceivable physical geometries.

Whether or not one accepts the soundness of necessity arguments, their historical import is undeniable. They at least convinced their authors of the truth of the derived theories. They sometimes produced new theories or new theoretical structures. Their genesis informs us on the style and cultural immersion of the author. Their adoption or their rejection may be used to characterize schools of thought. In addition, their ingeniousness is an abundant source of pleasure for the connoisseur. This book offers a rich sample of such arguments, selected according to their historical impact or to their intrinsic merit.[8] It deals successively with a few important theories or theoretical frames: mechanics, general principles of mechanical origin, physical geometry, spacetime, numbers and math, classical field theories, and quantum mechanics. The order of the chapters roughly corresponds to the chronology of necessity arguments. For example, the chapters on mechanics come first because proofs of necessity of some laws of mechanics are as old as Greek science; and the chapters on the more fundamental topics of geometry, numbers, and mathematics come fairly late in the

[7]On the cognitive importance of multiple formulations, cf. Vorms 2011. On their heuristic value, cf. Feynman 1965, pp. 53–5; de Courtenay 1999.

[8]Of course, I could not treat every necessity argument of interest. Among my many omissions are derivations of the dimension of space, or uses of the anthropic principle. On the latter, cf. Kragh 2011, Chap. 9.

book because no one seriously attempted to justify the axioms of physical geometry or the mathematical character of physical theory before the nineteenth century. For each theory or chapter, a historical survey of the relevant arguments precedes a critical analysis. This analysis should not be confused with history proper; its purpose is to stimulate the construction of necessity arguments that resist the corrosion of time. Each chapter can be read independently of the others, except perhaps the chapter on spacetime which should best be read together with the chapter on space.

Chapter 1 is a history of necessity arguments in classical mechanics, including the seventeenth-century opposition between René Descartes's rationalism and Isaac Newton's empiricism, the rationalism of earlier statics since Archimedes, the varieties of rationalism defended by Gottfried Wilhelm Leibniz, Leonhard Euler, and Jean le Rond d'Alembert in the eighteenth century, and the nineteenth-century persistence of a few deductive arguments against the empiricist tide. Some of the rational derivations of the laws of mechanics relied on theological arguments; others were of a more conceptual nature. In Chapter 2, which is again about classical mechanics, I combine some ingredients of the conceptual arguments (the impossibility of perpetual motion, the causal relation between force and motion, the relativity principle, and what I call the secular principle) with the definitions of four classes of ideal mechanical systems (connected systems, particles acting at a distance, continuous media, colliding particles) in order to derive the laws of equilibrium and motion of these systems. The usual laws of mechanics are retrieved in each case. I conclude that classical mechanics is the only theory of motion complying with a few natural requirements on the comprehensibility of motion at the macroscopic scale.

Chapter 3 is devoted to a few general principles that have received mechanical or dynamical justifications: the principle of energy conservation, the principle of least action, and the two laws of thermodynamics. I first recount how the evolution of mechanical reductionism and a new emphasis on conversion processes led to the energy principle. Then I discuss the reductionist arguments for the necessity of this principle and the more satisfactory arguments by Ernst Mach and Max Planck based on the impossibility of perpetual motion. I next show how the principle of least action, as a general principle of physics, resulted from the blackboxing of mechanism, and I deplore the lack of a non-reductionist foundation of this principle. The two laws of thermodynamics, discussed in the last section of this chapter, have a not purely rational but still persuasive kind of necessity as they express the impossibility of processes or devices never seen in nature. They receive further justification through statistical mechanics, as long as we accept the possibility of a dynamical reduction compatible with the existence and uniqueness of thermodynamic equilibrium under given macroscopic conditions. Altogether, the faith of physicists in the general, frame-like quality of thermodynamical and thermostatistical principles seems reasonable.

Chapter 4 deals with the foundations of geometry, with special emphasis on Helmholtz's approach to this problem in the late 1860s. I first recall how the ancient belief in the apodictic truth of the axioms of geometry came under attack at the turn of the eighteenth and nineteenth century, making room for the new geometries of Carl Friedrich Gauss, János Bolyai, Nikolai Ivanovich Lobachevski, and Bernhard Riemann. This opening of possibilities raised the question of the necessary structural kernel of any physical geometry. I give a detailed account of Helmholtz's answer to this question, according to which the locally Euclidean character of space follows from its measurability by freely mobile rigid bodies.

There are difficulties in this argument, due to a seeming circularity in the definition of rigid bodies and to the strictness of the assumed rigidity, which requires spaces of constant curvature. I show how these difficulties can be circumvented, leading to the necessity of the Riemannian structure of space (with any curvature) for measurable space.

Chapter 5 is an attempt to extend Helmholtz's theory of space to spacetime, preceded by a history of derivations of the structure of spacetime. I first recall how the Newtonian concept of absolute time gave way to Henri Poincaré's and Albert Einstein's concepts of apparent or relative time and to relativity theory in Einstein's form, based on the relativity principle and the light principle; and I describe some of the heuristics that led Einstein to general relativity. I mention Woldemar von Ignatowski's group-theoretical derivation of the special theory (1910), which has similarities with Helmholtz's doctrine of space; and I describe Alfred Robb's attempt to found Minkowskian spacetime on a conical order of events (1914), with the conclusion that Robb's axioms are too numerous and too artificial to pass for a rational foundation of relativity theory. A large section of this chapter is devoted to three ambitious and highly seductive attempts at rationally deriving the structure of spacetime and more. The first is Hermann Weyl's (1918), based on the intuition of a true infinitesimal geometry in which a connection is needed not only to compare the direction of vectors in successive tangent spaces but also to compare their lengths. The second is Arthur Eddington's (1923), in which the affine connection comes first and the observed metric is a derived construct. The third is the more Helmholtzian approach of Jürgen Ehlers, Felix Pirani, and Alfred Schild (1972), in which the Weyl structure of spacetime is derived from natural axioms for geodetic surveying by light signals and free-falling particles. These three impressive derivations share the defect of giving up a basic requirement of measurability: the independence of measuring standards on their history. This is why in the last section I develop a more Helmholtzian approach in which the possibility of optically controlling the rigidity of small bodies is assumed from the start. The end result is the pseudo-Riemannian, locally Minkowskian manifold of spacetime.

Chapter 6 addresses the broader question of the necessity of mathematics in physics. It begins with Descartes's and Kant's answers, completed by a historical reminder of the eighteenth-century progress in the mathematization of nature. It goes on with a historical sketch of the ways in which quantitative relations were expressed in physics, first through the Euclidean theory of ratios, then through numerical equations, and lastly through relations between dimensioned quantities. The next section is an analysis of Helmholtz's memoir of 1887 on counting and measuring, in which the arithmetic structure of physical theories is associated with the satisfaction of measurability criteria. The remainder is devoted to Poincaré's explanations of the success of mathematical analysis and group theory in theoretical physics. According to the French mathematician, idealized measurement is the source (though not the foundation) of the mathematical concept of continuum, and this genesis partly explains the success of mathematical analysis in physical theory. As for group theory, it is the mathematical tool for combining elementary operations, as is needed in the resolution of complex or extended phenomena into elementary phenomena. The chapter closes with an assessment of the relative merits of the ontological, transcendental, empiricist, and Poincarean views on the success of mathematics in physics.

Chapter 7 develops the consequences of a specific condition of measurability for fields, namely and roughly: the physically relevant properties of fields should all be testable by point-like particles. I call field theories satisfying this condition *Faradayan theories*, because Michael Faraday implicitly used it in his description of fields through lines of force. For such theories, and if the field dynamics derives from a Minkowski-invariant action principle, the only possible field theories are Nordström's theory for scalar fields, Maxwell's electromagnetic theory for vector fields, and Einstein's gravitation theory for tensor fields. Nordström's theory can be eliminated by slightly sharpening the measurability condition. In the tensor case, the assumed Minkowskian metric is a mere fiction, as the tensor field affects geodetic measurements in a manner compatible with the pseudo-Riemannian metric of general relativity. A further section of this chapter explains the relationship of the Faradayan derivation of Einstein's equations with the somewhat similar derivations by Suraj Gupta (1954), Richard Feynman (1963), and Stanley Deser (1970). Lastly and most daringly, I show that a sharpened Faradayan principle and the action principle directly lead to Einstein's theory of gravitation and electromagnetism when applied to a pre-metrical continuum of events.

Chapter 8 is devoted to several attempts at showing the necessity of quantum mechanics. I first examine whether necessity arguments can be extracted from the history of quantum mechanics, both on the matrix-mechanics side and on the wave-mechanics side. Then I recount the mathematical discovery (1979) that the only one-parameter deformation of the Lie algebra of classical Hamiltonian mechanics is the Moyal algebra, which is isomorphic to the algebra of infinitesimal evolutions of quantum mechanics. Next, I retrace some of the history of quantum logic as an attempt to justify the density-matrix representation of quantum states and the unitary evolution of these states by a natural logic of Yes–No experimental questions. Lastly, I discuss Lucien Hardy's (2001) and more recent proposals of "natural axioms" for statistical measurement correlations, leading again to the matrix-density formulation of quantum mechanics. To conclude this chapter, I ponder the extent to which these various arguments prove the necessity of quantum mechanics.

In all of these chapters, I have taken the unusual stand that some of the necessity arguments propounded by physicists from Euler to the present are valid or at least can be made valid after some improvements. Most of these arguments rely on falsifiable assumptions about the way in which phenomena are comprehensible. They therefore resist the traditional attacks against pure rationalism. They may seem more vulnerable to the objection that their deployment implies notions that are incompatible with any reasonable conception of physical theory. For instance, the Helmholtzian reliance on measurability criteria could be rejected as a naive operationalism in which the definition of physical quantities by concrete operations would determine their theoretical interrelations. In Chapter 9, the last of this book, I defend a view of physical theory that counters such potential objections in two manners: by allowing idealized measurements whose concrete realization may vary, and by integrating the variety of inter-theoretical connections used in comprehensibility arguments. Essential in this regard are the notion of *interpretive schemes* linking the symbolic universe of the theory to concrete experiments, and the notion of *modules* defined as components of a theory that are themselves theories with a different domain of application. The modular structure of theories appears to be necessary to the effective application, comparison, construction, and communication of theories. In addition, it enables us to conceive the virtual applications and the correspondence relations implied in comprehensibility arguments.

One of the beauties of comprehensibility arguments is the contrast between the naturalness of the comprehensibility requirements and the abstractness or artificialness of the central concepts and laws of the derived theory. This is most striking in the case of quantum mechanics, whose state concept belongs to the farfetched mathematical construct of Hilbert spaces. There is a price to this beauty: the mathematics needed to cover the distance from the comprehensibility criteria to the usual formulation of a theory can be non-trivial. For this reason, I have endeavored to provide demonstrations of all important points with the simplest mathematics possible. In some cases, especially those regarding quantum mechanics, the original deductions found in the literature require a kind of mathematics (for instance, lattice theory or cohomology) that is unfamiliar to most physicists. I have replaced them with more elementary demonstrations. For the sake of brevity and legibility, I have adopted a moderate level of rigor, leaving definitional details to the more mathematically inclined readers.

For each of the necessity arguments discussed in this book, my accounts are meant to be full enough to profit physicists and philosophers interested in the foundations of physics, as well as historians interested in the ever-going competition between rationalism and empiricism in science. Some of my readers will think I have overestimated the success of rational derivations, others that I have underestimated it. A few may complain that I have ignored other cases and possibilities. My aim, however, is not to give an exhaustive and incontrovertible account of all rationalist enterprises in physics. Rather, I want to enliven a kind of argument that is too frequently brushed away on naive empiricist grounds. Hopefully, my reader will see the beauty and fertility of a moderate rationalism in which the necessity of some of our theories is derived from the contingent possibility of certain ways of understanding the physical world.

While preparing this work, I have benefited from the support of the REHSEIS and SPHERE research teams of the Centre National de la Recherche Scientifique, from the warm encouragements of their directors David Rabouin and Pascal Crozet, and from the hospitality of Michèle Leduc at the Institut Francilien de Recherche sur les Atomes Froids. The directors of UC-Berkeley's Office for History of Science and Technology, Cathryn Carson and Massimo Mazzotti, opened to me wonderful intellectual and documentary resources. Guido Bacciagaluppi, Harvey Brown, and Jan Lacki generously commented on portions of this text and suggested significant improvements. The criticism of a few anonymous reviewers helped me sharpen my arguments. Discussions with Nadine de Courtenay, Jordi Cat, Martha-Cecilia Bustamante, and Dennis Dang inspired me all along this project. I also received important advice from Anouk Barberousse, Catherine Chevalley, Hasok Chang, Alexandre Guay, Carl Ipsen, Edward Jurkowitz, Joël Merker, and Jürgen Renn. Many thanks to all these colleagues and friends.

CONTENTS

CONVENTIONS AND NOTATIONS

c	velocity of light
ds, dS, $d\tau$	elements of curvilinear abscissa, surface, and volume (respectively)
D_μ	covariant derivative
δ	variation
$\delta(x)$	Dirac's delta function
δ_{ij}	unit tensor (0 for $i \neq j$, 1 for $i = j$)
Δ	Laplacian operator, or increment of a quantity (according to the context)
$\eta_{\mu\nu}$	Minkowskian metric tensor, with the signature $(+, -, -, -)$.
g	acceleration of gravity on earth
$g_{\mu\nu}$	metric tensor
$\Gamma^\mu_{\nu\rho}$	affine connection
h	Planck's constant
$\hbar$	$h/2\pi$
i	$\sqrt{-1}$
$\dot{q}$	derivative of the function q with respect to the time t or the parameter τ
∂_i	partial derivative with respect to the coordinate x_i
∇	gradient operator
$\mathbf{a} \cdot \mathbf{b}$	scalar product of the vectors $\mathbf{a}$ and $\mathbf{b}$
$\mathbf{a} \times \mathbf{b}$	vector product of the vectors $\mathbf{a}$ and $\mathbf{b}$
$[\mathbf{f}, \mathbf{g}]$	commutator of the operators $\mathbf{f}$ and $\mathbf{g}$
$\{f, g\}$	Poisson bracket of the functions $f(q, p)$ and $g(q, p)$
$\otimes$	tensor product
$a \wedge b$	meet of the propositions a and b
$a \vee b$	join of the propositions a and b
$\bar{a}$	negation of the proposition a

- Citations are in the author–date format and refer to the appended bibliography. Square brackets around the date indicate manuscript sources. Page numbers refer to the last mentioned source in the bibliographical item. Abbreviations are listed at the head of the bibliography.
- Translations from French, German, and Latin are mine, unless the source given in the reference is itself a translation. In the latter case, I have freely modified the translation whenever I judged it necessary.

1

RATIONALISM IN THE HISTORY OF MECHANICS

> And the demonstrations [of the rules of collision] are so certain, that even if experience seemed to show us the contrary, we would nonetheless have to trust our reason more than our senses.[1] (René Descartes, 1644)

> As [the principles of mechanics] have heretofore been insufficiently established, I demonstrate them in such a manner that they will be understood to be not only certain but even necessarily true.[2] (Leonhard Euler, 1736)

> The laws of statics and mechanics, as expounded in this book, are those which result from the existence of matter and motion. Now experience proves that these laws are the ones observed on the bodies around us. Therefore, the laws of equilibrium and motion, as known from observation, have necessary truth.[3] (Jean le Rond d'Alembert, 1758)

The seventeenth century saw the rise of a mechanical philosophy that purported to reduce every natural phenomenon to mechanical processes obeying uniform laws of motion. Knowledge of these laws evolved together with this project, against the Aristotelian doctrine of natural and violent motions. An important source was Galileo's study of free fall, which implied astute experiments, the idealization of fall in a non-resistant vacuum, and geometric concepts of velocity and acceleration. On the one hand, the experimental grounding of the new philosophy invited suspicion toward rational deductions of the laws of physics. On the other, its heavy reliance on geometrical and mathematical concepts could suggest that it shared the apodictic certainty of geometry.[4]

The two most influential natural philosophies of the century, Descartes's and Newton's, were thoroughly geometrical and mechanical, and they both implied a blending of geometry and mechanics. As Descartes reduced matter to pure extension and employed a theological principle that made changes of motion depend on geometric configuration only, he could easily believe that the laws of mechanics were as much accessible to pure reason as the axioms of geometry were commonly thought to be. Newton, who rejected this geometric reduction of mechanics and physics, preferred to regard the laws of mechanics as contingent truths induced from experience. In his opinion, mechanics preceded geometry because it served to construct the figures that formed the basis of geometrical reasoning: "Geometry is founded in mechanical practice, and is nothing but that part of universal Mechanics which accurately proposes and demonstrates the art of measuring." Newton indeed invented his differential geometry in a mechanical context. In the aftermath

---

[1] Descartes 1644, §52.   [2] Euler 1736, p. vi.   [3] D'Alembert 1758, p. xxix.   [4] Cf. Dijksterhuis 1961.

Physics and Necessity. First Edition. Olivier Darrigol.
© Olivier Darrigol 2014. Published in 2014 by Oxford University Press.

of an impressive synthesis of calculus, geometry, and mechanics, he resisted the temptation of making mechanics and geometry share the same kind of necessity. Others did not.[5]

This chapter is an account of a few clever attempts at proving the rational and quasi-geometrical necessity of the laws of mechanics, before and after Newton. The first section describes Descartes's wonderful dream of a world so thoroughly geometrical as to share the certainty of geometry. And it shows how Newton, in contrast, defended the inductive origin of his laws of motion, although he sometimes flirted with rational deduction. The second section offers a pre-Newtonian history of rationalist arguments in three important mechanical contexts: the equilibrium of machines from Archimedes to Varignon, collisions between Descartes and Newton, and free fall according to Huygens. It explains the kind of rationalism that led Leibniz to the conservation of live force and to the general idea of extremum principles later developed by Maupertuis and Euler. The third section shows how Daniel Bernoulli, in the early eighteenth century, defended the contingency of the laws of motion and the necessity of the laws of equilibrium. It details Euler's and d'Alembert's subsequent proofs of the complete necessity of the laws of mechanics, and it describes later exploitation and criticism of their arguments.

1.1   Descartes versus Newton

Descartes's dream

A few simple, clear, and distinct notions resist the systematic doubt on which René Descartes founded his philosophy: substance, duration, order, and number for all things; and extension for things corporeal. Other qualities of bodies, such as color or flavor, are not purely corporeal; they have an intellectual component and they are liable to confusion. This is why Descartes makes the physical world out of pure spatial extension. He divides space into rigid figures of various sizes and shapes, which form the particles of his three elements. The particles of the third element, or ordinary matter, are the grossest and have an arbitrary shape; those of the second element, or subtle matter, are round and fill as much as they can of the space between the former particles. Those of the first element are arbitrarily small and they fill the remaining interstices. Any change in the material world amounts to a rearrangement of the particles of the elements. As a corollary to this view, there is no vacuum in Descartes's world; every extension is matter; and there is no absolute motion: the motion of a body is always defined with respect to the contiguous bodies.[6]

Descartes's idea of God's perfection implies the conservation of the total amount of motion in the universe: "God . . . being omnipotent, created matter with the motion and the rest of its parts, and . . . now conserves, by his ordinary concourse, as much motion and rest in the universe as he introduced during creation." For the measure of motion assumed in this statement, Descartes forms the product of extension by velocity modulus, which

---

[5]Newton 1687, preface. Cf. Hankins 1970, p. 23.

[6]Descartes 1644, part 1, §48 (simple notions), §69 (privileging extension); part 2, §28 (relative motion); part 3, §52 (three elements). On Descartes's physics and natural philosophy, cf. Garber 1992; Gaukroger, Schuster, and Sutton 2000.

is the simplest choice in harmony with his identification of matter with extension. From God's immutability he also derives the tendency of individual bodies to preserve their rest or their motion. He expresses this tendency through three "laws of nature":[7]

1. Every thing remains in its state of being as long as nothing changes it.
2. Every moving body tends to continue its motion on a straight line.
3. If a moving body encounters a stronger body, it does not lose any of its motion; if a moving body encounters a weaker one that it may move, the former body loses as much motion as it gives to the latter.

The first law implies both the preservation of rest and the preservation of motion. It should not be confused with the modern concept of inertia, because the preserved rest is defined with respect to the surrounding bodies, not with respect to empty space. In particular, Descartes uses the preservation of rest to explain the rigidity of bodies whose parts are originally in mutual rest: the relative position of the internal parts of such bodies cannot change, because their surroundings are invariable: "There is nothing joining the parts of hard bodies but their being mutually at rest." Conversely, Descartes explains the fluidity of a body by the agitation of its parts.[8]

Descartes's second law is another consequence of God's immutability, following which not only the amount (modulus) of motion of the unperturbed body is preserved, but also the direction of motion. This idea is somewhat obscure. A first difficulty comes from the conflict between assuming the lack of obstacles to the motion and referring the motion to the surrounding bodies. This conflict may be avoided by noticing that the condition of unimpeded motion is a fiction that Descartes corrects by combining the inertial motion with the motion induced by the impinging bodies. A more serious difficulty comes from ambiguity in the definition of the true motion. In general, the moving body is surrounded by other bodies that differ in their motions. This is for instance the case for a rigid body immersed in a fluid necessarily made of agitated corpuscles. In one occasion, the motion of a hard body through a (macroscopically) stagnant fluid, Descartes solves this difficulty by taking the average relative velocity of the moving body with respect to the various surrounding bodies. Unfortunately, this subterfuge contradicts his derivation of a centrifugal force for the particles of subtle matter in the vortex that surrounds the sun in his theory of planetary motion. Indeed the velocity of one of these particles with respect to the neighboring particles is zero on average; the motion that Descartes considers in his justification of centrifugal force is the rotation of the subtle matter around the sun.[9]

The third law is even stranger in modern eyes, because it implies that a smaller hard body hitting a larger hard body at rest rebounds without communicating any of its motion to the larger body (Descartes's fourth rule of collision). Like the first law, this law assumes a special persistency of rest. Descartes avoids the obvious contradiction with experience by assuming that in observed collisions the air around the larger body increases its mobility. From his three laws and from the global conservation of motion, he derives seven

---

[7] Descartes 1644, part 2, §36 (motion conserved), §§37–41 (laws of nature).

[8] Descartes 1644, part 2, §§54–5.

[9] Descartes 1644, part 2, §62; part 3, §§54–5. A similar criticism can be found in Newton [c.1668?].

rules of collision corresponding to different kinds of initial conditions. In some cases, the collision involves bouncing (as they would in the modern elastic case); in others, the two bodies remain attached to each other after the shock (as they would in the modern soft case). Descartes's rules are evidently incompatible with common observation and with two principles that Huygens and Leibniz later brought to bear on this matter: the relativity principle and the principle of continuity. Descartes need not worry about these incompatibilities because his ideal collisions concern ideally hard bodies fictitiously extracted from the medium in which they necessarily bathe. His trust in the necessity of the rules of collisions is complete: "And the demonstrations [of these rules] are so certain, that even if experience seemed to show us the contrary, we would nonetheless have to trust our reason more than our senses."[10]

For Descartes, every natural phenomenon ought to be reducible to collisions or contact between the particles of the space-filling elements. Therefore, his principles or laws of nature determine the evolution of the universe from a given initial state. This God-chosen state does not matter much to the global appearance of the present universe; because for any initial state compatible with a given amount of motion, Descartes's universe passes through any state compatible with this amount (ergodicity *avant la lettre*). Observation is only necessary to select the form to be explained among the many other possible forms. Arguably, Descartes also needs observation and experiments to fill gaps in his deductions, and to relate the distribution of his elements to our sensations (as he does in his theory of the rainbow). There is no doubt, however, that he prefers rational deductions to sense-based inferences.[11]

To summarize, Descartes believed he could discover the laws of nature by mere reasoning, with no appeal to experience or experiments. For reasons of clarity and simplicity, he identified matter with mere extension. Any change in the world then became a rearrangement of geometrical figures. Descartes derived the laws of this change from the immutability of God. He meant them to account for every physical phenomenon, no matter how much conceptual distance there was between the hidden motion of the elements and the humanly accessible appearances. There is no need here to describe the precise way in which he explained light, colors, gravitation, fire, magnetism, and many other phenomena. The important point is that he believed that his mechanics of pure extension ruled the material word, and that the laws of this mechanics derived from his idea of God's perfection. There could not be any other laws of motion, and any experimental suggestion to the contrary was a delusion.

Even those of Descartes's contemporary who shared his antipathy with Aristotelian physics and his sympathy with mechanical philosophy could still object to his system for theological, internal, and empirical reasons. In the theological objections, the way in which Descartes appeals to God to justify his system is criticized, or his system is regarded as leading to atheism. In the internal objections, logical flaws are identified in his deductions. In the empirical objections, his laws of motion are declared incompatible with observation.

---

[10] Descartes 1644, part 2, §§46–52 (citation from §52).

[11] Descartes 1644, part 3, §47 (ergodicity). On empirical inputs in Descartes's philosophy, cf. Morrison 1989.

Newton's mechanics

Isaac Newton expressed objections of the three kinds. In the theological register, he argued that the identification of matter with extension led to atheism, because God, being immaterial, would be nowhere in Descartes's world and therefore would not exist. In contrast, Newton cleanly separated space from matter and identified his absolute space with an emanation of God. His most serious objections to Descartes were of the internal and empirical kinds. He, for instance, argued that Descartes's identification of light with a pressure transmitted by subtle matter would make any moving body shine, or would make light penetrate the shadow of any interposed screen. He also pointed to the aforementioned contradiction between Descartes's definition of true motion and his application of the principle of inertia.[12]

In his earliest extant notes on natural philosophy, written in the 1660s, Newton espoused a kind of atomism in which matter and light were made of immutable and invisible corpuscles occupying variable locations in an immaterial spatial frame. This absolute space plays an essential role in his *Principia* of 1687, as it provides the reference frame in which his three laws of motion are valid. These laws read:[13]

1. Every body perseveres in its state of rest, or of uniform motion in a right line, unless it is compelled to change that state by forces impress'd thereon.
2. The alteration of motion is ever proportional to the motive force impress'd; and is made in the direction of the right line in which that force is impress'd.
3. To every action there is always an opposed and an equal reaction: or the mutual actions of two bodies upon each other are always equal, and directed to contrary parts.

Although the first law resembles Descartes's first and second laws, it differs from them in referring motion to absolute space. Newton's two other laws are essentially different from Descartes's, since they make mechanics depend on a concept of force whereas Descartes makes it depend on encounters between ideally rigid bodies. Newton defines motive force as any action that alters the inertial motion, and measures it by the change in the quantity of motion (momentum) that it produces in a unit of time. Although this definition seems to imply the circularity of the second law, it truly does not because this law further implies that the momentum variation is the same whenever the circumstances that produce the force are the same. In modern terms, the force field can first be measured by a test particle, and then the motion of any other body in this field can be derived. In Newton's broader terms: "For all the difficulty of philosophy seems to consist in this, from the phænomena of motions to investigate the forces of Nature, and then from these forces to demonstrate the other phænomena." A last precision is in order: in corollaries to his second law, Newton makes clear that he means the forces to be applied impulsively (as they are in the case

---

[12] Newton [*c*.1664], p. 32 (light from moving body); [*c*.1668?], p. 25 (space and God); 1687, p. 369 (no shadow); 1713, *scholium generale* (space and God). Cf. Janiak 2004, 2009.

[13] Newton 1729, vol. 1, pp. 19–21.

of collisions). A continuous action, such as the attraction of a celestial body, is to be understood as a rapid succession of impulses.[14]

There is no doubt that Newton regarded the law of gravitation, which gives the expression of the gravitational force between two masses, as an empirical law. In the *Scholium generale* appended to the 1713 edition of the *Principia*, he famously rejected the deductive approach to gravitation and claimed that the basic concepts and laws of mechanics had an inductive origin:

> But hitherto I have not been able to discover the cause of those properties of gravity from phænomena, and I frame no hypotheses. For whatever is not deduc'd from the phænomena, is to be called an hypothesis; and hypotheses, whether metaphysical or physical, whether of occult qualities or mechanical, have no place in experimental philosophy. In this philosophy particular propositions are inferr'd from the phænomena, and afterwards render'd general by induction. Thus it was that the impenetrability, the mobility, and the impulsive force of bodies, and the laws of motion and of gravitation, were discovered. And to us it is enough, that gravity does really exist, and acts according to the laws which we have explained, and abundantly serves to account for all the motions of the celestial bodies, and of our sea.

In his preface to the same edition of the *Principia*, Newton's collaborator, Roger Cotes, explained the theological dimension of experimental philosophy:

> Without all doubt this World, so diversified with that variety of forms and motions we find in it, could arise from nothing but the perfectly free will of God directing and presiding over all. From this fountain it is that those laws, which we call the laws of Nature, have flowed; in which there appear many traces indeed of the most wise contrivance, but not the least shadow of necessity. These therefore we must not seek from uncertain conjectures, but learn them from observations and experiments. He who thinks the true principles of physics and the laws of natural things by the force alone of his own mind, and the internal light of his reason; must either suppose that the World exists by necessity, and by the same necessity follows the laws proposed; or if the order of Nature was established by the will of God, that himself, a miserable reptile, can tell what was fittest to be done. All sound and true philosophy is founded on the appearances of things.

In brief, physical laws result from the free will of God; they are not necessary truths and therefore they can only be discovered by observation and experimentation.[15]

In a scholium to his laws of motion, Newton explained that the two first laws could be inferred from Galileo's experiments on the fall of bodies, and that the validity of the third law in collisions had to be assumed, in addition to the two first laws, in order to retrieve the rules of collisions established by Wallis, Wren, Huygens, and Mariotte.[16] Despite this

---

[14]Newton 1729, vol. 1, p. iii (citation), p. 7 (motive force defined), p. 21 (impulsive force). The impulsive character of Newton's forces explains why he needs the law of inertia (first law) in addition to the second law; whereas in modern formulations of Newtonian mechanics, the proportionality of force and acceleration implies the law of inertia.

[15]Newton 1729, vol. 1, p. 392; Cotes (ed.), Newton 1729, pp. xxxiii–xxxiv (Cotes).

[16]Newton 1729, pp. 31–6.

professed inductivism, Newton did not always resist the charms of rational deduction. A first example of this kind is the derivation of the parallelogram of forces found in his two first corollaries to the laws of motion. According to this rule, two forces acting conjointly on the same material point have the same effect as a single force obtained by taking the diagonal of the parallelogram formed by these two forces. In his demonstration, Newton regards the two forces as two impulses measured by the motion they would induce on a free material point. Since these two impulses can be regarded as successive (with negligible delay) and since, by the second law, the motion after the second impulse is obtained by compounding the motions that the two impulses would separately impart, forces must combine as velocities do in the superposition of motions. Newton implicitly generalizes the result to continually acting forces by regarding them as rapid successions of impulses. The overall deduction is not purely a priori, for it presupposes the second law of motion, which Newton regards as empirical. The deductive thread goes on, however, with Newton's derivation of the law of the lever and other truths of statics from the parallelogram of forces.[17]

For the angular lever MON of Fig. 1.1, Newton seeks the condition of equilibrium of the suspended weights A and P. The letters K and L denote the intersection of the horizontal line through O with the suspending threads MA and NP; and the point D of the thread KA is such that DO = OL. Newton notes that the equilibrium is unchanged by rigidifying KMONLD. If DA represents the force acting on D, by the parallelogram rule this force is equivalent to the forces DC and CA acting in the directions perpendicular and parallel to OD. The force AC (acting on D) has no effect on the lever since it passes through the center O; and the force DC is balanced by an equal force applied vertically on L, since OD = OL. Therefore, the lever is in equilibrium if and only if the ratio of the weights A and P is equal to the ratio DA/DC. Using the similitude of the triangles OKD and DCA, the latter ratio is also equal to OK/OD or OK/OL. Newton thus retrieves the well-known condition of equilibrium of the lever (equality of moments). After discussing a couple of other examples involving inclined planes, pulleys, and screws, he propounds that the conditions of equilibrium of any machine derives from the parallelogram of forces:[18]

> Therefore the use of this corollary spreads far and wide, and by that diffusive extent the truth thereof is farther confirmed. For on what has been said depends the whole

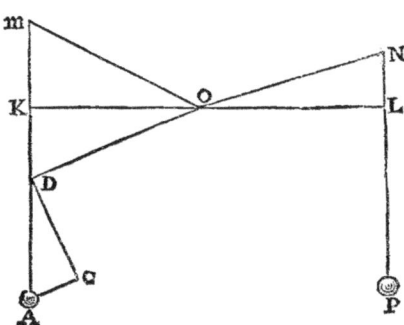

Fig. 1.1. Newton's diagram for his derivation of the law of the angular lever MON. From Newton 1687, p. 14.

[17] Newton 1729, pp. 21–2.   [18] Newton 1729, pp. 22–4 (citation from p. 24).

doctrine of Mechanics variously demonstrated by different authors. For from hence are easily deduced the forces of Machines, which are compounded of Wheels, Pulleys, Leavers, Cords and Weights, ascending directly or obliquely, and other Mechanical Powers; as also the force of the Tendons to move the Bones of Animals.

Another example of Newton's deductive flights is his justification of the third law in the case of (gravitational, electric, or magnetic) attractions. If the attraction of a body A by a body B were larger than the reciprocal attraction, Newton reasons, a rigid body C interposed between A and B would assume a forever accelerating motion since the net force acting on C would not vanish. The result would be both "absurd and contrary to the first law." Again, the deduction is not purely a priori since it assumes the first law or the impossibility of perpetual motion. It nonetheless conveys an air of necessity to the third law, because the impossibility of perpetual motion, be it of empirical origin or not, is very easily admitted.[19]

In this reasoning, the equilibrium of the rigid body C requires the equality and opposition of the forces acting on its extremities. In a more complicated mechanical machine for balancing an action and a resistance, such as a lever or a system of pulleys, it was known since the ancient Greeks that the ratio of the acting force and the resisting force should be equal to the ratio of (the projection of) the velocities of the corresponding parts of the machine (on the direction of the forces). This is a particular case of what is now called the principle of virtual work. Newton enunciated this principle right after his justification of the third law, for he regarded it as an extension of the equality of action and reaction. For the direct interaction of two bodies (by contact or by attraction), he simply measured the action and reaction by the corresponding forces or momenta; for interaction through a mechanism, he measured them by what is now called their virtual work. Although Newton did not intend to write a theory of mechanics in this machinist sense, he was eager to show some necessity of the laws of mechanism in the light of his fundamental laws of motion.[20]

1.2 Before Newton

Machines

There were many attempts to derive the law of the lever before Newton's. Archimedes' antic demonstration is based on the replacement of each of the balancing weights on a straight horizontal lever with a series of equal, equidistant weights. Consider, for instance, the lever of Fig. 1.2 symmetrically loaded with nine equidistant weights. According to Archimedes, the equilibrium is unaffected when weights 1 to 6 are replaced with a sextuple weight placed at their center of gravity A and when weights 7 to 9 are replaced by a triple weight placed at their center of gravity B. The ratio (2) of the resulting weights is equal to the inverse ratio of their distances from the suspension point. Archimedes thus established the law of

---

[19] Newton 1729, p. 37.

[20] Newton 1729, pp. 38–40. Newton's reasoning, unfortunately, seems to depend on the then widespread confusion between mass and weight, which makes virtual work (velocity × weight) more analogous to momentum (velocity × mass) than it should be.

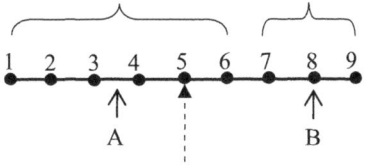

Fig. 1.2. Diagram for Archimedes' derivation of the law of the lever.

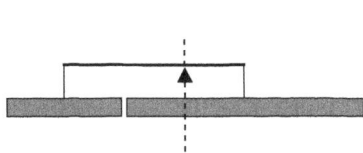

Fig. 1.3. Equilibrium of an asymmetric lever according to Stevin.

the lever for any pair of commensurable weights, and generalized it to incommensurable weights through the method of exhaustion.[21]

Archimedes' replacement of a set of weights by their sum amounts to assuming that their rotational effect is proportional to their distance from the axis, which is nearly the same as assuming the law of the lever from the start. Simon Stevin's sixteenth-century variant of Archimedes' demonstration does not have this defect. As is shown in Fig. 1.3, it is based on replacing the two balanced weights by two suspended bars of uniform section and of such length that the right end of the left bar nearly touches the left end of the right bar (for a given weight, this substitution can be justified by superposing to the original equilibrium the equilibrium of a symmetric lever loaded with this weight on one side and the replacement weight on the other). Stevin next assumes that solidifying the suspending threads preserves the equilibrium. For reasons of symmetry, the resulting system is in equilibrium if and only if the left end of the left bar and the right end of the right bar are at equal distances from the axis of the lever. This implies that the weights of the two bars should be in inverse proportion to their distance from the axis, in agreement with the law of the lever.[22]

As Guidobaldo del Monte partially indicated in his *Mechanicorum liber* of 1577, the law of the angular lever and the law of moments can be derived from the law of the rectilinear lever through the principle of superposition of equilibria and from the assumption that a force applied at a given point of rigid body has the same effect when applied on any other point on the line of action of the force. Firstly, the superposition of the equilibrium of the angular lever with the equilibrium of a rectilinear lever leads to the condition of equilibrium of the former lever in the case of applied forces perpendicular to the arms of the lever (see Fig. 1.4). Then for arbitrarily directed forces applied to the extremities of a straight lever, this lever can be replaced with an angular lever whose extremities are the projections of the fulcrum over the lines of actions the forces (see Fig. 1.5). The resulting law of equilibrium is the law of moments. As del Monte and his followers strove

---

[21] Cf. Jouguet 1908, vol. 1, pp. 7–15. On the history of statics in general, cf. Duhem 1905–6.

[22] Stevin 1586, pp. 11–13. Cf. Jouguet 1908, vol. 1, pp. 16n–17n. For the criticism of Archimedes, cf. Mach 1883, pp. 13–14.

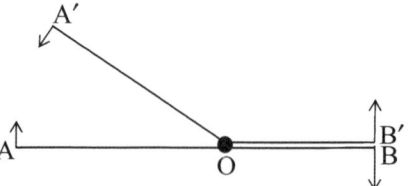

Fig. 1.4.   The law of the angular lever deduced from the law of the straight lever. The equilibrium of the lever
A′OB′ is undisturbed by rigidly connecting it with the lever AOB subjected to the same forces. The forces
on B and B′ being opposed to each other, the combined lever is equivalent to the lever AOA′, which is in
equilibrium for obvious symmetry reasons.

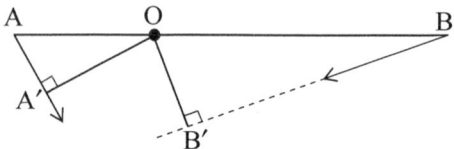

Fig. 1.5.   Equilibrium of a straight lever subjected to oblique forces. The equilibrium of the lever AOB is unchanged by rigidly attaching to it the lever A′OB′ and shifting the points of application of the two forces
from A to A′ and from B to B′.

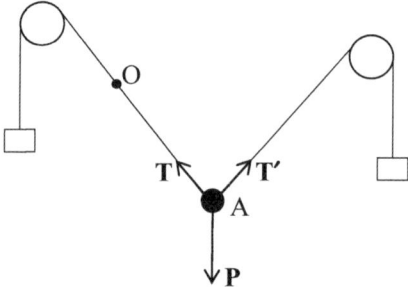

Fig. 1.6.   Equilibrium of a body A subjected to its weight **P** and to the tensions **T** and **T′** of two threads.

to demonstrate, every case of equilibrium of machines made of pulleys, ropes, and levers
can be derived from this law.[23]

As we saw, Newton derived the law of the lever from the parallelogram of forces.
Leonardo da Vinci and Gilles de Roberval had earlier done the reverse. A simplified and
somewhat modernized version of their reasoning goes as follows. In Fig. 1.6, the body
A is held in equilibrium through its weight **P** and through the tensions **T** and **T′** of the

---

[23]Del Monte 1577, p. 40. Cf. Jouguet 1908, vol. 1, pp. 24–7. Del Monte only treats the case in which the angle of
the lever is 180° (true lever, in which an upwards force is used to lift a weight, instead of balance). The generalization
to any angle is found in Lagrange 1811, pp. 6–7.

two suspending threads. The equilibrium persists if the portion OA of the first suspending thread is solidified to make a lever rotating around O. By the law of moments, the equilibrium of this lever requires that the vector sum $\mathbf{T}' + \mathbf{P}$ should be parallel to $\mathbf{T}$. Similarly, $\mathbf{T} + \mathbf{P}$ should be parallel to $\mathbf{T}'$. Therefore, the vector $\mathbf{T} + \mathbf{T}' + \mathbf{P}$ is parallel to both $\mathbf{T}$ and $\mathbf{T}'$ and it must vanish. This is the modern vector expression of the fact that the forces $\mathbf{T}$ and $\mathbf{T}'$ are conjointly equivalent to the force obtained by combining them through the rule of the parallelogram.[24]

Another possible foundation of statics is the condition of equilibrium of weights free to slide on two inclined planes (Fig. 1.7). Stevin ingeniously obtained this condition by starting with the case of Fig. 1.8, in which a heavy chain hangs on the double inclined plane. Any spontaneous movement of the chain would leave the global distribution of weights unchanged. Therefore, the tendency to motion would be preserved through this movement, and the chain would keep moving even in the presence of friction: a perpetual motion would occur. In order to exclude this possibility, the chain must be in equilibrium. For reasons of symmetry, the freely hanging part of the chain can be cut off without the equilibrium being disturbed. This means that weights proportional to the lengths of the inclined planes should be in equilibrium. In the analogous case of Fig. 1.7, the weights of the two stones must be proportional to the lengths of the planes on which they can slide.[25]

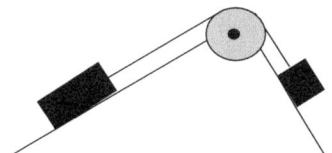

Fig. 1.7. Equilibrium of two weights on the two sides of an inclined plane.

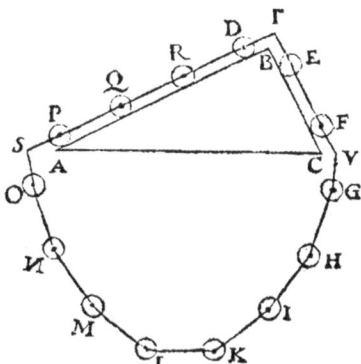

Fig. 1.8. Stevin's figure for his derivation of the rule of the inclined plane. From Stevin 1586, p. 41.

[24]Roberval 1636, pp. 21–4. Cf. Jouguet 1908, vol. 1, pp. 53–46.

[25]Stevin 1586, pp. 41–2. Cf. Jouguet 1908, vol. 1, pp. 45–53.

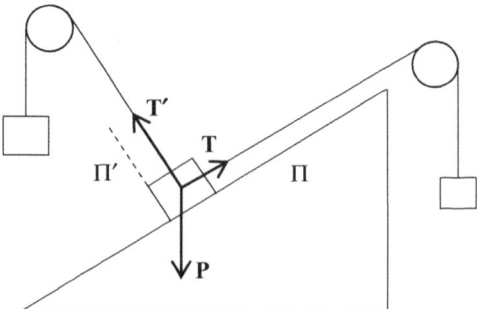

Fig. 1.9.   Figure for deriving the parallelogram of forces from the law of the inclined plane.

The law of the inclined plane evidently derives from the parallelogram of forces since the weight of each of the suspended bodies can be decomposed into a normal component counterbalanced by the underlying plane and a parallel component balanced by the tension of the thread (the equality of tensions on the two sides of the pulley being a direct consequence of the law of the lever). Stevin followed the reverse route, deriving the parallelogram of forces from the law of the inclined plane. In a simplified version of his reasoning, consider the weight $\mathbf{P}$ of rectangular shape suspended by two threads from its center of gravity, as in Fig. 1.9. The equilibrium persists if the inclined plane Π is brought to touch the lower right side of the weight and if the tension $\mathbf{T}'$ is removed. Therefore, the tension $\mathbf{T}$ and the weight $\mathbf{P}$ obey the rule of the inclined plane Π, following which $\mathbf{T} + \mathbf{P}$ must be parallel to $\mathbf{T}'$. Similar reasoning with a plane Π' brought underneath the lower left side of the weight implies that $\mathbf{T}' + \mathbf{P}$ should be parallel to $\mathbf{T}$. Therefore, $\mathbf{T} + \mathbf{T}' + \mathbf{P}$ must vanish, and the rule of the parallelogram must hold for the jointly applied forces $\mathbf{T}$ and $\mathbf{T}'$.

To sum up, the parallelogram or forces, the law of the lever, and the law of the inclined plane are interchangeable as a foundation for statics; and they can be derived from the impossibility of perpetual motion. The proofs appeal to at least two additional principles: the superposition of equilibria, and the principle of solidification. These principles were judged evident, and they were often used unconsciously, so that the laws of mechanism acquired an air of rational necessity. While the historian of mechanics Emile Jouguet denounced this illusion, he also valued the attempted deductions for their role in strengthening the empirical basis and the logical cohesion of the theory of mechanics:[26]

> I would almost say, in a paradoxical form slightly beyond my thought, that in the demonstrations at the basis of physical theories, the fact that an idea is introduced in a surreptitious manner is excellent, because it is a proof that these demonstrations involve notions that our experimental intuition forces onto us.
>
> Reasoning about the principles at the basis of a physical science does not so much consist in demonstrations as in logical analyses in which one scrutinizes various notions and tries to relate them to one another or to distinguish them.

Another approach to the equilibrium of machines (seen as the transfer of force through machinery) originated in the pseudo-Aristotle's *Mechanica*, with the statement that the

---

[26] Jouguet 1908, vol. 1, p. 23.

equilibrium of the lever requires the applied forces to be in inverse proportion to the velo-cities of the points of application. The rule appears in the writings of several authors from the thirteenth century to the Renaissance, either as a true principle (for Jordanus, Vinci, and Cardano) or as a corollary of other laws (for Galileo). In these texts it is generalized to other mechanical systems including pulleys and inclined planes, and to forces applied in a direction oblique to the motion of the point of application.[27]

The velocities used in the statement of this principle of equilibrium are fictitious since equilibrium or slow machine-mediated displacement, not motion per se, is the object of the principle; they are related through the physical connections of the machinery. In a letter to Varignon written in 1717, Johann Bernoulli introduced the name "virtual velocities" for the projection of the fictitious velocities on the direction of the forces, and asserted the vanishing of the algebraic sum of the "energies" of the applied forces in any machine in equilibrium, energy being defined as the product of force by virtual velocity. This is the first general statement of what Lagrange later called "principle of virtual velocities." The name "principle of virtual work" emerged in the mid-nineteenth century after the French engineering concept of work had become a central concept of mechanics.[28]

In a manuscript of 1637 entitled *Explanation of the machines through which one can lift a heavy load with a small force*, Descartes relied on Jordanus's principle that the elevation of the weight P at the height αh is equivalent to the elevation of the weight αP at the height h for any number α. Descartes regarded the latter principle as a necessary truth:

> This principle cannot fail to be admitted if one considers that the effect must always be proportional to the action that is necessary to produce it. Thus, if the action through which 100 pounds are raised by 2 feet is needed to raise another weight by 1 foot, this weight must equal 200 pounds. For it is the same thing to raise 100 pounds by one foot and again 100 pounds by one foot than to raise 200 pounds by 1 foot, and still the same to raise 100 pounds by 2 feet.

Descartes here implicitly assumes the uniformity of gravitation, since he does not see any difference in hoisting a body from the height H to the height $H + h$ and hoisting it from the height $H + h$ to the height $H + 2h$.[29]

In any mechanism with one degree of freedom, the two balanced forces can always be thought as generated by weights and pulleys (see Fig. 1.10) when they are not directly caused by weights. This is why Descartes's principle has the same scope as the principle of virtual work (limited to two forces). In the system comprising the mechanism and the force-generating pulleys and threads, this principle implies that the center of gravity of the two balanced weights remains at the same height during an infinitesimal displacement of the system around equilibrium. Several authors had used the center of gravity in the discussion of equilibrium before Descartes. For instance, it was well known that a rigid body free to rotate around a fixed point was in equilibrium if and only if its center of

---

[27] Cf. Jouguet 1908, pp. 57–67. For a thorough history of the principle of virtual work, cf. Capecchi 2012.

[28] Johann Bernoulli to Varignon, Jan. 26, 1717, in Varignon 1725, vol. 2, pp. 174–6. Varignon provided a proof of the principle based on the parallelogram of forces for all the machines included in his treatise. "Virtual work" occurs in Coriolis 1829, p. 12; "theorem of virtual work" in Delaunay 1856, p. 333.

[29] Descartes [1637]. Cf. Jouguet 1908, vol. 1, pp. 68–72.

Fig. 1.10.   Lever subjected to forces produced from pulley-weight systems.

gravity lied on the vertical through the fixed point. In 1644, Galileo's disciple Evangelista Torricelli enunciated the following principle as the basis of his theory of equilibrium:[30]

> Two weights, connected to each other, cannot move spontaneously unless their center of gravity goes down. Indeed, when two weights are related to each other in such a manner that the motion of the one implies the motion of the other, this connection being produced by means of a lever, a pulley, or any other mechanism, these two weights will behave as a single weight made of two parts; however, such a weight will never begin to move unless its center of gravity goes down. Consequently, when the [global] weight will be built up in such a manner that its center of gravity cannot descend in any manner, the weight will surely remain at rest in its original position.

Torricelli did not relate his principle to the principle of virtual work. In his times, there were at least four competing approaches to statics, based on the lever (del Monte), on the inclined plane (Stevin), on the principle of virtual work (Descartes and Wallis), and on the descent of the center of gravity (Torricelli). As we saw, in his *Principia* Newton sketched two approaches: a traditional one based on the principle of virtual velocities, and a novel one based on the parallelogram of forces. In his *Projet d'une nouvelle méchanique* published in the same year 1687, Pierre Varignon similarly based the theory of equilibrium on the parallelogram of forces. Having defined force by impressed motion, he fell into the common error of regarding the composition of forces as a trivial consequence of the composition of motions. His derivation of the law of the lever goes as follows.[31]

Two forces measured by the segments AS and AR are applied on the extremities M and N of the lever MBN represented in Fig. 1.11. The equilibrium of this lever is unchanged when it is attached to a rigid plane that can rotate around the pivot B. It is still unchanged if the two forces are applied on the intersection point A of their lines of action. Their action is therefore equivalent to that of the force AG, forming the diagonal of AR and AS, and acting on the point A of the rigid plane. Evidently, the equilibrium requires that the line of action of the latter force should pass through the pivot B. This is so if and only if the ratio BD/BP is the inverse of the ratio AS/AR, in conformity with the law of the lever. This reasoning excludes the case of parallel forces applied to M and N. Varignon treats this case as the limit in which the intersection point A is thrown to infinity.[32]

To summarize, Newton and Varignon continued the ancient tradition of deriving the laws of equilibrium of machines from a single principle. This principle could be the law of the lever, the principle of virtual work, the law of inclined planes, the necessary descent

[30]Torricelli 1644, p. 99.   [31]Varignon 1687.   [32]Ibid., pp. 58–60.

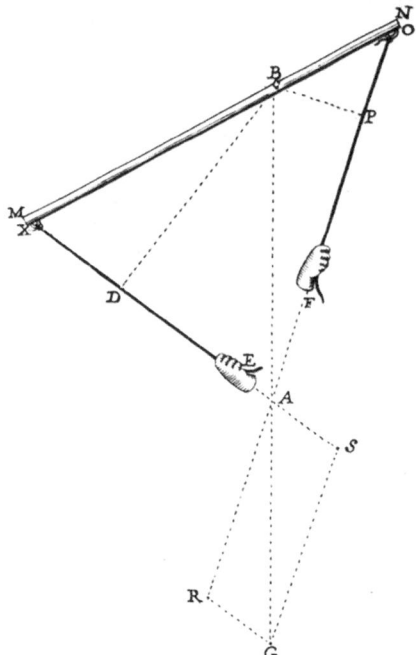

Fig. 1.11.   Varignon's figure for deriving the law of the lever. From Varignon 1687, plate 6.

of the center of gravity, or the parallelogram of forces. For Newton, this highly deduct-ive character of statics did not imply its rational necessity because the unifying principle derived from his second law of motion, which he regarded as contingent. Other authors regarded their founding principle as self-evident or as deriving from another self-evident principle. Archimedes' followers reduced the law of the lever to the evident equilibrium of a symmetric lever; Descartes reduced the principle of virtual work to the evident con-stancy of the effort needed to hoist a body to a height inversely proportional to its weight; Stevin deduced the law of the inclined plane from the impossibility of perpetual motion; and Varignon derived the parallelogram of forces from the composition of motions. These justifications of the founding principle, as well as the deductions based on this principle, typically involved implicit assumptions or even flawed reasoning. This explains the per-sistent diversity of choices for the founding principle. This does not imply, however, that the rationalist claims of their inventors were totally misguided. They globally indicated the genuine difficulty of imagining laws of statics differing from those verified in experiments.

Collisions

Another important background for Newton's mechanics and for the question of the ne-cessity of its laws was the problem of collisions. The glaring incompatibility of Descartes's collision rules with common observation prompted a few experimental and theoretical studies of the collisions of bodies. In 1668, after Robert Hooke agitated this question in the name of the Royal Society, John Wallis, Christopher Wren, and Christiaan Huygens published the now-accepted rules for elastic and binding frontal collisions. The first in print was the English mathematician John Wallis, who treated the case of "absolutely hard" bodies for which the two colliding bodies move together after their encounter. Wallis's rules were based on what we now call the conservation of momentum, namely:

$$m_1 \bar{v}_1 + m_2 \bar{v}_2 = m_1 \bar{v}_1' + m_2 \bar{v}_2', \tag{1}$$

where m_1 and m_2 are the masses of the two bodies, $\bar{v}_1$ and $\bar{v}_2$ their original (algebraic) velocities, and $\bar{v}_1'$ and $\bar{v}_2'$ their final (algebraic) velocities. For binding collisions, this implies that

$$\bar{v}_1' = \bar{v}_2' = \frac{m_1 \bar{v}_1 + m_2 \bar{v}_2}{m_1 + m_2}. \tag{2}$$

Wallis's intuition of the conservation of momentum derived from analogy with the principle of virtual motion, according to which the "force" of a weight acting on a machine is to be measured by the product of the weight with the (virtual) velocity of the part of the machine to which it is applied. As Wallis confused weight and mass, he measured the "force" or "momentum" of a moving body by the product of mass and velocity. Unlike Descartes, he regarded collisions (and the equilibrium of machines) as processes involving the mutual impeachment of two motions. This is why he assumed a complete destruction of motion in the case of a symmetric hard-body collision. He nevertheless shared Descartes's belief in the rational necessity of the laws of motion.[33]

The English architect Christopher Wren approached the collision problem in an experimental manner, observing collisions between the bobs of two pendula. It is not clear to what extent the rules he formulated derived from these experiments or from analogy with the equilibrium of a lever. Stated in modern terms, these rules amount to a sign change for the velocity of each body with respect to the center of gravity of the two bodies:

$$\bar{v}_1' - \bar{v}_G = -(\bar{v}_1 - \bar{v}_G), \quad \bar{v}_2' - \bar{v}_G = -(\bar{v}_2 - \bar{v}_G), \quad \text{with } \bar{v}_G = \frac{m_1 \bar{v}_1 + m_2 \bar{v}_2}{m_1 + m_2}, \tag{3}$$

They are identical to the modern rules for the frontal collisions of perfectly elastic bodies.

For initial velocities such that $m_1 \bar{v}_1 + m_2 \bar{v}_2 = 0$, the collision amounts to a sign change of the velocity of each body. Wren justified this result by arguing that in this case the two bodies had the "proper velocities" and therefore retained these velocities (save for the sign change) after the shock. He then reduced the general case to the former case by subtracting from each velocity the velocity $\bar{v}_G$ such that $m_1(\bar{v}_1 - \bar{v}_G) + m_2(\bar{v}_2 - \bar{v}_G) = 0$. He illustrated these relations through the equilibrium of a lever, again using a loose analogy between momentum and virtual work.[34]

Of all contemporary derivations of the rules of collision, Huygens's came closest to a rational deduction. Namely, he derived these rules from three principles whose necessity

---

[33]Wallis 1668.

[34]Wren 1668. On his subsequent *De motu* (Wren 1669–71), cf. Jouguet 1908, vol. 1, pp. 114–25.

became more and more commonly accepted: the principle of inertia, the Galilean principle of relativity, and a variant of Torricelli's principle according to which the center of gravity of a system of bodies cannot rise in a spontaneous manner. Huygens started with a restricted class of collisions in which the rules were evident enough, and extended this class by making the restricted collisions happen on a moving ship and considering them from ashore. A first obvious case is the symmetric collision between two equal bodies. Seen from the ship, the collision amounts to a sign change of the two equal velocities. Seen from the shore, it provides the rule for any collision between equal bodies, since any values of the initial velocities can be obtained by varying the ship velocity and the common relative velocity. More generally, Huygens observed that for any collision between two different bodies, there is a velocity u of the ship for which the relative velocity of the first body simply changes sign:

$$\bar{v}_1' - u = -(\bar{v}_1 - u), \text{ if } u = \frac{1}{2}(\bar{v}_1 + \bar{v}_1'). \tag{4}$$

In this case, the intuition of an elastic collision implies that the other body should also experience a mere sign change of its velocity:

$$\bar{v}_2' - u = -(\bar{v}_2 - u). \tag{5}$$

Consequently, the relative velocity of the two bodies simply changes its sign

$$\bar{v}_2' - \bar{v}_1' = -(\bar{v}_2 - \bar{v}_1). \tag{6}$$

This rule does not suffice to determine the outcome of a generic collision. Huygens supplemented it with an appeal to Torricelli's principle.[35]

Suppose that the velocities of the two colliding bodies have been obtained by making them fall from a certain height and then turn their vertical motions into horizontal ones by reflection on ideal elastic planes. After collision in a horizontal plane, let the final velocities be turned into elevations (with no residual motion). According to Torricelli's principle, the center of gravity at the end of this series of operations cannot be higher than it was at the beginning. Reversibility being assumed for elastic collisions, it must remain at the same height. Since, according to Galileo's laws of free fall, the equivalent height of a body is proportional to the square of its velocity, the following relation must hold:

$$m_1 v_1{}^2 + m_2 v_2{}^2 = m_1 v_1'^2 + m_2 v_2'^2. \tag{7}$$

Together with the reversal of the relative velocity, this relation implies the conservation of momentum in Wallis's form (1):

$$m_1 \bar{v}_1 + m_2 \bar{v}_2 = m_1 \bar{v}_1' + m_2 \bar{v}_2'.$$

It also implies Wren's expression of the final velocities.

In retrospect, we can imagine a simpler way of reasoning from Huygens's principles. Torricelli's principle (completed with the uniform acceleration of gravity and with the

[35] Huygens 1669, 1703. Cf. Jouguet 1908, vol. 1, pp. 139–55; Vilain 1996.

reversibility of elastic collisions) leads to Eq. (7). Together with the relativity principle, this equation leads to

$$m_1(\bar{v}_1 + u)^2 + m_2(\bar{v}_2 + u)^2 = m_1(\bar{v}'_1 + u)^2 + m_2(\bar{v}'_2 + u)^2 \qquad (8)$$

for any value u for the velocity of the ship. Consequently, the conservation of momentum must hold, and so too must the reversal of the relative velocity. Alternatively, Huygens could have based his theory of collisions on this reversal (itself derived by applying the relativity principle to the special case in which both bodies rebound with the same velocity modulus) combined with the conservation of momentum. However, unlike Wallis, he did not regard this conservation law as an evident truth.

Ten years later Edme Mariotte published his *Traité de la percussion*, in which he distinguished between "flexible bodies with spring" and "flexible bodies without spring" in conformity with the modern distinction between elastic and soft collisions. His approach being resolutely experimental, he ignored the ideal notion of perfectly hard body (without spring), and he based the various rules of collision on a series of *Principes d'expérience*, all drawn from precise, effective experiments. His success did not prevent Newton from deriving the rules of collision from more fundamental laws of motion. Nor did it discourage future attempts at a rational foundation of these rules.[36]

Huygens on falling bodies

The rational tendency of Huygens's derivation of the rules of elastic collisions may seem limited by his relying on the laws of free fall, which Galileo claimed to have inferred from experiments. This defect is corrected in Huygens's celebrated *Horologium oscillatorium* of 1673, a work meant for the improvement of pendulum clocks. There Huygens derives the laws of free fall from three "hypotheses":

1. If there were no gravity and if the air did not impede the motion of bodies, any body, once set into motion, would go on moving with an equal velocity on a straight line.
2. The action of gravity, whatever be its origin, is to make bodies assume a motion composed of the uniform motion they have in such-and-such direction and of the downward motion impressed by gravity.
3. These two motions can be considered independently of each other, and do not interfere with each other.

Taken together, these hypotheses imply that the motion of a body falling with a given initial velocity is obtained by compounding the motion it would have if it were originally at rest with the uniform motion it would have if gravity did not act. This rule is in harmony with the principle of relative motion, which Huygens had already used in his derivation of the rules of collision. Yet it does not quite result from this principle, because it further assumes that the motion of the body depends only on its initial velocity and not on the higher-order derivatives of the motion.[37]

[36] Mariotte 1679. Cf. Jouguet 1908, vol. 1, pp. 127–36.   [37] Huygens 1673, p. 21.

Let a body be dropped at time zero with no original velocity. At time τ, it has traveled the distance a and it has acquired the velocity b. According to Huygens's hypotheses, its motion between the times τ and 2τ is obtained by compounding the former motion with a uniform downward motion of velocity b. Therefore, at time 2τ the body has traveled the distance $2a + b\tau$ and it has acquired the velocity $2b$. By iteration of this reasoning, at time $n\tau$ the body has traveled the distance $na + n(n-1)b\tau/2$ and its velocity has acquired the value nb. Since the value of τ is arbitrary, this means that the velocity of the body increases linearly in time, and that the traveled distance increases quadratically in time. Consequently, this distance is proportional to the square of the acquired velocity.[38]

The necessity of this reasoning depends on the necessity of the hypotheses on which it is based. Since Huygens did no comment on the nature of these hypotheses, the best we can do is to guess his opinion on them by tracing their origins. They are all found in Galileo's and Descartes's works, though in different form and scope. As Huygens broadly accepted Descartes's variety of mechanical philosophy, he is likely to have shared his belief in the necessity of defining motion in a relative manner and in tracing any alteration of motion to the impact of bodies. It is therefore tempting to regard his derivation of the laws of fall as a mostly rational derivation. We will see that Euler later used similar reasoning to prove the necessity of Newton's second law. Newton himself plausibly drew on the *Horologium oscillatorium* in his formulation of his two first laws of motion. His second law can indeed be seen as a discrete version of Huygens's derivation of the laws of free fall.

The bulk of Huygens's treatise was of course devoted to the constrained fall that occurs in pendulum motion. In order to determine the period of oscillation of a rigid pendulum of any shape and loads, Huygens assumed the following extension of Torricelli's principle: at any given instant of the spontaneous motion of a rigidly connected system of (point-like) weights the center of gravity of these weights would always rise to the same height if the connections between them were suddenly removed and if their velocities were turned into additional elevations through properly inclined reflecting planes. In anachronistic symbols, this statement reads

$$\sum_{\alpha} m_{\alpha}(h_{\alpha} + v_{\alpha}^2/2g) = \text{constant}, \qquad (9)$$

where h_{α} and v_{α} denote the height and velocity of the weight α at the given instant, m_{α} denotes its mass, and g denotes the acceleration of gravity. For simplicity, suppose that all the weights of the pendulum are on the same plane and that the pendulum oscillates in its own plane. If θ denotes the angle by which the center of gravity of the pendulum deviates from the vertical, and if r_{α} denotes the distance between the weight α and the suspension point, Eq. (9) leads to

$$-\sum_{\alpha} m_{\alpha} r_{\alpha} g \cos\theta + \sum_{\alpha} m_{\alpha} r_{\alpha}^2 \dot{\theta}^2 = \text{constant}. \qquad (10)$$

---

[38] Huygens 1673, pp. 21–6. The reasoning implicitly assumes that a and b do not depend on the original height of the body (uniform gravity).

Therefore, the period of oscillation is the same as that of a simple pendulum of length:[39]

$$l = \sum_{\alpha} m_\alpha r_\alpha^2 \Big/ \sum_{\alpha} m_\alpha r_\alpha. \qquad (11)$$

After multiplication by g, Eq. (9) agrees with our modern expression of the conservation of the sum of the potential and kinetic energies of the system of weights. Huygens regarded his own statement as a consequence of the impossibility of perpetual motion, completed by the reversibility of the motion of the (undamped) pendulum. He did not mean to enunciate a new general law of mechanics. Rather, he meant to compensate for the contemporary ignorance of the mutual actions of the connected weights by appeal to principles that no one would deny. Similarly, in his earlier derivation of the conservation of the sum of the products of mass by squared velocity in elastic collisions, he had not given a name to this quantity and he had not tried to generalize its conservation to other mechanical processes.

Leibniz's dynamics

It was Gottfried Wilhelm Leibniz, who had benefited from Huygens's teachings during a Parisian sojourn, who truly perceived the necessity of a new conservation law. Despite his early embrace of Descartes's philosophy, Leibniz had come to believe that the re-duction of matter to pure extension was impossible. The collision between two bodies, he argued, could never imply bouncing if size was the only property that characterized them.[40] Moreover, the rigidity assumed by Descartes for the colliding bodies could not exist, for there was nothing to keep the parts of these bodies together in a purely geo-metrical concept of matter. Matter and its evolution had to be essentially continuous; and its substance had to be endowed with force, namely, with a God-given ability for change and action. Descartes's rules of collision contradicted the principle of continuity, since the slightest difference of size between the colliding bodies could imply a finite change in the outcome of the collision; they were obviously contrary to experience; and they relied on a conservation principle that led to the possibility of perpetual motion.[41]

In 1686, Leibniz published a brief essay on this last point. Inspired by Huygens's pen-dulum studies, he measured the "motive force" of a moving body by the height it would reach if its velocity were turned into an elevation (as occurs in a pendulum's swing or when a body is reflected by an elastic plane). Two bodies that have the same quantity of motion according to Descartes generally have different motive forces. Consequently, a perpetual motion can be built: Let a body of mass 2 acquire the velocity 1 by falling from a height of

---

[39] Huygens 1673, part 4, pp. 93–100. I have modernized Huygens's reasoning.

[40] Being given by God, Cartesian inertia is not purely geometric; still, it is purely passive, whereas action is needed for the bouncing of a body on another.

[41] On the evolution of Leibniz's views on matter, motion, and force, cf. Leibniz 1695, pp. 240–2; Costabel 1960; Duchesneau 1994; Fichant 1994; Garber 2009, pp. 129–55. I leave aside the most exotic aspects of Leibniz's doc-trine: the idea that any element of substance contains its own determination for change, irrespective of external actions on it; and the pre-established harmony which explains the compatibility of the internal determinations of the various bodies in nature.

1; by Descartes's measure of motion, it can be replaced by a body of mass 1 with velocity 2; the velocity of this new body can raise it to height 4, from which it can be made to act on a lever to raise the former body to height 2; the net result of the whole process is the raising of a body without any effort, which implies the possibility of perpetual motion. Conversely, the impossibility of the perpetual creation or destruction of motion implies the conservation of motive force defined as the sum of equivalent heights or the sum of the products of mass by velocity squared. This quantity is what Leibniz later called "live force" (*vis viva*), in opposition with the Newtonian "dead force" (*vis mortua*) impressed by a body on another. Of course, Leibniz knew that in the real world live force seemed to continually decrease. He solved this contradiction by assuming that any apparent loss of live force corresponded to its dissemination in small invisible parts of the interacting bodies.[42]

There were important theological stakes in Leibniz' introduction of force at the heart of what he called "dynamics" (from the Greek δύναμις for capacity or power). In his and many others' opinion, Descartes's system dangerously bordered on atheism by making the states and evolution of the world independent of divine providence. Leibniz was especially shocked by the section of the *Principia* in which Descartes claimed that the world could reach any given state from any initial state. In addition, Leibniz agreed with the Cartesian philosopher Nicolas Malebranche that the outcome of collisions could not be determined by purely geometrical considerations. God had to intervene in a direct or indirect manner. In his "occasionalism," Malebranche favored the direct manner of divine intervention, in which the encounter of two bodies provides God the opportunity (*occasion*) to intervene (according to stable rules). Leibniz preferred the indirect manner, in which God endowed matter with a force that determined its capacity for change. In brief, Leibniz shared Descartes's ambition of a mechanical explanation of the world, but his reducing dynamics, unlike Descartes's mechanics, involved a concept of active substance that could not be reduced to pure extension. Leibniz also differed from Descartes in his higher tolerance of final causes (what we would now call aim or purpose). He believed that such causes, for instance the principle that light takes the easiest path in a series of transparent bodies, were often more accessible to human inquiry and that they offered a pleasant demonstration of God's wisdom: God selects the best of all possible worlds.[43]

Leibniz famously introduced a sharp distinction between necessary and contingent truths. In his definition, necessary truths are those which can be reduced to an identity through analysis; in other words, their truth is a logical consequence of the definition of the terms they contain. In contrast, a contingent truth can only be shown to be the consequence of another truth. The reduction to identity is still possible, but only through an infinite chain of antecedents that is accessible to God only. For humans, contingent truths can only be known a posteriori, through experience, unless they can be traced to a natural expression of God's wisdom. As there are two kinds of truths, there are two basic principles

---

[42] Leibniz 1686; [1692], prop. 4 (Descartes leading to perpetual motion); 1695, p. 238 (*vis viva*); [1698], pp. 229–31 (dissemination of live force).

[43] Leibniz to Philippi, 1679–80 (dangers of the *Principia*), in Leibniz 1875–90, vol. 4, pp. 281–7; Leibniz 1695, p. 242 (occasionalism vs. Leibnizian force), pp. 242–3 (mechanical reduction), p. 243 (final causes). Cf. Garber 2009 (on active substance); Hankins 1967 (on Malebranche).

of reasoning: the principle of contradiction, from which necessary truths are derived; and
the principle of sufficient reason, which defines contingent truths:

> There are two first principles of all reasoning, the principle of contradiction, accord-
> ing to which every identical proposition is true and its contrary is false; and the
> principle that a reason must be given [*principium reddendae rationis*], according to
> which every true proposition that is not known per se has a proof from prior elements,
> or that a reason can be given for every truth, or, as is commonly said, that nothing
> happens without a cause. Arithmetic and geometry do not need this principle, but
> physics and mechanics do, and Archimedes employed it.

In his reference to Archimedes, Leibniz was alluding to the equilibrium of a symmetric
lever, which Archimedes could have derived by arguing that such a lever had no more
reason to move one way than the other (he, in fact, did not). Leibniz gave a general
description of this negative use of the principle of sufficient reason:[44]

> When two incompatible things are equally good, and when neither in themselves, nor
> by their combination with other things, has the one any advantage over the other,
> God will produce neither of them.

As we will see in a moment, this usage of the principle of sufficient reason became very
popular in eighteenth-century natural philosophy. It lends a sort of rational necessity to
some contingent truths, even though they do not result from the principle of contradiction.
According to Leibniz, some other contingent truths have a "moral" necessity: they find
their reason in God's wisdom. This is the case for the conservation of live force and for the
optical principle of the easiest path. In the best possible world, some quantities should be
conserved and some others should be a minimum. Unfortunately, this metaphysical argument does not tell which the conserved or minimized quantities are. In order to arrive at
his expression of live force, Leibniz had to appeal to Galileo's empirical laws of free fall.[45]

1.3   After Newton

Leibniz's and Newton's decisive contributions to the science of mechanics and the
needed calculus were roughly contemporary. The major Swiss and French contributors to
eighteenth-century mechanics combined them in diverse proportions. They were aware of
the conflict between Newton's empiricism and the rationalisms of Descartes and Leibniz,
and they addressed it in their reflections on the foundations of mechanics.

---

[44] Leibniz [*c*.1688], p. 309; 4th letter to Clarke, in Leibniz 1875–90, vol. 7, p. 374. On Leibniz and Archimedes'
lever, see Leibniz 1875–90, p. 301. The main purpose of Leibniz's distinction between contingent and necessary
truths was the conciliation of human freedom with God's perfection.

[45] Leibniz [*c*.1686]. Leibniz later tried rational deductions of the quadratic form of live force: cf. Garber 2009,
pp. 149–50.

Daniel Bernoulli on necessity and contingency

In 1726 the young Daniel Bernoulli graced the inaugural volume of the proceedings of the Petersburg Academy with an "Examination of the principles of mechanics," the purpose of which was to decide whether two principles of mechanics, Newton's second law and the parallelogram of forces, were contingent or necessary. The recent publication of an amplified version of Varignon's *Nouvelle mécanique*, which made the parallelogram of force the basis of all statics, presumably motivated Bernoulli to reflect on the nature of this principle. As Bernoulli knew, Newton and Varignon derived the parallelogram of forces from the vector composition of motions and from the relation between force and velocity. In the first part of his article, Bernoulli argued that Newton's second law, according to which force is proportional to acceleration, only was a "contingent truth," because other laws such as the proportionality between force and the square or the cube of the acceleration were equally conceivable. Only experience could decide between these various possibilities.[46]

In contrast, Bernoulli asserted that the parallelogram of forces was a "necessary truth" of the same sort as the truths of geometry. Specifically, he proved that this principle derived from three "hypotheses":

1. Two forces are "equivalent" if and only if they are parallel and have the same intensities and directions.
2. The joint application of two parallel forces is equivalent to a third parallel force whose intensity is the sum or the difference of the intensities of these forces, according as their directions are identical or opposite.
3. The joint application of two forces of the same intensity is equivalent to a force acting on the line bisecting the angle that they make.

Bernoulli justified the second and third hypotheses by appeal to Leibniz's principle of sufficient reason, according to which every truth must have a reason. If two forces are equal and opposite, there is no reason why the body to which they are applied would move one way or the other. Therefore, the body does not move. This result, together with the principle that a whole is the sum of its parts, implies hypothesis (2). Similarly, a body subjected to two forces of equal intensity has no more reason to move on one side of the bisecting line than it has on the other side. Therefore, it must move on the bisecting line. Since Bernoulli regarded the principle of sufficient reason as a necessarily true "metaphysical axiom," he also regarded its consequences as necessary true.[47]

---

[46] D. Bernoulli 1726, pp. 126–34. Cf. Dhombres 1987. As Bernoulli showed, the mentioned alternatives to Newton's second law entail a violation of the parallelogram of forces. Contrary to the opinion of some modern commentators, this does not imply that Newton's second law follows from the rule of the parallelogram. Indeed, laws in which the force is proportional to a linear combination of derivatives of the motion are compatible with the parallelogram of forces. At any rate, Bernoulli and Varignon both erred in asserting that the proportionality between force and velocity or acceleration implied the parallelogram of forces.

[47] D. Bernoulli 1726, pp. 134–42. According to Leibniz [c.1686], the principle of sufficient reason is implanted by God in our mind and confirmed by reason. Truths justified by sufficient reason (other than non-contradiction) are contingent. It is not clear, however, which status Leibniz accorded to truths derived by negative use of the principle of sufficient reason.

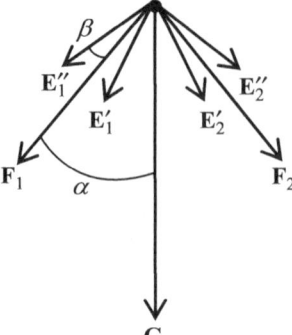

Fig. 1.12.   Diagram for proving the parallelogram of forces in
     the case of forces of equal intensity.

The most ingenious part of Bernoulli's article is his proof that the three hypotheses
imply the rule of the parallelogram. The following is a simplified and modernized version
of this proof. In Fig. 1.12, the force $\mathbf{G}$ is the resultant $\mathbf{F}_1 \oplus \mathbf{F}_2$ of two forces $\mathbf{F}_1$ and $\mathbf{F}_2$
of equal intensity F; the force $\mathbf{F}_1$ is the resultant of the forces $\mathbf{E}_1'$ and $\mathbf{E}_1''$; the force $\mathbf{F}_2$ is
the resultant of the forces $\mathbf{E}_2'$ and $\mathbf{E}_2''$; and the forces $\mathbf{E}_1', \mathbf{E}_1'', \mathbf{E}_2'$ and $\mathbf{E}_2''$ all have the same
intensity E. Dimensional homogeneity requires the relations

$$G = f(\alpha)F \quad \text{and} \quad F = f(\beta)E, \tag{12}$$

wherein f is a dimensionless function, α the angle that $\mathbf{F}_1$ and $\mathbf{F}_2$ make with $\mathbf{G}$, and β
the angle that $\mathbf{E}_1'$ and $\mathbf{E}_1''$ make with $\mathbf{F}_1$ (also the angle that $\mathbf{E}_2'$ and $\mathbf{E}_2''$ make with $\mathbf{F}_2$).
Now, by associativity and commutativity of the composition of forces (which Bernoulli
implicitly assumes), the force $\mathbf{G}$ can be regarded as the resultant of the forces $\mathbf{E}_1' \oplus \mathbf{E}_2'$ and
$\mathbf{E}_1'' \oplus \mathbf{E}_2''$, which are both parallel to $\mathbf{G}$. The intensities of these two forces are $f(\alpha - \beta)E$
and $f(\alpha + \beta)E$, since they make the angles $\alpha - \beta$ and $\alpha + \beta$ with $\mathbf{G}$. Consequently, we
have the identity

$$f(\alpha)f(\beta) = f(\alpha - \beta) + f(\alpha + \beta), \tag{13}$$

wherein the angles α and β can take any value.

Bernoulli did not have the modern concept of function, let alone the idea of a functional
equation. Instead he started from the evident case $f(\pi/2) = 0$, and obtained other values
of f for a sequence of other angles by repeated consideration of figures of the type of Fig.
1.12. The modern approach, inaugurated by d'Alembert and perfected by Poisson, consists
in solving the functional equation (13). Granted that the function f is differentiable, this
equation can be developed in powers of β to give

$$f(\alpha)\left[f(0) + \beta f'(0) + \frac{1}{2}\beta^2 f''(0) + \ldots\right] = 2f(\alpha) + \beta^2 f''(\alpha) + \ldots \tag{14}$$

Equating the terms of order 0, 1, and 2 yields

$$f(0) = 2, \ f'(0) = 0, \text{ and } f(\alpha)f''(0) = 2f''(\alpha). \tag{15}$$

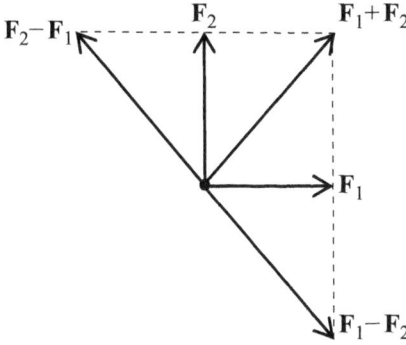

Fig. 1.13.   Diagram for proving the parallelogram of forces
in the case of perpendicular forces.

The solution of this system is

$$f(\alpha) = 2 \cos k\alpha, \text{ with } 2k^2 + f''(0) = 0. \tag{16}$$

The special case $\alpha = \pi/2$, for which f evidently vanishes, leads to the value 1 of the constant k. Consequently, the resultant of two forces of equal intensities is the diagonal of the parallelogram made by these two forces.[48]
    Now assume that the forces $\mathbf{F}_1$ and $\mathbf{F}_2$ have any intensity but make a right angle (see Fig. 1.13). Then the forces associated with the vectors $\mathbf{F}_1 + \mathbf{F}_2$ and $\mathbf{F}_1 - \mathbf{F}_2$ have equal intensity, so that

$$(\mathbf{F}_1 + \mathbf{F}_2) \oplus (\mathbf{F}_1 - \mathbf{F}_2) = (\mathbf{F}_1 + \mathbf{F}_2) + (\mathbf{F}_1 - \mathbf{F}_2) = 2\mathbf{F}_1. \tag{17}$$

Similarly, we have

$$(\mathbf{F}_1 + \mathbf{F}_2) \oplus (\mathbf{F}_2 - \mathbf{F}_1) = 2\mathbf{F}_2. \tag{18}$$

Consequently, we have

$$2\mathbf{F}_1 \oplus 2\mathbf{F}_2 = [(\mathbf{F}_1 + \mathbf{F}_2) \oplus (\mathbf{F}_1 - \mathbf{F}_2)] \oplus [(\mathbf{F}_1 + \mathbf{F}_2) \oplus (\mathbf{F}_2 - \mathbf{F}_1)] = 2(\mathbf{F}_1 + \mathbf{F}_2), \tag{19}$$

which means that the parallelogram rule applies to perpendicular forces. Then the rule holds for two arbitrary forces, because these forces can each be regarded as the resultant of two component forces directed along two given orthogonal axes.
    Has Bernoulli truly proven the necessity of the parallelogram of forces? The deduction from his three hypotheses is fairly solid, despite a few implicit additional assumptions of continuity, commutativity, and associativity. In his justification of the three hypotheses, the recourse to the principle of sufficient reason is unproblematic, because it boils down to considerations of symmetry: the effect must have the symmetry of the cause, if no asymmetric circumstance (for instance, a heterogeneous or anisotropic space) intervenes. More questionable are the definition of the equivalence of forces (hypothesis 1) and the

[48] Poisson 1811, pp. 11–19. On d'Alembert's contribution, cf. Dhombres and Radelet 1991.

additivity of intensities for parallel forces. Here Bernoulli seems to be confusing the properties of purely geometric objects with the properties of concrete objects, without realizing that this conflation presupposes the satisfaction of certain empirical criteria. As Hermann Helmholtz argued in the 1880s, the measurability of a quantity requires that concrete equality and concrete addition, defined through conceivable laboratory operations, share some properties of arithmetic equality and addition such as symmetry, transitivity, commutativity, and associativity. Bernoulli's hypotheses assume the measurability of forces through some unspecified procedure. Whether the criteria for this measurability are met is contingent.[49]

Euler's Mechanica

In 1736 Leonhard Euler published his *Mechanica*, aimed at clarifying the foundations and extending the applications of mechanics. In his introduction, he announced: "As [the principles of mechanics] have heretofore been insufficiently established, I demonstrate them in such a manner that they will be understood to be not only certain but even necessarily true." Somewhat like Newton, Euler assumed an absolute space conceived as a fictitious rigid body of infinite extension. He then derived the principle of inertia from the principle of sufficient reason, arguing that a body originally at rest in empty space could not start moving because it had no more reason to move one way than the other; that a body originally moving in empty space had to conserve the direction of its motion, since it had no more reason to be deflected one way than the other; and that its velocity modulus could not change since it had no more reason to decrease than to increase.[50]

Euler measured forces by static means, through their mutual compensation and addition in concrete devices. He regarded the parallelogram of forces as a necessary law, presumably because he agreed with Daniel Bernoulli's derivation. Like Huygens, he assumed that the effect of a given force on a body moving with a given original velocity could be obtained by compounding a uniform motion at this velocity with the motion that the body would acquire if it were starting from rest.[51] Consequently, for a given body there must be a one-to-one correspondence between its acceleration and the force to which it is subjected. For a given acceleration, the force must be proportional to the mass of the body defined as its "number of points." Euler's justified the latter law by replacing the mass M subjected to the force F with n masses M/n, each subjected to the force F/n. Lastly, Euler claimed that the parallelogram of forces implied the proportionality of force and acceleration for a given mass. His proof goes as follows.[52]

In Fig. 1.14, the body A is subjected to the equal forces AE and AF, whose resultant is AC. Euler mentally splits this body into two equal parts when the forces start to act; he

[49] On Helmholtz's reflections, see below, Chapter 6, pp. 195–7.

[50] Euler 1736, p. vi. Euler was not quite satisfied with this way of reasoning, because it implies the complete removal of all external bodies. In order to preserve the inertia of a body in the presence of distant matter, he made it an inherent property of the body, its "inertial force." He later himself denounced the obscurity of this concept.

[51] This assumption was part of Euler's definition of *vis absoluta* as a force whose effect does not depend on the state of motion of the body.

[52] Euler 1736, pp. 41 (measurement of forces), 46–52 (force and acceleration), 54–5 (force and mass), 58–60 (proportionality between force and acceleration).

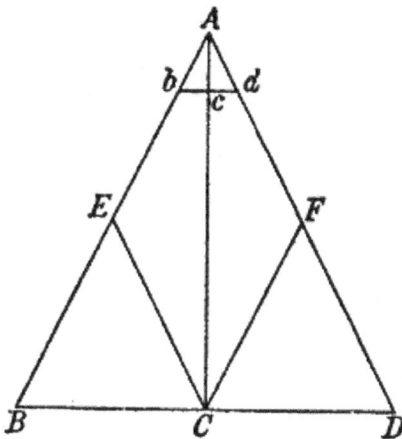

Fig. 1.14.   Diagram for Euler's derivation of the proportionality between force and acceleration. From Euler 1736, table II, fig. 3.

applies the force AE to one part, and the force AF to the other. After a given time, these two parts reach the points b and d. Euler then mentally reunites the two parts through an infinite cohesive force. The reconstituted A is found in the middle c of bd because it has no more reason to be found on one side of this point than on the other. Since the motion of a body is the same when mass and force are multiplied by the same number, the body A would move to b if it were subjected to the force AB, which is double of AE. Therefore, the distance Ac is to Ab what the force AC is to AB, which implies that the acceleration of a body is proportional to the force to which it is subjected.

Putting together Euler's various partial results, the acceleration of a body at a given time depends only on the applied force and on the mass of the body, and it is proportional to the former and inversely proportional to the latter. Taking into account the evident parallelism of force and acceleration, this is equivalent to the modern statement of the fundamental law of the mechanics of material points. Euler insisted that this law was "necessarily true" and that the alternative laws imagined by Bernoulli involved contradictions. The weakest point of his demonstration seems to be the derivation of the proportionality between force and acceleration, which implicitly assumes that the distance traveled by a body during a given time interval only depends on the resultant of the external forces applied to this body, whatever be the internal evolution of the body in this interval; the body can even be split and reunited through internal forces. This seems to be a strong assumption, tantamount to assuming that the center of gravity of a system of bodies (material points) always moves as if the total mass was concentrated on this center and the resultant of all external forces was applied to it.[53]

The principle of least action

In sum, Euler believed that the principles of mechanics derived from proper definitions of space and forces supplemented with the principle of sufficient reason. In the 1740s, he

[53]Euler 1736, p. 62. Cf. Jammer 1957, pp. 215–17; Pulte 1989, pp. 108, 115.

began to argue for a more metaphysical necessity of the laws of mechanics, based on the principle of least action. The idea of expressing a law of mechanics by the condition that a certain quantity is an extremum can be traced to the principle that a connected system of weights is in equilibrium if and only if the height of its center of gravity is a minimum. This principle has roots in the Aristotelian notion of a natural motion toward the center of the universe; it was exploited by Renaissance mechanicians like da Vinci and del Monte; and Torricelli generalized it to indifferent equilibrium.[54]

In 1740, the early French Newtonian Pierre-Louis Moreau de Maupertuis offered another sort of generalization, in which the forces acting on the material points of the connected system are any central forces (expressible as a power of the distance from fixed centers). In modern parlance, the quantity to be minimized is the total potential V of the forces. In the first case treated by Maupertuis, that of a rigid body able to rotate around a fixed point, the variation of the position $\mathbf{r}_\alpha$ of the point α of the body is restricted to the form $\delta\mathbf{r}_\alpha = \delta\boldsymbol{\theta} \times \mathbf{r}_\alpha$. The resulting variation of the potential V,

$$\delta V = -\sum_\alpha \mathbf{f}_\alpha \cdot \delta\mathbf{r}_\alpha = -\delta\boldsymbol{\theta} \cdot \sum_\alpha (\mathbf{r}_\alpha \times \mathbf{f}_\alpha), \text{ with } \mathbf{f}_\alpha = -\nabla_\alpha V, \tag{20}$$

vanishes if and only if the sum of the moments of the forces vanishes, in conformity with the general law of the lever.[55]

Four years later, Maupertuis introduced another minimum principle in the context of the dynamics of light corpuscles. From Fermat and Leibniz he knew that during refraction light takes the path that makes the sum $n_1 AI + n_2 IB$ a minimum, where n_1 and n_2 are the optical indices of the two media, A and B are the fixed extremities of the path, and I is its inflection point at the interface between the two media. In conformity with Newton's corpuscular theory of light, Maupertuis made the index n in a medium proportional to the velocity v of the particles of light in this medium. The minimized quantity then becomes the "quantity of action" $v_1 AI + v_2 IB$. As Leibniz had done for his principle of least resistance, Maupertuis regarded the truth of the principle of least action as a proof of God's wisdom:

> One cannot doubt that every thing is ruled through a supreme Being who, while he endowed matter with forces that denote his power, has destined it to execute effects that show his wisdom; and the harmony of these two attributes is so perfect that every effect of Nature could probably be deduced from each of them separately.

Maupertuis nonetheless deplored the difficulty of identifying the quantity that God had chosen to minimize, as witnessed by Fermat's and Leibniz's conflicting interpretations of the optical principle.[56]

---

[54] On this prehistory, cf. Jouguet 1908, vol. 1, pp. 59–61. On the history of variational principles in mechanics, cf. Lanczos 1966; Szabò 1979; Fraser 1983; Pulte 1989.

[55] Maupertuis 1740. Cf. Pulte 1989, pp. 46–9.

[56] Maupertuis 1744, p. 425. Cf. Pulte 1989, pp. 49–56; Panza 1995, pp. 461–5. Leibniz introduced the name action for the product of mass, length, and velocity. He may even have anticipated Maupertuis's principle. Samuel König alleged so much in 1707, thus initiating a famous controversy.

To a treatise on problems of maximum and minimum published in the same year 1744, Euler appended a proof that the trajectory of a particle subjected to central forces was such that the integral $\int mv\,ds$ over a path of fixed extremities was a maximum for a given value of the first integral of motion

$$E = \frac{1}{2}mv^2 + V(\mathbf{r}). \tag{21}$$

In a modernized proof, the variation that must vanish reads

$$\delta \int v\,ds = \int \delta v\,ds + \int v\delta\,ds, \tag{22}$$

with $\delta v = -(\nabla V/mv) \cdot \delta\mathbf{r}$ from varying Eq. (21), and $\delta\,ds = (d\mathbf{r}/ds) \cdot \delta\mathbf{r}$ from varying $ds^2 = d\mathbf{r} \cdot d\mathbf{r}$. Integrating by parts the second integral and equating the varied action to 0 yields

$$-\frac{\nabla V}{mv} - \frac{d\mathbf{v}}{ds} = 0 \quad \text{or} \quad \mathbf{f} = -\nabla V = m\frac{d\mathbf{v}}{dt}, \tag{23}$$

which is the fundamental law of dynamics.[57]

In the same treatise Euler asserted a Leibnizian faith in final causes:

> As the construction of the world is most perfect and as it is the doing of the wisest Creator, nothing happens in the entire world that does not result from some property of minimum or maximum. Consequently, there is no doubt that every effect of the world can be determined from final causes through the method of maxima and minima with as much success as it can be from efficient causes.

Like Leibniz, Euler anticipated that in many cases the efficient causes would remain hidden so that the method of maxima and minima would be the only available one. He nonetheless recognized that the expression of the quantity to be minimized was hard to reach by metaphysical means. For a particle in a force field, his guess was that the *motus collectivum* (total motion), which is the line integral of the momentum, should be a minimum. Euler later credited Maupertuis for introducing the principle of least action, after Maupertuis congratulated him for generalizing this principle to the motion of planets. On this occasion, Maupertuis offered a proof that both in hard and in elastic collisions a certain quantity was a minimum and he interpreted this quantity as the quantity of action. He also claimed that his principles amounted to a proof of the existence of God:[58]

> What a satisfaction for the human mind, in contemplating these laws, which are the principle of Motion and of Rest of all Bodies in the Universe, to find therein a proof of the existence of the One who rules it!

[57] Euler 1744, pp. 318–20. Cf. Panza 1995, pp. 465–94.

[58] Euler 1744, pp. 245 (citation), 309 (difficulty of finding the minimized quantity); Euler 1748; Maupertuis 1746, p. 276. In the notation of p. 16, the minimized quantity in collisions is $m_1(\mathbf{v}'_1 - \mathbf{v}_1)^2 + m_2(\mathbf{v}'_2 - \mathbf{v}_2)^2$. Cf. Szabó 1979, Chap. 2; Pulte 1989, pp. 70–5; Panza 1995.

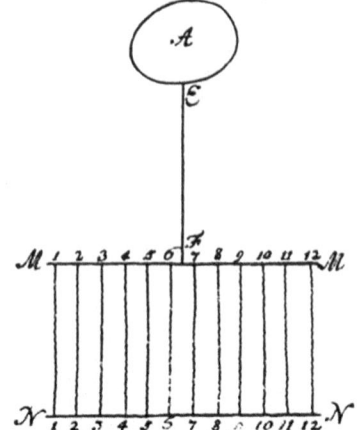

Fig. 1.15.   Euler's force synthesizer. The elastic threads 1–1 to
12–12 pull the rigid bar MM and the attached body A with a
force proportional to their number. From Euler 1751b, p. 246.

Notwithstanding with his earlier cited reference to God's wisdom, Euler usually avoided
theological considerations in his physics. Instead, in 1750 he offered what he called a
"metaphysical" derivation of Maupertuis's principles of equilibrium and motion. The basis
of this derivation was a conceptual analysis of the basic concepts of force and motion.
Euler there defined force as the cause of a change in the state of motion of a body, and
measured its intensity F in a semi-concrete manner by the number of threads of unit tension necessary to produce it (see Fig. 1.15). Assuming that this tension does not depend
on the length x of the threads, the force tends to increase the product Fx, more generally
the integral $\int F dx$ when the force is variable. When a (connected) system of bodies is subjected to a number of antagonistic forces, the sum V of these integrals for all the applied
forces should be a minimum because the generating threads contract as much as they can.
In the case of motion from a given configuration to another, Euler found it "very natural"
to require that the sum $\int V dt$ of all the intermediate values of the function V should be a
minimum. For a conservative system, $V = E - T$, where E is a constant and T is half the
live force of the system (our kinetic energy). Therefore, the integral $\int T dt$ should be a maximum. In the case of a single particle, $T = (1/2)mv^2$, and this integral is proportional to the
integral $\int mv ds$ that Euler had made a minimum in his treatise of 1744. He brushed away
criticism of the transition from a minimum to a maximum property as "pure chicanery,"
since the mathematical condition (vanishing variation) was the same in both cases.[59]

Thus, Euler had two ways to show the necessity of the laws of mechanics. In the first, he
derived the law of inertia and the relation between force and acceleration from the principle
of sufficient reason and from his definition of force. In the second, he derived the principles
of minimal potential and least action from a (meta)physical interpretation of the potential
as a sum of competing endeavors. Both approaches involve a few implicit assumptions and
some hazardous reasoning. For instance, the first approach implicitly assumes the relativity

---

[59] Euler 1748b, pp. 205–7 (threads); 217–18 (*très naturel*); 1751a, p. 159 (*chicanes*); 1751b (*démonstration
métaphysique*). Cf. Pulte 1989, pp. 188–92.

principle and the absence of higher order derivatives in the equation of motion. The reasoning behind the principle of least potential is merely intuitive, and a naïve analogy is involved in going from the principle of least potential to the principle of least action. We will later see whether, despite these shortcomings, something can be saved of Euler's rationalism.

In the *Mechanica* and in his memoirs on least action, Euler did not dwell on the true origin of forces. There is no doubt, however, that he preferred Cartesian contact action to Newtonian action at a distance. For optics, gravitation, and magnetism he defended ether theories that had a partial resemblance with the speculations of Descartes and contemporary French Cartesians. In a memoir of 1750 on the origin of forces, he argued that every force should be traced to the inertia and impenetrability of matter. This is not to say that he wished to eliminate force from the edifice of mechanics. Unlike Descartes, he did not try to derive the outcome of collisions from merely geometric considerations of ideally hard bodies. On the contrary, he introduced short-distance forces that expressed the impenetrability of the colliding bodies in conformity with Maupertuis's principle. In his pioneering investigations of the motion of rigid, elastic, and fluid bodies, he developed a Newtonian approach in which each element of the body obeyed the fundamental law of dynamics if only the applied force included internal forces of pressure or cohesion. As a result of this force-based practice, Euler's Cartesianism remained very discreet.[60]

D'Alembert's dynamics

It was the French philosopher-mathematician Jean le Rond d'Alembert who, in his influential *Traité de dynamique* of 1743, revived the Cartesian dream of deriving the laws of mechanics from the mere impenetrability of bodies. Being a fervent supporter of Newton's celestial mechanics, d'Alembert could not follow Descartes's philosophy in all respects. He rejected the materiality of pure extension, and defined motion as Newton and Euler had done: as the displacement of bodies in absolute space. He did not pursue the reduction of action at a distance to contact action through a medium; like Newton, he contented himself with a mathematical expression of this action. And he did not require the conservation of motion; on the contrary, we will see that the destruction of motion was the central concept of his dynamics.[61]

D'Alembert nevertheless followed Descartes and his disciple Malebranche in their rejection of forces at the foundational level of mechanics. In Descartes's system, direct contact and inertia determined all motion in the universe. For Malebranche, the contact of colliding particles was the "occasion" for God to intervene in their motion (according to rule); there still was no force in this mechanics, because force or power was a divine attribute, not to be found in matter. D'Alembert's own rejection of forces rested on his criticism of the vagueness of metaphysical principles usually evoked in this context, such as the proportionality between cause and effect. In his eyes the quarrel of live forces, which had

---

[60]Euler 1736, pp. 11–12 (all forces from motion); 1750a (Newtonian approach); 1750b (origin of forces). On Euler's Cartesianism, cf. Pulte 1989, pp. 108–10, 117–18, 170–81.

[61]D'Alembert 1743. Cf. Hankins 1970; Firode 2001. D'Alembert made clear that his use of the world "dynamics" for the general theory of motion only was a concession to contemporary usage, and did not imply a force-based approach.

long opposed supporters of the Leibnizian and Newtonian definitions of force, derived from the metaphysical hardening of concepts that were merely derivative.[62]

D'Alembert prided himself on having banished the causal relation between force and acceleration from his mechanics. In his system, the formula $\mathbf{f} = m\mathbf{a}$ is only a definition of force, the only one available since the true cause of the motion is forever unknown; force has no independent existence. Thus, the gravitational force acting on a part of a connected system of masses is defined through the acceleration that this part would acquire if it were disconnected from the other parts of the system. *Pace* d'Alembert, it should be noted that the contrafactual character of this definition implies the existence of a cause of motion that exists independently of its being effective. In the end, d'Alembert did not truly differ from Newton in his treatment of distance forces, and he implicitly assumed Newton's second law. The true originality of d'Alembert's dynamics lies elsewhere: in the elimination of all contact forces in hard-body collisions and in connected systems.

This elimination enabled d'Alembert to derive the fundamental laws of motion "from the mere consideration of motion." He had three such laws: the law of inertia, the law of compound motion, and the law of equilibrium. His statement and justification of the law of inertia did not significantly differ from Euler's: he appealed to the principle of sufficient reason, in a manner he believed to be more rigorous than Euler's. The law of compound motions states that the motion caused by the simultaneous application of two impulsive forces (*puissances*) to a material point (*corps*) is obtained by forming the parallelogram of the motions caused by each of these forces separately. This is the rule of the parallelogram of forces, which Newton had derived from his second law of motion (remember that in this derivation, the second impulse comes right after the first and increases the momentum vector by a specific amount). Since d'Alembert did not assume Newton's second law, he needed to retrieve its basic content, namely: the momentum change induced by a given impulse does not depend on the initial velocity of the body. He did so in a tortuous reasoning involving a material plane sliding at the velocity induced by the first impulse through coulisses themselves sliding at the velocity induced by the second impulse. This argument seems to implicitly assume the relativity principle. D'Alembert was also aware of Daniel Bernoulli's "highly ingenious" derivation of the parallelogram of forces, which he later simplified by introducing functional methods.[63]

Even though d'Alembert's second law is usually identified with the parallelogram of forces, it truly amounts to an indirect formulation of the original form of Newton's second law because both of these laws concern the application of successive impulses to a body. As for d'Alembert's third law, the law of equilibrium, it is closely related to Newton's third law. It states that two colliding hard bodies come to rest if and only if their velocities are in inverse proportion to their masses. As Newton explained, the latter law is equivalent to balancing the impulsive momentum changes of the two bodies, as required by the equality of action and reaction (Newton's third law). D'Alembert's derivation starts with the case of two equal bodies with opposite velocities, for which equilibrium results from symmetry.

---

[62] D'Alembert 1743, pp. xi–xii, 18.

[63] D'Alembert 1743, pp. xvi (citation), 3–8 (inertia), 22–5 (compound motions); D'Alembert 1758, *Avertissement* (praising D. Bernoulli). D'Alembert calls his three fundamental laws *principes* in his foreword; in the main text the first law is a *loi*, the second and third are *théorèmes*.

For him as for most of his contemporaries, the mass of a body is defined by quantity of matter, that is, by the volume of its material parts (which have unit density by definition). Consequently, two bodies of commensurable masses can be regarded as made of a different numbers of blocks that are all equal. Call n_1 and n_2 two numbers of blocks, $\mathbf{v}_1$ and $\mathbf{v}_2$ their common velocities. If the global momenta of the two sets of blocks are equal and opposed, then $\mathbf{v}_1/n_2 = -\mathbf{v}_2/n_1 = \mathbf{u}$. Each block of the first body acts like n_2 blocks of velocity $\mathbf{u}$, and each block of the second body acts like n_1 blocks with the velocity $-\mathbf{u}$, so that $n_1 n_2$ blocks of velocity $\mathbf{u}$ act on $n_1 n_2$ blocks of velocity $-\mathbf{u}$. Granting this sort of decomposition, equilibrium must occur between the two bodies. The case of incommensurable masses is easily obtained by rational approximation.[64]

As d'Alembert noted in the second edition of his treatise, this argument only proves that the equality of momenta is a sufficient condition of equilibrium. In order to prove the necessity of this condition, he introduced (by means of a concrete sliding plane) the reference frame that moves at the velocity of the center of gravity of the two bodies. In this frame, the momenta of the two colliding bodies are equal and opposite; therefore, they are in equilibrium and their final relative velocity must vanish; hence they cannot be in equilibrium in absolute space unless the velocity of the center of gravity vanishes.[65]

D'Alembert claimed that he could derive the entire theory of mechanics from his three principles. Let us first see how he derived the law of the lever. An obvious case of equilibrium is that of a straight lever subjected to equal and opposed forces on the line of the lever. Now consider the lever ACB represented by the curved line in Fig. 1.16, and subjected to the forces AH and BE. According to d'Alembert's second principle, the force AH can be replaced by the forces AG and AK, which act in the directions of AC and AB. Similarly, the force BE can be replaced by the forces BF and BP. The forces AG and BF have no effect on the lever, since their lines of action pass through the fulcrum C of the lever. The forces AK and BP are in equilibrium if and only if they have equal intensity, because they have the same effect on the curved lever ACB as they would have on the straight lever AB. Consequently, the condition of equilibrium is the equality of the moments AH × CM and BE × CV. Although d'Alembert was aware of Newton's and Varignon's earlier derivation

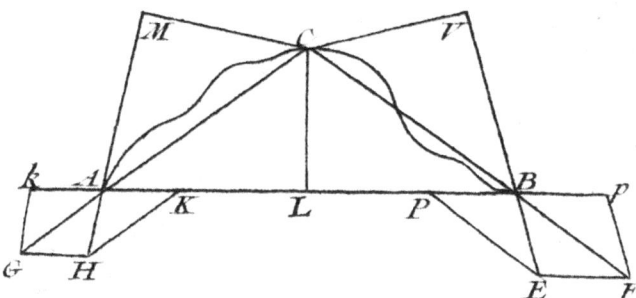

Fig. 1.16.   D'Alembert diagram for his derivation of the law the lever. From d'Alembert 1743, plate.

[64]D'Alembert 1743, pp. 37–40.   [65]D'Alembert 1758, pp. 53–4.

of the law of the lever from the parallelogram of forces, he preferred his own for being based on his definition of equilibrium through opposed momenta.[66]

D'Alembert did not dwell on statics, for he judged that his readers would already know that the rules of equilibrium of any machine derived from the law of the lever or from the parallelogram of forces. He contented himself with enunciating the general law that "the forces should be as the inverse of the velocities, estimated in the direction of these forces," which is a succinct formulation of the principle of virtual work.[67]

D'Alembert's statics is a statics inasmuch as it deals with the traditional problems of the equilibrium of machines. Its generating concept is the dynamical concept of destroyed motion owing to the impenetrability of bodies. Equilibrium amounts to the mutual destruction of the motions impressed on the various parts of the system. When this destruction is incomplete, motion ensues. In this case, d'Alembert may have found inspiration in Newton's following commentary to his third law: "If a body impinge upon another, and by its force change the motion of the other; that body . . . (because of the equality of the mutual pressure) will undergo an equal change, in its own motion, towards the contrary parts." In modern terms, this means that even when the bodies are not brought to rest by the collision, the momentum variations of the two bodies are equal and opposite. As indicated in the parentheses, Newton derived this rule from the equality of the action and reaction. In d'Alembert's philosophy of motion, there is no force acting between the two frontally colliding hard bodies, and the rule becomes the principle that the destroyed motions (momenta) of the two bodies must balance each other. In symbols,

$$(m_1\mathbf{v}_1 - m_1\mathbf{v}'_1) + (m_2\mathbf{v}_2 - m_2\mathbf{v}'_2) = \mathbf{0}, \text{ (with } \mathbf{v}'_1 = \mathbf{v}'_2). \tag{24}$$

This is a simple case of d'Alembert's general idea that the motion of a system of bodies is obtained by applying the equilibrium condition of the system to the motions destroyed by the contact of its rigid, impenetrable parts.[68]

Another simple case, the first encountered in d'Alembert's treatise, is the impact of a hard ball on an infinite rigid plane of infinite mass. If the incidence of the ball is normal to the plane, by the principle of sufficient reason it cannot move sideways from the impact point. By the same principle, it cannot rebound because there would be no preferred value for a non-zero rebound velocity. Therefore, the ball must come to rest. In other words, the ball and the plane are in mutual equilibrium for any initial velocity of the ball. In the case of oblique incidence, d'Alembert's second principle allows him to decompose the initial motion of the ball into a motion perpendicular to the plane and a motion parallel to the plane. The former motion is destroyed by the previous argument, and the former is unaffected because it is obviously unimpeded by the plane. Varignon had used a similar argument in his treatise.[69]

In the general case of a system of bodies on which a given motion $m_\alpha \mathbf{v}_\alpha$ is externally impressed on the body α of mass m_α, d'Alembert decomposed this motion into the

[66]D'Alembert 1743, pp. 43–4. D'Alembert treated the case of a flat angle ACB by taking the limit of a nearly flat angle.

[67]D'Alembert 1743, p. 47.   [68]Newton 1687, p. 13; 1729, p. 20; D'Alembert 1743, p. 138.

[69]D'Alembert 1743, pp. 31–3; Varignon 1725, vol. 1, p. 25. Cf. Schmit 2011.

motion $m_\alpha \mathbf{v}'_\alpha$ effectively acquired by the body and the motion $m_\alpha \mathbf{v}_\alpha - m_\alpha \mathbf{v}'_\alpha$ destroyed by its connection or collision with the other bodies of the system. Then he subjected these destroyed motions to the conditions of equilibrium of the system. This is what is now called d'Alembert's principle. D'Alembert believed the principle to be a trivial consequence of his definition of equilibrium by the mutual destruction of motion. Thereby he implicitly assumed that the rules for the mutual destruction of motion in the system did not depend on whether motion occurred in the original and final states of the system.[70]

There are essentially two classes of applications of d'Alembert's principle. In the first class, the bodies are disconnected and they collide. The impressed motion is the initial momentum of the bodies, and the acquired motion is their momentum after collision. Two simple members of this class have already been treated. The second class is that of machines made of threads, pulley, levers, pivots, etc., that is, the "connected systems" of modern analytical mechanics. Of course, all rigid components must be regarded as ideally rigid, all threads as perfectly inextensible and flexible, and all pivoting and all sliding as entirely frictionless. If, at a given time, the parts of the machine initially move at the velocity $\mathbf{v}_\alpha$ and if they are subjected to the external impulses $\mathbf{I}_\alpha$, the impressed motion is the momentum $m_\alpha \mathbf{v}_\alpha + \mathbf{I}_\alpha$ that the particle α would acquire if it were free to change its motion under the sole effect of the impulse $\mathbf{I}_\alpha$. The acquired motion is the momentum $m_\alpha \mathbf{v}'_\alpha$ that the particle takes when the connections of the system are taken into account, and the destroyed motion is $\mathbf{I}_\alpha - m_\alpha \mathbf{v}'_\alpha - m_\alpha \mathbf{v}_\alpha)$. If a power such as gravitation is continually acting on the system, d'Alembert follows Newton in regarding this power as the dense repetition of impulses. During a time interval dt large enough to contain a large number of impulses and small enough for the change in the configuration of the system to remain small, the net destroyed motion is $\mathbf{f}_\alpha dt - d(m_\alpha \mathbf{v}_\alpha)$, wherein $\mathbf{f}_\alpha dt$ is the momentum that the particle α would acquire during the time dt if it were free to move under the acting power. Consequently, if $\mathbf{a}_\alpha$ denotes the acceleration of the particle α at a given instant, d'Alembert requires that the conditions of equilibrium of the system should be met when it is subjected to the forces $\mathbf{f}_\alpha - m_\alpha \mathbf{a}_\alpha$ at this instant.

The first example of application that d'Alembert gave of this principle is the compound pendulum, which had been the object of numerous studies since Huygens's pioneering solution. The compound pendulum is a particular case of the problem of a rigid body, free to rotate around a fixed point O, and subjected to forces acting on its particles. In the former notation, the destroyed motion of the particle α of this body is $\mathbf{f}_\alpha - m_\alpha \mathbf{a}_\alpha$. According to d'Alembert's principle, the body must be in equilibrium with respect to these destroyed motions. Therefore, the sum of their moments must vanish:

$$\sum_\alpha \mathbf{r}_\alpha \times (\mathbf{f}_\alpha - m_\alpha \mathbf{a}_\alpha) = 0. \tag{25}$$

This may be rewritten as

$$\sum_\alpha \mathbf{r}_\alpha \times \mathbf{f}_\alpha = \sum_\alpha \mathbf{r}_\alpha \times m_\alpha \mathbf{a}_\alpha = \frac{d}{dt} \sum_\alpha \mathbf{r}_\alpha \times m_\alpha \mathbf{v}_\alpha, \tag{26}$$

[70] D'Alembert 1743, pp. 49–51.

which is the now classical theorem according to which the moment of forces applied to the rotator is equal to the time derivative of its angular momentum. In the case of the pendulum, $\mathbf{v}_\alpha = \boldsymbol{\omega} \times \mathbf{r}_\alpha$, where $\boldsymbol{\omega}$ is the angular velocity. Therefore, the equation of motion of flat pendulum oscillating in its own plane reduces to

$$\sum_\alpha \mathbf{r}_\alpha \times m_\alpha \mathbf{g} = \dot{\boldsymbol{\omega}} \sum_\alpha m_\alpha r_\alpha{}^2. \tag{27}$$

Comparison with the equation of a simple pendulum leads to the expression (11)

$$l = \sum_\alpha m_\alpha r_\alpha{}^2 \Big/ \sum_\alpha m_\alpha r_\alpha$$

for the length of the simple pendulum that has the same period of oscillation, in conformity with Huygens's result.[71]

Some forty years earlier Jacob Bernoulli (Johann's elder brother) had published a very similar derivation of this equivalent length, based on the idea that the discrepancies between the impressed and actual motions of the various masses of the pendulum should balance each other according to the law of the lever. Jakob Hermann, Johann Bernoulli, and Leonhard Euler later used slightly different and less justified procedures, by which they equated the sum of the moments of the weights $m_\alpha \mathbf{g}$ to the sum of the moments of the accelerative forces $m_\alpha \mathbf{a}_\alpha$. D'Alembert seems to have been aware of the latter derivations only. Possibly, he arrived at his principle of dynamics by reformulating this condition as the vanishing of the sum of the moments of the differences $m_\alpha \mathbf{g} - m_\alpha \mathbf{a}_\alpha$. More plausibly, he followed the consequences of his definition of equilibrium as the mutual destruction of motion and the hints provided by Newton's and Varignon's understanding of momentum variation in collisions.[72]

Modern Newtonian solutions of the compound pendulum problem involve internal forces of cohesion which are eliminated by taking the sum of moments. The great advantage of d'Alembert's method is that it gives an unambiguous prescription for finding the equation of motion directly in terms of physically accessible quantities, without introducing internal forces that are entirely fictitious in ideal connected systems. To this day, it remains the most general foundation of dynamics. Other principles, be they Newton's laws, the principle of least action, or Lagrange's equations, apply only to restricted classes of mechanical systems.

Here is briefly how other mechanical principles can be derived from d'Alembert's for properly restricted classes of system. The most efficient way of reasoning appeared in 1788 in the *Méchanique analitique* of the Franco-Italian mathematician Joseph Louis Lagrange. It is based on d'Alembert's principle in conjunction with an analytical formulation of the principle of virtual work. According to the latter principle, a connected system subjected to the forces $\mathbf{F}_\alpha$ is in equilibrium if and only if the sum of the virtual works $\mathbf{F}_\alpha \cdot \delta \mathbf{r}_\alpha$ vanishes for any virtual displacement $\delta \mathbf{r}_\alpha$ (i.e., for any infinitesimal displacement compatible with the connections). D'Alembert's principle then implies that for any possible motion of the system,

---

[71] D'Alembert 1743, pp. 69–70.   [72] Jacob Bernoulli 1703. Cf. Vilain 2000.

$$\sum_{\alpha} (\mathbf{F}_\alpha - m_\alpha \ddot{\mathbf{r}}_\alpha) \cdot \delta \mathbf{r}_\alpha = 0 \tag{28}$$

at any time and for any virtual displacement $\delta \mathbf{r}_\alpha$. As a special case of virtual displacement we may take the actual displacement $\mathbf{v}_\alpha dt$ of the system. This yields

$$\sum_{\alpha} (\mathbf{F}_\alpha - m_\alpha \ddot{\mathbf{r}}_\alpha) \cdot \mathbf{v}_\alpha = 0, \tag{29}$$

or

$$\sum_{\alpha} \mathbf{F}_\alpha \cdot \mathbf{v}_\alpha = \frac{dT}{dt}, \tag{30}$$

with

$$T = \sum_{\alpha} \frac{1}{2} m_\alpha v_\alpha^2. \tag{31}$$

In modern parlance, the work of the applied forces is equal to the variation of the kinetic energy of the system. In the absence of applied forces, the kinetic energy of a connected system is a constant. If the forces $\mathbf{F}_\alpha$ all derive from the potential $V(\mathbf{r}_1, \mathbf{r}_2, \dots \mathbf{r}_\alpha, \dots)$ (so that $\mathbf{F}_\alpha = -\nabla_\alpha V$), as is the case if they are gravitational forces, then the quantity $T + V$ is conserved.[73]

This derivation presupposes the continuity of motion. When hard-body collisions occur within the system, d'Alembert's principles lead to a loss of energy. D'Alembert illustrated the relation between continuity and conservation in the simple case of a particle moving along a polygonal wall. At each corner of the wall, there is a loss of motion. This loss is an infinitesimal quantity of second order when the sides of the polygon become vanishingly small. Consequently, in this limit the velocity of the particle is conserved (its energy as well).[74]

For a free rigid body, equilibrium under the forces $\mathbf{F}_\alpha - m_\alpha \ddot{\mathbf{r}}_\alpha$ requires that

$$\sum_{\alpha} \mathbf{F}_\alpha = \sum_{\alpha} m_\alpha \ddot{\mathbf{r}}_\alpha = \frac{d}{dt} \sum_{\alpha} m_\alpha \dot{\mathbf{r}}_\alpha, \tag{32}$$

and

$$\sum_{\alpha} \mathbf{r}_\alpha \times \mathbf{F}_\alpha = \sum_{\alpha} \mathbf{r}_\alpha \times m_\alpha \ddot{\mathbf{r}}_\alpha = \frac{d}{dt} \sum_{\alpha} \mathbf{r}_\alpha \times m_\alpha \dot{\mathbf{r}}_\alpha. \tag{33}$$

In common parlance, the resultant of all applied forces is equal to the rate of variation of the momentum of the solid; and the moment of all applied forces is equal to the rate of variation of the angular momentum of the solid.[75]

---

[73] Lagrange 1788.   [74] D'Alembert 1743, pp. 34–7.

[75] *Pace* Truesdell 1968, the principle of angular momentum only is a derived theorem in the d'Alembert–Lagrange approach.

Now consider an evolution $\mathbf{r}_\alpha(t) + \delta\mathbf{r}_\alpha(t)$ of a connected system, close to its actual motion $\mathbf{r}_\alpha(t)$ and compatible with its constraints. At every instant t, the displacement $\delta\mathbf{r}_\alpha(t)$ is a virtual displacement. Therefore, we have

$$\sum_\alpha [\mathbf{F}_\alpha(t) - m_\alpha \ddot{\mathbf{r}}_\alpha(t)] \cdot \delta\mathbf{r}_\alpha(t) = 0 \tag{34}$$

at every instant. Integrating between the times t_0 and t_1, we also have

$$\int_{t_0}^{t_1} \sum_\alpha \mathbf{F}_\alpha \cdot \delta\mathbf{r}_\alpha \mathrm{d}t = \int_{t_0}^{t_1} \sum_\alpha m_\alpha \ddot{\mathbf{r}}_\alpha \cdot \delta\mathbf{r}_\alpha \mathrm{d}t. \tag{35}$$

Anyone familiar with the calculus of variations recognizes in the right-side integral the variation

$$-\delta\int_{t_0}^{t_1} T \mathrm{d}t = -\delta\int_{t_0}^{t_1} \sum_\alpha \frac{1}{2} m_\alpha \dot{\mathbf{r}}_\alpha^2 \mathrm{d}t = -\int_{t_0}^{t_1} \sum_\alpha m_\alpha \dot{\mathbf{r}}_\alpha \cdot \delta\dot{\mathbf{r}}_\alpha \mathrm{d}t = \int_{t_0}^{t_1} \sum_\alpha m_\alpha \ddot{\mathbf{r}}_\alpha \cdot \delta\mathbf{r}_\alpha \mathrm{d}t \tag{36}$$

for fixed values of $\mathbf{r}_\alpha(t_0)$ and $\mathbf{r}_\alpha(t_1)$. If the forces derive from the potential V, the integral on the left-hand side of Eq. (35) is the sign-reversed variation of its time integral. Hence the variation of the quantity

$$S = \int_{t_0}^{t_1} (T - V) \mathrm{d}t \tag{37}$$

vanishes for any variation of its trajectory (compatible with the constraints) if and only if the equations of motion are satisfied. This is the principle of least action in William Rowan Hamilton's formulation.[76]

Let us now restrict the attention to connected systems such that the constraints are integrable. This means that the positions $\mathbf{r}_\alpha$ of the particles of the system can be expressed as functions of independent coordinates $q_1, q_2, \ldots q_i, \ldots q_n$, wherein n is the number of degrees of freedom of the system. Call L the difference $T - V$ expressed as a function of the variables $q, \dot{q}, t$. Varying the action (37) leads to

$$\delta S = \int_{t_0}^{t_1} \sum_i \left(\delta q_i \frac{\partial L}{\partial q_i} + \delta\dot{q}_i \frac{\partial L}{\partial \dot{q}_i} \right) \mathrm{d}t = \sum_i \int_{t_0}^{t_1} \delta q_i \left(\frac{\partial L}{\partial q_i} - \frac{\mathrm{d}}{\mathrm{d}t} \frac{\partial L}{\partial \dot{q}_i} \right) \mathrm{d}t. \tag{38}$$

The coordinates q_i being independent, the vanishing of this variation requires that

$$\frac{\mathrm{d}}{\mathrm{d}t} \frac{\partial L}{\partial \dot{q}_i} - \frac{\partial L}{\partial q_i} = 0 \tag{39}$$

for every coordinate. This is the Lagrangian form of the equations of motion.[77]

---

[76] On the history of variational principles in mechanics, cf. Lanczos 1966; Szabó 1979; Fraser 1983; Pulte 1989.

[77] When the forces $\mathbf{F}_\alpha$ no longer admit a potential, one can still introduce the generalized forces Φ_i such that $\sum_\alpha \mathbf{F}_\alpha \cdot \mathrm{d}\mathbf{r}_\alpha = \sum_i \Phi_i \mathrm{d}q_i$. The equations of motion then read $\frac{\mathrm{d}}{\mathrm{d}t} \frac{\partial L}{\partial \dot{q}_i} - \frac{\partial L}{\partial q_i} = \Phi_i$.

These deductions show the great power of d'Alembert's principles. They do not prove the necessity of these principles. In 1756, the Berlin Academicians opened a prize competition on the question "whether the principles of statics and dynamics were necessary or contingent." D'Alembert published his answer to this question two years later, in the foreword to the second edition of his *Traité de dynamique*. After summarizing the deductions given in the first edition of his treatise and completing them with the proof of uniqueness of the condition of equilibrium, he concluded:

> The laws of statics and mechanics, as expounded in this book, are those which result from the existence of matter and motion. Now experience proves that these laws are the ones observed on the bodies around us. Therefore, the laws of equilibrium and motion, as known from observation, have necessary truth.

To his religious reader, d'Alembert conceded that God had the power to make the empirical motions differ from the theoretical ones; but experiment showed that God refrained from using this power.[78]

Mechanics thus joined the ranks of algebra and geometry in "bearing the stamp of evidence." In his *Discours préliminaire* of 1751 for the *Encyclopédie*, d'Alembert had made clear that physics in general did not have this certainty: it was an application of mathematics to enlightened observation. Even in mechanics, experience had to be consulted in order to determine the expression of impressed forces of gravitational, magnetic, electric, or cohesive nature; only the relation between these forces and the resulting motion was necessary.[79]

After d'Alembert

It is not difficult to find flaws in d'Alembert's proof of the necessity of the laws of mechanics. For example, his derivation of the composition of motions is unclear and seems to involve an implicit use of the principle of relativity. Also, d'Alembert's principle of dynamics is not a straightforward consequence of his definition of equilibrium, despite his opinion to the contrary. We will return to this difficulty in Chapter 2. Among d'Alembert's readers, opinions varied as to his success in establishing the necessity of the laws of mechanics. The astronomer Jérôme Lalande approved d'Alembert's claim. Lagrange seems to have shared d'Alembert's opinion that the principle of dynamics resulted from the definition of equilibrium. In 1798, he offered an influential proof of the principle of virtual work, of which an improved version will be given in Chapter 2. However, he did not insist on the necessity of the laws of mechanics as much as d'Alembert had done; his main purpose was to ease their development in setting them in a most elegant and powerful analytical framework.[80]

---

[78] D'Alembert 1758, pp. xxiv–xxviii, 53 (citation from p. xxix). The prize was never attributed, despite the large number of submissions (cf. Winter 1957, pp. 224, 239, 257). The choice of the question may have been related to Maupertuis 1756, in which Maupertuis rejects (neo-)Cartesian derivations of the necessity of the laws of mechanics and traces these laws to God's free choice (cf. Panza 1995, pp. 518–20).

[79] D'Alembert 1751, p. xi.

[80] Lalande, in Montucla 1799–1802, vol. 3, p. 628; Lagrange 1788, pp. 180, 193; 1798 (see also Lagrange 1811–15, vol. 1, pp. 23–6).

In contrast, the engineer Lazare Carnot strongly denied the necessity of the laws of mechanics, despite his adoption of d'Alembert's concepts of matter and motion and his shared emphasis on discontinuous impacts. In his *Essai sur les machines* of 1786, Carnot asserted that the two laws on which he based his system, namely, the law of reaction and the binding character of hard-body collisions, were "purely experimental truths" and he denied the possibility of a purely rational definition of the basic concepts of mechanics:

> A more detailed explanation of these principles . . . would only have made things more embroiled. A science is like a beautiful river, whose course is easy to follow once it has acquired some regularity; but if we look for the source, we do not find it anywhere; it is as if the source were lying on the entire surface of the earth. Similarly, if we attempt to trace the origin of a science, we only find obscurity, vague ideas, vicious circles, and we get lost in the primitive ideas.

In his *Principles* of 1803, Carnot agreed with the empiricist philosopher John Locke that every science, including mathematics, had its origin in experience, and he judged it vain to try to separate the rational from the empirical in the foundations of a science. The basic facts of mechanics, he argued, "are too familiar for us to know to what extent, without them, reason alone could establish its definitions." In reality, Carnot did not differ as much from d'Alembert as these statements would suggest. In his *Discours préliminaire* d'Alembert asserted that "sensations were the principle of all our knowledge" and traced the basic concepts of matter and motion to sensory experience. His rationalism was limited to the claim that these concepts, once given, implied the laws of equilibrium and motion.[81]

The last great eighteenth-century exposition of the foundations of mechanics was the first book of Pierre-Simon Laplace's *Mécanique céleste*, published in 1798. There the French astronomer adopted a partially rationalist position in which he regarded some of the laws of mechanics as results of observation, and others as necessary true. In the first category he placed the principle of inertia, the proportionality between impulse and velocity, and the equality of action and reaction; in the necessary-true category, he placed the parallelogram of forces, the principle of virtual work (granted the equality of action and reaction), the relation $\mathbf{f} = m\mathbf{a}$ (granted the proportionality of impulse and velocity in the discrete case), and d'Alembert's principle (so obvious to him that he did not even name its author). His derivation of the parallelogram of forces was a clever variant of Bernoulli's. His derivation of $\mathbf{f} = m\mathbf{a}$ rested on the analysis of the force $\mathbf{f}$ as a series of impulses and implicitly relied on the relativity principle.[82]

The influential treatise of Laplace's disciple Siméon Denis Poisson, first published in 1811, harbors a similar mixture of contingent and necessary truths in the foundations of mechanics. Another shared characteristic is the recourse to two sorts of intuitions: some deriving from the impenetrability of hard bodies as in d'Alembert's dynamics, others deriving from the Newtonian concept of force and from the Laplacian concept of bodies as made of disjoint molecules interacting through central forces. For instance, while Laplace

---

[81] Carnot 1786, p. 107; 1803, p. 5; D'Alembert 1751, p. ii. On Carnot's mechanics, cf. Drago, Manno, and Mauriello 2001.

[82] Laplace 1796, pp. 4–7 (parallelogram of forces), 14 (law of inertia most natural and most simple), 15 (velocity proportional to impulse), 19–20 ($\mathbf{f} = m\mathbf{a}$), 38 (action = reaction), 38–41 (virtual work).

and Poisson relied on d'Alembert's definition of equilibrium through hard-body collision, they derived the equilibrium of machines and the principle of virtual work by consideration of internal forces of tension or cohesion. Whereas such forces were alien to d'Alembert's system, they naturally occurred in the molecular concept of bodies that subtended most of Laplacian physics.[83]

For the sake of consistency with the Laplacian theory of matter, it would seem better to define all the basic concepts of mechanics at the level of interacting material points. Poisson thought so much, as can be judged from the following extract of his memoir of 1828 on elasticity:[84]

> It would be desirable that geometers reconsider the main questions of mechanics un-
> der this physical point of view which better agrees with nature. In order to discover
> the general laws of equilibrium and motion, one had to treat these questions in a quite
> abstract manner; in this kind of generality and abstraction, Lagrange went as far as
> can be conceived when he replaced the physical connections of bodies with equations
> between the coordinates of their various points: this is what *analytical mechanics* is
> about; but next to this admirable conception, one could now erect a *physical mech-
> anics*, whose unique principle would be to reduce everything to molecular actions
> that transmit from one point to another the given action of forces and mediate their
> equilibrium.

The engineer Adhémar Barré de Saint-Venant implemented this program in the 1820s and 1830s, in a series of manuscripts whose contents were published in a partial and be-lated manner. Out of concern for concrete engineering problems, Saint-Venant rejected the rational mechanics of ideally perfect hard bodies, inextensible threads, rigid rods, etc. Like most of his contemporaries, he regarded the porous and molecular character of matter as proven by compressibility and other physical properties. In his opinion, d'Alembert's claim that he could derive the laws of mechanics from the principle of sufficient reason was "one of the mistakes of this great genius," for it betrayed "the arrogance of a mind that would substitute its limited wisdom to the infinite wisdom [of God]." The true origin of our knowledge of the world was God, who in his goodness had allowed the success of experimental induction by making phenomena depend on sufficiently simple laws.[85]

Saint-Venant based his mechanics on two principles:

1. In a system made of two atoms only, these atoms undergo equal and opposite accelerations on the line joining them, with an intensity depending on their distance
   only.
2. In a system made of several atoms, the acceleration of a given atom is the geometric
   sum of the accelerations it would take if it were subjected separately to the actions of
   each of the other atoms.

[83] Poisson 1811, vol. 1, pp. 11–19 (parallelogram of forces), 238–53 (virtual work), 278 (motion proportional to impulse as the "simplest law."

[84] Poisson 1828, p. 361.

[85] Saint-Venant [1835], pp. 3 (citation), 5, 14–15, 35. See also Saint-Venant 1851. Cf. Darrigol 2001.

Saint-Venant implicitly adopted Newton's absolute space, and he explicitly regarded the atoms as point-like, all identical, and discretely distributed in space, as André Marie Ampère had already done for mostly chemical reasons. Saint-Venant regarded the point-like character of the atoms and the discreteness of matter to be inevitable consequences of the existence of rigid bodies. Indeed a body made of a continuous homogeneous distribution of material points interacting according to Saint-Venant's two principles cannot have any rigidity because the net interaction remains unchanged during a shearing deformation of the body. Saint-Venant carefully avoided the concept of force in his two principles, because he regarded the search for true causes as "arrogant curiosity, bordering on the negation of God." In his view, forces were a remnant of paganism, a confusion between the material and spiritual orders; mathematics had to be our true guide in the world of God's wonderful creations:[86]

> In order to reach the beautiful, the sublime, the immaterial, feelings, love, and God at last, mathematics must banish the profane poetry, the sensual poetry of imagination and enthusiasm, must expel *forces* and the qualities of matter through the same exit as the sylphs, the nymphs, the gnomes, and the elves; then mathematics shall reveal the Sublime Truth, which has its own, intense poetry.

This is why Saint-Venant's basic laws only involved the kinematic concepts of distance and acceleration. He derived all other concepts and principles of mechanics from these two laws. For example, he defined the mass of a body as the number of atoms, the accelerating force of a body as the sum of the accelerations impressed on its atoms. He regarded rigid bodies as combinations of atoms interacting in such a way that their mutual distance could not be altered without much external effort. More generally, he assumed that connected systems could be approximately synthesized by proper combinations of atoms interacting through short-range forces. Call $\phi_{\alpha\beta}(r_{\alpha\beta})$ the potential of this interaction for the atoms α and β separated by the distance $r_{\alpha\beta}$. In addition, any given atom is subjected to long-range forces of gravitational, electric, or magnetic origin; and some of the atoms may be interacting with the atoms of another system that is in contact with the system under consideration. Call $\mathbf{F}_\alpha$ the sum of all these long-range and external forces for the atom α. Saint-Venant's two principles lead to the equation of motion

$$\ddot{\mathbf{r}}_\alpha = \mathbf{F}_\alpha - \sum_\beta \nabla_\alpha \phi_{\alpha\beta}(r_{\alpha\beta}), \tag{40}$$

For a virtual displacement $\delta\mathbf{r}_\alpha$ of the atoms, we have

$$\sum_\alpha (\mathbf{F}_\alpha - \ddot{\mathbf{r}}_\alpha) \cdot \delta\mathbf{r}_\alpha = \delta \left[\frac{1}{2} \sum_{\alpha\beta} \phi_{\alpha\beta}(r_{\alpha\beta}) \right]. \tag{41}$$

The right-hand side of this equation represents the variation of the total potential of the internal short-range atomic forces.[87]

---

[86] Saint-Venant [1835], pp. 4 (citation), 15 (citation), 40 (principles).   [87] Saint-Venant [1834].

Within a solid, an inextensible thread, or an incompressible fluid, and for atomic displacements compatible with the constraints of solidity, inextensibility, or incompressibility, the variation of the potential is negligible because the distances of the contiguous pairs of atoms that contribute to it vary very little. For the rolling or the sliding of solid over solid, there are contact forces involving hybrid pairs of atoms. The contribution of these pairs to the total potential is also negligible, because the relative configuration of relevant atoms remains globally the same (up to a relabeling of the atoms). In general, it can be seen that the variation of the total potential is negligible (compared to the work of applied forces) for any virtual displacement of a connected system. Together with Eq. (41), this result implies the approximate validity of both the principle of virtual work and d'Alembert's principle. Indeed, for groups $\bar{\alpha}$ of $m_{\bar{\alpha}}$ atoms having very nearly the same position $\mathbf{r}_{\bar{\alpha}}$ at any time, it implies Lagrange's general equation of dynamics

$$\sum_{\bar{\alpha}} (\mathbf{F}_{\bar{\alpha}} - m_{\bar{\alpha}} \ddot{\mathbf{r}}_{\bar{\alpha}}) \cdot \delta \mathbf{r}_{\bar{\alpha}} = 0, \tag{42}$$

which is d'Alembert's principle combined with the principle of virtual work.[88]

This route to the fundamental principles of dynamics has three defects. It admits without rigorous proof the existence of atomic interactions such that stable solids of various densities, inextensible threads, and incompressible fluids are possible. It relies on a case-by-case description of a connected system rather than on a general, abstract definition. It only implies an approximate validity of the dynamical principles. On the brighter side, it admits only one very simple kind of force, and it predicts precisely how real systems depart from ideal connected systems. Navier, Cauchy, Poisson, and Saint-Venant were indeed able to derive the fundamental equations of (rari-constant) elasticity and the equations of motion of elastic and viscous fluids on the basis of a similar model.

As we saw, Saint-Venant condemned d'Alembert's rationalism and regarded his own principles as contingent truths. He nevertheless came close to a new proof of the necessity of the laws of mechanics by making them derive from two very simple principles. This simplicity invites a rational justification, which will be attempted in Chapter 2.

Similar comments can be made about the French nineteenth-century textbook writers who relied on variants of Huygens's hypotheses. In 1856, the most rigorous of them, the astronomer Charles-Eugène Delaunay, based the mechanics of material points on four principles: (1) the principle of inertia, (2) the equality of action and reaction, (3) "the independence between the effect of a force and the motion previously acquired by the material point on which this force acts," and (4) "the mutual independence of the forces that act simultaneously on the same material point." Principle (4) is just another statement of the parallelogram of forces. The Huygensian principle (3) means that the motion induced by a force acting on an already moving body is the same as the motion obtained by compounding the original velocity of the body with the motion that the force would impart on it if it were originally at rest; it implies the relativity principle and the additional assumption that the motion of the body only depends on its original velocity. By reasoning similar to Huygens's, these principles imply the law $\mathbf{f} = m\mathbf{a}$. Although Delaunay must have known that principles (1), (2), and (4) had received rational justifications in Euler's *Mechanica*

---

[88] A similar derivation is found Saint-Venant [1834], and in Coriolis 1835.

and elsewhere, he preferred to regard his four principles as "originating in the observation of facts." Similar but less rigorous derivations of the law of acceleration could be found in the treatises of Gaspard Coriolis (1829), Jean-Marie Duhamel (1845), Charles Briot (1861), and Jules Violle (1883), to name but the most important ones. These authors mistook principle (3) for a statement of the principle of relative motion, whose necessity could be more easily argued than the additional assumption that the motion of the body only depends on its original velocity. Yet they all shared Delaunay's opinion that the laws of mechanics were contingent truths.[89]

So far I have only considered French treatises on mechanics, because they were the most influential ones from the mid-eighteenth to the mid-nineteenth century at least. In Britain, Euler's and d'Alembert's rationalism could not displace an empiricism that bore the stamp of Newton's and Locke's authority. The Germans, who drew their mechanics from Euler and the French, and who were strongly exposed to Descartes's and Leibniz's philosophies, were more likely to embrace a form of rationalism. As is well known, a central question of Immanuel Kant's philosophy was why certain truths could be established by a priori means and yet concern our experience of the world. According to the *Critique of Pure Reason*, first published in 1781, there are a priori conditions for the possibility of knowledge, grounded on the representative and legislative faculties of the mind. *Sensibility* and *understanding* are the names Kant gave to these two faculties. On the representative side, time and space are the necessary forms of inner and outer sense. One part of mechanics, which Kant calls *Phoronomie* (Ampère's later *cinématique*), draws its necessity from the doctrine of sensibility, for it describes motion irrespective of its causes.[90]

In his *Metaphysical foundations of natural science* of 1786, Kant further argued the rational necessity of three laws of mechanics: the conservation of mass, the law of inertia, and the equality of action and reaction. After showing that matter, for the sake of impenetrability and stability, could only be a continuous distribution of centers of force, he derived the three laws by applying to this notion of matter three categories of the understanding: substance, causality, and community. In a brief summary of Kant's convoluted reasoning, mass is conserved because matter is a substance; uniform motion cannot be altered without a cause; action must be equal to reaction because in binding collisions there is a reference frame in which the colliding bodies cancel each other's motion and because the interaction can only depend on the relative motion. Although Kant implicitly relied on the relation between force and momentum change, he did not make it a law of mechanics, presumably because he interpreted this relation as a definition of force.[91]

---

[89]Delaunay 1856, pp. 114–15 (inertia), 119 (reaction), 120 (3rd principle), 126–7 (4th principle); Coriolis 1829, p. 3; Duhamel 1845–6, vol. 1, p. 232; Briot 1861, pp. 72–4; Violle 1883–92, vol. 1, pp. 97–100.

[90]Kant 1781; 1783, §10 (motion); 1786, pp. 1–30 (*Phoronomie*). The *Phoronomie* is essentially limited to the motion of material points (for Kant does not admit rigid bodies in his dynamics). On the history of kinematics, cf. Martínez 2009.

[91]Kant 1786, pp. 106–37. Kant was not too clear about inertia. In the *Phoronomie* (part 1) he regards absolute space as unobservable and conventional. In the *Mechanik* (part 3), he derives the principle of inertia without linking it to a more physical notion of absolute space (pp. 119–20). However, his derivation of the law of reaction (in which the center-of-mass frame is called absolute space) and the "general remark" (pp.145–58) to the *Phänomenologie* (part 4) suggest a privileged role for the reference frame attached to the center of mass of all matter. For a detailed study, cf. Friedman 2013.

Although Kant's considerations were more obscure than Euler's and d'Alembert's, they could impress philosophically inclined physicists by their being anchored on a professedly rigorous philosophical system. Most important, they replaced the mysterious self-evidence of premises of earlier rationalist arguments with the powerful notion of the conditions of possibility of experience: some laws of nature are accessible by pure reason because they reflect the structure of the mental apparatus through which we grasp natural phenomena.

Conclusions

A large proportion of the inventors of modern mechanics believed that its foundations could be reached by pure reasoning, without appeal to experience. In this view, the laws of mechanics were necessary; experiments could not contradict them. Roughly, there were two kinds of rationalist arguments: theological and conceptual. In the theological kind, the laws were the direct expression of God's perfection, which could entail simplicity, conservation, or optimization. In the conceptual kind, the laws were logical consequences of the definition of the object of mechanics and of its basic quantities; the deductions often relied on the principle of sufficient reason, in a manner that we would now assimilate with considerations of space and time symmetry; they also involved supposedly self-evident principles such as the impossibility of perpetual motion. Descartes, Leibniz, and Maupertuis favored theological arguments; Huygens, Euler, and d'Alembert favored conceptual arguments.[92]

Another useful distinction concerns the concepts of interaction on which the rationalist deductions relied. For Descartes and d'Alembert, this concept was the contact or collision between ideally hard bodies; for all other rationalists, force played a role, to a variable degree. Huygens and Euler regarded force as a convenient way to represent mechanical actions without describing their true, contact-driven cause. Leibniz made it an essential component of his definition of substance. In eighteenth- and nineteenth-century natural philosophy there was a third, a "Newtonian" concept of force as direct action between two distant bodies. Although Bošković, Laplace, Saint-Venant, and Helmholtz famously argued for the necessity of this concept, they did not believe the laws of mechanics to be themselves necessary. With Kant's exception, none of the authors of necessitarian arguments accepted this concept, perhaps because they were, in some sense, heirs of Descartes.[93]

The adversaries of the rationalists were the contingentists who believed that the principles of mechanics could only be discovered by observation and experimentation, or at least that the only justification of these principles was the experimental validity of their consequences. Galileo, Newton, Carnot, and Saint-Venant all were contingentists, though for different reasons. Galileo was attacking the dogmatism of inherited scholasticism; Newton and Saint-Venant were denouncing the rationalists' pretense to read in the Creator's mind; and Carnot was expressing the engineer's common sense. Yet the declared contingentists occasionally flirted with rationalism. Galileo did so in his attempts to prove laws by

---

[92] Theological necessity roughly corresponds to Leibniz's "moral necessity"; the overlap of conceptual necessity with his "absolute necessity" is only partial, for the latter excludes appeal to the principle of sufficient reason.

[93] Bošković 1758, §74 (inductive origin of the law of reaction), §76 (inductive origin of the law of acceleration). As we saw, Saint-Venant avoided the name "force" for the direct action at a distance that he believed to be necessary.

thought experiments; Newton, Carnot, and Saint-Venant derived some of their principles from premises that could easily be (mis)taken as necessary, such as the impossibility of perpetual motion or the principle of relative motion. Some of the chief theorists of mechanics, for instance Daniel Bernoulli, Laplace, and Poisson, adopted an intermediate position in which they regarded some laws of mechanics as necessary and others as contingent.

At any rate, the authors of the best treatises on mechanics presented this science as a set of rigorous deductions from a minimal number of principles of maximal simplicity. First and foremost, Newton defined "Rational Mechanics" as "the Science of Motions resulting from any forces whatsoever and of the forces required to produce any motions, accurately proposed and demonstrated." Although this rationality should not be confused with rationalism, it invited and still invites speculation on the necessity of the few simple premises of the rational deduction.[94]

Euler and d'Alembert, arguably the greatest mechanical philosophers of the mid-eighteenth century, indulged in such speculation. Although they had a few adepts, they failed to set a trend: in the second half of the century, Cartesian rationalism followed the Cartesian worldview to the grave. With the partial exceptions of Kant and Helmholtz, the following generations of natural philosophers professed the empirical origin of any knowledge of nature and denounced the arrogance of a mind that would derive the laws of nature by mere introspection.

[94]Newton 1687, preface.

2

THE NECESSITY OF CLASSICAL MECHANICS

As we learned from the previous chapter, historical attempts at deriving the necessity of some laws of mechanics relied on a number of presuppositions that were judged self-evident or metaphysically necessary: the principle of sufficient reason, the impossibility of perpetual motion, the relativity principle, the uniformity of space and time, some sort of causality, the measurability of force and mass, the superposition of equilibria, the preservation of equilibrium by partial solidification of the system, and the equivalence of a continuously applied force with a rapid succession of impulses. In this long and yet incomplete list of presuppositions, the three first were the only ones to be commonly stated by their users; the others were often used in an unconscious manner, or they were judged so evident as not to require any statement. This silencing of premises implied gaps in the demonstrations. For instance, Newton's and Varignon's derivations of the laws of statics from the parallelogram of forces silently relied on the principles of superposition and solidification, whose obviousness can easily be contested. Other derivations were hampered by erroneous statements of their premises or errors of reasoning. For instance, a popular nineteenth-century derivation of the law of acceleration depended on confusion between the relativity principle and a stronger principle introduced by Huygens. In addition, the a priori necessity of even the most clearly formulated principles could be questioned. Was not the impossibility of perpetual motion an induction from experience? Was there not some empirical input in the applications of the principle of sufficient reason to mechanical problems?

In sum, historical demonstrations of the necessity of the laws of mechanics suffered three kinds of flaws: incomplete statement of the premises, errors in the deductive chain, and insufficient justification of the necessity of the premises. This is why historians have usually condemned these demonstrations as naïve, erroneous, and irreparable. Yet, I believe they provide a reservoir of ideas for designing proofs of necessity that resist the usual objections. This chapter provides such proofs for four sorts of mechanics: the theory of connected systems in the first section, molecular mechanics à la Saint-Venant in the second, continuum mechanics in the third, and collisions in the fourth and last section. The relevant ideas are eclectically borrowed from Huygens, Newton, d'Alembert, Lagrange, and Cauchy. As we will see, the implied kind of necessity is not quite the necessity dreamt by the founders of modern mechanics. It nonetheless excludes possible worlds in which the laws of mechanics would differ from those we know.

Physics and Necessity. First Edition. Olivier Darrigol.

2.1 Connected systems

The counter-weight concept of force

In order to move a body or to alter its motion at a given instant we must exert an effort that results in muscular sensations. We roughly know that such actions have degree and direction. They define a proto-concept of force. A more precise concept must appeal to ideal measurement. For this purpose I use the ideal pulley, although any other simple device whose equilibrium can serve for the comparison of forces, for instance the lever, would do.[1]

The direction of a force is tentatively defined as the direction of a thread that can be kept tense by this force. Two forces are said to be of equal intensity if they can equilibrate each other through the thread-pulley mechanism of Fig. 2.1.

As a consequence of this definition, a (vertical, downward) force can be produced by suspending a weight. The numerical value of the intensity of a force is then defined as the number of equal unit weights it can balance, with a precision given by the size of the unit. Just as in the case of length measurement, subunits can be introduced to refine the precision. In a convenient idealization, we may imagine the precision to be indefinitely improvable and represent forces by vectors in a three-dimensional, real vector-space. Note that this step also requires ideal, perfectly circular, rigid, and centered pulleys; and perfectly inextensible, flexible, and non-sliding threads.[2]

This way of measuring forces should only be regarded as tentative and provisional. Its acceptability depends both on empirical and on theoretical constraints. On the empirical side, it is not a priori obvious that our definition of the equality of the intensities of two forces is concretely transitive: two forces could be found equal to a third one without being equal to each other. On the theoretical side, it is not evident that our weight standard for the measurement of force is uniform. In fact, we will have to give up this standard as soon as the variation of gravity from place to place is no longer negligible.

Despite this limitation, our counter-weight concept of force is precise enough to allow a quantitative discussion of the conditions of equilibrium of various connected systems.

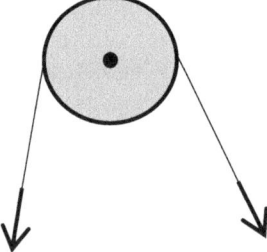

Fig. 2.1.   The equilibrating of two forces through a pulley-thread
            mechanism.

---

[1] I choose pulleys because they are well adapted to Lagrange's derivation of the principle of virtual works. The general idea is that knowing the equilibrium law of a very simple class of systems (pulleys or levers) is sufficient to derive the equilibrium law of any connected system under a few natural presuppositions.

[2] The general connection between ideal measurement and mathematical definition of quantity will be discussed in Chapter 6.

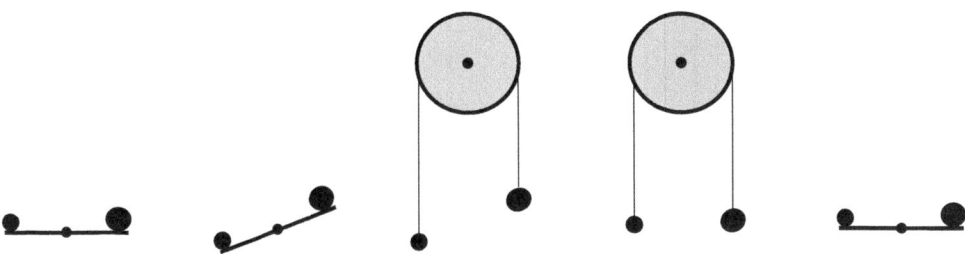

Fig. 2.2. Perpetual motion by means of a magic lever.

That of the pulley-thread system results from our definition of force. That of the lever, known as the law of moments, requires further considerations. A seemingly obvious case of equilibrium is that of the symmetrically loaded lever. Disequilibrium could still occur in this case when the two arms of the lever are built differently. What truly excludes disequilibrium is not symmetry, but the impossibility of perpetual motion: if it were possible to lift a heavier weight by means of a lighter weight applied symmetrically to a lever, then these two weights could be subsequently attached to the threads hanging from a pulley, and would induce a rotation of the pulley while returning to their original height (see Fig. 2.2). By repeating this double process, the pulley could be kept rotating even in the presence of friction. In order to block this possibility of a perpetual motion (of the first kind), the lever must be in equilibrium.[3]

The principle of virtual work

From this simple case, we will directly move to a generic system, intuitively defined as a set of pulleys, inextensible threads, solids that can slide or roll on each other without losing contact, and perfect liquids. In a more rigorous and abstract definition, a *connected system* is a (discrete or continuous) system of material points subjected to constraints that limit the possible changes of configuration of these points, and meeting the two following conditions:

1. For any infinitesimal change of configuration compatible with the constraints (called *virtual displacement*) in which the position of the material point α changes by $\delta\mathbf{r}_\alpha$, the opposite change $-\delta\mathbf{r}_\alpha$ is also possible.
2. Arbitrarily small forces acting in the direction of mutually compatible displacements suffice to break equilibrium.

The first condition implies the permanence of contacts between rigid bodies. The second excludes solid friction (caused by roughness, for example).

Calling $\mathbf{F}_\alpha$ the force acting on the material point α, the principle of virtual work stipulates that *the system is in equilibrium if and only if* $\sum \mathbf{F}_\alpha \cdot \delta\mathbf{r}_\alpha = 0$ *for any virtual*

---

[3]The change of horizontal distance of the weights during the rotation of the lever can be compensated without effort (by having them slide on a frictionless horizontal plane).

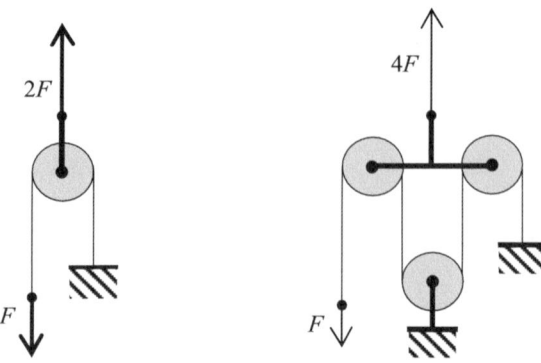

Fig. 2.3.   Simple tackle.   **Fig**. 2.4.   Triple tackle.

displacement $\delta\mathbf{r}_\alpha$. The forces involved in this statement do not include the internal contact forces used in elementary expositions of statics. They may include internal forces of
elastic, gravitational, or electric origin.

We will now follow an improved version of the demonstration of the principle of virtual
work that Lagrange gave in 1798. The basic idea is to synthesize the forces $\mathbf{F}_\alpha$ through a set
of tackles, a single rope running through them, and a weight. The simple tackle of Fig. 2.3
yields the force $2F$ under the tension F of the rope. Indeed if the force acting on the axis of
the pulley differed from $2F$, a perpetual motion could be generated by connecting this axis
and the two ends of the rope to the same rigid frame. Similarly, the triple tackle of Fig. 2.4
yields the traction $4F$, and so forth. As the intensities F_α can all be regarded as even multiples $2N_\alpha F$ of the same small intensity F (with a precision increasing with the smallness
of F), they can be generated by a properly arranged system of tackles through which the
same rope runs (see Fig. 2.5). The tension F of the rope is produced by a weight W.[4]

The virtual displacement $\delta\mathbf{r}_\alpha$of the material point α on which the force $\mathbf{F}_\alpha$ is acting
induces a shift $2N_\alpha(\mathbf{F}_\alpha/F_\alpha) \cdot \delta\mathbf{r}_\alpha$ of the rope, as an obvious consequence of the make-up of
the tackles. The resulting shift of the end of the rope is $F^{-1} \sum \mathbf{F}_\alpha \cdot \delta\mathbf{r}_\alpha$.

If there exists a virtual displacement such that $\sum \mathbf{F}_\alpha \cdot \delta\mathbf{r}_\alpha \neq 0$, this displacement or
the opposite displacement (warranted by property (1) of connected systems) is such that
$\sum \mathbf{F}_\alpha \cdot \delta\mathbf{r}_\alpha > 0$. The weight W therefore pulls the rope in the direction of a possible displacement. According to property (2) of connected systems, the rope must move no matter
how small this weight is. Therefore, the system is not in equilibrium. By contraposition, the
virtual work $\sum \mathbf{F}_\alpha \cdot \delta\mathbf{r}_\alpha$ must vanish for the system to be in equilibrium.

Reciprocally, the vanishing of $\sum \mathbf{F}_\alpha \cdot \delta\mathbf{r}_\alpha$ for any virtual displacement implies equilibrium. We will prove this *ad absurdum*. Suppose that the system is not in equilibrium under
the forces $\mathbf{F}_\alpha$. Then equilibrium can be restored by applying additional forces $\mathbf{X}_\alpha$ directed

---

[4]Lagrange 1798, also in Lagrange 1811–15, first section. In a concrete connected system on earth, the $\mathbf{F}_\alpha$ values
would include the weight of the various components of the system, so that Lagrange's construction can only be an
imaginary one (the more so because W itself is subjected to gravitation).

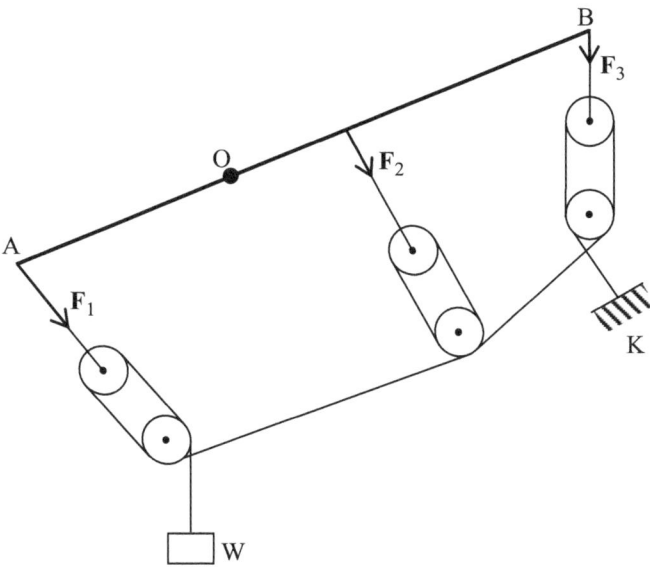

Fig. 2.5.   Lagrange's contraption for a proof of the principle of virtual velocities in the case of a lever. The forces $\mathbf{F}_1$, $\mathbf{F}_2$, $\mathbf{F}_3$, acting on the lever AOB, are produced by three rigidly held tackles through which a single rope runs from the anchor K to the suspended weight W.

against the initial displacements $\mathrm{d}\mathbf{r}_\alpha$ of the material points α. Otherwise, the same weight W and the same resulting displacements $\mathrm{d}\mathbf{r}_\alpha$ could be used to lift arbitrary heavy weights through a pulley-rope mechanism and perpetual motion would be possible. On the one hand, the counterbalancing forces $\mathbf{X}_\alpha$ verify the inequality $\sum \mathbf{X}_\alpha \cdot \mathrm{d}\mathbf{r}_\alpha < 0$. On the other hand, the restored equilibrium requires that $\sum (\mathbf{F}_\alpha + \mathbf{X}_\alpha) \cdot \mathrm{d}\mathbf{r}_\alpha = 0$. These two relations contradict the vanishing of $\sum \mathbf{F}_\alpha \cdot \mathrm{d}\mathbf{r}_\alpha$.[5]

Lagrange's own proof that the vanishing of the virtual work $\sum \mathbf{F}_\alpha \cdot \delta\mathbf{r}_\alpha$ implies equilibrium was based on Torricelli's principle. Namely, he argued that the vanishing of the virtual work implied that the weight W could not descend under any small displacement of the system, which is Torricelli's condition of equilibrium applied to the global system including the tackles and the weight W. This proof is erroneous, because the condition $\sum \mathbf{F}_\alpha \cdot \delta\mathbf{r}_\alpha = 0$ only implies that the weight W does not descend for any *infinitesimal* displacement of the system. In cases of unstable equilibrium, a small, *finite* displacement of the system can induce a descent of the weight W, as can be seen in Fig. 2.6.[6]

[5] Similar reasoning is found in Cornu 1875.

[6] Cf. Joseph Bertrand's note in Lagrange 1867–92, vol. 11, p. 24. The impossibility of perpetual motion in a mechanism subjected to friction implies the inequality $\sum \mathbf{F}_\alpha \cdot \mathrm{d}\mathbf{r}_\alpha > 0$ for the (real) displacement $\mathrm{d}\mathbf{r}_\alpha$ of the system caused by any forces $\mathbf{F}_\alpha$ that disturb equilibrium. This means that the system cannot start moving without a descent of the weight W of the same infinitesimal order as the displacement of the system (in the case of Fig. 2.5, the displacement is of second order with respect to the descent). Torricelli's principle does not imply this result, since it does not involve any restriction on the infinitesimal order of the descent.

Fig. 2.6.   Unstable equilibrium of a marble over a sphere.

The principle of virtual work only applies to the ideal connected systems of the above-given definition. Experimental realization is obtained by exploiting the operational meaning of the basic concepts of space, material point, and force, and by seeking or building concrete systems that approximately meet the definitional requirements of perfect solidity, inextensibility, perfect smoothness, etc. Such systems are rarely found in nature, and their construction may require special skills. Here we have an example of a physical theory that mainly applies to an artificial world, designed so as to enjoy simple regularities. Its laws are nonetheless necessary inasmuch as they logically follow from the definition of the systems to which it applies and from the fairly evident principle of the impossibility of perpetual motion. Empirical knowledge is only involved in judging the constructability of connected systems, for instance in ascertaining Euclidean geometry, the existence of rigid or inextensible bodies, the concrete possibility of smoothness, and the reliability of weight as a standard of force.

The relation between force and acceleration

The simplest case of accelerated motion is that of a particle subjected to a constant force and disconnected from other bodies. We assume the homogeneity of space and time, according to which the relation between force and motion should be invariant under spatial and temporal translations, and the isotropy of space, according to which this relation is also invariant under rotation. We further adopt the relativity principle, according to which the relation between force and motion should be the same in two reference systems that are in uniform rectilinear motion with respect to each other; and we assume that the transformation from a system of reference to another is Galilean.

If a free body is moving at a given time, the isotropy of space requires the subsequent motion to be rectilinear, and the uniformity of space and time requires this motion to be uniform (the distance traveled in a time interval cannot depend on the original time and position). The relativity principle excludes any further restriction on the motion of a free body: if a free body had to remain at rest (as Aristotle would have it), then it would have to do so in any reference frame, which is obviously impossible. To sum up, the assumed symmetries imply the principle of inertia, according to which a free particle is either at rest or in rectilinear, uniform motion.[7]

This fairly traditional introduction of space, time, and inertia raises a number of questions. Does it make sense to define time before motion? Should inertial motion be derived from the concept of time or should time be derived from inertial motion? Since motion can

---

[7]In modern terms, we adopt the Galilean spacetime endowed with both chronogeometric and affine (inertial) structure.

only be inertial in a limited class of reference frames, why does this class have a privileged role? Assuming that no quantitative notion of time is yet given and that the absolute simultaneity of events has been defined through infinite-speed signaling, why are there reference frames for which the distances traveled by any two free particles between two instants are always proportional? Does not this amount to a strange pre-established harmony between the motions of a priori independent particles? I leave all these questions to Chapter 5, and I henceforth take the Galilean (or Minkowskian) spacetime as a given.

As we saw in the previous chapter (p. 44), a few nineteenth-century physicists believed that the principle of inertia and the principle of relativity together implied the proportionality between force and acceleration. Counter-examples of this implication are easy to find. For instance, the force could be the product of the acceleration by a never-vanishing function of the acceleration or of any higher derivative of the motion. A third principle is necessary to further restrict the form of the equation of motion. A reasonable option is to posit the *secular principle* according to which two time-dependent forces $F_1(t)$ and $F_2(t)$ cause the same average motion of the particle if their secular averages are the same. In particular, a smooth force function $F(t)$ can be replaced with a dense series of impulses of equal intensity and direction within any small time interval, the density of the impulses in the vicinity of the time t being proportional to the intensity of the force $F(t)$.[8]

According to the principle of inertia, the velocity of the particle is constant between two impulses. By some sort of causality, each impulse causes the same increment of the velocity. According to the relativity principle, this increment does not depend on the original velocity (before the impulse). Consider a time interval much longer than the time between two impulses and much smaller than the time during which $F(t)$ varies significantly. The velocity variation in this time interval is proportional to the number of impulses, which means that the force is proportional to the acceleration.

In a more mathematical vein, the (Galilean) relativity principle and time locality imply that $F(t)$ should be a function of the derivatives of the motion of order superior or equal to 2. Euclidean invariance and the principle of inertia further imply that $F(t)$ should be the product of the acceleration **a** with a scalar function m of **a** and higher derivatives. The secular principle implies that the functional relation between $F(t)$ and the motion $r(t)$ remains the same after secular averaging, that is, after convolution with any smoothing scalar function of time. This can only happen if $F(t)$ is a linear combination of the derivatives of the motion on which it depends.[9] Therefore, the scalar function m must be a constant, and the force must be proportional to the acceleration. This constant is what is called the *mass* of the particle.[10]

---

[8] This is Newton's conception, traces of which can be found in d'Alembert 1743 and in L. Carnot 1803.

[9] For a rectilinear motion on the x axis, the functional relation Φ between the Fourier transforms $\tilde{F}$ and $\tilde{x}$ must verify $\Phi(\tilde{\sigma} \, \tilde{x}) = \tilde{\sigma} \, \Phi(\tilde{x})$, if σ denotes the smoothing function. The Fourier transform $\tilde{\sigma}$ is any function that vanishes for frequencies ω such that $|\omega|$ exceeds the cutoff Ω. For a smooth motion, $\tilde{x}$ is such a function. Therefore, $\Phi(\tilde{x}) = \Phi(\tilde{x} \cdot 1_\Omega) = \tilde{x}\Phi(1_\Omega)$, wherein 1_Ω is the function of ω that takes the constant value 1 for $|\omega| < \Omega$ and vanishes for $|\omega| > \Omega$. Developing $\Phi(1_\Omega)$ into powers of ω and taking the reverse Fourier transform of the former expression of $\Phi(\tilde{x})$, we find that the restriction of Φ to smooth motions is a linear combination of the derivatives of the function $x(t)$ with constant coefficients. By rotational invariance, the constancy of these coefficients for any rectilinear motion implies their constancy for an arbitrary motion.

[10] In the relativistic case, the secular averages would have to be performed over the proper time.

There is a weak point in these demonstrations: they assume that forces are invariant under a change of inertial frame.[11] This invariance is far from obvious, because the static devices through which we have defined and measured force no longer are static devices when seen from a moving frame. One way to escape this difficulty is to replace the force by a dense series of impulses and to imagine that each impulse is caused by the normal reflection of a very fast particle of very small mass. In any frame moving at a velocity very small compared to that of the particles, the reflection process is very nearly the same in the moving frame as in the original frame. Therefore, the effect produced by the impulse should be the same.[12]

The dynamical definition of force

So far we have defined force by static means, and thereupon built a theory in which force ends up being proportional to the acceleration of the bodies on which it is impressed. This approach has limited validity. The static definition of force involves a device, the pulley, which can only be used to compare forces acting in the same neighborhood. And it relies on a local standard of weight that may vary from place to place. Consequently, the weight of a standard body cannot be taken as a proper unit of force if mechanics is to be applied at large scale. A simple way to remedy this defect is to adopt a more reliable standard of force, for instance the repulsion between two electrons separated by a unit of length. But this would involve more advanced, extra-mechanical physics.

Better, we may follow d'Alembert in measuring a force by applying it to an otherwise free material point and taking the product of the mass of this point by its acceleration. In this approach, mass must be defined before force, for example by means of a pulley-thread balance (Fig. 2.7) whose dimensions are small in comparison with the distance over which gravity varies appreciably.[13]

Two masses are said to be equal when they equilibrate each other in this device, and the mass of a compound body is taken to be the sum of its parts (remember that we are in a non-relativistic context). The numerical value of a given mass can then be determined by the number of identical unit masses it can balance, with a precision depending on the smallest subunit used. The measures of space, time, and mass being now defined, the relation $\mathbf{F} = m\mathbf{a}$ provides the measure of force.

---

[11] In a non-relativistic context, this invariance does not exclude electromagnetic forces: the force acting on a moving charge particle has the same expression in any inertial frame although its field-based expression is frame-dependent.

[12] This argument only works if arbitrarily high velocities can be conceived. As ought to be, it fails in the relativistic context in which force no longer is an invariant. A more rigorous consideration of collision dynamics is given below, pp. 67–72.

[13] This definition of mass presupposes the equality of inertial and gravitational masses in order that the associated definition of force be compatible with the laws of statics. Less concretely but less contingently, one could define the mass of a body as the inverse of the velocity it acquires under a given impulsion (elastic bouncing of a light and fast particle). Then the secular principle and the principle of inertia would make the product $m\mathbf{a}$ a well-defined measure of the force acting on a particle. This measure would be compatible with the laws of statics because it would agree with the parallelogram of forces, which by Varignon's reasoning implies the law of the lever and the balancing of equal forces by the pulley-thread mechanism.

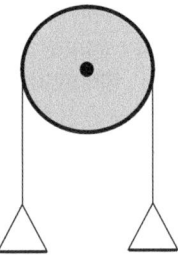

Fig. 2.7.   Pulley-thread balance.

There is something strange about this approach. It almost looks like $\mathbf{F} = m\mathbf{a}$ has become a definition of force, in which case it would not have any empirical content. This cannot be so, for the following reasons. Firstly, the product $m\mathbf{a}$ is an intrinsic characteristic of a material point (in a given class of reference systems), whereas force should remain an external circumstance on which the motion of a body depends, for instance contact with another body, gravity, or electric attraction. Therefore, the product $m\mathbf{a}$ can only be a *measure* of force, not its *definition*. Secondly, this way of measuring force is only available if the acceleration of a given material point is always the same in fixed external circumstances.[14] This is a consequence of the secular principle and the principle of inertia, by reasoning similar to that used in the earlier proof of the proportionality of force and acceleration.

Thirdly, this way of measuring force must be consistent with the validity of the laws of statics. By the new definition of mass, two equal masses equilibrate themselves in a pulley-thread mechanism placed in a uniform gravity. Since by Galileo's law of free fall these masses would be equally accelerated if they were disconnected from the mechanism, the corresponding forces or weights are equal by the new measure of force. The balancing of masses thus implies the balancing of forces, and the reasoning that leads to the principle of virtual work remains valid. In other words, the forces entering the principle of virtual work are the same as the forces entering the relation $\mathbf{F} = m\mathbf{a}$, whether the measure or force is static or dynamic.

The situation we encounter here is quite general. In order to arrive at a physical theory, one may start with some operational definitions and some empirical relations between the operationally defined quantities. Early in this procedure, idealizations lead to a mathematical representation of the class of systems under consideration. The theory is then expressed as a set of rules for deriving relations between the quantities occurring in this representation. These relations are exactly true in the mathematical representation. In the case of the mechanics of connected systems, the representation of the relevant class of systems is based on Euclidean space, the one-dimensional vector spaces of time and mass, the vector space of forces, and the abstract definition of a connected system. Basic rules are the principle of virtual work and the law of acceleration (more generally: d'Alembert's principle). The theory can be developed at a purely mathematical level, just like geometry or arithmetic. At that level, all relations among quantities are on the same footing, because the quantities belong to abstract spaces.

---

[14] In the case of an electric charge interacting with a magnet, the relative velocity of the charge and magnet must be counted as an external circumstance.

In order to connect this theory with the empirical world, one must identify concrete systems that fit the mathematical representation, and measurement procedures that relate to some of the theoretical quantities in such a manner that theoretical relations among them are approximately verified. The most evident identification is the one suggested by the first observations that motivated the theory. Alternative identifications may be more advantageous, either because they permit more precise measurements or because they allow applications of the theory to a wider empirical domain. How a given quantity should be measured, whether it should be directly measured or rather inferred from other quantities depends on the selected identification. Different identifications share the same empirical content as long as the various ways to measure or infer the same quantity lead to the same results.

D'Alembert's principle

The relation $\mathbf{F} = m\mathbf{a}$ only determines the motion of an isolated material point. For an arbitrary connected system whose material points are subjected to the forces $\mathbf{F}_\alpha$, the motion is determined by d'Alembert's principle, which may be stated as follows:

> At any instant, the system must be in equilibrium with respect to the fictitious forces $\mathbf{F}_\alpha - m_\alpha \ddot{\mathbf{r}}_\alpha$, if $\mathbf{r}_\alpha(t)$ denotes the position of the particle α of the system at time t.

D'Alembert's justification of this powerful and elegant principle rests on the following intuition.

First suppose that the system is in equilibrium under the forces $\mathbf{K}_\alpha$. If the particle α of the system was disconnected from the rest of the system, its momentum $m_\alpha \dot{\mathbf{r}}_\alpha$ would increase at the rate $\mathbf{K}_\alpha$. In reality, this momentum remains zero. Therefore, motion is destroyed by the connections, at the rate $\mathbf{K}_\alpha$ for the particle α. These rates are interrelated through the condition of equilibrium of the system, that is, $\sum \mathbf{K}_\alpha \cdot \delta \mathbf{r}_\alpha = 0$ for any virtual displacement $\delta \mathbf{r}_\alpha$. Now suppose that the system moves under the forces $\mathbf{F}_\alpha$. If the particle α were disconnected from the rest of the system, its momentum would increase at the rate $\mathbf{F}_\alpha$. In reality, this momentum increases at the rate $m_\alpha \ddot{\mathbf{r}}_\alpha$. Therefore, motion is destroyed by the connections at the rate $\mathbf{F}_\alpha - m_\alpha \ddot{\mathbf{r}}_\alpha$. Granted that the interrelation between the destroyed motions is the same whether or not the destruction is complete, we must have $\sum (\mathbf{F}_\alpha - m_\alpha \ddot{\mathbf{r}}_\alpha) \cdot \delta \mathbf{r}_\alpha = 0$ for any virtual displacement $\delta \mathbf{r}_\alpha$. In other words, the system must be in equilibrium with respect to the fictitious forces $\mathbf{F}_\alpha - m_\alpha \ddot{\mathbf{r}}_\alpha$.

Being based on the elusive concept of destroyed motion, this argument carries little conviction. In modern treatises of dynamics, d'Alembert's principle is given as an axiom. Yet it is possible to derive it from more obvious principles. Call $\mathbf{r}_\alpha(t)$ the actual motion of the particles of the system and suppose that the system has been at rest until the forces $\mathbf{F}_\alpha(t)$ began to act at $t = 0$ (which does not restrict generality, since systems are only observed for a finite stretch of time). Imagine that all connections among the particles are removed, and apply to the particles the forces $m_\alpha \ddot{\mathbf{r}}_\alpha$ (instead of the forces $\mathbf{F}_\alpha$) from the origin of time. Obviously, the motion of the particles will still be $\mathbf{r}_\alpha(t)$. Now re-establish the connections and keep the same applied forces $m_\alpha \ddot{\mathbf{r}}_\alpha$. This operation does not alter the motion, since the motion is compatible with the connections (intuitively, the connections are lax).

Consequently, the forces $\mathbf{F}_\alpha(t)$ and $m_\alpha \ddot{\mathbf{r}}_\alpha(t)$ induce the same motion of the connected system. As we will see in a moment, this implies that the system remains at rest when it is subjected to the forces $\mathbf{F}_\alpha(t) - m_\alpha \ddot{\mathbf{r}}_\alpha(t)$ from the origin of time. Hence, the system is in equilibrium under these forces at any instant, as was to be proved.

That a system cannot be set into motion by the forces $\mathbf{F}_\alpha(t) - \mathbf{G}_\alpha(t)$ if the forces $\mathbf{F}_\alpha(t)$ and $\mathbf{G}_\alpha(t)$ separately yield the same motion is not as obvious as it would seem. At a given positive time t, we only know that the forces $\mathbf{F}_\alpha(t)$ and $\mathbf{G}_\alpha(t)$ cause the same change of velocity for the specific value $\dot{\mathbf{r}}_\alpha(t)$ of the velocity at this time. They do not necessarily do so if this original value vanishes. In order to circumvent this difficulty, we can appeal to the secular principle and to the resulting equivalence between a continuously varying force and a dense series of impulses. Specifically, we replace $\mathbf{F}_\alpha(t)$ with a series of impulses $\mathbf{I}_\alpha^n$ occurring at the times $2n\tau$, and $\mathbf{G}_\alpha(t)$ with a series of impulses $\mathbf{J}_\alpha^n$ occurring at the times $(2n + 1)\tau$ (with gradually varying intensity and direction of the impulses). When the system is subjected to one of these series of impulses, the velocity of its parts undergoes a series of jumps, separated by intervals of gradual, inertial variation (the velocity of the particles of a freely moving connected system can vary, owing to centripetal acceleration for instance). The inertial motions only depend on the configuration and velocities at the instants at which they begin to occur, and are therefore approximately the same for two series of impulses that yield the same motion. Consequently, the individual impulses $\mathbf{I}_\alpha^n$ and $\mathbf{J}_\alpha^n$ must cause nearly the same velocity jump, say $\Delta \mathbf{v}_\alpha^n$; and the series of impulses mimicking the force $\mathbf{F}_\alpha(t) - \mathbf{G}_\alpha(t)$ induce the velocity jumps $\Delta \mathbf{v}_\alpha^0, -\Delta \mathbf{v}_\alpha^0, \Delta \mathbf{v}_\alpha^1, -\Delta \mathbf{v}_\alpha^1$, and so forth, separated by inertial velocity variations of second order with respect to the initial velocity (for example, $v^2\tau/R$ in the case of centripetal acceleration, if v is the initial velocity, and R the distance of the particle from the center). At the secular scale, this implies that the system remains at rest.

In sum, d'Alembert's principle can be derived from the following assumptions:

1. The motion of a disconnected material point subjected to a force satisfies the law of acceleration;
2. The forces acting on a system can be replaced by a dense series of impulses without significantly altering its secular motion;
3. The inertial motion of a connected system is completely determined by the initial configuration and the initial velocities;
4. Inertial accelerations are of second order with respect to the initial velocities;
5. The velocity increments caused by a given set of impulses only depend on the configuration of the system (not on its velocities); and
6. Two sets of opposite impulses yield opposite velocity increments when applied to the same configuration of the system.

As we saw, the first assumption results from the principle of inertia, the relativity principle, the secular principle, and some causality assumption. The second derives from the secular principle. The third follows from the observation that the initial state of motion can always be generated from rest by applying the impulses $m_\alpha \mathbf{v}_\alpha$ to the various particles of the system; the ulterior motion can only depend on this impulse, whose precise time structure is irrelevant according to the secular principle. The fourth assumption can be derived from a dimensional consideration: the inertial acceleration must be proportional to the square of the initial velocity, because the only other parameters of the system have the dimension of length. The fifth holds because no inertial motion can occur during the very short action of the impulsive forces. The sixth derives from the secular principle, applied to a dense series of impulses of equal intensity and alternating signs. To sum up, all the assumptions made in the present derivation of d'Alembert's principle can be reduced to three more basic principles: the principle of inertia, the principle of relativity, the secular principle, and a zest of causality.

As was already the case for statics, the dynamics of connected systems deals with ideal representations of concrete systems that are never strictly encountered in nature or even in man-made devices. In actual systems, there always is some friction, rigidity is only approximate, threads have some elasticity, parts can break, etc. In order to improve the match between theory and experiment, we may do our best to reduce these effects to a minimum (smooth surfaces, lubrication, improved materials, etc.), or we may try to include them in the theory. Viscous friction lends itself most easily to the latter strategy, for it can be represented by additional forces that follow simple empirical regularities. Departure from perfect rigidity is harder to take into account: it requires loosing up the kinematic constraints, and it dramatically increases the number of degrees of freedom of the system. We will now see how this can be done.

2.2 Molecular mechanics

The necessity of Saint-Venant's assumptions

We may now follow Saint-Venant in defining a different sort of mechanics, based on the following assumptions:

1. All matter is made of identical point atoms.
2. There exists a class of reference frames in which the motion of any isolated atom is rectilinear and uniform.
3. In an isolated pair of atoms, each atom undergoes an acceleration directed on the line joining the two atoms, opposite to that of the other atom, and depending only on the distance of the two atoms.
4. The acceleration of an atom is the vector sum of the accelerations it would take if it were subjected to the action of every other atom separately.

Although Saint-Venant regarded these assumptions as mostly contingent, we may try to derive them by reasoning similar to that given for the dynamics of connected system. The first assumption seems to be the necessary result of any ultimate reduction. The second, the principle of inertia, has already been discussed. The fourth immediately derives from the secular principle. The third can be derived from the relativity principle, the secular principle, some causality principle, and the impossibility of perpetual motion. This can be done as follows.

The secular principle and the causality principle allow us to replace the mutual interaction by a series of periodic impulses that cause well-defined velocity jumps. For symmetry reasons these impulses can be chosen to be simultaneous, opposed, and directed along the line joining the two atoms. The intensity of one of these impulses can only depend on the distance between the two atoms and on their inertial motion in the period preceding this impulse. Consequently, the intensity of the secular acceleration can only depend on the distance and the instantaneous velocities of the two atoms. The relativity principle further implies that the velocity dependence only involves the relative velocity of the two atoms. We will now see that the impossibility of perpetual motion excludes the latter dependence.[15]

Owing to rotational invariance, the acceleration of one of the atoms must have the form

$$\mathbf{a} = \mathbf{r}f(r, u, \mathbf{r} \cdot \mathbf{u}), \tag{1}$$

wherein $\mathbf{r}$ is the vector joining the two atoms and $\mathbf{u}$ their relative velocity. Consequently, the variation in time of the kinetic energy $T = \frac{1}{2}(v_1{}^2 + v_2{}^2)$ has the form

$$\mathbf{a} \cdot \mathbf{u} = (\mathbf{r} \cdot \mathbf{u})f(r, u, \mathbf{r} \cdot \mathbf{u}) = \dot{r}F(r, \dot{r}, u). \tag{2}$$

In order to avoid the possibility of perpetual motion, this kinetic energy must return to its original value after any cycle of the variable r (granted that perpetual motion remains impossible at the atomic scale, and granted that atomic processes are reversible). Hence $\mathbf{a} \cdot \mathbf{u}$ must be the total derivative (with respect to time) of a certain function of the variables $(r, \dot{r}, u)$ on which it a priori depends. This can only happen if the unknown function F depends on r only. Consequently, the acceleration of the two atoms cannot depend on their relative velocity. Another consequence is the existence of a potential from which this acceleration derives.[16]

The total exclusion of velocity-dependent forces has the inconvenience of excluding electromagnetic forces. However, we may follow Helmholtz and Maxwell in taking the velocity dependence of electromagnetic forces as a sign that they involve a third mechanical entity, the ether. In this case, an atom is no longer the direct cause of the motion of another atom. If electromagnetism is to be reduced to a pure Saint-Venant mechanics, the set of basic atoms must include the set of mass-points of which the ether would be made.

As we saw in the previous chapter, connected systems can be approximately synthesized from Saint-Venant atoms. Consequently, the general laws of statics and dynamics are (approximate) consequences of Saint-Venant's principles, which are themselves necessary consequences of the principle of inertia, the secular principle, the impossibility of perpetual motion, and a causality principle. Not surprisingly, the same principles are used in the direct and indirect proofs of the necessity of the laws of connected systems.[17]

The proofs of necessity are perhaps less convincing in the case of molecular mechanics, because they involve the now implausible reduction of matter to point atoms, and because they imply the application at the atomic level of principles whose intuitive evidence only

---

[15] Helmholtz 1847 believed he could derive the central character of molecular forces from the impossibility of perpetual motion alone. He later conceded that there was a gap in his derivation.

[16] If the secular principle did not hold, the accelerations could depend on higher derivatives of r, and the cyclic conservation of the kinetic energy could hold by compensation of the contribution of the various derivatives. A simple example of such a compensation is provided by Weber's law, for which $a = r^{-2}(1 - \dot{r}^2/2c^2 + r\ddot{r}/c^2)$.

[17] See Chapter 1, pp. 42–3.

holds at the macroscopic level. For example, the impossibility of building macroscopic machines that indefinitely produce or annihilate work without compensation is not necessarily incompatible with the existence of non-conservative atomic processes. This deficiency in the proofs of necessity of molecular mechanics is compensated by its larger scope: as Navier, Poisson, and Cauchy realized in the 1820s, this mechanics provides a powerful model for deriving the laws of deformable media of various kinds, including elastic fluids and solids.

Elasticity and viscosity

In order to derive the equations of elasticity, we only need to compute the resultant of all the molecular forces acting on a given molecule when the position of the molecules is modified by a large-scale deformation $\mathbf{u}(\mathbf{r})$. Alternatively, we may obtain these equations by computing the variation $\delta\Phi$ of the total potential of atomic or molecular forces

$$\Phi = \frac{1}{2} \sum_{\alpha\beta} \varphi_{\alpha\beta}(r_{\alpha\beta}) \tag{3}$$

under a variation $\delta\mathbf{u}(\mathbf{r})$ of the large-scale deformation and injecting this variation in the Lagrangian equation[18]

$$\sum_{\alpha} (\mathbf{F}_{\alpha} - \ddot{\mathbf{r}}_{\alpha}) \cdot \delta\mathbf{r}_{\alpha} = \delta\Phi \tag{4}$$

The value Φ of the total potential for the deformation $\mathbf{u}$ has the form

$$\Phi \approx \Phi_0 + \frac{1}{4} \sum_{\alpha\beta} \varphi''(r_{\alpha\beta})(\Delta r_{\alpha\beta})^2, \tag{5}$$

where $\Delta r_{\alpha\beta}$ denotes the variation of the distance between the molecules α and β owing to the deformation $\mathbf{u}$. This variation satisfies

$$r_{\alpha\beta} \Delta r_{\alpha\beta} = \mathbf{r}_{\alpha\beta} \cdot [\mathbf{u}(\mathbf{r}_{\beta}) - \mathbf{u}(\mathbf{r}_{\alpha})] \approx (\mathbf{r}_{\alpha\beta} \cdot \nabla)(\mathbf{r}_{\alpha\beta} \cdot \mathbf{u}_{\alpha}), \tag{6}$$

whence follows

$$\delta\Phi = \frac{1}{2} \sum_{\alpha\beta} r_{\alpha\beta}^{-2} \varphi''(r_{\alpha\beta}) [(\mathbf{r}_{\alpha\beta} \cdot \nabla)(\mathbf{r}_{\alpha\beta} \cdot \mathbf{u}_{\alpha})] [(\mathbf{r}_{\alpha\beta} \cdot \nabla)(\mathbf{r}_{\alpha\beta} \cdot \delta\mathbf{u}_{\alpha})]. \tag{7}$$

Suppose that the original state of equilibrium of the body is isotropic. Then, for any molecule α (except for those situated next to the surface of the body) the sum over β is proportional to the value of $\partial_i u_j \partial_i \delta u_j + \partial_j u_i \partial_i \delta u_j + \partial_i u_i \partial_j \delta u_j$ at the point $\mathbf{r}_{\alpha}$ (summation

---

[18] See Eq. (41) of Chapter 1, p. 42. This strategy is more powerful because it also yields the boundary condition. On the foundational works by Navier, Cauchy, and Poisson, cf. Darrigol 2002 and further reference there.

over the repeated space indices i and j being understood). Granted that the original state of the body is homogenous, summation over α next leads to

$$\delta\Phi = \int d\tau \sigma_{ij}\partial_i\delta u_j, \tag{8}$$

with

$$\sigma_{ij} = K(\partial_i u_j + \partial_j u_i + \delta_{ij}\partial_k u_k). \tag{9}$$

Injecting this result into Eq. (4) and assuming that the only external forces are the constant gravity $\mathbf{g}$ and the pressure $\mathbf{P}$ (obliquely) applied on the surface of the body, we get

$$\int \rho(\mathbf{g} - \ddot{\mathbf{u}}) \cdot \delta\mathbf{u} d\tau + \int \mathbf{P} \cdot \delta\mathbf{u} dS = \int d\tau \sigma_{ij}\partial_i\delta u_j = -\int d\tau(\partial_i\sigma_{ij})\delta u_j + \int n_i dS\sigma_{ij}\delta u_j, \tag{10}$$

where ρ is the density (number of atoms per unit volume) and $\mathbf{n}$ is the normal unit vector on the surface of the body. This relation holds for any $\delta\mathbf{u}$ if and only if

$$\rho\ddot{\mathbf{u}} = \rho\mathbf{g} + K[\Delta\mathbf{u} + 2\nabla(\nabla \cdot \mathbf{u})] \tag{11}$$

within the body, and

$$P_j = n_i\sigma_{ij} \tag{12}$$

on the surface of the body. These are the fundamental equations of elasticity in the simplest case in which there is only one elastic constant. Any problem of elasticity can be solved by means of more general equations of the same type.

In a fluid, molecules can freely slide on each other but resist closer packing. Consequently, under a small, large-scale displacement $\delta\mathbf{u}(\mathbf{r})$ of the molecules the total potential of the molecular forces in the volume element $d\tau$ varies by $-\lambda(\nabla \cdot \delta\mathbf{u})d\tau$, wherein λ is a positive coefficient and $(\nabla \cdot \delta\mathbf{u})d\tau$ is the volume increase of the portion of fluid that was originally in the element $d\tau$. Equation (4) then gives the equilibrium condition

$$\int \rho\mathbf{g} \cdot \delta\mathbf{u} d\tau + \int \mathbf{P} \cdot \delta\mathbf{u} dS = -\int \lambda\nabla \cdot \delta\mathbf{u} d\tau, \tag{13}$$

where $\mathbf{g}$ is the acceleration of gravity, ρ the density of the fluid, and $\mathbf{P}$ an external pressure applied on its free surface. The validity of this equation for any $\delta\mathbf{u}$ implies the relations

$$\rho\mathbf{g} - \nabla\lambda = 0 \text{ (within the fluid) and } \mathbf{P} = -\mathbf{n}\lambda \text{ (on the fluid surface)}. \tag{14}$$

If the fluid is moving, the distance of two neighboring molecules may be further affected by this motion. Following Navier, let us assume that this modification is proportional to the projection of the relative velocity of the two molecules onto the line that joins them:

$$\Delta r_{\alpha\beta} \propto \mathbf{r}_{\alpha\beta} \cdot [\mathbf{v}(\mathbf{r}_\beta) - \mathbf{v}(\mathbf{r}_\alpha)] \approx (\mathbf{r}_{\alpha\beta} \cdot \nabla)(\mathbf{r}_{\alpha\beta} \cdot \mathbf{v}_\alpha). \tag{15}$$

In analogy to the case of an elastic solid, this leads to the condition

$$\int \rho(\mathbf{g} - \ddot{\mathbf{u}}) \cdot \delta\mathbf{u}d\tau + \int \mathbf{P} \cdot \delta\mathbf{u}dS = -\int \lambda\nabla \cdot \delta\mathbf{u}d\tau - \int d\tau(\partial_i\sigma_{ij})\delta u_j + \int n_i dS\sigma_{ij}\delta u_j \tag{16}$$

with

$$\sigma_{ij} = \eta(\partial_i \upsilon_j + \partial_j \upsilon_i + \delta_{ij}\partial_k \upsilon_k). \tag{17}$$

For a nearly incompressible fluid, the resulting equation of motion is Navier's

$$\rho\ddot{\mathbf{u}} = \rho\left[\frac{\partial \mathbf{v}}{\partial t} + (\mathbf{v} \cdot \nabla)\mathbf{v}\right] = \rho\mathbf{g} - \nabla\lambda + \eta\Delta\mathbf{v}. \tag{18}$$

The readers who judge Navier's assumption (15) too artificial may prefer Poisson's assumption that every fluid has some rigidity, but that the stresses induced by shearing deformations rapidly decay owing to rearrangements of the molecules. This intuition leads to (secular) stresses that depend on the fluid velocity $\mathbf{v}$ as the stresses in a solid depend on its deformation $\mathbf{u}$.[19]

The Laplacian molecular theories of elasticity have restricted validity: they concern only the so-called rari-constant media for which the elastic behavior of isotropic media depends on one elastic constant only; and they hold only to the extent that macroscopic mechanics can be separated from thermal phenomena. As is well known, the explanation of the latter phenomena prompted mid-nineteenth-century physicists to replace the Laplacian static model of matter with a kinetic model in which a disordered motion of the molecules is associated with heat. Still, the partial success of the Laplacian theories is a valuable indication that the laws of the mechanics of deformable media can be derived from a molecular mechanical basis. It also teaches us that these laws do not depend on the details of the molecular model: there are infinitely many choices of the molecular distributions and forces that lead to the same macroscopic equations of motions.

2.3   Continuum mechanics

Concepts of pressure

Although static and kinetic molecular theories go a long way explaining the mechanical (and thermal) properties of macroscopic matter, it is desirable to have a purely macroscopic continuum mechanics for at least two reasons. Firstly, the classical molecular models on which these theories once depended are now obsolete. Secondly, *pace* Saint-Venant, a

---

[19]On Navier's and Poisson's relevant works, cf. Darrigol 2002.

deformable medium need not have a molecular structure. What makes pure continuum mechanics possible is the concept of internal pressure, foreshadowed in Newton's *Principia* and later developed by Euler and Cauchy. So far, the basic concepts of force we have used are the force applied to a material point of a connected system, and the force applied to a point atom. For a continuous body, two derived concepts are usually introduced: the force $\mathbf{f} d\tau$ applied to the volume element $d\tau$ of the body, and the force $\mathbf{P} dS$ applied to the surface element dS of the body. Discrete distributions are thus turned into continuous ones, and discrete sums into integrals. Although this is a standard procedure of the physics of continua, it relies on a subtle series of idealizations.

Remember that we started with the operational idea of a force as something that can be balanced by a weight through a pulley-thread system. This leads to the idealization of a force applied to a material point. We can then imagine a dense set of material points and try to take the limit of a distribution of force over these points when the density becomes infinite. This limit may lead to a discrete distribution *à la* Dirac, to a continuous distribution, or to a combination of both. These idealizations allow for rigorous mathematical constructs by means of Laurent Schwarz's theory of distributions. Their relevance to the physics of continua depends on their compatibility with basic principles and on the empirical success of the theories that rely on them. It is not a priori obvious that a mechanical action on a tiny part of a body can be replaced by a single force acting on a single point of the body. Neither is it obvious that a distributed action over a large part of a body can be replaced by a continuous distribution that ignores microscopic fluctuation. To assume so much is to accept some spatial counterpart to the secular principle.

Granted that force densities can be defined, we are in possession of the concept of force per surface unit or pressure. So far, this concept only concerns the surface of a body and the external actions on this surface. As history shows, some intellectual audacity was needed to move from this operationally meaningful notion to the concept of internal pressure. Euler (in the case of hydrodynamics) and Cauchy (in the case of elasticity) both assumed that an internal portion of a body could be regarded as subjected to two sorts of forces: long-range forces of gravitational origin, and pressures exerted on the surface of this portion by the rest of the body. This idealization can be taken as an axiom, whose worth is judged through its consequences; or it can be derived from a discrete model of matter such as Saint-Venant's. In the latter approach, the pressure across an internal surface element of a body is defined as the resultant of the forces acting from any molecule on one side of the element to any molecule on the other side such that the line joining the two molecules crosses the element.[20]

The Cauchy stresses

Cauchy further assumed that a sufficiently small volume element of the body behaved as a rigid body with regard to dynamical principles.[21] Accordingly, the element is in equilibrium if and only if the resultant and the resulting torque of all applied forces

---

[20] On the history of stress, cf. Truesdell 1968a.

[21] The sole purpose of this assumption is to justify the application of the principles of linear and angular momentum to the elements of the continuum. It may be discarded by taking these principles as axioms, as recommended by Truesdell 1968b and by most modern mechanicians. Or it may be replaced with a molecular justification along the lines of the previous section.

vanishes. In particular, the resultant of the pressures on the faces of the pyramid $x \geq 0$, $y \geq 0$, $z \geq 0$, $\alpha x + \beta y + \gamma z \leq \varepsilon$ must be balanced by the weight of the pyramid (x, y, z are Cartesian coordinates from an arbitrary origin within the body). As the surface of the faces is of order ε^2 and the weight of order ε^3, this is only possible if the pressure on the four faces balance each other at the order ε^2. Consequently, the internal pressure on the surface element $\mathrm{d}\mathbf{S}$ has the form $\sigma_{ij}\mathrm{d}S_j$, where σ_{ij} is called the stress tensor.

For any internal portion of the body, the resultant of the pressure forces is

$$\int \sigma_{ij}\mathrm{d}S_j = \int \partial_j\sigma_{ij}\mathrm{d}\tau. \tag{19}$$

The internal equilibrium of the body under the external force density $\mathbf{f}$ (usually $\rho\mathbf{g}$) requires that

$$f_i + \partial_j\sigma_{ij} = 0. \tag{20}$$

The other condition of equilibrium is the balance of moments:

$$x_if_k - x_kf_i + \partial_j(x_i\sigma_{kj} - x_k\sigma_{ij}) \equiv x_i(f_k + \partial_j\sigma_{kj}) - x_k(f_i + \partial_j\sigma_{ij}) + \sigma_{ki} - \sigma_{ik} = 0. \tag{21}$$

Combined with the first condition this gives the symmetry condition

$$\sigma_{ij} = \sigma_{ji}. \tag{22}$$

Most generally, the motion of a continuous body verifies the internal condition

$$f_i + \partial_j\sigma_{ij} = \rho\ddot{x}_i, \tag{23}$$

where σ_{ij} is a symmetric stress system. At the border of the body, the dynamical equilibrium of a thin, tangent volume element further requires the balance of any external pressure with the internal pressure.

Elasticity and viscosity

In order to determine the motion from these conditions, more must be known about the stress system. In a perfect fluid, this tensor has the isotropic form

$$\sigma_{ij} = -P\delta_{ij}, \tag{24}$$

which leads to Euler's equation

$$\rho\left[\frac{\partial\mathbf{v}}{\partial t} + (\mathbf{v}\cdot\nabla)\mathbf{v}\right] = \rho\mathbf{g} - \nabla P. \tag{25}$$

If the fluid is incompressible, this equation completely determines the motion under given boundary conditions. In the contrary case, the pressure must be known as a function of other physical variables such as density and temperature. This may require leaving the narrow context of mechanics.

In a viscous fluid, we intuitively expect shear stresses proportional to the shearing velocity. For example, a plane parallel motion implies a stress $\sigma_{xy} \propto \partial v_x / \partial y$ if the x axis is parallel to the motion and the y axis is perpendicular to the planes of equal velocity. The three-dimensional, isotropic, and symmetric generalization of this special case leads to

$$\sigma_{ij} = \eta \left(\partial_i v_j + \partial_j v_i - \tfrac{2}{3} \delta_{ij} \partial_k v_k \right), \tag{26}$$

to which we must add the hydrostatic pressure $-\delta_{ij} P$. The Navier–Stokes equation immediately results from this stress system.

In an elastic solid, we expect the stress system to depend on the local deformation of the system. In a natural generalization of Hooke's law, the dependence should be linear for a small deformation. The local deformation is given by the variation of the square of the distance between the material points originally situated at $\mathbf{r}$ and $\mathbf{r} + \mathrm{d}\mathbf{r}$. Calling $\mathbf{u}(\mathbf{r})$ the shift of the material point originally at $\mathbf{r}$, this variation is $e_{ij} \mathrm{d}x_i \mathrm{d}x_j$ with

$$e_{ij} = \partial_i u_j + \partial_j u_i. \tag{27}$$

In a homogenous, isotropic, and linear medium, the relation between the stress tensor σ_{ij} and the deformation tensor e_{ij} is restricted to the form

$$\sigma_{ij} = K' e_{ij} + K'' e_{kk} \delta_{ij}. \tag{28}$$

The corresponding equations of motion are more general than Navier's, since the coefficient of the $\nabla(\nabla \cdot \mathbf{u})$ term can now have any value. This derivation of the equations of elasticity is much simpler than Navier's. It also is better fitted to the elastic behaviors found in nature, for there are bodies for which the constants K' and K'' do not satisfy Navier's relation $K' = 2K''$.

To sum up, the concept of internal pressure provides a much easier access to the dynamics of continuous media than any molecular model could do. However, the laws of this dynamics only derive from the necessary laws of simpler dynamics through further idealization of notions obtained either in the connected-system approach or in the molecular approach. This route also requires some empirical input, such as Mariotte's law for gases, Newton's law of shear stress for viscous fluids, or Hooke's law for elastic solids. The only exception is the perfect liquid, which could already be treated as a connected system.

Internal pressures in connected systems

Any connected system can be regarded as made of material volume elements whose motion obeys

$$f_i + \partial_j \sigma_{ij} = \rho \ddot{x}_i. \tag{29}$$

Suppose the motion is caused by the force density **f** and by the external pressures **P** applied on the exposed surface S_e of the bodies of which the system is made. This does not restrict generality, because localized forces can be regarded as pressures applied to tiny surfaces. The virtual work of all impressed and inertial forces under the virtual displacement $\delta\mathbf{u}(\mathbf{r})$ is

$$\delta W = \int (\mathbf{f} - \rho\ddot{\mathbf{u}}) \cdot \delta\mathbf{u}\,d\tau + \int_{S_e} \mathbf{P} \cdot \delta\mathbf{u}\,dS = -\int (\partial_j\sigma_{ij})\delta u_j\,d\tau + \int_{S_e} \mathbf{P} \cdot \delta\mathbf{u}\,dS. \qquad (30)$$

Partial integration leads to

$$\delta W = \int \sigma_{ij}\partial_j\delta u_i\,d\tau - \int_{S_e \cup S_c} \sigma_{ij}\delta u_i\,dS_j + \int_{S_e} \mathbf{P} \cdot \delta\mathbf{u}\,dS, \qquad (31)$$

wherein S_c denotes the contact surfaces between the various bodies of which the system is made. Granted that the boundary condition

$$\sigma_{ij}dS_j = P_i dS \qquad (32)$$

holds on the surface S_e, the virtual work also reads

$$\delta W = \int \sigma_{ij}\partial_j\delta u_i\,d\tau + \int_{S_c} \mathbf{P} \cdot \delta\mathbf{u}\,dS. \qquad (33)$$

The first integral represents the work done by the stresses under the deformation caused by the virtual displacement, as can be verified by cutting up each body into cubic volume elements. This work always vanishes for the rigid bodies, inextensible threads, and perfect liquids of which connected systems are made. In the case of a rigid body, this vanishing is obvious because there is no deformation.[22] In the case of a perfect liquid, there are no shear stresses, and the internal isotropic pressure does not work since there is no compression. In the thread case, the longitudinal stress does not work because of the inextensibility of the thread, and there are no transverse stresses because of the flexibility of the thread.

The second integral in Eq. (33) represents the work of the contact pressures between any two bodies of the system. In case of a rolling contact, this work vanishes because the relative displacement vanishes on the small contact surface and because the pressures of the two bodies on each other are opposite and equal. This work also vanishes for sliding contacts, because smooth sliding requires the vanishing of the pressure component parallel to the slide. To sum up, the virtual work of all inertial forces and impressed forces must vanish during a virtual displacement of the system.

---

[22]This can be verified directly because the most general deformation of a solid, $\delta u_i = a_i + \varepsilon_{ikl}(\omega_k x_l - \omega_l x_k)$, makes $\partial_j\delta u_i$ a skew-symmetric tensor, whose contraction with the symmetric tensor σ_{ij} vanishes.

The principle of virtual work and d'Alembert's principle can thus be derived by regarding connected systems as made of material volume elements subjected to both internal pressure forces and long-range forces and obeying Newton's laws of motion. The dynamics of connected systems thus becomes a special case of the mechanics of continua. Reciprocally but somewhat artificially, any continuous mechanical system can be regarded as a connected system in which internal stresses are treated as impressed forces. Indeed the vanishing of the virtual work of all applied forces,

$$\int \mathbf{f} \cdot \delta \mathbf{u} d\tau + \int \mathbf{P} \cdot \delta \mathbf{u} dS - \int \sigma_{ij} \partial_j \delta u_i d\tau = 0, \tag{34}$$

yields, after partial equation of the last integral, the internal condition of equilibrium (20) and the boundary condition (32).

The main interest of this remark is the connection it reveals between the equations of elasticity and the principle of least action. Suppose, as was done in the old elastic-solid theories of light, that space is filled with various elastic bodies. Then the only impressed forces are the internal stresses and the inertial force density $-\rho\ddot{\mathbf{u}}$. Owing to the linear dependence between stress and strain, the virtual work of stresses takes the form

$$\delta V = \delta \int \frac{1}{4} \sigma_{ij} e_{ij} d\tau. \tag{35}$$

Consequently, the equations of motion derive from the principle of least action, with the Lagrangian

$$L = \int \frac{1}{2} \rho \dot{\mathbf{u}}^2 d\tau - \int \frac{1}{4} \sigma_{ij} e_{ij} d\tau. \tag{36}$$

Unfortunately, the principle no longer applies in non-elastic bodies in which deformation implies the production of heat. General validity of this principle only obtains in molecular theories in which heat is understood as a mode of motion.

To summarize, continuum dynamics can be developed through a system of purely macroscopic idealizations involving the concept of internal stress. It is fairly autonomous, since it can be applied without consulting molecular mechanics or the dynamics of connected system. However, it requires the laws of motion of a mass element of the continuum, which it must borrow from one of these competitors. Two bridges can been thrown between the various approaches: Continuum dynamics can be regarded as a large-scale consequence of molecular mechanics, with some restrictions on the resulting stresses; and the dynamics of connected systems can be derived from the continuum dynamics of a restricted class of systems.

2.4 Collisions

We have so far addressed three ways of founding mechanics: through connected systems, through point-like atoms and central forces, and through internal stresses. There is still a third way, through collisions, which has played a significant, historical role. Descartes,

Huygens, Newton, Leibniz, and d'Alembert all gave a central role to collisions in their quest for the true laws of mechanics. Johann and Daniel Bernoulli tried to reduce every force, including gas pressure, to molecular impacts. Later in the eighteenth century, another Swiss geometer, Georges-Louis Lesage, nearly succeeded in reducing gravitational forces to the impact of *corpuscules ultramondains* freely moving between gravitating bodies and impacting them. In the late nineteenth century, Osborne Reynolds attempted a "submechanics of the universe" based on the close packing and collisions of hard spherical molecules of ether.[23]

None of these reductions truly worked; and I do not intend to offer a new one of the same kind. There are several reasons, however, to preserve a privileged role of collisions in mechanics. Collisions are the simplest mechanical process: they only involve discrete jumps of a few physical quantities (attention being focused on asymptotic states), and the relevant interaction may result from mere contact (as is approximately the case for billiard balls). Owing to this simplicity, it is tempting to bring collision dynamics to bear on the foundations of other forms of dynamics, as Newton did through the analogy between a continuously acting force and a rapid succession of impacts. Earlier in this chapter, we encountered derivations of Newton's second law and of d'Alembert's principle that exploit the same analogy in the name of the secular principle. Last but not least, modern particle physics largely is a physics of scattering processes. For all these reasons, it is important to discuss the general characteristics of collision dynamics.

One way to do that is to assume a given theory of the matter and forces of the colliding bodies and to explore its consequences. In his dynamics, d'Alembert assumes ideally rigid bodies, whose mutual collisions destroy motion. In this view, the symmetric, frontal collision of two equal rigid bodies brings the two bodies to rest and permanent contact (this is more like a soft-body collision in the modern conception). D'Alembert then defines mass in such a manner that momentum is conserved in any collision. In contrast, Saint-Venant assumes that collisions involve central forces between the atoms of the two colliding bodies that come close to each other. This mechanism (or any mechanism implying forces that satisfy the principle of reaction) implies the conservation of momentum in general, and the conservation of kinetic energy in cases for which the molecular configuration of the colliding bodies is asymptotically unchanged by the collision.[24]

To this constructive approach to collisions process, one can oppose a principle-based approach that refrains from any particular assumption on the colliding mechanism. We only assume, for every collision, the invariance of quantities of the form

$$I = \sum_i F_i(\mathbf{v}_i), \tag{37}$$

wherein $\mathbf{v}_i$ is the velocity of the particle i and F_i is a continuous, differentiable function of this velocity depending on the kind of particle.[25]

---

[23] Reynolds 1903. On Lesage, cf. Chabot 2004.

[24] On d'Alembert's and Saint-Venant's views, see Chapter 1, pp. 32–4, 41–2.

[25] Comte 1986 is a clever implementation of this kind of approach. His results agree with those given below, although the derivations differ somewhat.

The one-dimensional case

We first consider the case of a one-dimensional space. As the following considerations lend themselves to relativistic generalization, we leave the Galilean framework in which we were so far confined. Call rapidity the function φ of the velocity v that is additive with respect to the composition of inertial boosts. In these terms, a possible invariant has the form

$$I = \sum_i f_i(\varphi_i). \tag{38}$$

Owing to left–right symmetry, the odd and even parts of this invariant must be conserved separately. Without loss of generality, we may therefore assume that there exists an invariant with definite parity (f is either an odd or an even function of φ). The relativity principle implies the conservation of

$$I_a = \sum_i f_i(\varphi_i + a) \tag{39}$$

for any value of the rapidity a of an inertial boost. Consequently, the quantities

$$I^{(n)} = \sum_i f_i^{(n)}(\varphi_i) \tag{40}$$

should be conserved for any value of the order n of derivation.

This proliferation of conserved quantities should be avoided, because there cannot be more than two independent conservation laws in a one-dimensional space. Indeed, for a binary elastic collision, in which the particles remain the same, there are only two unknowns, the final velocities. A first way to avoid the proliferation is to assume the constancy of the second derivatives f_i''. This yields the conserved quantities

$$M = \sum_i m_i, \; P = \sum_i m_i \varphi_i, \text{ and } E = \sum_i \frac{1}{2} m_i \varphi_i^2 \tag{41}$$

(if f_i'' does not vanish). In the case of binding collisions, only the two first invariants can exist.

If the derivatives f_i'' are not constants, the non-proliferation of invariants in binary elastic collisions requires that the invariant $f_1''(\varphi_1) + f_2''(\varphi_2)$ should be a function of the invariants $f_1(\varphi_1) + f_2(\varphi_2)$ and $f_1'(\varphi_1) + f_2'(\varphi_2)$. Since φ_1 and φ_2 are independent variables, this function can only be an affine function. Consequently, f_i'', f_i, and f_i' are related by the same affine function. As f_i has a definite parity, f_i'' should in fact be proportional to f_i. Consequently, f_i is proportional to the sine or the cosine of the rapidity (a numerical coefficient in the sine or cosine can be absorbed in the definition of the rapidity). In order to avoid conservation laws involving periodic functions of the rapidity, the sine and the cosine must be hyperbolic. Hence there should be two invariants:

$$E = \sum_i m_i \cosh \varphi_i, \tag{42}$$

and

$$P = \sum_i m_i \sinh \varphi_i. \tag{43}$$

So far we have left the function $\varphi(v)$ undetermined. The first case leads to the non-relativistic invariants

$$M = \sum_i m_i, \ P = \sum_i m_i v_i, \ \text{and } E = \sum_i \frac{1}{2} m_i v_i^2 \tag{44}$$

granted that $\varphi(v) = v$. The second case leads to the relativistic invariants

$$E = \sum_i m_i (1 - v_i^2/c^2)^{-1/2}, \tag{45}$$

and

$$P = \sum_i m_i v_i (1 - v_i^2/c^2)^{-1/2}, \tag{46}$$

granted that

$$\varphi(v) = \tanh^{-1}(v/c). \tag{47}$$

In a one-dimensional world, these choices of the function $\varphi(v)$ remain unjustified.

Rapidity determined

We will now see that in a three-dimensional world, the non-proliferation of conservation laws completely determines the function $\varphi(v)$ as well as the form of the possible invariants. The quantity (42)

$$E = \sum_i m_i \cosh \varphi_i$$

lends itself to a three-dimensional generalization in which $\varphi_i = \varphi(|\mathbf{v}_i|)$. A boost of infinitesimal velocity $\mathbf{u}$ leads to a variation of the rapidity φ such that

$$d\varphi = (\mathbf{u}/c) \cdot (\mathbf{v}/v), \tag{48}$$

because this is the only expression compatible with the additivity of rapidities (when $\mathbf{u}//\mathbf{v}$), the isotropy of space, linearity with respect to $\mathbf{u}$, and the equivalence of velocity and rapidity for small velocities ($\varphi(u) \approx u/c$). The relativity principle therefore leads to the conservation of the quantity

$$\sum_i m_i \sinh \varphi_i \, d\varphi_i = \sum_i (\mathbf{v}_i/v_i) \cdot (\mathbf{u}/c) \, m_i \sinh \varphi_i \qquad (49)$$

for any choice of $\mathbf{u}$, which means that the vector quantity

$$\mathbf{P} = \sum_i m_i(\mathbf{v}_i/v_i) \sinh \varphi_i \qquad (50)$$

is conserved.

The non-proliferation of conservation laws requires that the application of a second boost to the latter quantity should not produce any new conservation law. Suppose that all the velocities $\mathbf{v}$ are parallel and that the infinitesimal boost $\mathbf{u}$ is perpendicular to them. Then the velocities $\mathbf{v}$ vary by $d\mathbf{v} = \mathbf{u}$, and v and φ do not vary at this order (the normal component of $\mathbf{v}$ cannot vary for symmetry reasons, and the generated parallel component is obviously equal to the velocity of the boost). Consequently, the quantity

$$E' = \sum_i m_i v_i^{-1} \sinh \varphi_i \qquad (51)$$

is a conserved scalar quantity, which must be proportional to the already known invariant of the same kind (42)

$$E = \sum_i m_i \cosh \varphi_i,$$

in order to avoid proliferation. This implies the relation

$$\tanh \varphi = v/c, \qquad (52)$$

which is thus retrieved without recourse to relativistic kinematics! The resulting invariants are the total energy

$$E = \sum_i m_i(1 - v_i^2/c^2)^{-1/2} \qquad (53)$$

and the total momentum

$$\mathbf{P} = \sum_i m_i \mathbf{v}_i(1 - v_i^2/c^2)^{-1/2}. \qquad (54)$$

More general boosts applied to more general collisions do not generate new invariants. This can be verified directly by differentiating the total momentum and injecting the expression

$$d\mathbf{v} = \mathbf{u} + (1 - v^2/c^2)(\mathbf{u} \cdot \mathbf{v})\mathbf{v}/c^2 \qquad (55)$$

of the variation of the particle velocity $\mathbf{v}$ owing to the infinitesimal boost $\mathbf{u}$ (the latter expression is the only one that complies with linearity in $\mathbf{u}$, space isotropy, and compatibility with the earlier given relations $v d\varphi = \mathbf{u} \cdot \mathbf{v}$ and $\tanh \varphi = v/c$). The same conclusion can be reached in a more concise manner by noting that the total energy and the total momentum are the components of a 4-vector, and therefore transform linearly under an inertial boost.

Similar reasoning applied to the presumed invariant $\sum_i m_i \varphi_i{}^2$ leads to $\varphi \propto \upsilon$ and to the non-relativistic expressions of energy and momentum. To sum up, the non-proliferation of conservation laws implies that the rapidity should be proportional either to the velocity or to its hyperbolic arctangent. The possible invariants are mass and the classical energy and momentum in one case, the relativistic energy-momentum in the other case. Note that in the relativistic case, energy conservation applies to binding collisions or to disintegrations (momentum conversation cannot occur without energy conservation, since both conservations intermix under Lorentz transformations). Consequently, the mass of a compound particle exceeds the mass of its disintegration products by an amount given by their total kinetic energy (in the frame of the center of mass) divided by c^2. This is a particular case of Einstein's general statement that the mass of a body depends on its energy content.

To summarize: if we assume the existence of additive invariants in collision processes, the principle of relativity completely determines the form of these invariants in conformity with the known expressions of energy and momentum either in relativistic or in non-relativistic mechanics. The knowledge of these invariants does not exhaust collision dynamics. Both in classical and in quantum mechanics, the probability of a given final state (efficient cross section) can be determined as a function of the initial states. The derivation of this probability requires knowledge of the laws of molecular dynamics, whose necessity in the classical case has already been discussed in section 2.2.

Conclusions

The necessity and unity of mechanics

We may now judge to what extent and in which sense the laws of classical mechanics can be regarded as necessary truths. The form of the invariants in collision processes result directly from the spacetime structure. The derivation of the laws of the full mechanics of connected systems, molecular systems, and continua requires additional assumptions. The first of these assumptions is a causality principle according to which forces (at least impulses) alter the motion in a well-defined manner. The second is the secular principle following which any action on a given body is equivalent to a rapid succession of impulses. The third is the impossibility of perpetual motion.

With these assumptions and for a given structure of spacetime, the laws of mechanics follow from the very definition of a mechanical system, to which an operational definition of force must be added in the case of connected systems. The question of the necessity of mechanical laws is thus reduced to the question of the status of the assumptions and definitions from which we have derived them. I leave aside the status of the spacetime structure, which will be discussed in Chapter 5. The causality principle and the secular principle may be seen as resulting from the comprehensibility of the motion of mechanical systems. This comprehensibility entails that well-defined causes have well-defined effects, and that inaccessible details of impressed forces do not affect the observed motion (more generally, the motion at the scale of interest should not depend on the details of forces at a finer scale). The impossibility of perpetual motion is also a matter of comprehensibility, because its violation would imply the existence of mechanical effects without a cause: for instance, a weight could be elevated without compensatory change in the lifting agency. Consequently, a reasonable claim of necessity can be phrased as follows: if classical mechanics is defined as a theory that makes evolution in space and time comprehensible, there can only be one classical mechanics in a given spacetime and for a given definition of its ideal object.

This conclusion does not entail the empirical validity of classical mechanics, because neither the comprehensibility of motion nor our ideal definitions of mechanical systems need have empirical counterparts. Firstly, the comprehensibility of nature is not a transcendental truth: whether a certain class of phenomena is comprehensible in a certain sense can only be decided empirically. In particular, the validity of the causal and secular principles is to be decided empirically. Secondly, the existence of real systems satisfying the ideal definitions of mechanical systems is not granted a priori. So it cannot be said that a mechanical essence of nature has been proved. What has been proved in this chapter is that the laws of mechanics can be derived from more obvious assumptions regarding the comprehensibility of motion.

A weak point in this moderate necessity argument is the arbitrariness in the definition of mechanical systems. We have the choice between various intuitions of interaction (contact, action at a distance, internal pressure, collisions) from which the ideal definition of mechanical systems derives. So far I have only argued the necessity of the laws of mechanics *for a given definition of its object*. A critical reader could reproach me with having adjusted these definitions so that the usual mechanical laws follow from them. The suspicion seems high in the case of the statics of connected systems, for which the principle of virtual work is a mere consequence of the two definitional properties (and of the impossibility of perpetual motion). In defense of the necessity of mechanical laws, it can still be argued that the definitions of ideal systems naturally follow from the intended applications of mechanics.

Better, it can be argued that the various definitions of mechanical systems lead to the same mechanical laws in different guises, because there are bridges between the corresponding forms of mechanics. The existence of such bridges should not be a surprise since all these forms satisfy the same principles of the comprehensibility of motion and fit in the same spacetime structure. The following diagram indicates the bridges, some of which have already been discussed in the sections on molecular mechanics and on continuum mechanics.

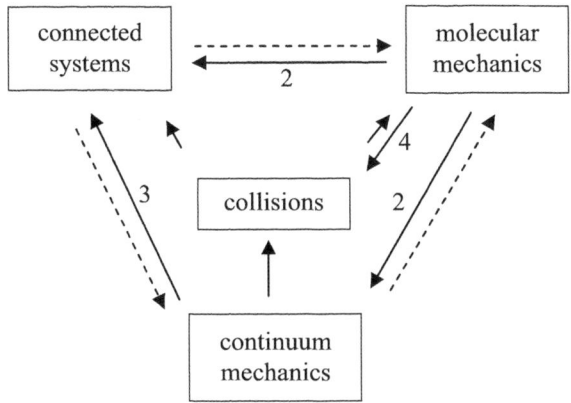

The solid arrows indicate well-established connections, with numbers indicating the relevant sections of this chapter; the dotted arrows stand for tentative connections; the short arrows stand for a heuristic analogy. There are enough arrows to contemplate the reduction of any of the four mechanics to another of them.[26]

Molecular mechanics can be used to construct approximations of connected systems (section 2.2), and also to define the internal stresses of large-scale continua (section 2.3). Its concept of interaction is directly adapted to a detailed description of collision processes (beginning of section 2.4). The stress-based mechanics of continua can be used to construct approximations of connected systems (section 2.3), as is implicitly done in elementary theories of these systems based on internal contact forces. As Faraday and Maxwell showed in the case of electromagnetism, action at a distance (assumed in molecular mechanics) can be mimicked by stress systems in the intervening medium. The mechanics of continua can also be used to study the collisions between elastic bodies.[27]

More speculatively, some of the greatest physicists of the nineteenth century tried to reduce every mechanics to the dynamics of connected systems. This is the "Matter and Motion" program. For example, Kelvin tried to derive elastic stresses from turbulent motion in a perfect liquid (which is a connected system). Maxwell hoped to derive electromagnetic stresses from a hidden field mechanism. Late in the century, Joseph Larmor claimed to have achieved so much with his concept of gyrostatic ether. Heinrich Hertz based his own mechanics on the possibility of a general reduction to connected systems. Whether this is truly possible remains an open question.[28]

Collisions can be made responsible for stresses, as in the kinetic theory of gases, or even for action at a distance, as in Lesage's theory of gravitation. The latter attempt having failed and there being little chance to explain cohesion in this manner, collision-based reduction seems to be the least promising option.[29]

To sum up, mechanics is deeply unified despite the variety of guises in which it has been developed. Its basic laws are essentially a necessary consequence of the comprehensibility of motion in terms of causal evolution in space and time, even though they may take different forms in different sorts of mechanics. Again, this necessity does not imply the empirical success of mechanics. It does not even clearly delineate the sort of system to which mechanics should tentatively be applied, as can be inferred from the rise and fall of mechanical reductionism. However, from past successes and failures of mechanical explanation, we may judge to what extent the ideal of a causal description of phenomena in space and time can be realized in natural and artificial phenomena.

[26] Roughly, I consider one mechanics to be reduced to another when selected features of a subclass of systems of the latter approximately enjoy the definitional properties of the systems of the latter and approximately obey the same laws. The existence of such reductions does not imply that all forms of mechanics are equally convenient in solving a given mechanical problem. Marion Vorms (2011) has argued that form is, in fact, essential to problem-solvers who have a finite inferential capacity.

[27] Maxwell 1873, §§105–11, 641–5.

[28] On Matter and Motion, cf. Stein 1981. On Hertz, cf. Lützen 2005. Euler anticipated this neo-Cartesian program in his conception of the principle of least action: cf. Pulte 1989, pp. 108–10, 117–18, 170–81.

[29] Cf. Chabot 2004.

The necessity of least action

The purest theories of motion are the mechanics of connected systems without impressed forces, and the mechanics of systems of particles directly acting at a distance. Indeed these theories contain only one kind of interaction, and do not require an explicit concept of force. For these reasons, Kelvin, Hertz, and Larmor dreamt of a world entirely made of perfectly connected masses, while Saint-Venant and Helmholtz dreamt of a world made of point atoms directly acting on each other. Despite strong conceptual dissimilarities, these two approaches lead to equations of motion that make a certain quantity, the action, stationary. Reciprocally, the equations of motion can be obtained by requiring that the action should be stationary under infinitesimal variations of the successive configurations of the system. The action is the integral over time of a Lagrangian function that does not explicitly involve time. Through Emmy Noether's theorem, this symmetry implies the existence of an invariant, the energy. Other continuous symmetries imply other invariants. For example, the translational invariance of a system of point atoms implies the conservation of momentum.[30]

Because of this relation between dynamical invariants and Lagrangian symmetries, and because of many other mathematical niceties of theories based on extremum principles, the principle of least action is often regarded as the best foundation of mechanics. Is this belief rational? Historically, the principle has been justified in four different manners: by theological arguments (Maupertuis), by analogy with optics (Hamilton), by reduction to connected systems (Maxwell, Kelvin), and by reduction to systems of particles interacting through central forces (Helmholtz). Perhaps the analogy with optics is the deepest, for it anticipates the physical interpretation of the action as a quantum phase. It is nonetheless worth trying to justify least action within the context of elementary mechanics, as Maxwell and Helmholtz tried to do.[31]

What is it exactly that leads to least action in the purest forms of mechanics? In the approach based on connected systems, there are no forces *stricto sensu*, there are only inertial forces. The principle of least action results from the principle of virtual work and from d'Alembert's principle. The former principle derives from the definition of connected systems and from the impossibility of perpetual motion; the latter derives from the secular principle and from Newton's second law, which itself derives from the secular principle, the principle of inertia, and the relativity principle. In molecular mechanics, the same principles imply that the acceleration of a molecule is proportional to the gradient of a function of the molecular configuration. Consequently, the equations of motion can be obtained by minimizing the time integral of the sum of this function and of the kinetic energy.

In both approaches, Newton's second law or some generalization of it is assumed, whatever the precise meaning of the force in it may be. Although there could obviously

---

[30]On the relation between conservation law and invariance of the Lagrangian density, see (e.g.) Landau and Lifshitz 1960, Chap. 2.

[31]On least action in general, cf. Dugas 1950, Lanczos 1966. On Maupertuis's (and Euler's) motivations, see Chapter 1, pp. 28–31. On Hamilton's, cf. Hankins 1980.

be least action involving higher derivatives of motion than the second, the acceleration law clearly is a basic ingredient of the non-relativistic derivations of least action. It can be justified through the principle of inertia, the relativity principle, the secular principle, and some causality. For the rest, the derivation of the principle of least action must rely on the special simplicity of the forces entering Newton's law. In one case they express the effect of constraints; in the other they result from additive action at a distance. In both cases their general form derives from the impossibility of perpetual motion. The moral seems to be that the secular and causal principles, the impossibility of perpetual motion, and the structure of spacetime together lead to the least action principle whether contact action or distance action is regarded as more primitive. In this sense, the principle least action appears to be almost as necessary as the other laws of mechanics. However, it does not directly apply to macroscopic mechanics in general. It only applies to a submechanics to which macroscopic mechanics can always be reduced. Extensions of the principles of least action and energy conservation beyond ordinary mechanical phenomena will be discussed in the next chapter.

3

FROM MECHANICAL REDUCTION TO GENERAL PRINCIPLES

> The qualities of bodies which cannot be intended and remitted, and which apply to all bodies on which it is possible to set up experiments, are qualities of all bodies universally... We are certainly not to relinquish the evidence of experiments for the sake of dreams and vain fictions of our own devising; nor are we to recede from the analogy of Nature, which uses to be simple, and always consonant to itself.[1] (Isaac Newton, 1713)

> We might reason, *a priori*, that... absolute destruction of living force cannot possibly take place, because it is manifestly absurd to suppose that the forces with which God has endowed matter can be destroyed any more than they can be created by man's agency.[2] (James Joule, 1847)

> The theoretical science of nature shall have fulfilled its mission when the reduction of phenomena to elementary forces is accomplished and when this reduction is known to be the only possible one. Then the reduction would be proved to be the necessary comprehending form of our conception of nature, and it would be regarded as objectively true.[3] (Hermann Helmholtz, 1847)

> A theory is the more impressive the greater the simplicity of its premises is, the more different kinds of things it relates, and the more extended is its area of applicability. Therefore the deep impression which classical thermodynamics made upon me. It is the only physical theory of universal content concerning which I am convinced that within the framework of the applicability of its basic concepts, it will never be overthrown.[4] (Albert Einstein, 1946)

Descartes's rationalism associated the necessity of the laws of mechanics with the necessity of a mechanical reduction of the world. Although this connection seems natural, it is not logically necessary. One may very well subscribe to a mechanical worldview and yet ascribe an empirical origin to the laws of mechanics. So did Newton, Laplace, Saint-Venant, Helmholtz, and nearly every nineteenth-century physicist. Consequently, the question of the necessity of mechanical reduction should be treated separately from the question of the necessity of the laws of motion. From a historical point of view, the former question can be divided into two subquestions: (1) Why did almost every physicist from the seventeenth century to the late nineteenth century believe that the physical world was inherently and fundamentally mechanical? (2) Were some features of this mechanical reduction deemed necessary?[5]

---

[1]Newton 1713, p. 357.   [2]Joule 1847, p. 271.   [3] Helmholtz 1847, p. 13.   [4]Einstein 1949, p. 33.

[5]"Mechanical" is here used in the broad sense it had in the nineteenth century, although in earlier times "mechanical" often referred to clockwise or corpuscular mechanism.

The first question is relatively easy to answer. As we saw, Descartes argued that extension and duration, being the only attributes of corporeal things that resisted systematic doubt, had to be the sole foundation of the physical world. In practice, he also needed a theologically justified concept of inertia; other mechanical reductionists included mass and force in the list of basic mechanical concepts. With this latitude, most natural philosophers came to agree with Descartes that mechanical concepts were most basic, and that secondary qualities such as color, flavor, and pitch had to be reduced to the primary qualities of extension and motion. Phenomena that were not mechanical in a macroscopic sense, such as optics, electricity, and magnetism, were believed to be explicable by a variety of mechanisms, although there was much disagreement on the means and timeliness of such explication. While the strictly Cartesian ideal of reduction to contact mechanisms gradually lost its appeal, the Newtonian reduction to elements of matter interacting through distance forces or the Leibnizian reduction to matter endowed with force took over. Each variety of mechanical reduction was accompanied with a metaphysical proof of the priority of its mechanical concepts. Beyond these controversial arguments, there was a consensus that the comprehensibility of nature implied a strong analogy between the inner mechanism of physical phenomena and the mechanics of macroscopic bodies. This is Newton's "analogy of nature," conjugated with the idea that motion is the simplest kind of change that we can imagine. The 1713 edition of his *Principia* contains the following rule of philosophizing:[6]

> The qualities of bodies which cannot be intended and remitted, and which apply to all bodies on which it is possible to set up experiments, are qualities of all bodies universally: For since the qualities of bodies are only known to us by experiments, we are to hold for universal all such as universally agree with experiments; and such as are not able to diminution, can never be quite taken away. We are certainly not to relinquish the evidence of experiments for the sake of dreams and vain fictions of our own devising; nor are we to recede from the analogy of Nature, which uses to be simple, and always consonant to itself.

The second question, about the necessary features of the mechanical basis, has a more complex answer, because it implies the various ways in which mechanics was itself founded. It is worth exploring in some detail because it has strong implications on the way mechanical reducibility affects the form of macroscopic laws. One important issue was whether mechanics, at its most fundamental level, conserved the total measure of motion. This issue will be treated in the first section of this chapter, as historical background for the introduction of the energy principle in the first half of the nineteenth century. The removal of motion-destroying processes from the foundations of mechanics in the early nineteenth century indeed was a crucial precondition for conceiving energy conservation.

The second section is a history of various foundations of the energy principle by Saint-Venant, Joule, Thomson, Helmholtz, and Mayer, followed by a sketch of historico-critical investigations of this principle by Mach, Planck, and Poincaré. It involves a variety of necessity arguments including the perfection of God, the equality of cause and effect, the unity of force, the impossibility of perpetual motion, the transcendental necessity of central

---

[6]Newton 1713, p. 357. On Newton's analogy of nature, cf. McGuire 1970.

forces, and the reducibility of dissipative processes to conservative ones. I will try to decide which of these arguments may survive the modern evolution of physics.

The third section of this chapter is devoted to the principle of least action, as a means to express the mechanical reducibility of a system without specifying the reducing mechanism. The heroes of this section are Maxwell, Helmholtz, and Poincaré, who suggested that the principle should become the general foundation of physics. More broadly, Poincaré perceived a global evolution of physics from naïve mechanism to a "physics of principles" in which the mathematical structure of phenomena was directly revealed, without the subterfuge of hidden mechanisms.[7] Thermodynamics, which Poincaré regarded as a paradigm of this new physics, is presented in the fourth and last section of this chapter, together with its later statistico-mechanical foundation. The necessity of its two laws is discussed both in the phenomenological and in the reductionist approach.

3.1 Varieties of mechanical reduction

Early conservative reduction

As we saw, Descartes regarded the conservation of total motion, measured by the product of mass (bulk) and velocity modulus, as a central pillar of his system. Leibniz ascribed equal metaphysical importance to the conservation of motion, although he rejected Descartes's measure and replaced it with the product of mass by squared velocity. As Leibniz framed this reinterpretation as an opposition between two measures of force, the Newtonian "dead force" and his "live force," few Newtonians followed him. Important exceptions were Johann Bernoulli and his son Daniel, who conciliated the two kinds of force and insisted on the conservation of live force. They supplemented Leibniz's concept of live force with the potential contribution of compressed springs and elevated bodies, arguing like Huygens that elasticity was reducible to the agitation of entrapped subtle matter. They used the conservation of the total live force, including the potential contribution, to solve mechanical problems of various kinds, for example the efflux of a liquid through an opening on its container. Like Leibniz, the Bernoullis argued that any apparent loss of live force in collision, friction, or abrupt hydraulic change, was due to the dissemination of motion into small invisible parts of the system or to irreversible micro-compressions.[8]

In his foundational *Hydrodynamica* of 1738, Daniel Bernoulli not only derived the laws of efflux but also pioneered what we would now call the energetics of hydraulic machines. He defined the effect of a machine as the product of elevated weight by elevation, and the "moral" contribution to the "efforts" of the operators, or *absolute power*, as the product of force by the displacement of the point of application. In general, concern with the efficiency of various machines brought eighteenth-century mechanicians and natural philosophers to think about ways to measure input and output and about the causes of the difference between input and output. Besides Antoine Parent's and Daniel Bernoulli's pioneering reflections, the most influential work of this kind was the series of experiments performed

---

[7] Poincaré 1900b, 1904.

[8] J. Bernoulli 1724, 1735; D. Bernoulli 1738. Cf. Hass 1909, p. 22; Scott 1970, pp. 22–3; Truesdell 1954, XXIII–XXXI; Truesdell 1968b; Séris 1987, pp. 235–64; Darrigol 2001, pp. 289–94.

by the British engineer John Smeaton in the 1750s on small- and full-scale waterwheels. Smeaton measured the input by the amount of live force of the incoming water, and the output by the *mechanical effect*, or elevated weight times elevation. In Leibnizian spirit, he explained the limited efficiency (ratio between output and input) of waterwheels by the "change of figure" of the incoming water during its collision with the paddles of the wheel. He also experimented to confirm the Leibnizian idea that apparent losses of live force in collisions were caused by diverted motion or permanent deformation of the colliding bodies.[9]

Hard bodies, machines, and losses of live force

Although the theoretical importance of Daniel Bernoulli's *Hydrodynamica* and the practical importance of Smeaton's waterwheel studies were quickly recognized, the Leibnizian conservation of force remained unpopular until the early nineteenth century. There were two reasons for this neglect. In the wake of the famous quarrel of live forces that opposed the supporters of Newton's and Leibniz's definitions of force, any recourse to live force was usually regarded as an offence to Newton. Consequently, the tolerance for this concept decreased with the spread of Newton's philosophy. Another reason for denying the conservation of live force originated in the fact that Newton and early students of the collision problem admitted ideally rigid bodies whose collisions implied the annihilation of motion. The conflict became acute after d'Alembert, in total disrespect for Leibniz's principle of continuity, made hard-body collisions the very foundation of the theory expounded in his *Traité de dynamique* of 1743. The most influential authorities on mechanics at the end the century, Lagrange, Lazare Carnot, and Laplace, maintained this essential annihilation of motion at the core of their mechanics.[10]

This is not to say that live force played no role in French rational mechanics. D'Alembert and his heirs argued that live force was a perfectly well-defined and even useful concept once stripped of its metaphysical garment. Moreover, they proved that in any mechanical system in which motion proceeded gradually and smoothly, without impact between hard parts and without friction between sliding parts, live force was conserved. More generally, d'Alembert and Lagrange enunciated the "theorem of live forces" according to which the live force of a (frictionless) connected system whose masses are subjected to central forces is the same whenever the system returns to the same spatial configuration (we would now say that the kinetic energy is the same whenever the potential energy is the same). The theorem was useful as a first integral of the equations of motion; and it had a potentially wide range of application in fluid mechanics because according to d'Alembert the motion of fluids was inherently continuous.[11]

In the last third of the eighteenth century, a couple of learned French engineers hybridized the d'Alembertian and Leibnizian traditions by combining hard-body collisions and considerations of live force. First, the chevalier Jean-Charles de Borda corrected Daniel Bernoulli's evaluation of the loss of pressure owing to the sudden enlargement of a water

---

[9] D. Bernoulli 1738, pp. 163–5; Smeaton 1794. Cf. Séris 1987, pp. 299–331; Darrigol 2001, pp. 294–301.

[10] Cf. Scott 1970; Darrigol 2001.   [11] Cf. Darrigol 2001, pp. 301–5, and further reference there.

pipe by assuming a hard-body collision between successive slices of water in this case. Most impressively, Borda showed that Smeaton's experimental result for the maximal efficiency of an undershot waterwheel could be derived theoretically by assuming a discontinuous velocity change of the water impacting the paddles of the wheel. The engineer and statesman Lazare Carnot systematized the combined recourse to hard-body collisions and live-force considerations in his ingenious *Essai sur les machines* of 1683. As was mentioned, one of Carnot's aims was to clarify the foundations of d'Alembert's dynamics, in particular in showing the empirical origin of its basic laws. Another aim was to adapt it to the needs of the theory of machines.[12]

Carnot did so in two ways: by replacing d'Alembert's ideal bodies with natural bodies always endowed with inertia and some degree of malleability and elasticity, and by treating every continuous motion of a machine as the approximation of a discrete succession of mutual impacts of its parts. He managed to quantify the effect of an abrupt velocity change in a system of (not necessarily hard) bodies through the relation

$$\sum_{\alpha} m_\alpha (\mathbf{v}'_\alpha - \mathbf{v}_\alpha) \cdot \mathbf{V}_\alpha = 0, \tag{1}$$

where the sum is performed over the parts α of the bodies with the masses m_α, the initial velocities $\mathbf{v}_\alpha$, and the final velocities $\mathbf{v}'_\alpha$; the velocities $\mathbf{V}_\alpha$ are any virtual velocities that do not alter the connectivity of the system. In hard-body collisions, one may take $\mathbf{V}_\alpha = \mathbf{v}'_\alpha$, so that

$$\sum_{\alpha} m_\alpha {v'_\alpha}^2 - \sum_{\alpha} m_\alpha {v_\alpha}^2 = -\sum_{\alpha} m_\alpha (\mathbf{v}'_\alpha - \mathbf{v}_\alpha)^2, \tag{2}$$

which is the theorem now named after Carnot.[13]

In the continuous case, Carnot introduced the *moment of activity* of a force as the integral of its projection along the displacement of its point of application (now called *work*), and enunciated a second theorem:

> In every system of bodies in which the motion changes by insensible degrees, the sum of live forces increases during a given time by a quantity always equal to twice the moment of activity consumed in the same time by all the motive forces.

As Carnot explained, the moment of activity of the forces driving a machine is the basis for "estimating the labor of workers and for paying their salaries." The moment of activity of the forces applied by the machine measures its action. Consequently, there are three limitations to the efficiency of a machine: the moment of activity of internal frictional forces, the live force of motions that do not contribute to the desired effect, and collisional loss quantified by Carnot's first theorem. Carnot concluded his analysis with concrete prescriptions for optimizing the efficiency of a machine, and with a derivation of the impossibility of perpetual motion from the assumption that all active forces in nature depended only on mutual distances. The total live force diminishes forever in Carnot's world.[14]

---

[12] Borda 1766, 1767; Carnot 1783. Cf. Darrigol 2001, pp. 305–11, and further reference there.

[13] Carnot 1803, theorems XVI, XII. Cf. Drago, Manno, and Mauriello 2001.

Dissipation

Owing to the competition of Lagrange's *Méchanique analitique* on the mathematical side, and of Charles-Augustin Coulomb's memoirs on labor and machines on the engineering side, Carnot's treatise remained largely unnoticed until the publication of his fuller *Principes fondamentaux de l'équilibre et du mouvement* in 1803. By that time, the newly founded Ecole Polytechnique favored elaborate mixes of mathematical and practical considerations on machines. A few of its early students, including Claude Louis Navier, Gaspard Coriolis, Jean Victor Poncelet, and Adhémar Barré de Saint-Venant, read Carnot and Smeaton, and developed the energetics of machines with special emphasis on the concept of moment of activity, which Coriolis and Poncelet renamed *work*. In their view, this concept was more fundamental than the Leibnizian live force, because it was more directly related to the engineer's interest. Consequently, Coriolis and Saint-Venant redefined live force by the equivalent work, which is half the product of mass by squared velocity.[15]

None of these engineer-physicists adopted Carnot's concept of hard-body collisions and the concomitant loss of live force. There are several reasons for that: their interest in Smeaton's work on waterwheels and collisions, their desire to purge mechanics from unrealistic idealizations, and, most importantly, the contemporary attractiveness of Laplacian physics. Impressed by late-eighteenth-century successes in quantifying electricity, magnetism, and heat, Laplace wanted to subject the entire domain of physics to the mathematical and experimental precision of his original field, astronomy. On the theoretical side, he developed a grand unified picture in which matter and the imponderable fluids of electricity, magnetism, and heat were made of quasi-punctual molecules interacting through distance forces, following the model of gravitation theory. Unlike earlier molecular conceptions, which had remained vague and qualitative, this picture permitted precise calculations based on summing the forces acting on a given molecule and approximating the sums with integrals. The only arbitrary input was the force law between the molecules. Experiments recommended $1/r^2$ laws for gravitation, electricity, and magnetism; while commonsense required small-range forces for capillarity, cohesion, elasticity, and thermal effects. In the latter case, the precise force law did not matter to the results in the continuum approximation.[16]

Although Laplace's closest disciples, Poisson and Cauchy, were the furthest from the world of engineering, the aforementioned engineer-physicists all contributed to Laplacian physics, sometimes in a decisive manner. Navier developed the first molecular theories of elasticity and viscous flow, which were the starting point of Poisson's and Cauchy's more genuinely Laplacian theories in this domain. Granted that central forces are solely responsible for the molecular interactions, the theorem of live forces trivially applies to any closed system so that any variation of the global live force must be compensated by a change in the relative configuration of the molecules of the system. Consequently, seeming losses of live force in collision processes must be related to permanent alterations of the molecular configurations. Navier asserted so much in 1819, and so too did Coriolis, Poncelet, and Saint-Venant. Poncelet was most eloquent in his description of the various ways in which

---

[14]Carnot 1803, pp. 181–2, 231.   [15]Cf. Grattan-Guinness 1984; Séris 1987; Darrigol 2001, pp. 311–17.

[16]Cf. Crosland 1967; Fox 1974; Heilbron 1993.

live force was "dissipated" in molecular deformation and agitation, or in small-scale fluid motion. In his *Mécanique industrielle* of 1829, he even described the cascading of motion from the macroscopic scale to the molecular scale:[17]

> In general, the production of eddies is one of the means which nature uses to extinguish, or rather to dissimulate live force during sudden changes in the motion of a fluid, as vibratory motions are themselves another cause of its dissipation, of its dissemination in solids. The careful observation of the facts sufficiently justifies the belief that independently of the gyratory motions shared by a whole portion of the fluid mass, there are also secondary or less apparent motions which concern a more or less large number of molecules and which distribute themselves in the intervals of the previous motions... But we can go further and assume without too much risk that such motions of rotation or oscillation imparted on the individual molecules or on the last groups of molecules are, besides the adherence and cohesion to which we shall presently return, one of the most important causes of the loss of motion in fluids and especially of the resistance that their streaming filaments experience when gliding on each other or on the surface of solids.

3.2 The energy principle

Saint-Venant's anticipation

For Navier, Coriolis, and Poncelet, any macroscopic decrease of live force is compensated by the production of invisible motion or by the work of internal molecular forces. There seems to be little conceptual distance between this statement and a statement of conservation for the sum of the total kinetic energy and the total potential energy of the molecular motion. In symbols, the former statement reads

$$\Delta T_{\text{macro}} = -\Delta T_{\text{micro}} - \Delta V \tag{3}$$

while the second reads

$$T_{\text{macro}} + T_{\text{micro}} + V = \text{constant}, \tag{4}$$

where T denotes a kinetic energy and V a potential energy. Yet none of these authors took this step. Several reasons can be given for this reticence: They were more concerned with observed losses of live force than with occult conservation (or with conversion processes); the theorem of live forces, as they knew it, stated the conservation of live force during cycles only; last but not least, Leibnizian insistence on conservation had an undesired theological flavor in post-revolutionary France.

This last prevention did not apply to the deeply religious Saint-Venant. In an unpublished memoir on general mechanics of 1834, this learned engineer defined the "patent

---

[17] Navier, in Bélidor 1819, pp. x–xi, 76n–80n; Poncelet 1829, pp. 529–30. Cf. Darrigol 2001, pp. 322–5.

capital" of a physical system as half the sum of the squared velocities of its atoms, and the "latent capital" as half the sum of the integrals

$$V_{\alpha\beta} = \int_0^{r_{\alpha\beta}} a_{\alpha\beta}(r)\mathrm{d}r, \tag{5}$$

where $r_{\alpha\beta}$ denotes the distance of the atoms α and β, and $a_{\alpha\beta}$ the acceleration of the atom α owing to the presence of the atom β (reckoned positively in the attractive case). For an isolated system, Saint-Venant derived the conservation of the sum of these two quantities, which he called "dynamical capital." He commented:

> Applied to the entire system of the world, [this theorem] gives, if one takes the dynamical capital as the common measure of the effective motion and the possible motion, this great principle invented by Descartes and elucidated by Leibniz, that *Motion* remains forever constant in the universe.

As can be judged from Saint-Venant's private writings, his reference to Descartes and Leibniz was not a mere gloss. He agreed with these illustrious forerunners that God had introduced motion in the world he created, and that the perfection of his oeuvre implied the conservation of this motion. In Saint-Venant's view, the new molecular mechanics, in the purest form of point-like atoms interacting through central forces, offered an opportunity for a precise mathematical formulation of Leibniz's intuition in terms of a first integral of the equations of motion. This is not to say that Saint-Venant agreed with the rest of Leibniz's philosophy. His condemnation of the concept of force as a remnant of paganism also applied to Leibnizian force. His own expression for live force, "patent capital," conveyed the economic importance that Navier and others had already emphasized. Unlike other Laplacian engineers, Saint-Venant did not fear melding the practical with the spiritual.[18]

Conversions and conservation

Despite Navier's and Ampère's approval, Saint-Venant's memoir remained buried in a *pochette de séance* of the French Academy of Sciences. Some fifteen years elapsed before others propounded similarly precise formulations of energy conservation. The main reason for this delay probably was the French engineers' lack of interest in the conversion processes that made energy conservation a more palpable notion. After Alessandro Volta's discovery of the electric battery at the close of eighteenth century, there were the discoveries of electrolysis (1800), electromagnetism (1820), thermoelectricity (1822), and electromagnetic induction (1831). Steam engines, which produce work from heat, were becoming very common in Britain. Early in the century, Count Rumford and Humphry Davy argued that heat could be created by friction. In an anti-vitalist crusade, some German physiologists came to regard the human body as a machine capable of converting chemical force into heat and mechanical force. The alleged discoverers of energy conservation benefited

---

[18] Saint-Venant [1834], p. XXXII. Cf. Darrigol 2001, pp. 325–35.

from their familiarity with these conversion processes. In addition, they were aware of the developments that had led Saint-Venant to his own doctrine of conservation.[19]

In early-nineteenth-century Britain, Smeaton's "mechanical effect" became more and more important in judging the efficiency of industrial machines. A few natural philosophers, including John Dalton's friend Peter Ewart, William Wollaston, and Thomas Young, braved the Newtonian prohibition of Leibnizian live force and accepted Smeaton's idea that the destruction of live force in collisions was only apparent. By 1840, the French energetics of machines was widely taught and appreciated. The Cambridge philosopher William Whewell and the London professor Henry Moseley largely drew on Poncelet's and Coriolis's writings in their mechanical engineering treatises of 1841 and 1843. So too did the engineering professor Lewis Gordon in lectures attended by his young friend William Thomson. The Manchester brewer, chemist, and experimental philosopher James Joule was well aware of this evolution. He was also familiar with the mechanico-molecular concept of matter of his friend John Dalton, based like Laplace's on mutually attracting molecules with self-repulsive atmospheres of heat. And he inclined toward the Newtonian, kinetic conception of heat, which Count Rumford, Humphry Davy, and Thomas Young defended against the then dominant caloric theory. These preconceptions informed and partially motivated his quantitative studies of the electric and mechanical production of heat in the 1840s. Together with the uniform conversions established in these experiments, they motivated his general statement of energy conservation in May 1847:

> The phenomena of nature, whether mechanical, chemical or vital, consist almost entirely in a continual conversion of attraction through space, living force, and heat into one another. Thus it is that order is maintained in the universe—nothing is deranged, nothing ever lost, but the entire machinery, complicated as it is, works smoothly and harmoniously.

Joule assorted this statement with a theological comment:[20]

> We might reason, *a priori*, that ... absolute destruction of living force cannot possibly take place, because it is manifestly absurd to suppose that the forces with which God has endowed matter can be destroyed any more than they can be created by man's agency.

Like Joule, the young William Thomson knew and approved Smeaton's considerations of "mechanical effect." He was also familiar with Poncelet's concept that live force was dissipated rather than destructed in natural processes. In his early theoretical works on electricity, he exploited what we now call the potential energy of the electric distribution. Worries about apparent losses of mechanical effect determined his acceptance of Joule's kinetic concept of heat. In the early 1940s, while he and his brother were watching with his brother James "the consumption of power in the flow of water into a canal lock," they wondered "what became of the power as we could not suppose the water *worn* and therefore altered as solids might be supposed to be when power is consumed in their friction." In

---

[19] Cf. Kuhn 1959; Harman 1982.   [20] Joule 1847, pp. 268–71. Cf. Cardwell 1989; Smith 1998, pp. 52–76.

1949, he described a similar paradox in the case of heat flowing from a hot body to a cold body:

> When 'thermal agency' is thus spent in conducting heat through a solid, what becomes of the mechanical effect which it might produce [by acting on a steam engine]? Nothing can be lost in the operations of nature –no energy can be destroyed.

As we will see in a moment, Thomson solved these paradoxes by accepting Joule's idea that motion could be converted into heat, and Clausius's idea that the work produced by a steam engine implied the consumption of an equivalent amount of heat. In private, Thomson accorded theological significance to the universal conservation of energy: "No destruction of energy can take place in the material world without an act of power possessed only by the supreme ruler."[21]

In sum, Joule and Thomson shared with Saint-Venant the theological motivation, the engine-efficiency context, and the belief that physical phenomena could be reduced to a conservative mechanism. There were important differences, however. Whereas Saint-Venant reasoned mathematically on point-like atoms and central forces, Joule and Thomson left the underlying mechanism largely undetermined. Whereas Saint-Venant was mainly concerned with macro-mechanical phenomena, Joule and Thomson performed or exploited quantitative experiments of conversion between various kinds of force.[22]

Closest to Saint-Venant was Hermann Helmholtz in his famous memoir of 1847 on "the conservations of force." Helmholtz imagined a world made of mass points interacting through central forces; he derived the theorem of live forces and the principle of virtual velocities on the basis of this picture; he defined the "mechanical work" (*mechanische Arbeit*) or "work quantity" (*Arbeitsgrösse*) as the capacity to elevate a weight; he noted the importance of this concept in the study of machines, and he measured live force via the equivalent work, which yields the 1/2 factor. Helmholtz identified the "total force" of a system with the "maximum quantity of work to be won" from this system; he introduced the "sum of tensional forces" (*Summe der Spannkräfte*), which is the exact counterpart of Saint-Venant's latent capital (our potential energy); and he expressed his fundamental principle as the conservation of the sum of the latter quantity and the live force for an isolated system.[23]

Like Saint-Venant, Helmholtz probably obtained the picture of point-masses interacting through central forces by simplifying the Laplacian scheme. In a philosophical introduction to his memoir, he justified this picture through a pseudo-transcendental argument. The comprehensibility of nature (*Begreiflichkeit der Natur*), Helmholtz tells us, implies the possibility of finding the ultimate, invariable causes of natural processes. This is done by means of the two inseparable abstractions of matter and force. By definition, matter is that which only changes by motion, and force is the cause of motion. Full comprehensibility implies reduction to forces that depend on spatial relations only, not on time. Every

---

[21] James Thomson to William Thomson, Aug. 13, 1863, quoted in Smith and Wise 1989, p. 286; Thomson 1849, pp. 118n–119n; Draft of Thomson 1851, quoted in Smith and Wise 1989, p. 329. See also Thomson's notes of 1839–40, quoted in Smith and Wise 1989, pp. 306–7: "All created things must be sustained," or "The continual presence of God is necessary to preserve as to create."

[22] Saint-Venant and Thomson were both familiar with Thomas Reid's variety of theological voluntarism.

[23] Helmholtz 1847, pp. 17 (*Arbeit*), 18 (live force), 22 (*Spannkräfte*), 28 (machines). Cf. Bevilacqua 1993.

interaction in an extended system can then be decomposed into time-independent forces between two material points. As these elementary forces can only depend on distance, they are central forces. Helmholtz concludes:[24]

> The theoretical science of nature shall have fulfilled its mission when the reduction of phenomena to elementary forces is accomplished and when this reduction is known to be the only possible one. Then the reduction would be proved to be the necessary comprehending form of our conception of nature, and it would be regarded as objectively true.

Like Kant in the dynamics of the *Metaphysical foundations of natural science*, Helmholtz decomposes matter into centers of force, and he seems to approve the definition of causality as a category of understanding, as well as the transcendental necessity of space and time relations. Yet there are important differences between his and Kant's reasoning. Whereas the categories of Kant's transcendental logic and the forms of intuition of his transcendental aesthetics are necessary features of any representation of the world we might have, causality and even the existence of space and time relations are only hypothetical for Helmholtz:

> Here we do not have to decide whether nature is fully comprehensible or involves processes that elude the law of a necessary causality and thus fall into the domain of spontaneity or freedom; at any rate, it is clear that science, which aims at understanding nature, must begin with the assumption of its comprehensibility, and proceed and investigate [the world] under this assumption, until perhaps she might be forced by incontrovertible facts to recognize her limits.

In the latter event, phenomena would partially resist reduction to central forces acting on pairs of material points. In Helmholtz's opinion, the theorem of live forces would then cease to apply and perpetual motion would be possible, against overwhelming empirical evidence.[25]

From his picture of a fully comprehensible world, Helmholtz deduced the conservation of what we would now call the sum of the potential and kinetic energies of a closed system. In some cases, such as gravitation and electrostatics, he regarded the full reduction to central forces as already accomplished. In others, such as electrodynamics, he contented himself with a macroscopic, phenomenological application of energy conservation, based on identifying a conserved quantity during the conversion of a kind force to another kind.[26]

In later writings, Helmholtz acknowledged gaps in his demonstration that matter could be decomposed into pairs of material points interacting through central forces. Firstly, this picture and its relation to energy conservation require Newton's laws of motion, which are empirical laws at least according to the later Helmholtz. Secondly, the force acting on a given material point is not necessarily the sum of forces separately emanating from every other material point. Thirdly, velocity-dependent forces, for instance magnetic forces, are compatible with energy conservation. Most fundamentally, Helmholtz emphasized that causality was only a hypothesis reducible to the regularity or lawfulness of phenomena.

---

[24] Helmholtz 1847, p. 17. Cf. Darrigol 1994, pp. 217–18; Jurkowitz 2010.

[25] Helmholtz 1847, p. 13. Cf. Heimann 1974; Wise 1981.   [26] Cf. Bevilacqua 1993.

In this view, force is nothing but law regarded as an objective power beyond us; and cause is a permanent force. Theorists and experimenters pursue the same aim of identifying the laws of nature by different means. Consequently, energy conservation is better discussed at the phenomenological level of conversion processes and machines, without speculating on an ultimate mechanical picture of the world:

> Any reduction of the phenomena to underlying substances and forces pretends to have found something unchanging and final. We are never entitled to such a claim. This is allowed neither by the imperfections of our knowledge, nor by the nature of the inductive inferences on which all our perceptions of the actual rest.

Helmholtz thereby denounced the futility of pursuing a specific, detailed mechanical reduction of phenomena. But he remained convinced that the world was essentially mechanical and that the laws of physics should be compatible with the possibility of a reduction to central forces acting in pair. As we will see in a moment, this belief motivated his late enthusiasm for the principle of least action.[27]

Although Helmholtz's essay on the conservation of force was presented as a "physical memoir" and although its main conceptual sources belonged to physics and philosophy, it originated in his interest in Justus von Liebig's theory of the chemical origin of animal heat. If muscular work is not a consequence of a specific vital force and if heat is nothing but motion, Helmholtz reasoned, it must imply the conversion of chemical heat into work; and this conversion must be uniform in order to prevent the possibility of perpetual motion. Similarly, a few years earlier the philosopher-physician Julius Robert Mayer had privately reasoned that the heat Q and work W produced by an animal, being both the result of an internal combustion process, had to be quantitatively related: since the work W can be entirely converted into the heat Q_W by friction, and since the maximal heat K that the combustion may produce should be independent of the details of this production, one must have $K = Q + Q_W$. Mayer famously related this reasoning to his accidental observation that venous blood had a lighter color under the hot climate of the Eastern Indies, which confirmed Lavoisier's combustion theory of animal heat: higher outside temperature means lower combustion and therefore cleaner blood.[28]

In 1842 Mayer obtained the heat/work conversion factor J by equating the heat C_V required to increase the temperature of a unit mass of a gas by one degree at constant volume, to the heat C_P required for the same temperature increase under constant pressure, minus the heat equivalent of the work done by the gas in the latter process. In symbols this gives the "Mayer equation"

$$J(C_P - C_V) = R, \tag{6}$$

where R is the constant of perfect gases. French calorimetric data gave Mayer $J = 360$ joule/calorie (in modern units). Mayer framed the equivalence between heat and work within a general principle of the "equality of cause and effect," through which causally interdependent "forces" such as work, live force, and heat possessed numerical

---

[27] Helmholtz 1878, p. 243. Cf. Darrigol 1994, pp. 219–21.

[28] Helmholtz 1845; Mayer 1851. Cf. Lenoir 1982; Caneva 1993.

equivalents. He left the nature of heat undetermined, and he avoided mechanical reduction. His foundation of energy conservation was partly metaphysical, partly empirical.[29]

Saint-Venant, Mayer, Joule, Helmholtz, and Thomson had unequal importance in the later development of the doctrine of energy conservation. Saint-Venant had none, since his memoir remained unpublished. Mayer had very little, because his articles of 1842 and 1845 remained unnoticed until he claimed priority over Joule and until Helmholtz defended him. Despite a poor initial reception, Joule's works were highly influential, owing to his precise, quantitative experiments about the conversion of mechanical, thermal, and electrical energies. Helmholtz's memoir had little impact until Thomson recognized its power. Thomson, even though he is usually not regarded as one of the discoverers of energy conservation, was most important not only through his approbation of Joule's and Helmholtz's results, but also through his systematic reliance on the energy concept in every domain of physics, through his integration of energy conservation in the new thermodynamics, and through his zeal in promoting energetics in physics and engineering communities. His friends Peter Guthrie Tait, William Rankine, and James Clerk Maxwell seconded him in this task.[30]

Crosbie Smith's recent history of energy conservation emphasizes the contribution of Thomson and his allies. In contrast, earlier histories of this subject considered a wide range of precursors, contributors, and promoters of energy conservation. There were two main reasons for this inclusiveness: nationalistic priority concerns, and, more importantly for us, an interest in the historical and philosophical necessity of the energy principle.

The necessity of the energy principle

From Leibniz to Thomson, highly diverse arguments contributed to the gradual form-ation of the doctrine of conservation: theological arguments for Leibniz, Saint-Venant, Joule, and Thomson; metaphysical arguments for Leibniz and Mayer; mechanical evalu-ation of the efficiency of engines for all actors; conservative mechanical reducibility for all of them except Mayer; uniform convertibility of various kinds of forces including heat and work for Joule, Mayer, Helmholtz, and Thomson; a kinetic concept of heat for Joule, Helmholtz, and Thomson; the impossibility of perpetual motion for Leibniz, Helmholtz, and Thomson. In addition, the young Helmholtz offered a pseudo-transcendental justific-ation of a conservative kind of mechanical reducibility.

Later in the century, when the energy principle had become the strongest pillar of the physical sciences, it became the object of thorough historico-philosophical investigations. In his first book, the *Conservation of work* of 1872, Ernst Mach attacked the mechanical-reductionist foundation of this principle by arguing that mechanical concepts, having their origin in the study of macroscopic motion, need not apply to hidden microphysical entit-ies; that the assumption of such entities contradicted the economy of thought; and that the impossibility of perpetual motion, far from being a consequence of the mechanical

---

[29] Mayer 1842, 1845. Cf. Caneva 1993. As Mayer explained in 1845, his calculation of J presupposes the pos-sibility of connecting the states of a gas after isochoric and isobaric heating without consuming heat or work; the Gay-Lussac expansion of a gas into a vacuum provides this connection, since it does not imply any temperature change.

[30] Cf. Smith and Wise 1989; Smith 1998.

nature of the world, was the origin of most mechanical laws. In the end, Mach distinguished empirical, logical, and formal components of the energy principle. First comes the empirical awareness of causal relations between physical quantities. In order to avoid the vicissitudes of absolute space and time, Mach defined this causality in an operational manner, as a functional relation between the measurable (mechanical, thermal, electrical, etc.) state variables of a closed system of bodies. This means that a given small enough subset of state variables is found to be a well-defined function of the complementary subset of variables. For instance, for a closed system including a vertically mobile weight, experience teaches us that the height of the weight is the same whenever the state of the other bodies of the system is the same. In other words, the height of a weight cannot change without compensation in its environment. According to Mach, experience further teaches us that every change of state of the global system is reversible.[31]

Next comes the logical step: the reversibility of transformations implies the possibility of cycles; and causality implies that a mechanical subset of variables of a closed system takes the same value whenever the other variables take the same values. This implies that no work can be produced (or consumed) during a cycle, in conformity with the impossibility of perpetual motion. The last, formal step is the introduction of the conservation of energy as the formal expression of this impossibility. If $\alpha, \beta, \gamma, \ldots$ denote the state variables of a closed system, causality implies the existence of relations of the form "$\alpha = f(\beta, \gamma, \ldots)$." This may be rewritten as "$F(\alpha, \beta, \gamma \ldots) = $ constant" without change of content. In sum, Mach saw nothing more in energy conservation than causality formally conceived as conservation, for the sake of representational economy. In his opinion, Mayer's *Causa aequat effectum* was nothing but this form of conception (*Form der Auffassung*), expressed without the arbitrary reduction of physics to mechanics.[32]

Although Mach's distinction between empirical, logical, and formal aspects of energy conservation may seem useful, its implementation is spoiled by two incompatible requirements: on the one hand, Mach assumes the reversibility of the transformations of closed systems; on the other hand, he purports to describe these systems through measurable quantities only. Another difficulty is the ambiguity in the choice of an invariant for a closed system. In the particular case in which a vertically mobile weight P of height H is a subsystem of the global, closed system, we all know that the proper invariant has the form $PgH + f(\alpha, \beta, \ldots)$, where g is the acceleration of gravity, and $\alpha, \beta, \ldots$ are the other state variables of the system. But could not there be other invariants?

These difficulties are solved in Planck's own history of energy conservation, a remarkably clear, prize-winning essay published in 1887. Like Mach, Planck used history as an antidote for the metaphysical hardening of our concepts. He denounced the inclination of contemporary Kantian philosophers and physicists to regard the energy principle as the result of a transcendental deduction:

> Recently, it has been proclaimed that a proof of the [energy] principle is neither possible nor necessary because this principle is valid a priori, namely, because it is a

---

[31] Mach 1872; 1896, pp. 324–7. Mach remarked that his concept of causality was purely relational and did not involve prior concepts of space and time.

[32] Mach 1896, pp. 326–7.

necessary, nature-given form of our intuition and understanding. In this case and for many other truths that have been conquered through century-long efforts, the truths are said to be evident and inborn after the fact, when the force of habitude has acted on us.

Although Planck did not question the mechanical world picture, he agreed with Mach that it was only a hypothesis, that it should not be made the foundation of the energy principle, and that the impossibility of perpetual motion provided a better foundation. In the usual formulation, the impossibility of perpetual motion states the impossibility of a cyclic machine that can produce an unlimited amount of work without compensation in the environment. Planck extended this impossibility to negative work, that is, to the consumption of work in an uncompensated manner. He then considered a system interacting mechanically and thermally with its environment, and defined the energy of a given state of this system as the work equivalent of the external actions needed to bring the system to this state from a fixed reference state. This definition, borrowed from Thomson, is consistent only if the work equivalent does not depend on the path from the reference state to the end state. This independence is implied in the generalized impossibility of perpetual motion, because otherwise the system could undergo cycles in which work is produced or consumed without thermal compensation.[33]

In sum, the inventors and commentators of the energy principle expressed four kinds of necessity arguments for this principle: theological, metaphysical, transcendental, mechanical, and empirical. Even if there would still be a physicist to embrace theological or metaphysical arguments, he or she would have to agree with Leibniz that these arguments are not sufficient to determine the expression of conserved quantities. Transcendental arguments in the Kantian sense, hypothetical causality in Helmholtz's sense, and empirical causality in Mach's sense still suffer from this difficulty. They may be used either to assert the existence of a deterministic evolution for a closed system or to assert the existence of strict correlations between empirically accessible variables of the system. In both cases, invariants can be defined but they are not unique. As Poincaré noted in his *Thermodynamique*, this is especially evident in the first case: time and initial conditions can be eliminated from the equations

$$x = f(x_0, y_0, \ldots; t), \quad y = g(x_0, y_0, \ldots; t), \ldots \qquad (7)$$

of the deterministic evolution to obtain $n - 1$ independent invariants if there are n degrees of freedom. Some means must be given to select a special invariant among all the possible ones.[34]

These means are found in mechanics regarded either as a reductive foundation for every physical phenomenon, or as a gauge of other phenomena. In the first option, the reducibility of physics to a variety of conservative mechanics is assumed. The conservative character

---

[33] Planck 1887, pp. 131 (citation), 138–42. Mach approved Planck's reasoning in later editions of his *Mechanics*. Implicit in this reasoning is the assumption that any two states of a system can be connected (both ways) by exchanging heat and work with the environment. When this was not concretely the case, Planck imagined ideal transformations. On Thomson's definition, see below, p. 99.

[34] Poincaré 1892, pp. ix–x.

of the fundamental mechanics of the world is thereby essential. As we saw, until the early nineteenth century (Newtonian) mechanical reducibility implied pervasive violations of energy conservation, because mechanics, at its most fundamental level, involved the destruction of motion in hard collisions. The gradual demise of ideally hard bodies, and accumulated evidence that motion was never lost without thermal or micro-mechanical compensation led to the conviction that mechanics, at its most fundamental level, should be conservative. The kind of necessity here evoked is partly empirical, partly economical: if dissipative processes can be traced to invisible conservative processes, it is simpler to assume that fundamental processes are all conservative.

A first variety of conservative mechanics, born in a Laplacian context, privileged direct action at a distance and assumed two fundamentally different forms of energy, potential and kinetic. Its most eloquent supporters were Saint-Venant and Helmholtz. A later form, promoted by British physicists like Thomson, Rankine, Maxwell, FitzGerald, and Larmor favored contact action and regarded every form of energy as ultimately kinetic (as the Bernoullis had done long ago). Potential energy had to be traced to hidden mechanism, as it was in the kinetic theory of gases, in Maxwell's vortex model of the magnetic field, in Thomson's vortex model of atoms and ether, and in Larmor's theory of the ether. In the 1890s, the continental predilection for direct action at a distance suffered from the experimental proof that electromagnetic interactions propagated at finite speed. The author of this proof, Heinrich Hertz, propounded a form of mechanics in which the interconnections of hidden masses were responsible for all distance action.[35]

Not all nineteenth-century mechanical reductionists wanted an explicit, unique reduction. Hertz never did, and, as we will see in a moment, Helmholtz and Maxwell did not in their maturity. However, they all agreed that a reduction was in principle possible, and that it should involve one of the former varieties of what we would now call classical mechanics. In retrospect, Mach was rightly suspicious of this reductionism since nineteenth-century mechanical reduction did not survive the advent of relativity and quantum theory. Nevertheless, an avatar of this reductionism is well alive in the assumption that every phenomenon, at least in the pre-quantum stage, is reducible to a relativistic, Lagrangian field dynamics. In this context, energy (and momentum) conversation are intimately related to spacetime structure through Noether's theorem. We will return to this point in Chapter 7.

The second option for defining energy, the one based on coupling non-mechanical phenomena with mechanical processes, was strongly supported by Joule, Thomson, and Planck. Its possibility derives from the fact that every physical phenomenon, being ultimately observed through induced mechanical changes, must be coupled to macroscopic mechanical phenomena. In the mature form Planck gave to this option, it builds on the impossibility of the uncompensated creation or destruction of mechanical work. As Poincaré remarked, this non-reductionist approach has a dangerous tendency to turn the energy principle into a mere convention. Indeed the principle only has precise experimental meaning insofar as the various sorts of compensation for missing or surging mechanical work can be identified and quantified, as was the case in his times for heat, electricity, and

---

[35] Cf. Stein 1981; Kragh 2011, Chap. 2; Lützen 2005.

chemical affinity. Whenever a new sort of uncompensated work appears, two attitudes are logically possible: one may decide that the energy principle is violated, or one may introduce a new invisible entity that saves the principle. Physicists since Poincaré have recurrently faced this dilemma, famously after the discovery of radioactivity and after the discovery of the continuous spectrum of β-decay. In the end, the second alternative always won, with the application of the mass-energy relation to the atomic nucleus in the case of radioactivity, and with the acceptance of Pauli's neutrino in the second case.[36]

To conclude, two necessity arguments have survived the intricate history of the energy principle: the reduction to a conservative dynamics itself deemed necessary, and the reduction to the simple fact of the impossibility of creating or annihilating observable motion without compensation.

3.3   The principle of least action

Hidden mechanism

One essential virtue of the energy principle, emphasized by Mayer and Mach, is that despite its mechanical origins it is able to relate measurable quantities without knowing the precise mechanism of this relation. Another principle of mechanical origin, the principle of least action, shares this "blackboxing" of mechanism with the energy principle. As we saw in Chapter 1, Leibniz and Euler both believed that extremum principles would be in many cases the only way to predict the motion of physical and non-immediately mechanical systems, because the description through efficient causes or Newtonian forces would remain inaccessible to us. Something similar can be said of Hamilton's version of the principle of least action, from which Lagrange's equations derive. In his *Mécanique analitique* of 1788, Lagrange showed that in a holonomic[37] connected system subjected to central forces the equations of motion could be derived from the expressions of the live force and total potential in terms of as few coordinates as there are degrees of freedom in the system. The material constitution of the system and the detailed play of force within it do not enter this deduction. Lagrange did not truly exploit this circumstance, since he only treated examples in which the mechanical system was fully described (albeit without figures). In his own eyes, his main achievement was to have reduced mechanics to "a branch of analysis."[38]

The first to exploit the blackboxing quality of the principle of least action were the British mathematicians and natural philosophers George Green and James MacCullagh. In the late 1830s, they based their optical ether theories on a Lagrangian function for the macroscopic displacements of the elastic ether, thus avoiding the intricacies of the molecular ethers earlier assumed by Fresnel, Poisson, and Cauchy. MacCullagh inferred the form $K(\nabla \times \mathbf{u})^2$ of the potential energy of the ether ($\mathbf{u}$ denotes the displacement) from known optical laws. He recognized that a mechanical realization of this potential, if there were

---

[36] Poincaré 1904, pp. 305–7.

[37] A holonomic system is a connected system for which the differential equations expressing the constraints are integrable, so that generalized coordinates can be introduced.

[38] Lagrange 1788, p. vi. On Leibniz, Euler, and Lagrange, see Chapter 1, pp. 21, 29, 38.

any, would widely differ from ordinary elasticity. Most influentially, in 1865 James Clerk Maxwell exploited the Lagrangian method to derive some of his electromagnetic equations from the mere assumption of a hidden mechanical motion in magnetic fields. The artificial character of his honeycomb model of 1861, and the necessity to consolidate the empirically verifiable consequences of this theory determined this move. In essence, Maxwell assumed that the electric currents were generalized velocities in Lagrange's sense, commanding the motion of a hidden mechanism in the magnetic field. For a system of linear currents i_α, the kinetic energy of the hidden motion has the form

$$T = \sum_{\alpha\beta} M_{\alpha\beta} i_\alpha i_\beta, \tag{8}$$

where the coefficients $M_{\alpha\beta}$ depend on the spatial configuration of the currents. The Lagrangian derivative $\mathrm{d}(\partial T/\partial i_\alpha)/\mathrm{d}t$ with respect to the generalized velocity i_α yields the electromotive force of induction in the circuit α, and the spatial derivatives of T yield the electromechanical forces acting on the circuits.[39]

Maxwell later captured the essence of this procedure through the analogy of a belfry, the machinery of which is controlled by a number of ropes. The belfry being originally at rest, finite velocities are impressed impulsively on the ropes. If the required impulses are measured for every possible value of the positions and final velocities of the ropes, the kinetic energy of the system can be computed as a function of generalized coordinates and velocities. Indeed the homogeneity of T implies that

$$2T = \sum_i p_i \dot{q}_i, \tag{9}$$

since by definition $p_i = \partial T/\partial \dot{q}_i$. Then the motion of the ropes for any applied force f_i is given by the Lagrange equations

$$\mathrm{d}(\partial T/\partial \dot{q}_i)/\mathrm{d}t = f_i \tag{10}$$

if there is no potential energy.[40]

In 1880 George Francis FitzGerald recognized that MacCullagh's optical ether theory and Maxwell's electromagnetic theory shared the same equations of motion, if the local velocity $\dot{u}$ of MacCullagh's ether was identified with the magnetic induction field **B**. Through this correspondence, FitzGerald inaugurated the practice of defining electromagnetic theory through a field Lagrangian. Although the best known of these Lagrangians is the one Karl Schwarzschild propounded in 1903, the most eloquent promoters of the field-Lagrangian method were Hermann Helmholtz and Henri Poincaré. In the 1880s, Helmholtz became convinced that the principle of least action should be the general

---

[39] MacCullagh 1848 [read 1839]; Maxwell 1865; 1873, §554. Cf. Buchwald 1985, pp. 20–3; Darrigol 2000, pp. 157–60, and further reference there. On MacCullagh, cf. Darrigol 2010.

[40] Maxwell 1879, pp. 783–4.

foundation of physics, for it drew all the benefits of the possibility of a mechanical reduction without falling into the arbitrariness of specific mechanical pictures:

> The general validity of the principle of least action seems to me so widely secured that it can have a high value as a heuristic principle and as a lead in our striving to formulate the laws of new classes of phenomena. Besides, the principle has the advantage of condensing in just one formula all the relations that are relevant to the investigated class of phenomena; which gives us a complete overview of everything essential.

Helmholtz himself applied the principle of least action to thermodynamics and electrodynamics. As this application required more general forms of the Lagrangian than encountered in ordinary mechanical theories (connected systems with central forces), he proved that the desired generality could be obtained by eliminating the coordinates of hidden or cyclic motions.[41]

Poincaré shared Helmholtz's enthusiasm about the milder form of mechanical reductionism implied in the principle of least action. In the foreword of his Sorbonne lectures of spring 1888 on electricity and optics, he defined Maxwell's "fundamental idea" as follows:

> In order to demonstrate the possibility of a mechanical explanation of electricity, we need not worry about finding the explanation itself. We only have to know the expression of the two functions T and U that are the two parts of the energy, to form the Lagrange equations, and to compare these equations with the empirical laws.

This interpretation of Maxwell's recourse to the Lagrangian method rested on the following theorem. Consider a physical system, the configuration of which is completely determined by empirically controllable coordinates $q = (q_1, q_2, \ldots q_n)$. By assumption, there is a function $T(q, \dot{q})$ and a function $V(q)$ such that the motion of the system is given by Lagrange's equations

$$\frac{\mathrm{d}}{\mathrm{d}t}\frac{\partial L}{\partial \dot{q}_i} - \frac{\partial L}{\partial q_i} = 0, \quad \text{with } L = T - V. \tag{11}$$

By definition, this system has a mechanical explanation if and only if there is a system of points $\mathbf{r}_\alpha(q)$ with masses m_α such that the kinetic energy T takes the form

$$T = \sum_\alpha \frac{1}{2} m_\alpha \dot{\mathbf{r}}_\alpha^2. \tag{12}$$

This will be true if the functions $\mathbf{r}_\alpha(q)$ satisfy the condition

$$\sum_\alpha \frac{1}{2} m_\alpha \left(\sum_i \frac{\partial \mathbf{r}_\alpha}{\partial q_i} \dot{q}_i \right)^2 = T(q, \dot{q}). \tag{13}$$

---

[41] FitzGerald 1880; Schwarzschild 1903; Helmholtz 1886, p. 210.

Since the number of mass points is as large as one wishes, this condition can always be satisfied in an infinite number of ways, and there are infinitely many mechanical explanations.[42]

Poincaré enunciated the following corollary:

> If a phenomenon allows of a complete mechanical explanation, it will allow for infinitely many other explanations which will equally well account for all particularities revealed by experiments.

This overabundance of possible explanations led Poincaré to imagine a future in which physicists would content themselves with the virtual explanation offered by the Lagrangian method:

> Among all possible explanations, how could we perform a selection for which experiments are of no avail? Perhaps the day will come when physicists will lose interest in such questions—which are inaccessible to positive methods—and will leave them to metaphysics. This day has not come yet; men do no easily resign themselves to forever ignore the essence of things.

At the turn of the century, Poincaré judged that history had taken this phenomenogical turn. Laplacian molecular theories had largely been supplanted by continuum theories. British models of the ether gave way to Hertz's formal definition of electromagnetic theory through a system of partial differential equations. The new, preferred physics was "a physics of principles" in which the energy principle, the principle of least action, and other general principles constrained and guided the evolution of physical theory without the need of a mechanical picture. Microphysical entities such as atoms, ions, and electrons were accepted only to the extent that their characteristics became accessible to experiments. "Indifferent hypotheses," such as the molecules of Laplacian physics or the vortices of Maxwell's ether model were left behind as naïve mechanism.[43]

The necessity of least action

We may now discuss the necessity of the principle of least action in a broader context than we did at the end of Chapter 2. As was told there, theological justifications of this principle are as vacuous as the similar justifications of the energy principle, for they do not determine the form of the quantity to be minimized. As long as the action remains deprived of direct empirical meaning, there is no counterpart to the empirical justification of the energy principle. One might think of interpreting the action as the phase of a wave phenomenon, as suggested by Hamilton's analogy between ray optics and particle mechanics and confirmed by quantum mechanics. This is probably the best option, although it only displaces the question of the necessity of least action to the question of the necessity of quantum mechanics. If one wishes to remain in a classical context, the only possibility is to relate the principle of least action to mechanical reducibility.

In the nineteenth-century, connected systems and point-atoms directly acting at a distance both offered plausible justifications for the validity of the principle of least action in

---

[42]Poincaré 1890, pp. xiv–xv.   [43]Poincaré 1890, p. xv; Poincaré 1900b, 1904.

any domain of physics. The choice between these two possibilities was a matter of taste, because the connection of the basic mechanical models to experiment could be very indirect (as in molecular models of matter or in mechanical models of the electromagnetic field). Maxwell, Kelvin, and Hertz favored (holonomic) connected systems; Saint-Venant and Helmholtz favored direct action at a distance in the wake of the Laplacian tradition. As we saw in Chapter 2, the two kinds of representations are mutually related through reduction and approximation. Both options now seem obsolete, because they are incompatible with relativity theory. Just as is the case for the energy principle, there remains the field-stress approach, which will be addressed in Chapter 7.

3.4   Thermodynamics and statistical mechanics

Perhaps the most enduring of all physical theories is thermodynamics, if properly defined or redefined. Since Rudolf Clausius and William Thomson founded it in the mid-nineteenth century, its two principles or "laws" (as English speakers prefer to say) have never failed and they are still commonly regarded as constraints to which the macroscopic consequences of any theory must be subjected. Let us see whether we can find clues for this exceptional longevity in the early history of thermodynamics.

Sadi Carnot's reflections

In 1824, Sadi Carnot, a son of Lazare, published his *Reflections on the motive power of fire*, which are rightly regarded as the foundation of a thermodynamic style of reasoning. A military engineer and former student of the Polytechnique school, Sadi had a special interest in steam engines and he was aware of their recent improvements by Thomas Newcomen and by James Watt. In his groundbreaking essay, he first remarked that the production of work by thermal means depended on a temperature difference that permitted "a fall of caloric." *Calorique* was the name that the French then gave to the substance of heat, in the wake of the Scottish and French calorimetry of the late eighteenth century. In Laplacian physics, the caloric was an elastic, self-repelling substance, the pressure of which represented temperature. Consequently, the transfer of heat from a higher to a lower temperature was analogous to the expansion of a spring or the fall of water. Analogy with hydraulic engines suggested that in an optimal engine there should be no temperature change without a corresponding change of volume of the working substance. Carnot described the simplest cyclic process that met this criterion: isothermal expansion of air by contact with a hot source, adiabatic temperature increase by compression, isothermal contraction by contact with a cold source, and adiabatic return to the original state. This reversible cycle is now called a Carnot cycle.[44]

Then came Carnot's most important theorem: a reversible engine has the highest possible efficiency among all engines that work between two given temperature sources. The proof is based on two axioms: the conservation of the caloric, and the impossibility of perpetual motion. Suppose, *ad absurdum*, that there exists an engine that is more efficient than the reversible engine. The work produced by this hypothetical engine is inferior to the

---

[44]Carnot 1824. On the history of thermodynamics, cf. Cardwell 1971; Truesdell 1980; Darrigol 2003b.

work needed to return the "fallen" caloric to its original level by operating the reversible engine in its reversed mode. The simultaneous operation of the two engines would then permit the indefinite production of work without any compensation. In order to avoid this consequence, Carnot's theorem must hold.

A corollary of this theorem is that all reversible bithermal engines have the same efficiency: the ratio of the produced work to the transferred heat is a function solely of the temperatures of the source and sink. Carnot tried to determine this universal function. For this purpose, it is in principle sufficient to know the elastic and thermal properties of a fluid on which a Carnot cycle can be performed. Carnot found contemporary data insufficient for a full determination. However, he realized that his theorem could be used to derive relations between the constitutive properties of fluids and the universal efficiency function.

For example, consider the heat and work exchanged in an infinitesimal, elongated Carnot cycle made of two isotherms at the temperatures θ and $\theta + d\theta$ and two adiabats at the approximate volumes V and dV, $d\theta$ being an infinitesimal of higher order than dV. The exchanged work is $P(\theta + d\theta, V)dV - P(\theta, V)dV$. The transferred heat is ldV by definition of the latent heat of expansion l. Equating the ratio of these two quantities to the universal efficiency $\mu(\theta)d\theta$ leads to the relation

$$\partial P/\partial \theta|_V = \mu\, l. \tag{14}$$

Although Carnot obtained only a particular case of this important relation, it deserves the name "Carnot's relation" because its demonstration only requires Carnot's theorem without recourse to the conservation of the caloric. Most of Carnot's other results did not survive the later rejection of caloric. His theorem and the styles of its derivation and application nevertheless had a brilliant future: ideal machines and processes, reversible cycles, *ad absurdum* reasoning, and organizing principles became the gist of modern thermodynamics.[45]

Clausius's and Thomson's thermodynamics

In the late 1840s, Joule, Thomson, and Helmholtz became aware of Carnot's reasoning and saluted it as an elegant application of the new doctrine of energy conservation. Indeed a thermal engine, in Carnot's view, is nothing but a device for turning the potential energy of an elastic fluid, the caloric, into mechanical work. The only problem with this view is that it contradicts Joule's equivalence of heat and work. As Joule pointed out, if heat is conserved in a Carnot cycle and yet can be created by friction, then the work provided by a Carnot engine could be turned into heat, and the total heat in the universe would thus be increased *ad libitum*. A young German physicist, Rudolf Clausius, solved this paradox in a crucial memoir of 1850 by assuming that part of the heat provided by the hot source was transformed into work. Carnot's theorem could then be maintained without contradicting Joule's statement of the equivalence between heat and work, although the theorem could no longer be based on the impossibility of perpetual motion. Imitating Carnot's *ad absurdum* reasoning, Clausius assumed the existence of a bithermal engine with a higher efficiency

[45] Cf. Truesdell 1980.

than a reversible Carnot engine, and used the work produced by this hypothetic engine to run a reversed Carnot engine between the same sources. The net result would be a transfer of heat from a cold source to a hot source without any compensation. The impossibility of such a transfer implies Carnot's theorem.[46]

This impossibility agreeing with "the known behavior of heat," Clausius took Carnot's theorem as his second principle. His first principle was a statement of Joule's equivalence between heat and work: "In all cases when work is produced by heat, a quantity of heat is consumed that is proportional to this work, and reciprocally the same quantity of heat can be produced by consuming an equal amount of work." Clausius then borrowed from Helmholtz a derivation of the universal Carnot function $\mu(\theta)$ based on this principle, on Carnot's theorem, and on the additional assumption that for a sufficiently dilute gas the energy of the gas only depended on its temperature. The latter assumption derives from the kinetic-molecular picture of the gas, since for a dilute gas the potential energy of the intermolecular interactions is negligible. It implies that the work equivalent $J l dV$ of the heat $l dV$ used to expand a dilute gas at constant temperature must be equal to the work $P dV$ provided by this expansion. Together with Carnot's relation $\partial P/\partial \theta|_V = \mu l$, this gives

$$\mu = J/\theta \qquad (15)$$

if the temperature θ is defined so that $PV = R\theta$ for a dilute gas.[47]

Clausius's memoir contained the two-principle foundation of thermodynamics, as well as the constraints on constitutive properties that result from these principles. His approach was mostly macroscopic, except for the kinetic-molecular justification of $l = P$ in a dilute gas. After reading Clausius, Thomson provided a similar foundation, with some differences. Most important, Thomson related the first law of thermodynamics to his principle of mechanical effect, introduced the modern meaning of the word energy, and defined the internal energy of a system. The principle of mechanical effect, he reasoned, implies that the total work of the external forces acting on a system must be equal to the variation of the kinetic energy of the system minus the work of the internal forces of the system. Consequently, this total work vanishes during a cycle of the system (granted that the internal forces derive from a potential). It comprises the work of macroscopic forces and that of the microscopic forces responsible for thermal exchanges. Therefore, the sum of exchanged work and heat (expressed in work units) vanishes during a cycle. This is Thomson's sharpened statement of Joule's equivalence law. As Thomson further noted, this law implies that the sum of the work and heat exchanged by a system during a non-cyclic process only depends on the initial and final states of the system. Assuming a fixed reference state, Thomson defined the "mechanical energy" of a system in a given state as the value of this sum for any process leading from the reference state to this state.[48]

Thomson further departed from Clausius in his statement of the second law. In his own variant of Carnot's *ad absurdum* reasoning, the heat taken from the hot source by the hypothetical better-than-reversible engine is regained by working a Carnot engine backward.

---

[46] Clausius 1850. Cf. Daub 1971; M. Klein 1969; Truesdell 1980.   [47] Clausius 1850, p. 373.

[48] Thomson 1851. Cf. Smith and Wise 1989.

The work required in the latter operation being less than the work required in running the hypothetical engine, it becomes possible to produce work by borrowing heat from a single source. Thomson's statement of the second law was the impossibility of such monothermal work production: "It is impossible by means of inanimate material agency, to derive mechanical effect from any portion of matter by cooling it below the temperature of the coldest of the surrounding objects."[49]

The other basic concepts of classical thermodynamics derive from Clausius's or Thomson's principle. In 1854, Thomson defined the "absolute temperature" T as the inverse of the Carnot function μ. Together with the first law, this choice implies the relation

$$Q_1/T_1 + Q_2/T_2 = 0 \qquad (16)$$

for a Carnot cycle in which the heats Q_1 and Q_2 are exchanged with sources at the temperatures T_1 and T_2. For a reversible cycle of a system exchanging heat with any number of sources, Thomson further derived the relation

$$\sum Q_i/T_i = 0, \qquad (17)$$

which he regarded as "the mathematical expression of the second principle." In the same year 1854, Clausius introduced the "equivalence value" $\int \delta Q/T$ of any transformation of the environment of a system, the heats δQ being exchanged with a continuous sequence of sources of temperature T. In these terms, he obtained "the analytical expression of the second principle," namely, the vanishing of this integral for a reversible cyclic transformation. This property implies that $\delta Q/T$ should be a differential. In 1865, Clausius introduced the integral of this differential, named it entropy (from the Greek τροπή for transformation), and showed that it necessarily increased during the spontaneous evolution of a closed system. He concluded this memoir with two pronouncements:[50]

(1) The energy of the world is constant.
(2) The entropy of the world tends to a maximum.

Concrete application of the principles of thermodynamics requires further assumptions on the state manifold of the system under consideration, mechanical concepts to describe work done on the environment, calorimetric concepts to describe thermal sources. These data are usually symbolized by the differentials δW and δQ of exchanged work and heat. Once these have been given, the necessity of classical thermodynamics clearly depends on the necessity of its two principles. The first principle, which is the vanishing of the sum of work and heat brought to a system in a cycle of operations, is a consequence of the energy principle and is therefore as necessary as we judge the latter principle to be. The second principle, which is the impossibility of a monothermal engine in Thomson's formulation and the impossibility of uncompensated heat transfer from a cold to a warm body in Clausius's formulation, is sometimes called the impossibility of a perpetual motion of

---

[49]Thomson 1851, p. 179.

[50]Thomson 1857 [read May 1854]; Clausius 1854; Clausius 1865, p. 400. Cf. Daub 1971.

the second kind because it forbids a sort of thing we would never expect to happen, for instance the propulsion of a ship by turning seawater into ice. This is an empirical sort of necessity, though of a kind that no sane person would doubt.

The kinetic theory of gases

In the phenomenological form later defended by Max Planck and by Pierre Duhem, classical thermodynamics makes no reference to sub-mechanical or kinetic-molecular foundations. Mechanics only intervenes in defining some of the macroscopic properties of the system and its environment; heat is defined empirically through thermometry and calorimetry; and the two principles are traced to the impossibility of perpetual motions of different kinds. Carnot, Thomson, and Clausius already understood the utility of developing thermodynamics as a direct theory of machines, mostly independent on molecular mechanics. However, the three founders embraced mechanical reductionism in general and a mechanical concept of heat in particular. Carnot's theory is essentially mechanical in its description of heat as a falling substance and in its compatibility with Laplacian physics. Clausius occasionally appealed to molecular arguments in his thermodynamics, for instance in his determination of the Carnot function, in his defining the work of internal forces, and in the related concept of disgregation. Thomson was more circumspect, although he clearly agreed with Clausius that the first principle of thermodynamic rested on heat being a mode of motion.

In 1857, building on earlier intuitions by Daniel Bernoulli, John Herapath, John Waterston, and James Joule, Clausius derived the equation of perfect gases by summing the impacts of mostly free-flying molecules on the walls of the container. The following year, he introduced the concept of the mean free path of the molecules, which controlled the speed of diffusion of the gas. In 1860, James Clerk Maxwell determined the distribution of velocities of the gas, established the equipartition of energy between different gases in a hard-ball model, and computed transport phenomena by the mean-free-path method. In 1866, he developed a deeper and more general approach based on a statistical formula for the number of binary collisions as a function of the initial velocities of the colliding molecules. So far the developments of the kinetic molecular theory of gases did not intersect much macroscopic thermodynamics. They completed the latter theory by determining relations and quantities that were otherwise left to experimenters, for instance the equation of state of dilute gases and the temperature-dependence of specific heats; and they transcended this theory by determining the speed of transport phenomena.[51]

Maxwell nonetheless argued that the molecular picture, if taken seriously, implied a merely statistical validity of the second law of thermodynamics. This is the substance of the "demon" argument found in a letter he wrote to Tait in December 1867: a "finite being" who could "see the individual molecules" would be able to create a heat flow from a cold to a warm gas without expense of work, for he could control a diaphragm on the wall between warm and cold gas, and let only the swifter molecules of the cold gas pass into the warm gas. In 1878, Maxwell remarked that the dissipation of work (or the mixing

---

[51] Clausius 1857, 1858; Maxwell 1860, 1867. Cf. Brush 1976.

entropy in Gibbs's terms) during the interdiffusion of two gases depended on our ability to separate them physically or chemically, and concluded:[52]

> The dissipation of energy depends on the extent of our knowledge... It is only to a being in the intermediate stage, who can lay hold of some forms of energy while others elude his grasp that energy appears to be passing inevitably form the available to the dissipated state.

Maxwell's insight was critical and qualitative; it did not provide a (statistico-) mechanical interpretation of quantitative thermodynamic relations. In Germanic context, Clausius's parallel between energy and entropy invited such interpretation. Starting with a memoir of 1866 by the Austrian physicist Ludwig Boltzmann, there appeared a few interpretations of entropy as a non-statistical attribute of certain classes of mechanical systems (periodic, or with cyclic variables). Of more lasting value were Boltzmann's attempts to give a quantitative, statistical interpretation of the entropy concept.[53]

Boltzmann's statistical interpretations of entropy

In 1868, in an attempt to generalize Maxwell's distribution, Boltzmann proved that for a conservative mechanical system such that its trajectory in phase-space filled the energy-shell after an infinite amount of time, the fraction of time spent by the system in an element of this shell was uniform with respect to the canonical measure $\mathrm{d}^N p \, \mathrm{d}^N q$ ($q_1, q_2, \ldots q_N$ and $p_1, p_2, \ldots p_N$ being the conjugate coordinates and momenta). In modern terms, he proved that an ergodic Hamiltonian system was microcanonically distributed. As a consequence, he showed that a given molecule of such a system obeyed the Maxwell–Boltzmann distribution law. Three years later, he showed that any small subsystem of the original system was distributed according to the canonical law

$$\rho = \mathrm{e}^{-\beta H}/Z, \tag{18}$$

wherein H is the Hamiltonian of the subsystem, β a constant parameter to be identified with the inverse of temperature, and Z the normalizing factor

$$Z = \int \mathrm{e}^{-\beta H} \mathrm{d}^N p \, \mathrm{d}^N q. \tag{19}$$

He then subjected a canonically distributed system to an infinitesimal change of external conditions (corresponding for instance to a macroscopic change of temperature and volume). Identifying the work provided to the system during this change with the canonical average $< \mathrm{d}H >$ of the resulting change of the Hamiltonian, and the internal energy with the canonical average $< H >$ of the energy, he obtained the expression

$$\delta Q = \mathrm{d} <H> \, - \, < \mathrm{d}H > \tag{20}$$

---

[52] Maxwell to Tait, Dec. 11, 1867, in Maxwell 1990–2002, vol. 2, pp. 331–2; Maxwell 1878. Cf. Klein 1970b.

[53] Cf. Klein 1972.

of the exchanged heat. The product $\beta\,\delta Q$ is easily seen to be the differential of

$$S = \beta < H > +\ln Z. \tag{21}$$

In other words, there exists an entropy function, which Boltzmann soon rewrote as

$$S = -\int \rho \ln \rho \, \mathrm{d}^N p \mathrm{d}^N q. \tag{22}$$

This is the first occurrence of a mathematical relation between entropy and probability, the probability ρ being here defined as the fraction of time spent by the system around a given point of phase space after a very long time has elapsed.[54]

In 1872, in an attempt to prove that Maxwell's distribution was the only equilibrium distribution in a gas, Boltzmann determined how an arbitrary initial distribution of velocities $f(\mathbf{v})$ evolved in time under collisions occurring at the rate given by Maxwell in 1866. The result was the Boltzmann equation, which implies an irreversible evolution of this distribution towards Maxwell's distribution. In order to demonstrate this irreversible behavior, Boltzmann formed

$$\mathrm{H} = \int f \ln f \, \mathrm{d}^3 v, \tag{23}$$

no doubt in analogy with the entropy formula (22), and derived its monotonous behavior from the Boltzmann equation. He naturally interpreted $-\mathrm{H}$ as the entropy of a gas out of equilibrium. In reply to the obvious objection that a reversible microscopic dynamics could not by itself lead to macroscopic irreversibility, he explained that Maxwell's collision number and the resulting evolution of the distribution f were valid only in a statistical manner. Namely, the Boltzmann equation or the time-reversed equation expresses the secular behavior of f (smoothing out small fluctuations) for most choices of the initial microstate (and for a reasonably long time, a Boltzmann later made clear in his reply to Ernst Zermelo's recurrence-based objection). In order to bring out the probabilistic dimension of his theorem, Boltzmann showed that the quantity $-\mathrm{H}$ represented the logarithm of the number W of microstates compatible with the distribution f. This is the origin of the formula which Planck later wrote as[55]

$$S = k \ln W. \tag{24}$$

As Planck noted, the fact that various macrostates of a system have different probabilities W does not imply that a system, left to itself, (most likely) evolves toward states of higher probability. This is only an intuition, which today we can easily verify by numerical simulation. The necessity of Boltzmannian irreversibility is a difficult question, which will

---

[54] Boltzmann 1868, 1871b, c. Cf. Klein 1973; Darrigol and Renn 2000.

[55] Botzmann 1872, 1877a, b. On Boltzmann's understanding of irreversibility, cf. Brush 1976, pp. 616–39; Klein 1970a; Brown, Myrvoll, and Uffink 2009. Boltzmann originally used the notation E instead of Burbury's later H (capital Greek eta).

not be addressed here. Worth mentioning, however, are the connections Boltzmann per-ceived between his three entropy formulas. He related the second to the first by considering a generalization of the Boltzmann equation to two interacting ensembles of systems; and the third to the first by showing that for a microcanonically distributed gas, the measure W of the domain of phase space defined by the velocity distribution $f(\mathbf{v})$ was approxim-ately proportional to e^{-H}. This suggests that equilibrium thermodynamics is intimately related to irreversible macroscopic evolution by a common measure of the probability of the macrostates of the system.[56]

A deeper understanding of the entropy concept and of the origins of thermodynamic irreversibility is not the sole merit of Boltzmann's and Gibbs's statistical mechanics. As Einstein fully realized, this theory provides a two-way bridge between invisible mi-crophysics and macroscopic thermodynamics. If the microphysics is known, as is today the case for sufficiently simple systems, the full range of thermodynamic properties can be derived from it, for instance by computing the associated partition function (the trace of $e^{-\beta H}$, if H is the quantum-mechanical Hamiltonian). If the thermodynamic properties are empirically known, some features of the underlying microphysics can be inferred. For instance, the high-temperature specific heat of a gas is related to the number of molecular degrees of freedom of a molecule. Moreover, statistical mechanics enables us to derive the probability of fluctuations of thermodynamic quantities around their equilibrium value, for instance the probability of the energy fluctuations of a small system coupled to a thermo-stat, simply by inverting Boltzmann's relation between entropy and probability. As Einstein argued, in some cases such as Brownian motion the fluctuations are observable and provide a measure of Boltzmann's constant. In other cases, the theoretical expression of the fluc-tuations, as derived from the macroscopic entropy function, provides information on the underlying microdynamics. This is how Einstein arrived at the lightquantum hypothesis and at the wave-corpuscle duality.[57]

In some sense, the existence of fluctuations contradicts ordinary thermodynamics, since the latter theory assumes a perfectly defined value of thermodynamic quantities at equilib-rium. Yet it does not contradict Thomson's two laws of thermodynamics, because energy conservation is exact for a closed system even in statistical thermodynamics and because the possibility of producing work by fluctuation in a single cycle of a tiny monothermal engine does not contradict the second law for the large number of cycles that would be necessary for macroscopic work production.[58]

The statistico-mechanical necessity of thermodynamics

We may now examine to what extent Boltzmann's connections between entropy and prob-ably bear on the question of the necessity of equilibrium thermodynamic. As Maxwell, Boltzmann, and Gibbs fully explained, the microcanonical and canonical distributions in phase space imply the usual thermodynamic relations between temperature, entropy,

---

[56] Planck to Graetz, cited in Kuhn 1978, p. 27; Boltzmann 1872, pp. 401–2; Boltzmann 1881a.

[57] Cf. Klein 1963, 1964, 1982; Renn 1997.

[58] In fact, the recent fluctuation theorem by Christopher Jarzinsky (1997) implies that the average work over many cycles must be positive (received): cf. Jarzinsky 2010.

energy, and other state variables.[59] The necessity of these relations therefore depends on the adequacy of stationary ensembles to describe a system in thermodynamic equilibrium. Boltzmann considered three possible justifications: ergodicity, collision formulas à la Maxwell, and the empirical existence of equilibrium. Boltzmann doubted ergodicity, and rightly so. Even today, it remains difficult to characterize the subclass of Hamiltonian systems that satisfy this property. At any rate, ergodicity is physically irrelevant because the real time scale at which equilibrium is reached and observed is much smaller that the ergodic time scale for which the system fills nearly all the accessible phase space. Typically, a system initially out of equilibrium reaches the region of most probable states before it has had time to explore much of phase space and it remains in this region for an extremely long time before it goes on exploring other regions.[60]

Boltzmann's second justification, which dates from 1871, rests on a generalization of Maxwell's collision number to giant molecules representing the whole system. By reasoning analogous to Maxwell, kinetic equilibrium requires that the giant molecules be distributed according to the Maxwell–Boltzmann law, so that the system they represent is canonically distributed. This justification requires much faith in Maxwell's collision number and its hidden probabilistic assumptions. The third and best of Boltzmann's justification is found in a memoir of 1881 in which he relates Maxwell's stationary ensembles to the equilibrium properties of a single system by assuming, as a well-established empirical result, that the average properties of a closed system over a sufficiently long time (the concrete equilibrium time) do not depend on the choice of the initial state (except perhaps for choices too rare to be ever observed) as long as this state belongs to an ensemble compatible with a given value of the dynamical invariants of the system.[61] Under this assumption, a time-averaged property of a single system can be replaced with the ensemble average of the time average, which is also the time average of the ensemble average; so that for a stationary ensemble, the ensemble average should be equal to the time average. This argument, rediscovered by Einstein in 1903, probably is the most convenient justification of the use of statistical ensembles as mechanical models of thermodynamics. In sum, the empirical existence of equilibrium implies the validity of its representation through stationary statistical ensembles; in turn, this representation implies the usual thermodynamic properties of equilibrium states.[62]

Thermodynamics is not only about equilibrium states. It also determines the possibility or impossibility of the spontaneous evolution of a system from a first quasi-equilibrium state to another equilibrium state. For instance, the theory is able to determine the possibility of chemical reactions through their free-energy or free-enthalpy balances (according

---

[59] Maxwell 1879; Boltzmann 1885; Gibbs 1902. Cf. Klein 1972; Gallavotti 1994; Renn 1997. Whether this implication provides an example of intertheoretical reduction has been abundantly discussed by philosophers: cf. Sklar 1993; Barberousse 2000, and further reference there.

[60] On recent attempts based on weakened ergodicity, cf. Uffink 2007; Frigg, Berkovitz, and Kronz 2011.

[61] This is related to the principle of equilibrium that is called "Minus First Law" in Brown and Uffink 2001.

[62] Boltzmann 1871a, 1881b; Einstein 1903. Recent justifications of the stationary ensembles based on information theory are too weak to be considered here: they rest on the gratuitous assumption that the lack of information on the physical system should be maximized. The Boltzmann–Einstein justification implicitly admits the existence of a Hamiltonian for which the desired property holds.

as volume or pressure is kept constant). The principles of thermodynamics, in Clausius's or in Thomson's form, contain the criteria for this discrimination. As is well known, these criteria are reducible to Clausius's condition that the entropy can only increase during the spontaneous evolution of a closed system (including the heat and work sources). Reciprocally, this condition, joined to the first principle, implies the two forms of the second principle. Indeed the global entropy variation during the uncompensated transfer of the heat Q from a T_1 source to a T_2 source is equal to $Q/T_2 - Q/T_1$, which is positive if and only if $T_2 < T_1$; and the global entropy variation during a cycle of a monothermal machine receiving the heat Q from the T source is $-Q/T$, which is positive if and only if the work $W = -Q$ is negative. Consequently, the principles of thermodynamics will result from statistical mechanics if the latter theory implies Clausius's entropy condition. For a closed system, the relevant ensemble is the microcanonical ensemble and the corresponding entropy is the logarithm of the surface of the energy shell in phase space. The initial state of quasi-equilibrium differs from the final state of equilibrium by an energy barrier whose removal increases the effective surface of the energy shell. Consequently, the microcanonical entropy increases during a spontaneous evolution of the system.

To sum up, the representation of thermodynamic equilibrium by stationary ensembles not only implies the existence of the energy and entropy functions, it also implies the law of entropy increase in the evolution of a system from a quasi-equilibrium state to full equilibrium. Consequently, the empirical fact of the existence and uniqueness of equilibrium and the Hamiltonian character of the underlying microscopic dynamics together imply the principles of thermodynamics. Ideally, in order to perfect the similarity between the reductionist justifications of the energy and entropy principles, one would wish to derive the existence of equilibrium from the microdynamics only. Can this be done without any arbitrary statistical assumption? One might perhaps prove in a precise mathematical sense that most initial microstates lead to the same equilibrium macrostate after a moderate time. However, we still would not know why nature has not selected the more exceptional microstates.[63]

Just as is already the case for the energy principle, the reductionist demonstration of necessity of the entropy principle hinges on the necessity of the reduction. Boltzmann, Maxwell, and Gibbs already speculated that molecular mechanics might differ from macroscopic mechanics; and we all know that it does. Then the survival of the reductionist approach depends on the availability of a new mechanics that shares some features of classical mechanics. Quantum mechanics answers the call, for it harbors a concept of stationary ensemble described by density operators depending only on operators that commute with the Hamiltonian. Altogether, the physicists' confidence in the eternity of thermodynamic and statistico-mechanical principles seems reasonable. The thermodynamic principles can be reduced to a basic empirical fact, the impossibility of two kinds of perpetual motion. They are further consolidated by their deduction from a microdynamics supplemented with the empirical existence of thermodynamic equilibrium.

---

[63] Cf. Sklar 1993. Perhaps quantum-theoretical blurring could be evoked, but then a similar problem would occur at the level of quantum states.

Conclusions

Physicists long believed in the necessity of a mechanical reduction of all physical phenomena. They now believe in their reduction to some basic (relativistic) field dynamics, quantized or not. Arguments for the necessity of the mechanical reduction, such as Newton's analogy of nature, have proved to be fragile. While reductionism remains, the basis of the reduction varies. Early in the nineteenth century, the privileged basis was the Laplacian picture of molecules acting through central forces. This reduction, or later reductions to ideal connected systems imply the conservation of energy, defined as the sum of the kinetic and potential energies of the mechanical basis. It also implies the principle of least action. These principles have proved to be far more general and far more stable than their reductionist justifications. Being more directly related to observation, they can be applied without knowing the detailed mechanism of phenomena. It is therefore desirable to justify them in a non-reductionist manner. This can be done for the energy principle, by relating it to Planck's axiom of the impossibility of producing or annihilating work without compensation in the environment. Unfortunately, no such justification exists for the principle of least action, due to the lack of a direct physical interpretation of the action (unless we introduce its quantum-theoretical interpretation as a phase).

The anchoring of principles on evident empirical truths is best illustrated in thermodynamics. The two laws or principles of this theory can be expressed as the impossibility of two kinds of perpetual motion. However, these principles also admit reductionist justifications. The first principle, which is the equivalence of the heat and work exchanged during a cycle of a machine, can be derived from kinetic-molecular reduction by identifying the exchanged heat with work done on the molecules. The second principle, or the impossibility of a monothermal engine, can be derived from the representation of systems in (quasi-)equilibrium by stationary statistical ensembles. This representation itself derives from the empirical existence of equilibrium combined with the microdynamical reduction. As was already the case for energy conservation and least action, the mechanical and statistico-mechanical reductions do much more than justifying a few general principles. They relate the micro-world and the macro-world in a detailed, reciprocal manner. The necessity of this relation hinges on the existence of thermodynamic equilibrium and on general features of the microdynamics, found both in classical mechanics and in quantum mechanics.

In sum, general principles and microphysical reductions appear to be intimately related, both historically and logically. In some situations, the principles are deduced from evident empirical truths and they are used to constrain the possible reductions of macroscopic phenomena. In other situations, the necessity of some features of the reductionist basis is first established, and the principles are derived from these features. In both cases the principles appear to be necessary, although the implied kind of necessity differs. In the macroscopic approach, the necessity hinges on manifest empirical impossibilities; in the reductionist approach, the necessity results from comprehensibility assumptions which evolved when smaller scales became empirically accessible.

4

GEOMETRY

> All the external objects of our sense-world must necessarily conform with the most complete accuracy to the propositions of geometry.[1] (Immanuel Kant, 1783)

> Space, regarded as the domain of measurable quantities, is not at all the most general concept of a manifold of three dimensions. On the contrary, it involves rules that result from the completely free mobility of rigid bodies without change of shape through any position or orientation.[2] (Hermann Helmholtz, 1870)

Geometry is often regarded as a purely mathematical theory. Its purpose can indeed be defined as the demonstration of propositions from other propositions called axioms; and this activity does not require experimentation. Yet the question of the necessity of geometry does not exclusively belong to the history and philosophy of mathematics, for two reasons: because geometry originated in surveying practices, as indicated by the Greek etymology; and because geometry, qua mathematical theory, has never ceased to be successfully applied to aspects of the physical world. In this state of affairs, there are two questions about the necessity of geometry: (1) Is there a purely intellectual necessity of the axioms of geometry? (2) Is there a necessary geometry of the world? Until the late eighteenth century, the usual answers to these questions were a resounding "Yes" to the first and a softer "Yes" to the second. Nineteenth-century developments led to a strident No to the first, and to a nuanced Yes to the second.

In a plain way, consider a concrete rectangular triangle drawn with a ruler and a set square, and measure its sides with a graduated ruler. No one will doubt that the result complies with Pythagoras' theorem within the uncertainty margin of the measurement. But is there a reason why it should be so? One answer could be: yes, because Euclid's axioms are approximately verified for concrete points and lines and because Pythagoras' theorem is a consequence of these axioms. This answer has two defects: it is not clear in what sense it can be said that Euclid's axioms are approximately valid, because they involve infinitely long lines that are never observed in nature; and their validity, to the extent that they can be concretely verified, seems purely contingent. A better answer, which Hermann Helmholtz provided in 1868, is that the concrete operations through which the triangle is constructed and measured require freely mobile rigid bodies (the ruler and the set square), whose existence in turn implies that Pythagoras' theorem holds for small enough triangles and that it must hold for any triangle (at the scale for which the free mobility of rigid bodies has been verified) if it holds for one given finite triangle.

---

[1] Kant 1783, p. 60.   [2] Helmholtz 1870, p. 19.

In brief, space must be locally[3] Euclidean as long as it is measurable. Mathematicians are of course free to define abstract spaces, such as projective spaces or fibered bundles, in which ideal measurability is not assumed; and physicists may find such geometries useful in the construction of their most advanced theories. However, such geometries do not directly concern ordinary space. We must stick to the locally Euclidean structure, as long as we assume the measurability of ordinary space. It could of course well be that Euclidean geometry principally fails at very small scales: Bernhard Riemann already conceived this possibility in 1854. However, this can only happen if measurement becomes inconceivable at such scales.

The first section of this chapter is devoted to the historical evolution of the question of the necessity of geometry from the ancient Greeks to Helmholtz and Poincaré. The second section is an attempt to consolidate those of Helmholtz's claims that survived the later evolution of physical geometry. The third and last is a brief attempt to delineate the conventional and necessary elements of any physical geometry.

4.1 From Euclid to Helmholtz

Euclid's Elements

The *Elements* that Euclid wrote around 300 BC is a set of deductive proofs of propositions based on three kinds of premises. The premises of the first kind are twenty-three definitions of basic geometric entities: points, lines, surfaces, planes, circles, right angles, triangles, parallel lines, etc. The first four definitions read:

1. A *point* is that which has no part.
2. A *line* is a breadthless length.
3. The extremities of a line are points.
4. A *straight line* is a line which lies evenly with the points on itself.

Then come the five "requests" (usually called postulates):

1. To draw a finite straight line from any point to any point.
2. To produce a finite straight line continuously in a straight line.
3. To describe a circle with any center and distance.
4. That all right angles are equal to one another.
5. That, if a straight line falling on two straight lines make the interior angles on the same side less than two right angles, the two straight lines, if produced indefinitely, meet on that side on which are the angles less than the two right angles.

Lastly, Euclid lists five "common notions":[4]

1. Things which are equal to the same thing are equal to one another.
2. If equals be added to equals the wholes are equal.

---

[3] Here and elsewhere, I use "local" to qualify a property that is approximately valid in small enough domains and rigorously valid in the limit of a vanishing small domain. Although this acceptance is common among physicists, it differs from the mathematicians' local/global distinction.

[4] Euclid, in Heath and Heiberg 1908, vol. 1, pp. 153–5.

3. If equals be subtracted from equals, the remainders are equal.
4. Things which coincide with one another are equal to one another.
5. The whole is greater than the part.

Euclid probably adapted his distinction between definitions, requests, and common notions from Aristotle's similar distinction between definitions, postulates, and axioms. According to Aristotle, definitions set the meaning of basic objects without necessarily asserting their existence; axioms are evident preconditions of knowledge in various fields; in contrast, the truth of the postulates does not jump to the eyes; it is conceded for the sake of demonstration; and it is finally accepted when the truth of the consequences is recognized. Like Aristotle's axioms, Euclid's common notions are general requirements for magnitudes of any kinds[5]; his requests concern the generic existence (lines and circles), uniformity (right angles), or uniqueness (parallel lines) of entities introduced in the definitions.[6]

Plausibly, Euclid and his forerunners arrived at these premises by idealizing intuitive geometric notions and by seeking the minimum set of assumptions needed to solve geometric problems. Although these problems had roots in surveying practice and although their formulation and solution usually appealed to figures, the purely deductive character of the solutions and the irrelevance of the materiality of the figures suggest that the premises were meant to be purely ideal. Euclid most likely thought so. In this case, his common notions were self-evident truths, and the truth of his requested propositions depended on the purely intellectual evidence of some of their consequences. Experiment played no role in asserting geometrical truths. This does not mean that Euclid had no interest in the application of geometry. His *Optics* is a proof of the contrary. However, he must have shared Plato's opinion that concrete geometrical problems were imperfect realizations of the ideal problems, for at least two reasons: concrete points, lines, etc. do not satisfy the ideal definitions, and the postulates require indefinite extensions of lines and figures beyond the finite domain of experience or beyond the finite world of contemporary cosmology.[7]

Whether Euclid's definitions carry any sense that is not implicitly contained in the postulates, whether the meaning of the postulates is clear enough, and whether postulates and common notions are truly sufficient to solve any geometrical problem are questions that need not concern us for the moment. Although in the course of time Euclid's premises were improved in several manners, it has never been doubted that Euclid had founded a deductive system in which every geometrical problem could be solved from a few premises. Until the nineteenth century these premises were regarded as necessary true.

The fifth postulate

Despite Euclid's prudent distinction between common notions and postulates, his successors usually regarded his postulates as self-evident, the fifth excepted. In an influential

---

[5] Yet they are not as evident as it would seem, for the implied "equality" is not identity: it refers to a method of comparison, namely, the congruence of figures. Compare with Helmholtz's approach, p. 118 below.

[6] Cf. Heath 1908, pp. 117–22.   [7] Cf. Lanczos 1970, p. 25; Torretti 1978, pp. 1–9; Greenberg 2007.

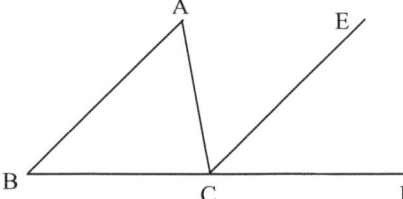

Fig. 4.1.   Figure for Euclid's proof that the sum of the
interior angles of triangle equals two right angles.

commentary, the fifth-century philosopher Proclus Lycaeus declared that the fifth postulate was actually a theorem that should be proved from Euclid's other premises, although
he was himself unable to mend Ptolemy's earlier proof of this kind. Proclus refused to regard the postulate as self-evident, because the intuitive convergence of the lines when the
interior angles that they make with a third line is less than two right angles does not necessarily imply that they intersect: they could just get indefinitely closer to each other. At any
rate, the fifth postulate differs from the other ones as it does not simply state the general
existence and uniformity of basic objects. Most likely, Euclid or one of his forerunners
introduced it for the sake of proving proposition I.32, according to which the sum of the
interior angles of a triangle is two right angles. Euclid's proof of this proposition goes as
follows.[8]

In addition to the triangle ABC of Fig. 4.1, let CE be drawn parallel to AB. Then, postulate 5 implies that the corresponding angles ABC and ECD are equal. Indeed if they were
not, the sum of the angles ABC and BCE would be inferior or superior to two right angles,
and the lines BA and CE would intersect by postulate 5. Similarly, postulate 5 implies the
equality of the alternate angles BAC and ACE. Therefore, the sum of the interior angles
BAC, ABC, and ACB is also equal to the sum of the angles ACE, ECD, and BCA, which
is evidently equal to two right angles.[9]

Postulate 5 is usually called the parallels postulate because it implies that there cannot be
more than one line passing through C and parallel to AB in the plane containing ABC (this
is the so-called Playfair postulate). Indeed, for any such parallel CE, the angle ECD must
be equal to the angle ABC; then CE is unique because there is only one line through C that
makes a given angle with CD. In contrast, the existence of at least one parallel (proposition
I.31) can be proved without recourse to postulate 5: the line CE such that ECD = ABC
must be parallel to AB, because if it were not, there would exist a triangle BCF for which
the exterior angle FCD is equal to the interior angle FBC, which is impossible according
to the construction of Fig. 4.2.

After Ptolemy's and Proclus' there were many attempts to prove the fifth postulate, either
from Euclid's other premises, or from a more evident postulate not contained in these
premises. A significant seventeenth-century example of the latter kind is John Wallis's
proof that the fifth postulate derives from the demand that "for every figure there exists
a similar figure of arbitrary magnitude." In the eighteenth century, a few philosopher-
mathematicians from Giovanni Girolamo Saccheri to Johann Heinrich Lambert tried to
prove that the negation of the fifth postulate led to absurdities. To this end, they explored

[8] Cf. Heath 1908, pp. 202–8.   [9] Euclid, in Heath and Heiberg 1908, vol. 1, pp. 316–17.

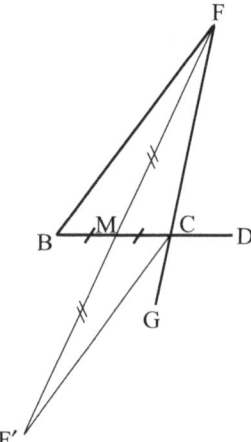

Fig. 4.2.   The point F′ is constructed so that FF′ bisects BC at M and so that FM = MF′. The triangles FMB
          and F′MC being similar, the angle BCF′ equals the angle FBC. Therefore, the inequality of the angles BCG
          and BCF′ implies the inequality of the angles FCD and FBC.

the consequences of this negation combined with those of Euclid's propositions that do not
depend on the fifth postulate. In this process they discovered what we would now regard
as theorems of Lobachevskian geometry: for example, the sum of the interior angles of a
triangle is inferior to two right angles. Lambert also realized that spherical geometry was
what we would now call a two-dimensional model for a system of axioms in which there
were no parallels; and he speculated that a sphere of imaginary radius would be a model
for a system in which there are infinitely many parallels. In the end he decided that these
systems were not genuine geometries, the first because it had straight lines intersecting in
more than one point; the second because it did not allow for similar figures at different
scales.[10]

Kant's transcendental aesthetics

The philosopher Immanuel Kant did not need these mathematical detours to demonstrate
the necessity of the propositions of geometry. The first part of his *Critique of pure reason*
of 1781, the *Transcendental aesthetics*, is devoted to the proof that space and time are pure
forms of outer and inner sensibility respectively. For space, this means that the perception
of external objects minimally requires the mind's ability to conceive spatial relations. Like
Descartes, Kant argued that any other property of sensed objects could be annihilated in
thought. Yet Kant's space, unlike Descartes's extension, is not purely ideal: it is a necessary
precondition of our perception of external objects. Neither can it originate in experience,
because that would jeopardize its necessity. Rather, all our experience of the outer world is

[10]Cf. Heath 1908, pp. 202–20; Torretti 1978, pp. 44–52; Gray 2007, Chap. 7.

necessarily spatial. Whether or not things in themselves are spatial is not an issue for Kant, since we only have access to the appearances of things.[11]

There is textual evidence that Kant's space is necessarily Euclidean. For instance, he writes: "All the external objects of our sense-world must necessarily conform with the most complete accuracy to the propositions of geometry"; and "All the propositions of geometry are valid . . . of all the objects of senses, and therefore in respect of all possible experience." However, sensibility or intuition *per se* does not require the Euclidean structure, it only requires an amorphous notion of space. This structure is brought by the understanding:

> Nature rests on laws which the understanding cognizes a priori, and indeed mainly on universal principles of the determination of space. Now I ask: Do these natural laws belong to space, and does the understanding learn them by merely seeking to investigate the abundant meaning contained therein, or do they belong to the understanding and to the manner in which it determines space according to the conditions of synthetic unity, on which all its concepts hinge? Space is something so uniform, and as regards all particular properties so indefinite, that certainly no one will seek any wealth of natural laws in it. What determines space to the circular form, to the figure of the cone or of the sphere, is the understanding insofar as it contains the ground of the unity of their construction. The mere universal form of intuition called space is the substratum of all particular objects of definable intuitions, and in this certainly lies the condition for the possibility and variety of these intuitions. But the unity of objects is determined simply by the understanding, according to conditions that lie in its own nature, and the understanding is thus the source of the universal order of Nature, since it comprehends all phenomena under its own laws.

Thus, it is the understanding that generates geometrical concepts, by allowing the construction and the comparison of figures. Kant did not doubt that these concepts and the resulting propositions were those of Euclidean geometry. Yet he made no attempt to derive the Euclidean premises from his categories of understanding. This leaves the possibility open that the categories would allow for a wider class of geometries. Although neo-Kantian philosophers later exploited this possibility, it was neither in Kant's mind nor in the mind of any of his early interpreters. Rather, his transcendental aesthetics was regarded as a further proof of the necessity of Euclidean geometry.[12]

A new geometry

The first thinkers to contemplate the possibility of geometries different from Euclid's were the mathematicians Carl Friedrich Gauss, János Bolyai, and Nikolai Ivanovich Lobachevsky in the early nineteenth century. They developed the consequences of the system obtained by removing the fifth postulate from Euclid's more thoroughly than Lambert and other forerunners had done, and found it to include an infinite number of similar theories depending on one parameter only (the constant negative curvature in the Riemannian

---

[11] Kant 1781, *Die transcendentale Aesthetik*, §§2–3.

[12] Kant 1783, pp. 60, 67, 116. In the pre-critical Latin dissertation (Kant 1770, §15), Kant regarded the Euclidean postulates as contained in the mere intuition of space, without appeal to the understanding.

interpretation) as well as Euclid's geometry (the limiting case of vanishing curvature). They did not doubt the consistency of these theories, presumably because they were obtained by removing one postulate from a set of premises that were already believed to be mutually consistent. And they dared regard the new class of theories as a genuine geometry that could well apply to the world at large scale. They even attempted to measure the parameter of this geometry (the curvature) through astronomic triangulation.[13]

This was an audacious move, as Euclid's geometry was universally believed to be the only true geometry, either in an ideal sense or in Kant's transcendental sense. Gauss probably had Kant in his collimator when, in 1817, he confided to a friend:

> I am ever more convinced that the necessity of our geometry cannot be proved, at least not *by*, and not *for* our HUMAN understanding. Maybe in another life we shall attain insights into the essence of space which are now beyond our reach. Until then we should class geometry not with arithmetic, which stands purely a priori, but, say, with mechanics.

Gauss did not publish his ideas, for fear of "the uproar of the Beotians." Lobachevsky and Bolyai, who lived in the outskirts of Europe, had less restraint and won the credit for the new geometry. They did not regard themselves as revolutionaries, however. On the contrary they believed to have isolated the true core of Euclidean geometry, as indicated by the qualification "absolutely true" that Bolyai gave to his new science of space. Far from reducing the premises of geometry to freely chosen abstract axioms in the modern sense, they maintained the traditional idea of their self-evidence.[14]

Riemann's manifolds

In his *Habilitationsvortrag* of 10 June 1854, the Göttingen mathematician Bernhard Riemann criticized the received approach to geometry in the following terms:

> It is known that geometry assumes, as things given, both the notion of space and the first principles of constructions in space. She gives definitions of them which are merely nominal, while the true determinations appear in the form of axioms. The relation of these assumptions remains consequently in darkness; we neither perceive whether and how far their connection is necessary, nor a priori, whether it is possible.

Riemann clearly understood that Euclid's definitions failed to fix the meaning of their objects, and that the axioms (requests and common notions in Euclid's terminology) determined them only as much as was necessary for proving the propositions of geometry. He saw that this lack of a cogent interpretation made it difficult to judge the independence and compatibility of the axioms, as witnessed by the endless debates on the status of the fifth postulate.[15]

---

[13] Cf. Torretti, pp. 53–66; Gray 2007, Chaps. 8–10.

[14] Gauss to Olbers, 28 Apr. 1817, and Gauss to Bessel, 27 Jan. 1829, both cited in Torretti 1978, pp. 55, 53. Gauss's statement shows that he did not believe in the a priori necessity of the laws of mechanics.

[15] Riemann 1867 [1854], p. 133. Cf. Scholz 1980, chap. 2; Torretti 1978, pp. 90–101.

Riemann's solution to this dilemma was to construct geometry from the better defined concept of numerical manifold (*Mannigfaltigkeit*). The idea that points in space can be represented by (three) numbers originates in Descartes's *Géométrie*, in which a plane curve is characterized by an algebraic equation between the segments x and y obtained by projecting the line OP on the axes OX and OY, if O denotes an arbitrary origin and P a point of the curve. This procedure is based on an algebra of geometric lines which is a natural extension of the Greek theory of ratios. In later times, after incommensurable ratios had become commonly accepted as genuine numbers, Descartes's method of coordinates became what we now think it to be: a numerical representation of points in space. This representation, however, remained subordinated to Euclid's geometry and its postulates. Riemann simply inverted the perspective, beginning with the general concept of a manifold, the elements of which are characterized by a finite sequence of numbers called coordinates. He distinguished discrete manifolds in which these numbers are integers, and continuous manifold in which they are what we would now call real numbers. He then introduced a measure on the manifold: the number of points in the discrete case, and an expression for the distance between two infinitely close points in the continuous case. Infinitesimal homogeneity requires this distance to be a homogenous form of first order in the differentials $dx_1, dx_2, \ldots dx_n$ of the coordinates. Although Riemann tolerated any homogenous form, he favored choices for which the square of the elementary distance was the quadratic form

$$ds^2 = \sum_{\mu\nu} g_{\mu\nu} dx_\mu dx_\nu \quad (\mu = 1, 2, \ldots n; \nu = 1, 2, \ldots n). \tag{1}$$

He thus generalized to higher dimensions the theory of curved surfaces that his mentor Gauss had given a few years earlier.[16]

Gauss's great discovery in this field was the *theorema egregium*, according to which the total curvature measured by the products of the two principal curvatures at a given point only depends on the metric coefficients (g_{11}, g_{12}, g_{22}) and not on the specific way in which the surface is immersed in three-dimensional Euclidean space. Any deformation of the surface that preserves the mutual distances of its points (bending without stretching) preserves the value of the Gaussian curvature.[17] This means that there is an intrinsic geometry of the surface, which could be explored by flat inhabitants of this surface. A remark of this sort probably motivated Riemann's new definition of space as a metric manifold.[18]

For a quadratic ds^2, Riemann managed to extend the concept of Gaussian curvature to higher dimensions by focusing on intrinsic geodesic properties. At a given point of the manifold, the geodesic lines passing through this point and tangent to a given (two-dimensional) plane of the (n-dimensional) tangent space form a curved surface whose Gaussian curvature has a well-defined value. As there are $n(n-1)/2$ independent directions for such planes, there are $n(n-1)/2$ independent curvatures in an n-dimensional manifold. For instance, there are three independent curvatures in three-dimensional space.

---

[16]Riemann 1867, pp. 135–41.

[17]For example, a cylinder has no Gaussian curvature since its unrolling onto a plane conserves all distances measured on it.

[18]Gauss 1828.

This connection between curvature(s) in n-dimensional manifolds and Gaussian curvature on ordinary surfaces enabled Riemann to intuitively grasp some properties of higher-dimensional metric manifolds. In particular, the Euclidean requirement that any figure can be transported without altering its metric relations implies that all the curvatures are equal and constant through the manifold. Indeed in the two-dimensional case it is intuitively clear that a flexible, inextensible membrane can slide on a curved surface only if its Gaussian curvature is a constant (the Pope's zucchetto can slide on his head only if is constantly curved).[19]

The simplest subcase of a Riemannian geometry obtains by setting the curvature to zero at every point of space. Then there exists a change of coordinates through which the metric is uniformly brought to the diagonal form

$$\mathrm{d}s^2 = \sum_\mu \mathrm{d}x_\mu{}^2. \tag{2}$$

This is the case of flat, Euclidean space. Constant positive curvature leads to (hyper) spherical geometry; and constant negative curvature leads to Lobachevskian geometry. In this manner, Riemann and his followers not only retrieved the postulates and propositions of the new synthetic geometries, but he also showed the possibility of these postulates. In modern terms, he provided a model for the system of axioms of these geometries.[20]

Riemann did not doubt that the simplest geometry, Euclid's, was the one best suited to the physics of his time. However, he regarded the choice between the different possible metrics of the space manifold as strictly empirical. Since available space measurements were only approximate and limited to moderate scales, space could well have a small constant curvature without contradicting observation. Riemann was also willing to relax the constancy of curvature, in which case curvature could undergo large fluctuations at the atomic scale. Most daringly, he added:[21]

> Still more complicated relations may exist if we no longer suppose the linear element expressible as the square root of a quadric differential. Now it seems that the empirical notions on which the metrical determinations of space are founded, the notion of a solid body and of a ray of light, cease to be valid for the infinitely small. We are therefore quite at liberty to suppose that the metric relations of space in the infinitely small do not conform to the hypotheses of geometry; and we ought in fact to suppose it, if we can thereby obtain a simpler explanation of phenomena.

In brief, Riemann did not believe in the necessity of any geometric postulates; and he did not believe in the necessity of geometry itself. In the order of increasing specificity, Euclidean geometry requires the existence of the continuous manifold of space, the measurability of space, the transportability of figures, and the vanishing of curvature. Riemann regarded all these requirements as "hypotheses" whose validity could be jeopardized in the

---

[19] Riemann 1867, pp. 141–2, 144.

[20] Riemann 1867, pp. 143–7; Beltrami 1868 for the connection with Lobachevskian geometry.

[21] Riemann 1867, pp. 147–50 (citation from p. 149).

physics of the future. Moreover, he did not believe that the metric of space was a purely geometric entity standing on its own:

> The question of the validity of the hypotheses of geometry in the infinitely small is bound up with the question of the ground of the metric relations of space. In this last question, which we may still regard as belonging to the doctrine of space, is found the application of the remark made above; that in a discrete manifold, the ground of its metric relations is given in the notion of it, while in a continuous manifold, this ground must come from outside. Either therefore the reality which underlies space must form a discrete manifold, or we must seek the ground of its metric relations outside it, in binding forces that act upon it.

As a free-thinking mathematician, Riemann found it his duty to denounce "traditional prejudices" on the nature of space and to open the spectrum of possibilities in theoretical physics. Yet he did not recommend a sudden revolution of physics. Rather, he suggested that new experimental findings would gradually lead to revisions of the received Newtonian theories; at the end of this process, geometry, dynamics, and their mutual relations would appear under a radically different light.[22]

Helmholtz's foundations

In the early 1840s, the young Hermann Helmholtz wrote a loosely Kantian essay on the foundations of the natural sciences. Somewhat like Kant, he argued that the theories of time, space, and mechanics were "pure sciences" whose concepts belonged to "general and necessary forms of our conception of nature [*Anschauung der Natur*]." Unlike Kant, however, he did not believe these pure sciences to imply any empirical law:

> The general concepts, which derive from the mere possibility of any conception of nature, need not further restrict the possibility of any empirical combination of perceptions, that is, we cannot derive from them any empirical fact or law; they can only give us a norm of our explanations [*eine Norm für unsere Erklärungen*].

This statement foreshadows Helmholtz's mature view that the comprehensibility of nature, if it is to imply any empirical law, must itself be a partially empirical notion.

The young Helmholtz further departed from Kant by blurring the border between understanding and intuition. This can already be seen from his placing more than Kant's intuition under the word *Anschauung*. Also, in his opinion arithmetic belonged to general logic and did not require any grounding in internal intuition. He was more faithful to Kant in his definition of time and space as necessary conditions for the differentiability of our internal and external sensations. In his view, this differentiability involved continuous manifolds, one-dimensional in the case of time, and multidimensional in the case of space. He thereby privileged the representation of space through a coordinate system. Unlike Riemann, he immediately assumed a metric significance of the space coordinates.[23]

---

[22] Riemann 1867, p. 149.

[23] Helmholtz, MS partially reproduced in Koenigsberger 1902–3, vol. 2, pp. 126–38 (citations from p. 127).

In the 1860s, Helmholtz returned to the problem of space through the physiology of sensations. In the wake of his physiological optics he associated different manifolds with different sensations: ordinary (tactile) space, visual space, and color space. He also reflected on how we infer the existence of external objects and how we develop spatial notions about them. His general approach was kinesthetic, that is, based on our awareness of relations between motor impulses and ensuing sensations. He concluded that geometric notions derived from our ability to obtain and observe the congruence of physical objects.[24]

This remark is the starting point of Helmholtz's important memoir of 1868 on the foundations of geometry. The realization of congruence implies the existence of freely mobile rigid objects. In his mathematical development of this notion, Helmholtz avoided the synthetic approach of traditional geometry, which he believed to be easily contaminated with unjustified intuitions. Instead he started with the general notion of a continuous manifold, which had served him well in his physiology of sensations. Altogether, his geometry rested on four "hypotheses" which may be briefly stated as the following:

    I. Space is a three-dimensional differentiable real-number manifold.
    II. There exist rigid bodies.
    III. The rigid bodies enjoy complete mobility.
    IV. Monodromy: the rotating motion of a rigid body around any given axis must lead back to its original configuration.[25]

In hypothesis I, Helmholtz includes the assumption that a point of space is determined by the measurement (*Abmessung*) of three independent coordinates. It is important to note that this concept of a manifold, unlike Riemann's, implies that coordinates have a metric significance by definition. Helmholtz's favorite example for the coordinates of a point is its distances from three reference points. His definition of the differentiability of the manifold seems circular, for it requires the position of a point to be a differentiable function of its coordinates.[26] In practice, however, he only requires the distance between two points to be a differentiable function of their coordinates.

By hypothesis II, Helmholtz means the existence of (non-trivial) transformations in which a certain function of the coordinates of any two points is preserved. Unlike the usual concept of rigidity, this concept does not imply prior knowledge of the preserved function. Helmholtz's aim is to determine this function from his hypotheses.

Hypothesis III states that any given point can be brought to any other point by such a transformation; that once the image of a first point is fixed, the image of a second point is only constrained by the invariance required in hypothesis II; and so forth. In the case of three-dimensional space, there are three degrees of freedom for choosing the image of a first point, two for choosing the image of a second, one for choosing the image of a

---

[24]Cf. Helmholtz 1856–67, vol.3; Helmholtz 1868a; DiSalle 1993, and further reference there.

[25]Helmholtz 1868b, pp. 618–20. Cf. Torretti 1984, pp. 155–71; Richards 1977; Stein 1977; DiSalle 1993; Hyder 2001, who rightly emphasizes the metrological character of Helmholtz's foundation of geometry and relates it to his earlier work on color theory. Friedrich Ueberweg's alleged anticipation (1851) rested on insufficient axioms for the free mobility of rigid bodies and flawed deductions of Euclid's axioms therefrom: cf. Torretti 1984, pp. 260–4; Schüller 1994.

[26]Klein (1890, p. 374) noted this defect.

third point, and none for choosing the image of any additional point. This implies that the configuration of a rigid body is determined by six parameters.

From these remarks and from some of Helmholtz's own commentary, it is clear that he thus meant to define a continuous group of transformations that acts transitively on the space manifold. His further determination of this group was based on the analysis of its action in an infinitesimal domain of space within which all transformations can be regarded as linear. Anticipating Sophus Lie's theory, he focused on the infinitesimal generators of this group of linear transformations. In particular, he showed that the rotations around a given axis (that is, the transformations that leave two separate points invariant) could be obtained by exponentiation of a single generator. In order to satisfy his axiom of monodromy, he required the eigenvalues of this generator to be purely imaginary. In this case, there exists a choice of coordinates for which the generator is a two dimensional rotation matrix. More generally, Helmholtz proved the existence of a choice (x, y, z) of co-ordinates for which the transformations preserve the quantity $x^2 + y^2 + z^2$. According to hypothesis III, any other invariant must be a function of this quantity. Homogeneity further requires the distance between two points to be given by $\sqrt{x^2 + y^2 + z^2}$, as is the case for the Euclidean distance in Cartesian rectangular coordinates.[27]

This result only holds in infinitesimal domains. It is compatible with any Riemannian metric for the global manifold. Helmholtz next examined the consequences of the free mobility of rigid bodies in finite domains. In agreement with Riemann, he asserted that this mobility implied the constancy of the Gaussian curvature. In this case there are only three sorts of geometry: Euclidean, spherical (with any positive curvature), and Lobachevskian (with any negative curvature). Helmholtz originally overlooked the third possibility, until he became aware of Eugenio Beltrami's contemporary interpretation of Lobachevski's axioms in terms of a Riemannian manifold with constant negative curvature. At any rate, Helmholtz was not willing to regard variable curvature as a viable option for a physical geometry. Nor was he willing to give up monodromy, although he briefly described two-dimensional geometries in which it would not hold, including the Minkowskian variety.[28]

Helmholtz's original aim seems to have been to justify Euclidean geometry on the basis of his four hypotheses and of the additional hypothesis that space is infinite. After becoming aware that these hypotheses did not exclude Lobachevskian geometry, he rather insisted on the impossibility of a complete a priori determination of geometry. Although he believed that the general form of our spatial intuition was a priori determined, he argued against Kant that only experience could decide the precise axioms of geometry. He thereby noted that this decision depended on the accepted theory of mechanics: "The geometrical axioms do not concern spatial relations *per se*, they also involve the mechanical behavior of our most rigid bodies when moved." In order that geometry truly bears on experience, Helmholtz explained, it needs to be completed by mechanical principles such as the principle of inertia or the uniformity of mechanical laws.[29]

---

[27] More details will be given below, pp. 126–9.

[28] Helmholtz 1868b, pp. 635–9; 1868a, p. 617 (on Beltrami); Beltrami 1868. Helmholtz read Riemann after he had developed his own view: cf. Helmholtz 1868a, p. 611.

[29] Helmholtz 1870, p. 30.

Altogether, Helmholtz's philosophy of geometry was far more conservative than Riemann's. Unlike Riemann, Helmholtz regarded the congruence of figures as a basic requirement for any geometry. He saw that this requirement implied the locally Euclidean character of space, as well as its constant curvature when applied through the entire space. He therefore excluded non-quadratic metrics and variable curvature. Since he further required space to be infinite, his geometry was necessarily Euclidean or Lobachevskian. Consequently, he agreed with Bolyai and Lobachevski that all of Euclid's postulates were necessary, except the fifth. The implied kind of necessity nonetheless differed: From the ancient Greeks to Bolyai, it had been a sort of self-evidence; for Helmholtz it resulted from the existence of freely mobile rigid bodies in nature.

After Helmholtz

Helmholtz's demonstration of the locally Euclidean character of space attracted the attention of Felix Klein, who advised his Norwegian friend Sophus Lie to work it out with continuous-group methods. Either directly or through Lie's interpretation, it probably was the main incentive for Henri Poincaré's definition of geometry through the group of displacements of rigid bodies. In his own distorted Kantianism, Poincaré regarded group theory as a necessary form of the understanding, pre-existing in our mind at least potentially. Groups indeed have to do with our ability to imagine the composition, inversion, and the indefinite repetition of operations in our mind, which is a necessity for extracting regularities from the complex flux of our internal and external sensations. Poincaré therefore regarded the general concept of measurable space as transcendental. In this regard, his position was similar to Helmholtz's insofar as Helmholtz's "fact" (*Thatsache*) of the existence of freely mobile rigid bodies could be interpreted as a requirement of our understanding.[30]

At the same time, Poincaré rejected the idea that experience allowed us to select a specific group among all the groups compatible with the general concept of measurable space. In his view, no such decision was possible because the choice of a class of rigid bodies was purely conventional. As Helmholtz had himself noted, our judgment of the rigidity of a body necessarily depends on the laws of mechanics. Experimentally, we can only test the consequences of the conjunction of the axioms of geometry and of the laws of mechanics. For a different set of axioms, these consequences can be preserved if the laws of mechanics are modified accordingly. From this fact, Poincaré inferred that the choice of the geometric group was purely conventional, although he allowed considerations of convenience to restrict this choice. It was convenient to regard as approximately rigid those bodies whose concrete realization was easy; in case of future contradictions between this choice and the laws of mechanics, it would be convenient to keep the geometry that this earlier choice had favored; and it was always more convenient to define rigid bodies so that the geometric group be very simple. As Euclidean geometry satisfied these three criteria of convenience, Poincaré did not doubt that it would forever remain the favorite geometry of physicists. In contrast, Helmholtz seems to have been unwilling to modify the most fundamental laws

---

[30] Klein 1890, p. 375n; Lie 1886; 1890, p. 416n (Klein's stimulus); Poincaré 1887 (refers to Lie); 1891 (refers to Helmholtz); 1902, p. 90 (on groups). Cf. Torretti 1984, 171–9 (on Lie); Carrier 1994; Heinzmann 2001.

of mechanics. In his replies to Kantian critics, he insisted on the reliability of empirical tests of congruence and on the resulting empirical determinability of the axioms of geometry.[31]

Morals

A first kind of necessity for the premises of geometry is the self-evidence partially assumed by Euclid and globally accepted by his followers. Although Euclid's fifth postulate lacked the desired evidence, geometers long hoped it would someday be derived from the other postulates. Alas, self-evidence is a vague and dangerous notion. The Euclidean dogma vacillated when Gauss, Bolyai, and Lobachevski imagined geometries in which the fifth postulate no longer held. It completely collapsed when Riemann, Helmholtz, and Poincaré recast geometry through the concept of metric manifold. In this approach, it becomes clear that Euclid's "common notions" are only a way of expressing the possibility of congruence, that lines and planes can be defined by congruence, and that the postulates derive from unlimited congruence, with spherical and Lobachevskian variants. Then it becomes possible to discuss what is necessary and what is not in geometry, with a precise definition of necessity bound to measurability. From a mathematical point of view, measurability (congruence) is defined *in abstracto* and we are not even bound to confine the definition of geometry to a theory in which congruence is possible. From a physical point of view, the laws of congruence-based geometry are necessary insofar as classes of rigid bodies can be concretely realized.

Riemann, Helmholtz, and Poincaré all agreed that ideal and unlimited congruence required space to be a locally Euclidean manifold of constant curvature. For Riemann, this congruence was only a hypothesis that could very well cease to hold at very small or very large scales; Helmholtz regarded it as a scale-independent fact empirically established at moderate scales; Poincaré regarded it as a rule of understanding. None of them believed in the necessity of a specific choice of the curvature. For Riemann and Helmholtz, this choice was a matter of empirical adequacy; for Poincaré, it was a matter of convention.

4.2 Improved foundations

I will now attempt a general construction of space based on Helmholtz's insight that the measurability of space implies the locally Euclidean character of space. In Helmholtz's original memoirs this insight is accompanied with more arbitrary considerations which have led many commentators to doubt its validity and generality. I list these weaknesses before developing an improved Helmholtzian approach to space.[32]

Helmholtz criticized

Helmholtz asserts that the measurement of space necessarily implies the assumption of freely mobile rigid bodies. In his memoir of 1868, he provides no justification for this claim.

---

[31] Poincaré 1891; Helmholtz 1870, p. 29; 1878. Cf. e.g., Friedman 1993, chap. 7; Paty 1993, pp. 250–63. In the 1920s, Hans Reichenbach extended conventionalism to the Riemannian concept of space, to time, and to spacetime (Reichenbach 1928).

[32] Cf. e.g., Paul Hertz and Moritz Schlick's commentary in Helmholtz 1921; Torretti 1978, pp. 157–8, 167–8; Ryckman 2005, 70–1.

Any of his readers could understand that the measurement of lengths and distances boils down to the transport and congruence of a rigid rod. But why require rigid *bodies*?

For the purpose of distance measurement, a linear, infinitely thin rod is sufficient. The rigidity and free-mobility of such a rod does not entail any restriction on the form of the distance element d*s*. According to Riemann, this element could be any homogenous function of the differentials of the coordinates. In the resulting geometry, freely mobile rigid linear bodies can be characterized as bodies that always fit portions of geodesic lines of constant length.[33]

The multidimensionality of Helmholtz's rigid bodies thus appears to be essential to the derivation of the locally Euclidean character of space. In a lecture of 1870, Helmholtz makes clear that by space measurement he not only means measurement of lengths, but also measurements of surfaces, volumes, and angles, which clearly require multi-dimensional standards. This does not answer the following question: Can the free mobility of *any* rigid body be justified without leaving the context of length and distance measurements?[34]

Another defect of Helmholtz's approach is that he assumes a concept of *infinitesimal* rigid body for which free mobility only implies the Riemannian structure of the space manifold, without any restriction on the curvature. Yet Helmholtz ultimately requires the free mobility of *finite* rigid bodies. Perhaps he believed that the former assumption implied the latter if a finite rigid body could be decomposed into an infinite number of infinitesimal rigid bodies. Or perhaps he was just following a Kantian requirement of homogeneity in our determinations of space. This leaves the following question open: Is there a consistent concept of freely mobile infinitesimal rigid body that does not imply the free mobility of finite rigid bodies? A positive answer to this question seems essential in order to save the Helmholtzian justification of the quadratic character of the metric in geometries of variable curvature.

A third defect of Helmholtz's approach is that he regards the coordinates of the real-number manifold as the results of surveys performed with freely mobile rigid bodies and yet defines rigidity in terms of the preservation of a certain function of the coordinates of two points of the body. He seems to be caught in a circle, for he cannot define coordinates without rigid bodies and he cannot define rigid bodies without coordinates. This raises the following question: Is there a definition of rigid bodies that does not require a prior concept of distance? Such a definition would avoid a paradox mentioned at the beginning of Helmholtz's memoir of 1868, namely that "we decide whether a body is rigid . . . only by means of the very laws [of geometry] whose factual truth should be established by experience." Helmholtz believed to have circumvented this difficulty by not using a specific expression of the distance of two points in his definition of rigid bodies. In reality, his definition of rigidity still eludes experimental testing, since it implies coordinate measurements done by means of rigid bodies.[35]

A more technical defect of Helmholtz's argument concerns the monodromy hypothesis. As Sophus Lie proved in 1886, non-monodromous geometries only exist in the

[33]Riemann 1867, pp. 138–40. These are the modern "Finsler geometries," named after Paul Finsler.

[34]Helmholtz 1870, pp. 19–20.   [35]Helmholtz 1868b, p. 619.

two-dimensional case. In three dimensions, Helmholtz's other axioms suffice to eliminate such geometries. Helmholtz was "very happy" to learn this from Felix Klein in a conversation held in 1893.[36]

Rigid rods

Following Helmholtz, I assume that geometry is about measuring distances by means of some gauge. For instance, we may count the minimum number of steps needed to go from one point to another; or, better, we may do the same with a rigid rod. The success and non-ambiguity of this procedure entails the following assumptions for the class of rigid rods:

(1) For any two rods, if an extremity of the first rod is kept in contact with an extremity of the second, the other extremity of the first rod can be brought in contact with at most one point of the second (no plasticity or elasticity).

(2) If coincidence can be obtained in one place and at one time between a pair of points of one rod and a pair of points of the other, this coincidence will be possible at any other place and time, no matter how variously and differently the two rods have traveled before meeting again (free mobility and stability).

This definition, unlike Helmholtz's definition of rigid bodies, does not entail any concept of distance. It permits a direct empirical test of rigidity and free mobility. Of course, there are infinitely many classes of rigid rods according to this definition. Rods obtained by subjecting the rods of a given class to a dilation that depends only on their location will form another class of rigid rods, no matter what the dilation law is. For instance, in a thermostatic universe with heterogeneous temperature, iron rods and copper rods form two distinct classes of approximately rigid rods.[37]

Although the above definition of rigidity is not to be found in Helmholtz's writings, it fits very well with Helmholtz's concept of measurement as expressed in his "Counting and measuring" of 1887. In this essay Helmholtz based measurement on the existence of concrete operations of comparison and composition that share the basic properties of arithmetic equality and addition. In particular, the concrete comparison must be transitive. As Helmholtz explains, in the case of distances this means that if two pairs of material points are congruent to a third, they must be congruent to each other, no matter when and where the two first congruences obtained. This condition trivially holds if there is no material connection between the points of a pair. When applied to a class of rods, it implies the rigidity of the rods.[38]

---

[36] Lie 1890; Klein 1890, p. 374n. Lie also noted Helmholtz's lack of rigorous justification of his use of a first-order expression of the free mobility of infinitesimal rigid bodies. Cf. Merker 2000.

[37] Reichenbach (1928, Chap. 1, §5) defines rigid bodies as solid-state bodies that are not affected by differential forces (forces that produce different effects on different materials). This definition is too weak to warrant definite length measurements, since it does not exclude flexible rods for instance. Moreover, the insensitivity to differential forces is unnecessary if these forces are kept constant, and it results from my definition if these forces are variable.

[38] Helmholtz 1887, pp. 378–9, 390. Cf. Darrigol 2003, and Chapter 6, pp. 194–6 below.

This implication can be understood by the following *ad absurdum* reasoning. Suppose the rods have some flexibility. Intuitively, this means that the extremities of a rod A can be constrained to fit the extremities of rods of different length, though not with those of any rod. Call B a rod of the minimal length, and C a rod of the maximal length for which congruence with A is still possible. The rods B and C cannot be mutually congruent, because their relative difference of length exceeds the limit beyond which their extremities can no longer be constrained to fit.

This reasoning is not completely adequate, for it relies on predefined notions of length and flexibility. Nevertheless, it strongly suggests that Helmholtz's requirement of transitivity is equivalent to my more natural conditions (1) and (2) of rigidity and free mobility. Helmholtz very much insisted that his requirement was an empirical criterion for selecting a class of rigid rods or compasses that permit distance measurement.

Once congruence has been defined by means of a class of rigid rods, the distance between two points (that is, two small objects) can be measured by means of chains of unit rods. At the precision of the unit, the distance is given by the minimal number of links of a chain joining the two points. This distance measurement can be refined by using smaller and smaller unit rods. In common practice, a sequence of subunits is used such that the lengths of two consecutive subunits differ by a factor ten (for instance). The outcome of the measurement is a decimal number whose last digit corresponds to the last subunit whose extremities can be distinguished.

So far, we have considered concrete objects and operations that can be realized in an approximate manner only. We may now leave the empirical world and take our flight to a mathematical set-theoretical world in which the properties (1) and (2) of rigid rods hold exactly, and the sequence of subunits can be pursued indefinitely. The usual sets of natural, rational, and real numbers can thus be engendered in harmony with the geometer's needs. Whether geometry truly motivated the historical introduction of these mathematical constructs, it is important to recognize that any theory of measurement requires these constructs or similar non-standard ones.[39]

The metric manifold

We now know how to measure distances with arbitrary precision. Suppose there exist three points A, B, and C whose mutual distances are found to be invariable. We know by experience that except for singular cases the location of any fourth point within a sufficiently small domain is determined by its distance from these three points. Moreover, the distance of a variable point M from a fixed point O varies linearly under small increments of its coordinates, except when M is originally at O. In the latter case, the variation cannot be linear since this would allow the distance between M and O to vanish without their coordinates

---

[39] As the ancient Greeks already knew, Pythagoras' theorem implies the existence of irrational ratios. This means that the idealization of arbitrarily precise measurement and the idealization of freely mobile rigid bodies can only be compatible if the results of measurement include irrational numbers. Indeed the former idealization makes any ratio a number, and the latter implies Pythagoras' theorem for infinitesimal rectangular triangles. I adopt the modern, Newtonian identification of ratios with numbers, although Helmholtz rather avoided it (1887, p. 385). On the history of this identification, cf. Roche 1998, and Chapter 6 below.

being equal. This variation must nonetheless be a homogenous function of first degree of the coordinate increments, because for a sufficiently small unit of length, a (reasonable) unit change implies a multiplication of all measured distances by the same constant.

These conclusions are only valid to a certain approximation, given by the precision of the distance measurement. Again we may jump to the ideal, mathematical level in which coordinates are sharply defined as real numbers. At this level, the distance OM should be a differentiable function of OA, OB, and OC whenever M differs from O. This implies the differentiability of any change of coordinates resulting from a different choice of the reference points A, B, and C. The resulting mathematical concept is that of a differentiable manifold in Riemann's original sense, namely, endowed with a metric that is not necessarily of the locally Euclidean form.

In order to further restrict the form of the metric, we need some additional condition. By experience we know that the position of a point M with respect to three rigidly connected reference points A, B, and C is completely determined by its distance from these three points. This implies that the distance between two points M′ and M″ is a function of their distances from the reference points. This function of course depends on the choice of the reference points. However, by experience we know that it only depends on the mutual distances of these reference points, as long as none of the involved distances is exceedingly large. This fact can be regarded as a precise expression of the homogeneity of space over moderate distances, since it means that the same relations between all measured distances can be used in surveys performed in a not too large domain.

This local homogeneity implies the existence of rigid bodies in Helmholtz's sense. Indeed it implies that the distances between any number of points remain the same when their distances to three points A, B, C and the distances between these three points are kept constant. In other words, there exist transformations that preserve the mutual distances of any number of points. Free mobility (Helmholtz's hypothesis III) also holds, since the choice of three new reference points A′, B′, C′ such that AB = A′B′, AC = A′C′, BC = B′C′ involves six degrees of freedom (nine coordinates minus three constraints).

Two demonstrations will now be given of the fact that the free mobility of rigid bodies implies the locally Euclidean character of space: one based on elementary geometrical considerations, the other on Helmholtz's (implicit) exploitation of the Lie-group structure. For the moment, the number of dimensions is limited to two.

Pythagoras' theorem

The assumption of freely mobile rigid bodies allows us to define an angle as a rigid connection of two straight segments of arbitrary length with a common extremity. The addition or difference of two angles is defined by making these two angles share one of their sides, and taking the angle made by the two remaining sides. The straight angle is the angle that makes a flat angle (a single straight line) when added to itself.

Call d the length of the hypotenuse of a rectangular triangle, x the length of one of its shorter sides, and α the (positive or negative) angle between this side and the hypotenuse. The intersection between two lines being unique (at least locally), and there being only one line perpendicular to a given line through a given point, the length d is a function of α

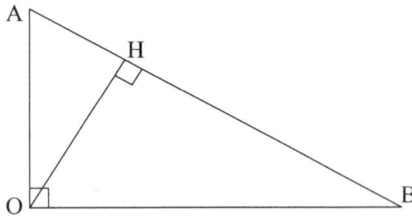

Fig. 4.3.   Figure for a proof of Pythagoras' theorem.

and x only. This function vanishes for $x = 0$, and it must be differentiable with respect to x because the space manifold is differentiable. Therefore, d is a linear function of x for small triangles. This means that the ratio between one of the shorter sides of a small rectangular triangle and its hypotenuse is completely determined by the angle that they make.

Owing to the free mobility of rigid bodies, this ratio cannot be altered by rigid displacement of the triangle. It may nonetheless depend on the sign of the angle α, because in two dimensions no displacement (translation or rotation) can turn a left-handed triangle into a right-handed one. In order to avoid this ambiguity, we temporarily immerse our two-dimensional space into a three-dimensional one, so that a flat-angle rotation of a rectangular triangle around one of its shorter sides can invert the sign of the angle it makes with the hypotenuse. If we remained strictly in a two-dimensional context, Pythagoras' theorem would no longer result from the free mobility of rigid bodies; the geometry would not need to be locally Euclidean.[40]

Now take a look at Fig. 4.3. The former theorem implies AH/OA = OA/AB as well as BH/OB = OB/AB. Since AB = AH+BH, we have the relation

$$AB^2 = OA^2 + OB^2, \tag{3}$$

which is Pythagoras' theorem. As is well known, Euclid's premises support the same proof. The reason is that his "common notions" contain the assumption of freely mobile rigid bodies, and his postulates turn the local validity of the theorem into a global one.[41]

For a deeper insight into the consequences of the assumption of freely mobile rigid bodies, we may follow Helmholtz and investigate the group of space transformations (isometries) that leave the distances between any two pairs of points invariant. The local homogeneity and the differentiability of the space manifold imply that the distance between two points of a small neighborhood is a homogenous function of the difference of their coordinates only. Therefore, a uniform (small) increment of the coordinates for each point of space leaves the mutual distances invariant. This special kind of isometry defines the group of (infinitesimal) translations of a rigid body.

Now consider the (positive) isometries that leave a given point of space invariant. This group only has one parameter, for instance the distance traveled by one of the displaced points (P). Indeed, the position of P is determined by this distance and its invariant distance

---

[40]This is why Helmholtz needs monodromy in his own treatment of the two-dimensional case.

[41]Helmholtz (1870, p. 7) remarked that Euclidean geometrical proofs were all based on congruence.

from the origin; and the position of any other point is determined by its invariant distance from the origin and its distance from P (for small enough transformations). In a given system of coordinates and in a small domain, the infinitesimal generator of this group is a 2×2 matrix. Call λ and μ the eigenvalues of this matrix. In the degenerate case $\lambda = \mu$, there is a choice (u, v) of coordinates for which the resulting transformations (obtained by exponentiation) have the form

$$u' = u + \alpha v, \quad v' = v. \tag{4}$$

This should be rejected, because these transformations can produce an arbitrary large increase of coordinates in conflict with Helmholtz's hypothesis IV.

In the non-degenerate case, the exclusion of arbitrary large increase of coordinates implies that the real parts of λ and μ vanish.[42] The real character of the generator further implies that $\lambda + \mu$ is real.[43] Consequently, there is a complex choice of coordinates (z, z^*) such that the generator has the form

$$z' = iz, \quad z'^* = -iz^*. \tag{5}$$

Exponentiation leads to the transformations

$$z' = e^{i\theta} z, \quad z'^* = e^{-i\theta} z^*, \tag{6}$$

wherein θ is the parameter. In terms of the real coordinates (x, y) such that $z = x + iy$ and $z^* = x - iy$, this gives the transformations

$$x' = x \cos\theta - y \sin\theta, \quad y' = y \sin\theta + x \cos\theta, \tag{7}$$

which leave the quadratic form $x^2 + y^2$ invariant. Consequently, the distance of a point from the origin must be a function of $x^2 + y^2$. Homogeneity further restricts this distance to the Euclidean form $\sqrt{x^2 + y^2}$. The parameter θ is easily seen to correspond to an angle in the above-defined sense, the lines $x = 0$ and $y = 0$ to two orthogonal axes, and the coordinates x and y of a point to the distances of this point from these two axes.

In three dimensions

In the three-dimensional case, Helmholtz considers the subgroup of transformations that leave two points (or a line) invariant. By hypothesis II, this subgroup is a one-parameter Lie group. By an argument similar to the one given in two dimensions, there is a generator

---

[42] The moduli $|e^\lambda|$ and $|e^\mu|$ of the exponentials of the eigenvalues must be 1, so that real powers of them cannot take arbitrarily large values.

[43] This is so because the trace is the same in any basis and must be real in a real basis.

of this subgroup with the eigenvalues $0, i, -i$, and there is a choice of real axes X, Y, Z for which it is represented by the matrix

$$G_Z = \begin{pmatrix} 0 & -1 & 0 \\ 1 & 0 & 0 \\ 0 & 0 & 0 \end{pmatrix}. \tag{8}$$

The generators G_X and G_Y of the transformations that leave the X and Y axes invariant have the form

$$G_X = \begin{pmatrix} 0 & a & d \\ 0 & b & e \\ 0 & c & -b \end{pmatrix}, \quad G_Y = \begin{pmatrix} a' & 0 & d' \\ b' & 0 & e' \\ c' & 0 & -a' \end{pmatrix} \tag{9}$$

since their trace must vanish in order that their eigenvalues be $0, i, -i$. Moreover, the existence of an invariant axis for the transformations generated by linear combinations of the three generators G_X, G_Y, and G_Z implies the vanishing of the determinant of any such combination. For the combination

$$\alpha G_Z + G_X = \begin{pmatrix} 0 & a+\alpha & d \\ -\alpha & b & e \\ 0 & c & -b \end{pmatrix}, \tag{10}$$

this condition gives

$$b(a+\alpha) + cd = 0, \tag{11}$$

whose validity for every α requires that $b = 0$ and $cd = 0$. The coefficient c cannot vanish, because otherwise G_X^3 would vanish and transformation generated by exponentiation of G_X would not meet the monodromy condition. Therefore, the coefficient d must vanish. The same argument for the combination $\alpha G_Z + G_Y$ leads to $a' = 0$ and $e' = 0$. Then the vanishing of the determinant of the combination

$$\alpha G_Z + \beta G_X + G_Y = \begin{pmatrix} 0 & \alpha + \beta a & d' \\ -\alpha + b' & 0 & \beta e \\ c' & \beta c & 0 \end{pmatrix} \tag{12}$$

leads to the condition

$$c'e(\alpha + \beta a) + cd'(b' - \alpha) = 0, \tag{13}$$

whose validity for any choice of α and β requires that $aec' = 0$, $cb'd' = 0$, and $c'e = cd'$.

For the sake of monodromy, e and d' cannot vanish. Therefore, a and b' must vanish. Using the condition $c'e = cd'$ and the positivity of $-e/c$ (needed for imaginary eigenvalues), we may rescale the Z axis by $\sqrt{-e/c}$ in order to reach the expressions

$$G_X = \begin{pmatrix} 0 & 0 & 0 \\ 0 & 0 & -1 \\ 0 & 1 & 0 \end{pmatrix}, \ G_Y = \begin{pmatrix} 0 & 0 & -1 \\ 0 & 0 & 0 \\ 1 & 0 & 0 \end{pmatrix}, \ G_Z = \begin{pmatrix} 0 & -1 & 0 \\ 1 & 0 & 0 \\ 0 & 0 & 0 \end{pmatrix} \qquad (14)$$

of the three generators. We recognize the expression of the generators of the rotations around three orthogonal axes in an orthonormal basis, for which we know the quadratic form $x^2 + y^2 + z^2$ to be invariant. This ends Helmholtz's proof of the locally Euclidean character of a manifold measurable by freely mobile rigid bodies.

Curved space

So far we have confined our investigation of the structure of space to domains of infinitesimal dimensions. Helmholtz's hypothesis of freely mobile rigid bodies, even limited to small bodies, further restricts the metric manifold to have constant intrinsic curvature as long as the rigidity is meant to be perfect. The derivation of the locally Euclidean character of space does not require perfect rigidity, however. It only requires the existence of transformations that do not change the distances between any given set of points up to corrections that are of second order with respect to the largest of these distances.

Relaxing the rigidity of bodies amounts to relaxing the local homogeneity that was earlier assumed. This can be done without giving up the perfect rigidity of linear rods. Even with imperfect rods, indefinitely precise distance measurement would remain possible if the error in appreciating the congruence of rods was of second order with respect to the length of the rod. Indeed the cumulated error would vanish in the limit of a vanishing rod length, as happens in the rectification of a circle by a polygon with a growing number of sides. There would remain the possibility that after traveling on a large loop a rod returns with a different length, so that length measurement would be path-dependent. This violation of the criterion (2) of rigidity runs against the basic concept of measurement, which implies unrestricted transitivity of the concrete operation of comparison of two quantities. Riemann and Helmholtz both excluded it. Hermann Weyl welcomed it, as we will see in the next chapter.

As long as we exclude this exotic possibility, the surveying of space by means of small rigid rods leads to a Riemannian manifold with a positive, definite local metric

$$\mathrm{d}s^2 = \sum_{\alpha\beta} g_{\alpha\beta}\mathrm{d}x_\alpha\mathrm{d}x_\beta \quad (\alpha = 1, 2, 3 \, ; \beta = 1, 2, 3). \qquad (15)$$

For consistency with the Helmholtzian justification of the locally Euclidean character, the Riemannian structure should be compatible with the local existence of freely mobile approximately rigid bodies. This is indeed the case because, as Riemann asserted in his habilitation lecture, for domains small compared to the local curvature there is a choice of coordinates for which the metric tensor $g_{\alpha\beta}$ is approximately uniform: the Euclidean character of space extends throughout any such small domain.[44] For larger domains, there is in general no reduction of this kind; homogeneity and similarity no longer apply to large figures. In a spherical geometry, two straight lines always intersect, and the sum of the angles of large triangles exceeds $180°$.[45]

4.3   Conventions and necessities

The measured curvature of space, which determines the extent to which the geometry departs from Euclid's, depends on the conventional choice of the class of rigid rods. Suppose that the geometry of a two-dimensional space is found to be spherical for one class of rods. We may fancy another convention according to which a rigid rod is a rod whose projection on a Mercator map of the sphere is rigid in the former convention. The geometry would be Euclidean (except at the poles) with respect to the new class of rods. Reciprocally, a Euclidean space can be mapped into a spherical one by changing the class of rigid rods. In practice we renounce this freedom, because we want rigid rods that can practically be used for surveying purpose. Natural candidates are rods made of any hard material, as long as we neglect physical effects that may destroy the congruence between two different rods made of different materials.

For precise length measurements, such effects are no longer negligible. Imagine a world in which each place has its own constant temperature. Owing to thermal dilation, there are as many classes of rigid rods (according to the earlier given definition) as there are materials with different dilation coefficients. For a given temperature distribution over space, the metric depends on the kind of rod used. There is a wood-based geometry and an iron-based geometry, for instance. As physicists do not want a temperature-dependent geometry, they make their rods out of the least dilatable material (invar); better, they keep them at a standard temperature. In more advanced physics, they may think of using the wavelength of a spectral line as a standard, but they only do so as long as this choice does not interfere with the global simplicity of the laws of physics. In the end, it is the theory which tells us how to select the class of rigid rods. Yet it still is the idealized concept of length measurement and the associated criteria (1) and (2) that tell us what rigidity is about: the congruence of rods of the same class must be transitive.

The conventionality of the metric structure is often said to result from the following circle: On the one hand, any experimental test of the rigidity of bodies must be based on the laws of geometry; on the other, the empirical determination of these laws requires bodies that are known to be rigid. The Helmholtzian approach reveals the partial fallacy of this argument. It is not true that *any* experimental test of the rigidity of bodies must be based on the laws of geometry. Helmholtz's transitivity criterion or my criteria (1) and (2) restrict the admissible classes of rigid bodies without appealing to these laws. What depends on the laws of geometry is the choice among the various admissible classes. Those

---

[44] In the geodesic central coordinates that Riemann introduced in his habilitation lecture, the ds^2 at the distance D from the center differs from its central value by terms of the order $(D/R)^2$, where $1/R^2$ denotes the Gaussian curvature (Riemann's more precise statement involves the Riemann tensor). Cf. Riemann 1867, p. 141.

[45] Torsion is excluded as long as we require all geometric properties, including the affine connection, to be uniquely determined by the metric.

who think this is only a minor corrective should consider the fact that the rigidity criteria and the local homogeneity and differentiability of space suffice to establish a very essential result: the locally Euclidean character of space.

Another conventional aspect of space derives from the choice of a reference system. In a given neighborhood, points of space are defined by their distances from three rigidly connected reference points. We may switch to three other reference points that are rigidly connected to the former points without truly changing the geometry. This change amounts to a relabeling of the points of space, which can be identified to the points of an extended rigid body to which the reference points belong. In a more significant change, we can refer the points of space to reference points that do not belong to this rigid body. Then there are as many spaces as there are independent rigid bodies.[46]

Despite the conventionalist aspects of its definition, physical space must have the locally Euclidean structure in order to be measurable. This necessity is only relative: it hinges on the empirical realizability of rigid bodies. Also, the importance of rigid bodies in Helmholtz's argument does not imply that they should be regarded as primitive objects in the global theory of space and matter. Rather, we should require any present or future theory of the constitution of rigid bodies to be compatible with the aforementioned rigidity criteria. This will happen, for instance, in a classical (Laplacian) molecular theory in which the forces responsible for the rigidity are Euclidean invariant central forces. All we can say is that the measurability of space, as long as it is empirically vindicated, has strong structural consequences on the structure of space and on the dynamics of processes occurring in space. There is no need to regard measuring rods as ontologically more primitive than the dynamics that allows their construction.

Conclusions

The Helmholtzian approach to space starts with the existence of a class of (small) freely mobile rigid *rods* that leads to a consistent surveying technique. Although the choice of this class is largely conventional, it must meet certain empirical conditions: my criteria (1) and (2) or Helmholtz's transitivity criterion. Together with the observed smoothness and local homogeneity of the space manifold, the existence of freely mobile rigid rods leads to the existence of freely mobile, approximately rigid *bodies*, from which the Riemannian character of physical geometry follows. The further determination of this geometry depends on the class of rigid bodies that has been conventionally adopted. For a given convention, systematic surveys determine the metric of the Riemannian manifold.

This approach involves a gradual sharpening of elementary spatial notions. Ultimately, it leads to a highly idealized theory in which the thickness of material points, errors in the appreciation of congruence, the imperfect rigidity of rods, and the rationality of the numbers yielded by actual measurements are ignored to permit exact relations and deductions. The outcome can be considered as a purely mathematical theory, as an abstract set-theoretical construct whose properties no longer depend on experience.

---

[46] For a given class of rigid rods, the existence of a rigid (reference) body is an empirical question. From everyday experience we know that such bodies approximately exist at our scale. In general relativity, extended rigid bodies exist only in exceptional cases.

The empirical motivation of this idealized geometry largely explains its success as a physical theory. The most important assumption, that of freely mobile rigid bodies, depends on a kind of invariance assessed in our experience of the world: the fact that congruence among a certain type of objects is constrained and reproducible. In concrete form, this assumption is the basis of our physical concept of space. In idealized form, it is the basis of the Riemannian concept of space. Although mathematicians are free to imagine more general concepts of space, the locally Euclidean property is essential to the measurability of physical space. This finding, important though it is, leaves several questions open. Does measurability through rigid bodies extend to arbitrarily small scales? Does the notion of a rigid reference body extend to arbitrarily large scales? What is the cause of the space metric? These questions will be addressed in the next chapter, in the broader context of relativistic spacetime.

5

SPACETIME

Everything real in the world is a manifestation of the world metric; the physical
concepts are none other than those of geometry.[1] (Hermann Weyl, 1918)

The whole of those laws of nature which have been woven into a unified scheme—
mechanics, gravitation, electrodynamics and optics—have their origin, not in
any special mechanism of nature, but in the workings of the mind.[2] (Arthur
Eddinton, 1920)

From Newton to Einstein, fundamental physics has moved from a concept of absolute
time to a concept of spacetime in which time and space are deeply interrelated. This history, which is reviewed in the first section of this chapter, shares at least one feature of
the history of space: it recurrently involved attempts at showing the necessity of some basic conceptual premises. For instance, Kant identified Newtonian time with an a priori
form of our inner intuition, in parallel with the outer intuition of space; Woldermar von
Ignatowski derived the Minkowskian structure of spacetime from the Lie-group structure
of changes of reference frames, just as Helmholtz derived the (locally) Euclidean structure of space from the Lie-group structure of the displacements of rigid bodies; or Alfred
Robb derived this structure from a partial ordering of events. In most cases, the rational
arguments faced criticism, and strategies were deployed to protect or modify them.

Some of the most vertiginous attempts at a purely rational foundation of physics occurred in the wake of Einstein's general relativity. They were Hermann Weyl's unified
theory of gravitation and electricity, and Arthur Eddington's subsequent world geometry.
These are recounted and commented in the second section of this chapter. Although few
physicists today would share the varieties of idealism defended by these luminaries, their
deductive flights had important byproducts, including modern gauge theories and an ingenious argument by Jürgen Ehlers, Felix Pirani, and Alfred Schild (the EPS argument)
for deducing the Weylian structure of spacetime from the incidence relations of light rays
and free falling particles. This argument ends the second section of this chapter.

The third and last section begins with my own attempt to extend the Helmholtzian derivation of the structure of space to the structure of spacetime. I argue that rigid bodies
and invariable clocks remain locally conceivable in a relativistic context, and that they can
be used to construct the concepts of local rigid frames, local inertial frame, inertial time,
and to derive the locally Minkowskian structure of spacetime. This construction has several advantages over the EPS argument: it is more directly related to the historical genesis
of relativity, it is more gradual and more elementary, and it harbors a concept of rigid

---

[1] Weyl 1918b, p. 2.   [2] Eddington 1920, p. 180.

bodies similar to the local frames that are commonly used in modern general relativity. This and other rational justifications of spacetime structure help understand this structure and answer classical philosophical questions about it. To illustrate this point, I briefly address the distinction between operations and idealizations, the small-scale and large-scale extrapolations, the conventionality of the metric, relationism versus absolutism or substantivalism, and the so-called hole argument. Lastly, I discuss the varieties of rationalism implied in the history of time and spacetime from Newton to the present, and conclude that the Helmholtzian variety remains a forceful justification of the pseudo-Riemannian structure of spacetime.

5.1 From time to spacetime

Time and times

The earliest thinking men must have been aware of an ordered succession of events in their consciousness, and they must have been able to count years, months, and days between two events. Thus they had a notion of time as a shared way of ordering events according to a growing sequence of numbers. This primitive experience defines time as a completely ordered continuum, since there is no apparent limit to the fineness of the ordering. As the ordering, in its quantitative version, relies on commonly witnessed periodic events such as the succession of days and nights, it is entirely objective. In other words, all men share or can be made to share the same time.

In agreement with this view, Aristotle defined time as an attribute of motion (in the broad sense of a continuous succession of states) shared by all motions, continuous, and expressible by number. In his *Principia*, Newton judged it unnecessary to define time, for it belonged to the elementary notions on which everyone agreed. In a *scholium* to his definition of dynamical concepts, he nonetheless specified:

> Absolute, True, and Mathematical Time, of itself, and from its own nature flows equably without regard to any thing external, and by another name is called Duration: Relative, Apparent, and Common Time is some sensible and external (whether accurate or unequable) measure of Duration by the means of motion, which is commonly used instead of True time; such as an Hour, a Day, a Month, a Year.

This distinction between absolute and relative time is quite necessary, because in Newton's mechanics the traditional ways of measuring time, through the motion of celestial bodies or through the oscillations of a pendulum, cannot be regarded as exact. Absolute time and space must be introduced in an abstract manner as the time and space for which Newton's laws hold, and it is only after deriving the consequences of these laws for the observable universe that one knows how to measure time.[3]

Newton's absolute time is clear in two respects: it has a definite mathematical representation as a succession of points on a line (the time axis); and its operational meaning is indirectly determined by the laws of motion. This probably explains why it went nearly

---

[3] Aristotle, *Physics*, IV, 10–14; Newton 1729, p. 9. On Aristotle, cf. Coope 2005.

uncriticized for about two centuries. It is not clear, however, what kind of concept Newton meant time to be. He rejected the direct empirical definition, and is not likely to have contemplated a purely mathematical definition in the manner of the modern semantic view of physical theories. More likely, his "true time" carried ontological weight as a fundamental mode of being. In his *Critique of pure reason*, Kant proposed a third interpretation, according to which time is the pure form of inner sense. This means that our self-awareness requires the one-dimensional continuum of time, just as our representation of the outer world requires the multi-dimensional continuum of space.[4] Time, as a pure form of inner intuition, is not yet a structured notion. Duration, ordering, and the notion of simultaneity are consequences of our subsuming time under three categories of the understanding: quantity, causality, community. Something like Newton's absolute time thus emerges as a consequence of the mind's demand for synthetic unity in the presentation of phenomena to our inner intuition. As has been done in the case of space, one could speculate that Kant misconceived or misapplied his table of categories and that a suitable correction of this defect would be compatible with relativity theory. This is not, however, how Kant and his early readers saw the matter.

The first influential critique of absolute time was the Austrian philosopher-physicist Ernst Mach, in his widely read *Mechanics* of 1883. Mach did not deny the existence of "forms of conception" (*Formen der Auffassung*) that shape our representation of the world. He nonetheless rejected the Kantian dogma of the complete rigidity of these forms. In his view, they are the contingent product of a Darwinian evolution; they can be revised when new domains of knowledge are explored; and they are nothing but partially subjective ways of improving the economy of our thinking. The true origin of the physical sciences is sensory experience, refined through measuring devices. Consequently, time can only be the comparison of various motions with a given motion, for instance the comparison of the motion of a pendulum with the rotation of the earth. Newton's absolute time, Mach tells us, is only an "idle metaphysical conception" reminiscent of medieval philosophy.[5]

Mach's operational definition of time has its own problems. It raises the question of the choice of the motion that serves as a reference. The simplest choice would be the motion of a free particle, which is rectilinear and uniform according to the principle of inertia. This choice raises the difficulty of the definition of the reference system for the inertial motion. Mach rejects Newton's absolute space, since such a space is unobservable. In his view, inertial motion can only be defined with respect to some distant bodies whose mutual distances are believed to be (nearly) constant. Ideally, the laws of mechanics should be formulated in terms of the relative distances of all bodies of the universe, without privileging a subclass of these bodies as a reference system. Although Mach made some efforts in this direction, it is only in recent years that Julian Barbour and Bruno Bertotti found a satisfactory reformulation of classical gravitation theory along these lines. This long delay and the conceptual subtlety of the latter theory can be taken as a suggestion that space,

---

[4] Kant 1781, *Die transcendentale Aesthetik*, §§4-7.

[5] Mach 1883, pp. 207–9. On *Formen der Auffassung* and *Denkweisen*, cf. Mach 1996, pp. 326–7, 380–90.

after all, might be a physical entity distinct from the imbedded matter. Mach himself had nothing against replacing Newton's space with a subtle all-pervading medium.[6]

In his discussion of Newton's absolute time and space, Mach tended to the conclusion that these concepts should be completely eliminated from the foundations of mechanics. However, his criticism is also compatible with the view that time can be defined conventionally by selecting a privileged class of motions, the best convention being that for which the formulation of the laws of mechanics is simplest. The latter view is the one that Henri Poincaré developed in his brilliant essay of 1898 on the measurement of time. Poincaré first reviewed the hierarchy of methods of time measurement, from the uncorrected pendulum to growingly precise definitions of the astronomical time; and he concluded that the astronomers implicitly defined time "so that the laws of mechanics would be the simplest possible." One could otherwise define time with respect to any well-defined motion, but the laws of mechanics would then lose their Newtonian simplicity:[7]

> In other words, there is not a manner of measuring time that is *truer* than others; the one that is generally adopted is only the most *convenient*. Of two clocks, we cannot say that one works well and that the other does not; we can only say it is advantageous to rely on the indications of the first.

Poincaré was most original in the second part of his essay, in which he criticized the notion of simultaneity. Our awareness of distant events, he argued, is necessarily indirect: it requires the transfer of signals. Since these signals travel at a finite speed, there is no way of timing distant events other than assuming a constant velocity for the signals and correcting for the delay. This is what astronomers do when they date astronomical events: they assume that light has traveled from the place of the event at the velocity c known from terrestrial measurements. There is no warranty, however, that this velocity is the same everywhere in the universe and that it does not depend on the direction of propagation. Distant simultaneity therefore depends on a conventional choice for the method of its determination. Poincaré gives the example of the sailors' determination of longitude, which requires a clock synchronized with a reference clock kept ashore. A first possibility is to carry on board a highly accurate and stable clock that was synchronized with the reference clock before departure. Although this works at the precision required for longitude measurement, there is no warranty that the beat of the clock is not altered by transportation. Another method is the observation of a celestial event; in all rigor, this requires a correction by the time that light takes to travel from the event to the sailor. A third method is to exchange telegraphic signals with the keeper of the reference clock; in all rigor, this requires a correction depending on the speed of the signals. Poincaré concludes that the rigorous definition of simultaneity depend on rules whose nature is open to question.[8]

---

[6] Mach 1883, pp. 218–22 (relative distances), 215–16 (subtle medium); Barbour and Bertotti 1982; Barbour 1999, pp. 113–20. For criticism of relative-distance mechanics, cf. Earman 1989, Chap. 5.

[7] Poincaré 1898, p. 6.   [8] Poincaré 1898, pp. 6–13. Cf. Galison 2003, Chap. 4.

Electrodynamics and relativity

Poincaré returned to time measurement in 1900, in the context of a criticism of the new electrodynamics of Hendrik Antoon Lorentz. According to a then nearly universal belief, the propagation of light or of electromagnetic interactions required a subtle medium, the ether, pervading the whole universe. It was therefore natural to expect that the outcome of optical experiments on earth should depend on the motion of the earth through the ether. Yet, attempts to detect this dependence had constantly failed since the beginning of the nineteenth century. At the time of Poincaré's writing, the most successful explanation of this failure was Lorentz's electromagnetic theory, in which electrons and ions interacted through a strictly stationary ether. Intuitively, in this theory the effects of the ether wind on the propagation of light are compensated by the reaction of the ions and electrons to the electromagnetic disturbance associated with light. Formally, Lorentz proved this compensation by showing that the Maxwell–Lorentz equations of electromagnetism were approximately invariant under the combination of two transformations. Calling u the velocity of the earth through the ether, t the absolute time, and x the abscissa in a direction parallel to the motion the earth, the first transformation is the Galilean transformation $x' = x - ut$ leading from the ether frame to the earth frame; the second is a purely formal transformation involving the "local time" $t - ux'/c^2$ and the dilated abscissa $x'' = x'/\sqrt{1 - u^2/c^2}$. The local time accounts for the absence of first order effect; and the dilation of the relative abscissa (with the concomitant contraction of lengths) accounts for Michelson and Morley's failure to detect the second-order fringe shift that the ether wind would imply in their interferometer.[9]

Lorentz believed this compensation between ether wind and electronic reaction to be true only to a certain approximation. In contrast, Poincaré judged the ether to be too immaterial a thing for its wind to have any detectable effect. In other words, he assumed the full validity of a relativity principle according to which physical phenomena are strictly the same in any reference frame moving at a constant velocity with respect to the ether frame. In particular, terrestrial physicists should not be able to trace their motion through the ether by optical means; and any attempt to measure the velocity of light on earth should yield a constant value independent of the earth's motion. Combined with Poincaré's earlier reflections on simultaneity, this implies that terrestrial physicists, when synchronizing their clocks by optical means, can do as if the velocity of light had the constant value c it truly has in the ether. To first order in u/c this procedure yields the expression $t' = t - ux/c^2$ for the time t' measured by these clocks.[10]

Poincaré thus provided a physical interpretation of the local time as the time effectively measured by terrestrial observers under a convention dictated by the relativity principle. Joseph Larmor, Lorentz, and Poincaré later modified their field and coordinate transformations to obtain the exact invariance of the Maxwell–Lorentz equations. In 1905, Poincaré interpreted these transformations as active transformations boosting the whole experimental setup at a constant velocity and formally expressing the invariance of the phenomena through this boost. If the setup includes the space and time measuring devices,

[9] Cf. Janssen 1995; Darrigol 2000, Chap. 8, and further reference there.   [10] Poincaré 1900, p. 272.

this invariance implies that the Lorentz transformations (as Poincaré named them) relate the time and space coordinates measured in the two systems.[11]

In this view, the equations of electromagnetism come first, and the Lorentz transformations are obtained through the invariance properties of these equations. Since a combination of the Maxwell–Lorentz equations implies the d'Alembertian operator $\partial^2/c^2\partial t^2 - \partial^2/\partial x^2 - \partial^2/\partial y^2 - \partial^2/\partial z^2$, the transformations leave the form $ct^2 - x^2 - y^2 - z^2$ invariant. The interpretation of these transformations in terms of effective space and time measurements then results from their being a formal expression of the relativity principle. A first weakness of this version of the theory of relativity is that it makes the Lorentz transformations depend on the electromagnetic nature of all physical processes, or at least on the assumption that all physical processes obey the same invariance properties as electromagnetic processes. Another weakness is that despite Poincaré's early criticism of the ether, his approach still maintains the ether as the carrier or electromagnetic energy and as the reference frame in which "true" space and time are measured. In Poincaré's words, the space and time measured in moving frames are only "apparent," even though the principle of relativity excludes any experimental means to decide which of all Galilean frames is the ether frame. This is the conservative side of Poincaré's conventionalism: as physical theory always requires conventions, we may as well pick the conventions that least disturb our mental habits. The choice of an ether frame is an arbitrary convention that enables us to preserve the traditional concepts of space and time as those defined in this frame.[12]

Although this is a logically possible option, the surviving ether makes it more difficult to disconnect the Lorentz transformations from electromagnetism and to perceive their kinematic necessity. A major step in this direction was accomplished by Albert Einstein, in his memoir of 1905 on the electrodynamics of moving bodies. Einstein based his theory on two principles:

1. The laws according to which the states of physical systems evolve do not depend on which of two systems of coordinates they are referred to, as long as these two systems are in uniform rectilinear translation with respect to each other.
2. Every light ray in a given system of coordinates moves with the well-defined velocity c, whether it is emitted by a body at rest in this system or by a moving body.

Taken separately, these two principles seem conservative. The first is the principle of relativity, well known to any student of mechanics. The second is a trivial consequence of the ether-based theory of light, as long as the given system of coordinates refers to the ether frame. Their iconoclastic character only emerges by combination: they together imply that the velocity of light is the same in two inertial frames, which obviously contradicts the absolute character of time. Granted a few natural assumptions of homogeneity and isotropy, they imply the Lorentz transformations.[13]

Einstein uses this capital result to define a "new kinematics" in which space and time measurements become entangled and frame dependent. He shows that moving rods seem shorter and the period of moving clocks seems larger when measured from the rest frame

---

[11] Poincaré 1905, 1906. On Poincaré's interpretation, cf. Bracco and Provost 2006.

[12] Cf. Paty 1993; Darrigol 2000, Chap. 9; Walter 2009.   [13] Einstein 1905, p. 895.

under natural conventions, and he gives the relativistic law for the composition of velocities. Turning to electromagnetism, he proves that the Maxwell–Lorentz equations are invariant under the space and time transformations and the associated field transformations, and he exploits this invariance to derive some results of the optics of moving bodies. Lastly, he shows that in order to comply with the relativity principle, the equations of mechanics must be modified in a specific manner.[14]

Einstein later described his theory as a "theory of principles," similar to thermodynamics and to be contrasted with "constructive theories" based on a detailed picture of the basic entities. This feature lends to Einstein's relativity a superior kind of necessity: it purports to be a necessary framework for any future particle or field dynamics, whatever the particles and fields may be. The implied kind of necessity hinges on the nature of the principles. Einstein regarded them as vast empirical generalizations, not as apodictically certain truths. To be true, the principle of relativity seems to be part of a general objectivity requirement: that the laws of physics should be the same for every observer, irrespective of his or her state of motion. But it truly is not so, because the invariance of the laws is restricted to a limited class of reference frames, and only experience tells us how to identify this class (before general relativity). As for the light principle, it seems to rest on the contingent fact that the velocity of light has been found to be the same under any circumstances of its emission; or it seems to depend on the contingent truth of the laws of electromagnetism. These contingencies can be reduced by replacing the light principle with the principle of the existence of an upper limit for the propagation velocity of all interactions, as Lev Davidovich Landau did in his lectures on field theory. If there is such a limit, the principle of relativity implies that it should be the same in every inertial system. In this approach, the only remaining contingency is the value of the limit. It can be finite, in which case Einstein's relativity applies; or it can be infinite, in which case Galilean relativity applies.[15]

Better: one could try to base relativity theory on the principle of relativity only. This is what the Germanized Russian physicist Vladimir Sergeyevitch Ignatowski managed to do in 1910. His argument, a variant of which will be given in a moment, is based on the group structure of the transformations linking two inertial frames. The general solution contains both the Lorentz group and the Galilean group as a singular limit of the former. Again, the group depends on one free parameter, easily interpreted as the maximal velocity that a frame can have with respect another. As Ignatowski identified the relativity principle with the denial of absolute motion, he must have felt close to have proven the necessity of relativity theory.[16]

Robb's conical order of events

In Einstein's approach to relativity theory, space comes before time, and the basic concepts are defined through ideal measurement procedures. Space and time coordination is described through frame-dependent coordinates. Emphasis is placed on the relativity of our time determinations. In particular, the simultaneity of two events is not conserved

---

[14] On kinematics and its earlier history, cf. Martínez 2009.   [15] Einstein 1919; Landau and Lifchitz 1959, §1.

[16] Ignatowki 1910. See below, pp. 174–5.

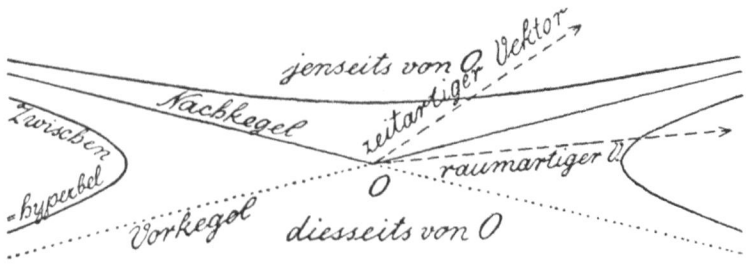

Fig. 5.1.   Minkowki's future and past light cones (*Vorkegel*, *Nachkegel*), and the distinction between time-like
and space-like vector in spacetime (*zeitartiger Vektor*, *raumartiger Vektor*). From Minkowski 1909, Fig. 2.

by a change of reference frame. It was Einstein's former professor, the Göttingen mathematician Hermann Minkowski, who in December 1908 introduced the idea that the
space and time coordinates referred to points of a four-dimensional spacetime, just as
Cartesian coordinates refer to points of a three-dimensional space. Minkowski thus developed an intrinsic geometry of spacetime, emphasizing the event-to-event relations that
were independent of the choice of a reference frame. In particular, he distinguished
between light-like, time-like, and space-like intervals, according as the squared 4-distance
$c(t_2 - t_1)^2 - (x_2 - x_1)^2 - (y_2 - y_1)^2 - (z_2 - z_1)^2$ between the two events was zero, positive, or negative. From a geometrical point of view, the time-like intervals from a given event
form a double cone centered on this event; the space-like intervals belong to the exterior
of this cone; the positive time-like intervals to the interior of the upper cone; the negative
time-like intervals to the interior of the lower cone (see Fig. 5.1). The physical meaning of
this intrinsic classification of events is the following: the time-like intervals are those engendered by light signals. The space-like intervals are those for which no causal connection
by a propagating particle is possible. The positive time-like intervals are those for which the
second event can be causally connected to the first. Hence come the names "future light
cone" and "past light cone" for the two parts of the double cone of time-like intervals.[17]

Somewhat like Minkowski, the Cambridge mathematician Alfred Robb wanted an intrinsic characterization of spacetime in which space and time coordinates would come last,
as a contingent way of labeling events. Robb began with an aversion for Einstein's concept
of distant simultaneity:

> From the first I felt that Einstein's standpoint and method of treatment were un-
> satisfactory, though his mathematical transformations might be sound enough, and I
> decided to proceed in my own way in search of a suitable basis for a theory. In particu-
> lar, I felt strongly repelled by the idea that events could be simultaneous to one person
> and not simultaneous to another; which was one of Einstein's chief contentions. This

---

[17]Minkowski 1909. Cf. Walter 2008. As Minkowski knew, Poincaré 1906 had already interpreted the Lorentz
transformations as rotations in a 4-space with a fourth imaginary coordinate.

seemed to destroy all sense of the reality of the external world and to leave the physical universe no better than a dream, or rather, a nightmare.

Robb's alternative was to purge the theory from any non-invariant notion, thus reaching an "objective standpoint." The simultaneity of distant events had to go, and the frame-dependent ordering of events had to be replaced by an "absolute relation of before and after" based on the Minkowskian cones. Being a Göttingen-trained mathematician, Robb sought an axiomatic formulation of this order, in analogy with the axiomatic formulations of geometry by Giuseppe Peano, Felix Klein, and Oswald Veblen. His great idea was to base the entire theory of time and space on axioms for the order of events. As he regarded this order as pre-metrical and pre-spatial, he claimed to have deduced the pseudo-Euclidean structure of spacetime from the mere ordering of events.[18]

In the first version published in 1914, Robb's theory spanned over 400 pages, and it included 30 postulates, 157 definitions, and 206 theorems. The slightly more compact version of 1921 has the advantage of revealing the heuristics behind this complex construction. Basically, Robb obtained his postulates by inspection of the order engendered by Minkowski's light cone. The first postulates are the usual axioms of an order relation, plus the explicit requirement that the order should only be partial:

> POSTULATE V. If A be any element [i.e., event], there is at least one other element distinct from A, which is neither before nor after A.

Robb's next step is the introduction of the future and past light cones, which he calls α and β subsets. This requires an additional postulate:

> POSTULATE VI. (a) If A and B be two distinct elements, one of which is neither before nor after the other, there is at least one element which is after both A and B, but is not after any other element which is after both A and B. [together with the symmetric statement (b)]

There follows the definition of the future light cone or "α sub-set."

> If A be any element of the set, then an element X will be said to be a member of the α sub-set of A provided X is either identical with A, or else provided there exists at least one element Y distinct from A and neither before nor after A and such that X is after both A and Y but is not after any other element which is after both A and F.

Thanks to a number of other ad hoc postulates, Robb was able to erect an affine structure, that is, a pre-metric geometry based on Peano's axioms of incidence and order and on the Euclidean axiom of parallels. His light cone induces the distinction between three kinds of line: light-like, space-like, and time-like (his "optical lines," "separation lines," and "inertial lines").[19]

Most impressively, Robb proves that the light-cone structure and the affine structure together imply a relation of congruence for two time-like intervals or for two space-like intervals. In his definition, two time-like intervals are said to be congruent if they are the

---

[18] Robb 1914; 1921, pp. v, vi. Cf. Briginshaw 1979 for a quick overview; Winnie 1977 for a simplified version of Robb's theory; Sklar 1977, pp. 208–10, for a criticism of Robb's anti-conventionalist agenda.

[19] Robb 1921, pp. 17, 19.

second diagonals of two light-parallelograms that share the same first diagonal (a light-parallelogram being a parallelogram whose four sides belong to light lines). Call x and x' the two intervals. This statement can be written as

$$x = k + l, \quad x' = k' + l', \quad \text{with} \quad k - l = k' - l', \tag{1}$$

wherein the intervals k, l, k', l' are all time-like. From a Minkowskian point of view, the success of this definition results from the algebraic implication

$$k^2 = l^2 = k'^2 = l'^2 = 0 \quad \Rightarrow \quad x^2 = x'^2, \tag{2}$$

the squares being computed from the Minkowskian distance. Robb needed additional postulates in order to be able to compare intervals of different sorts. He also had an Archimedean postulate, a postulate à la Dedekind for continuity on a line, and a postulate for the tridimensional character of space. In the end, Robb retrieved Veblen's axioms of Euclidean geometry for the space-like, three-dimensional hyperplanes of his spacetime; and he was able, by means of four reference intervals, to construct coordinate systems in which the length of an interval of coordinates (x_0, x_1, x_2, x_3) was given by the Poincaré–Minkowski metric $x_0{}^2 - x_1{}^2 - x_2{}^2 - x_3{}^2$.[20]

On the one hand, Robb's theory is impressive by its rigor, ingeniosity, and seeming ability to derive Minkowskian spacetime from the mere causal ordering of events. On the other hand, the theory is lengthy, difficult, and highly artificial, to such an extent that it may not have been published without the friendly support of another Ulsterman, Joseph Larmor. In order to be a convincing proof of the necessity of relativistic spacetime, Robb's theory should be based on postulates that can be reached without previous knowledge of Minkowskian spacetime. In reality, it owes most of its postulates from the will to integrate aspects of the previously known Minkowskian spacetime. Moreover, the compatibility of the postulates can only be judged by using the conical order of this spacetime as a model. These defects probably explain why Robb's theory has usually been ignored by physicists. They certainly make it difficult to regard it as a successful proof of necessity of the Minkowskian spacetime.[21]

General relativity

Soon after inventing the theory of relativity, Einstein began to worry about how to integrate gravitation in this new scheme. Newton's theory obviously needed to be modified, since it involved instantaneous action at a distance. The theory of relativity was itself unsatisfactory, since it suffered the defect that Mach had earlier identified in Newton's mechanics: the privileging of a class of reference frames without any physical reason. In 1907, Einstein hit upon "the most felicitous idea of [his] life," which he believed would direct him to a generalization of relativity including gravitation in a natural manner. This idea was the principle of equivalence, according to which the gravitational field can be locally eliminated

---

[20] Robb 1914, 1921, 1936.

[21] Recent attempts to simplify Robb's theory do not make it less artificial. For references, see Ehlers 1973, p. 37.

by adopting a free falling reference frame (Einstein's free falling elevator). This principle is a consequence of the equality between inertial mass and gravitational mass, which itself results from Galileo's observation that all bodies fall at the same speed irrespective of their size and mass. Einstein first tried to extend the transformations of space and time coordinates to uniformly accelerated frames. This did not lead him very far.[22]

From the principle of equivalence, Einstein concluded that all reference frames were equally valid, since it was not possible to decide whether the accelerations observed in a given frame were of gravitational or inertial origin. He also understood that the choice of an inertial frame was a local affair, since gravitation generally varied from place to place in a manner depending on the distribution of matter. Furthermore, by reasoning based on tangent inertial frames[23] he judged that space and time measurements in accelerated frames ceased to obey the usual rules; in particular, he argued that space measurement in a rotating frame led to a non-Euclidean metric. At the same time, he was aware of Minkowski's interpretation of special relativity as a pseudo-Euclidean geometry of spacetime; and he knew that Riemann and his followers had developed a generalized concept of space in which the choice of coordinates was arbitrary and in which the validity of Euclidean properties was limited to small regions of space.[24]

In these circumstances, Einstein, with the help of his mathematician friend Marcel Grossmann, tried to represent spacetime as a four-dimensional pseudo-Riemannian manifold with a metric of signature $(+, -, -, -)$. This means that at any given point of spacetime, a proper change of coordinates turns the metric

$$ds^2 = g_{\mu\nu} dx^\mu dx^\nu \tag{3}$$

into the Minkowskian form

$$ds^2 = dx_0{}^2 - dx_1{}^2 - dx_2{}^2 - dx_3{}^2. \tag{4}$$

Einstein regarded this mathematical feature as expressing the local validity of the special theory of relativity. In a flat spacetime, there is a choice of coordinates for which the Minkowskian form holds everywhere and the geodesics are straight lines in conformity with the principle of inertia. In a curved spacetime, the geodesics are curved lines which Einstein interpreted as the trajectories of free falling particles. In this view, gravitation becomes a geometrical effect; and the metric must depend on the distribution of matter since gravitation does. Two principles guided Einstein's determination of this dependence: the principle of general covariance which he regarded as expressing the complete freedom in the choice of reference frames, and the principle of correspondence according to which the predictions of Newton's theory of gravitation should be retrieved in the limit of small spacetime curvature.[25]

[22] Einstein 1907; [1920], p. 265 (*glücklichste Gedanke meines Lebens*). Cf. Norton 1989a; Stachel 1995.

[23] A tangent inertial frame is an inertial frame that has the same velocity as the accelerated frame in the vicinity of a given event.

[24] Cf. Stachel 1989a.

[25] Einstein and Grossmann 1912. Cf. Norton 1989b; Stachel 1989b; Pais 1982, pp. 201–6.

Einstein encountered difficulties on this glorious path, partly due to his persistent association of concrete reference frames with systems of coordinates. After temporarily retreating from general covariance, he obtained a satisfactory theory by choosing the simplest generally covariant generalization of Poisson's relation

$$\Delta\varphi + 4\pi\, k^2 \rho = 0 \tag{5}$$

between the gravitational potential φ, the mass density ρ, and the gravitational constant k^2. The equivalence between mass and energy makes the energy-momentum tensor $T_{\mu\nu}$ of matter (including electromagnetic fields) a natural generalization of the mass density. The equation of geodesics,

$$\ddot{x}^{\mu} + \dot{x}^{\nu}\dot{x}^{\rho}\Gamma^{\mu}_{\nu\rho} = 0, \tag{6}$$

with

$$g_{\mu\sigma}\Gamma^{\sigma}_{\nu\rho} = \frac{1}{2}(\partial_{\nu}g_{\mu\rho} + \partial_{\rho}g_{\mu\nu} - \partial_{\mu}g_{\nu\rho}), \tag{7}$$

makes the metric tensor $g_{\mu\nu}$ a natural generalization of the potential φ (the dot derivation refers to the proper time). Consequently, the generalization of $\Delta\varphi$ should be a generally covariant function of the second order derivatives of the metric tensor. The simplest such quantity (the one of lowest algebraic degree) is $R_{\mu\nu} + \alpha g_{\mu\nu}R + \beta g_{\mu\nu}$, where α and β are arbitrary constants and $R_{\mu\nu}$ is the Ricci tensor obtained by contracting the Riemann curvature tensor $R_{\mu\nu\rho\sigma}$. The relation between the metric and the distribution of matter should therefore read

$$R_{\mu\nu} + \alpha g_{\mu\nu}R + \beta g_{\mu\nu} = -k^2 T_{\mu\nu}. \tag{8}$$

Hilbert and Einstein showed that this equation could be obtained by minimizing an invariant action (the integral of the trace R of the curvature tensor in Hilbert's case) and led to the conservation of the total energy-momentum of gravitation and matter field if only $\alpha = -\frac{1}{2}$.[26]

The same restriction may also be reached as follows. By covariant generalization, the Minkowskian expression of the conservation of the energy-momentum of matter,

$$\partial^{\nu}T_{\mu\nu} = 0, \tag{9}$$

leads to the equation

$$D^{\nu}T_{\mu\nu} = 0, \tag{10}$$

---

[26] For a rigorous proof that R is the only invariant scalar that can be formed from the metric tensor and its first- and second-order derivatives and that is linear with respect to the second-order derivatives, cf. Weyl 1921a, pp. 287–8. The genuine history of Einstein's motivations is much more complex: cf. the thorough investigations by Michel Janssen, John Norton, Jürgen Renn, Tilman Sauer, Matthias Schemmel, and John Stachel in Renn 2007; and the brief account in Janssen 2005.

where D^ν is the contravariant derivative. Combining this equation with the field equation (8) and taking into account the contracted Bianchi identity

$$D^\nu R_{\mu\nu} = \frac{1}{2}\partial_\mu R, \tag{11}$$

we get

$$\left(\alpha + \frac{1}{2}\right)\partial_\mu R = 0, \tag{12}$$

which suggests the choice $\alpha = -\frac{1}{2}$. As Einstein originally set β (the cosmological constant) to zero, he obtained the fundamental equation[27]

$$R_{\mu\nu} - \frac{1}{2}g_{\mu\nu}R = -k^2 T_{\mu\nu}. \tag{13}$$

Einstein did not regard arguments of this sort as rigorous proof of the necessity of his theory. Rather, he introduced them as evidence for the "psychologically natural" character of the fundamental equations. Commentators have detected various flaws in his train of reasoning. For instance, the requirement of general covariance is in itself devoid of physical content.[28] Any field theory can be put in generally covariant form, just as the equations of mechanics can be expressed in curvilinear coordinates. What Einstein truly meant by requiring general covariance was that the fundamental equations of the theory should only imply covariant combinations of dynamic fields (the metric field, the electromagnetic field, and matter fields) and of their (covariant) derivatives. That is to say, the general equations of the theory should not favor any specific choice of the coordinates, although a specific solution of these equations may do so. The idea that the local validity of Minkowskian geometry leads to pseudo-Riemannian geometry at large is more vulnerable: more general kinds of geometry, for instance Weyl's geometry or geometries with torsion, are compatible with local Minkowskian structure. In addition, the correspondence principle does not uniquely lead to Einstein's equation (13). It must be supplemented with simplicity arguments excluding higher-order equations, scalar contributions to the gravitational field, etc.[29]

Yet there is a sense in which Einstein was right to be so confident about the truth of his theory, independently of any experimental test. The known validity of Minkowskian geometry at small scale, the local eliminability of gravitation by a change of reference frame, and the Newtonian dependence of gravitation on the distribution of matter strongly suggest that spacetime should be described by a geometry in which Minkowskian geometry is locally valid and in which the connection between the local geometries depends on the distribution of matter. The suggestion seems to become an implication if one excludes

---

[27] Einstein 1916. Cf. Renn and Sauer 2003.

[28] This was first observed by Erich Kretschmann in 1917. For a thorough review of this point, cf. Norton 1993.

[29] Einstein 1916, p. 177.

the conventionalist escape in which the geometry would be divorced from the measuring procedures that make Minkowskian geometry locally valid. As Jürgen Ehlers put it,[30]

> An attempt to construct a theory which contains in a consistent way the empirically supported structural ingredients of both Newtonian mechanics including gravity theory and of special relativity theory leads, with considerable force, to Einstein's theory of a curved spacetime.

Admittedly, there is some freedom in choosing the way in which the connection between the local geometries is done. The simplest way is the Riemannian one in which some spacetime measuring procedure must be the counterpart of Riemann's surveying space through invariable rods. Whether this is a necessary assumption will be discussed in a moment. Another freedom resides in the choice of the relation between metric field and matter distribution. A theorem by David Lovelock establishes, in conformity with Einstein's intuition, that Einstein's equations (including the cosmological term) are the only covariant, second-order partial-differential equation relating the metric field to the energy-momentum tensor of matter for which the divergence of the implied combination of the metric field and its derivatives vanishes identically. The latter condition results from the invariance of the gravitational field action from which the equation derives. Granted that such an action exists, the theorem excludes equations of higher algebraic degree. It does not exclude equations with higher order derivatives or equations involving scalar contributions to the gravitational field. Against this possibility, it is tempting to evoke Einstein's "Subtle is the Lord, but malicious he is not."[31]

A most frequent objection to Einstein's heuristics runs as follows. His reasoning implies clocks and rods whose rate and length are constants when they are attached to any inertial frame, and it even implies rigid frames of references. Do these notions retain a meaning in the curved spacetime of general relativity? Most modern commentators answer negatively, because rigid bodies are strictly incompatible with the variable curvature of spacetime. At most they assume, as John Lighton Synge did, that moving standard clocks preserve their rate, and they base spacetime surveying on geodesic triangles involving light rays and traveling clocks. If they are right, the usual deductions of the Lorentz group, based on rigid frames and rods and on faithful timekeepers fall through, and the local validity of Minkowskian geometry is in need of a new justification.[32]

5.2 Rational spacetime

Hermann Weyl

The first reader of Einstein's theory who formulated the latter kind of criticism was the Göttingen-educated mathematician Hermann Weyl. In order to understand Weyl's approach to the spacetime problem, it is necessary to focus on an essential result of Riemann

---

[30]Ehlers 1973, p. 3.

[31]Lovelock 1971. Cf. Ehlers 1973, pp. 39–40. Einstein's citation is from the German inscription: "Raffiniert ist der Herr Gott, aber boshaft ist er nicht" above the fireplace in Fine Hall at Princeton University.

[32]On clock-based surveying, cf. Synge 1960, Chap. 3.

that was briefly mentioned in the previous chapter: it is possible to choose the coordinates of the metric manifold so that at a given point the derivatives of the metric coefficients $g_{\mu\nu}$ vanish. In terms of these coordinates and in the vicinity of this point taken as an origin, the equation of a geodesic line takes the form $x_\mu = \alpha_\mu s + O(s^3)$ (the α_μ values being constants, and s being the curvilinear abscissa), which means that the same affine Euclidean structure is shared by the tangent plane of a given point and by the tangent planes at infinitesimally close points of the manifold. An important consequence of this fact, which Tullio Levi-Civita and Weyl developed in 1917–18, is the existence of a notion of parallelism for vectors belonging to the tangent spaces of two infinitesimally close points. Thanks to this notion, a vector can be parallely transported along any curve; parallel transport of a vector on a loop generally does not lead back to the original direction of the vector; and the departure from the original direction measures the curvature. The covariant derivative of a vector field can be defined by comparing the vector at point x_μ with the vector obtained by parallel transport of the vector at point $x_\mu + dx_\mu$ from the latter point to the former point.[33]

If $\Gamma^\mu_{\nu\rho}$ is the symbol entering the equation (6) of geodesics, the parallel transport of the vector a^μ over the length dx^μ implies the variation

$$\delta a^\mu = -\Gamma^\mu_{\nu\rho} a^\rho dx^\nu \tag{14}$$

of its components, in conformity with the expression

$$D_\nu a^\mu = \partial_\nu a^\mu + \Gamma^\mu_{\nu\rho} a^\rho \tag{15}$$

of the covariant derivative. As Weyl emphasized, the form of these relations and the symmetry of the Γ symbols with respect to permutation of the lower indices are mere consequence of the existence of a system of coordinates for which the components of a vector do not vary during parallel displacement ($\delta a^\mu = 0$). Indeed, if $\bar{x}^\mu$ denotes these coordinates, this condition reads

$$\bar{a}^\mu(\bar{x}) = \bar{a}^\mu(\bar{x} + d\bar{x}). \tag{16}$$

Through the coordinate change $x \rightarrow \bar{x}$, all vectors are transformed through the matrix $\partial \bar{x}^\mu / \partial x^\nu$. Therefore, the former relation can be rewritten as

$$\frac{\partial \bar{x}^\mu}{\partial x^\nu}(x) a^\nu(x) = \frac{\partial \bar{x}^\mu}{\partial x^\nu}(x + dx) a^\nu(x + dx) = \left[\frac{\partial \bar{x}^\mu}{\partial x^\nu} + \frac{\partial^2 \bar{x}^\mu}{\partial x^\nu \partial x^\rho} dx^\rho \right] a^\nu(x + dx), \tag{17}$$

which implies that

$$\delta a^\sigma = a^\sigma(x + dx) - a^\sigma(x) = -\frac{\partial x^\sigma}{\partial \bar{x}^\mu} \frac{\partial^2 \bar{x}^\mu}{\partial x^\nu \partial x^\rho} a^\nu dx^\rho. \tag{18}$$

---

[33]Riemann 1867 [1854], pp. 141–2. Levi-Civita 1917; Weyl 1918b, c. On Weyl's infinitesimal geometry and its context, cf. Ryckman 1994, 2005; Scholtz 2001; Afriat 2009; Bell and Korté 2011. For proofs of Riemann's assertions, cf. Heinrich Weber's comments to Riemann [1861] and Weyl 1919b.

This expression has the desired form (14), with[34]

$$\Gamma^{\mu}_{\nu\rho} = \frac{\partial x^{\mu}}{\partial \bar{x}^{\sigma}} \frac{\partial^2 \bar{x}^{\sigma}}{\partial x^{\nu} \partial x^{\rho}}. \tag{19}$$

As a mathematician, Weyl was a champion of the impoverishment strategy, which consists in seeking the minimal structure for which some mathematical properties or theorems survive. An elementary example of this strategy is found in the first chapter of his *Raum Zeit Materie* of 1918. There he begins with Euclidean space, defined intuitively by congruence and mathematically by the group of isometries. He then impoverishes this structure by extracting from this group the subgroup of translations, which transform every line into a parallel line. In the spirit of Felix Klein's Erlangen program, this subgroup defines the "affine geometry" in which the equality of parallel segments is defined without a distance being given. In modern terms, Weyl's affine space is a set of points on which the group of translations acts transitively; and this group is a vector space of finite dimension.[35]

The tangent spaces of a differentiable manifold are affine spaces, owing to the linearity of changes of coordinates in an infinitesimal neighborhood. Consequently, the parallelism of two vectors is well defined in every tangent space. In order to compare vectors belonging to neighboring tangent spaces, Weyl defines an "affine connection" (*affiner Zusammenhang*) through the existence of a local system of coordinates in which a small parallel displacement of vectors keeps the coordinates constant. As we just saw, this condition implies the form (14) of the connection:

$$\delta a^{\mu} = -\Gamma^{\mu}_{\nu\rho} a^{\rho} dx^{\nu},$$

with $\Gamma^{\mu}_{\nu\rho} = \Gamma^{\mu}_{\rho\nu}$; and the transformation rules for these coefficients are the same as in the case of Riemannian geometry. Covariant derivation can be defined through Eq. (15):

$$D_{\nu} a^{\mu} = \partial_{\nu} a^{\mu} + \Gamma^{\mu}_{\nu\rho} a^{\rho}.$$

An affine geodesic line can be defined as a line $x^{\mu}(\tau)$ such that the tangent vector $\dot{x}^{\mu} = dx^{\mu}/d\tau$ remains parallel to itself along the line

$$\dot{x}^{\mu}(\tau + d\tau) - \delta \dot{x}^{\mu} \propto \dot{x}^{\mu} \text{ with } \delta \dot{x}^{\mu} = -\Gamma^{\mu}_{\nu\rho} \dot{x}^{\nu} \dot{x}^{\rho} d\tau, \text{ or } \ddot{x}^{\mu} + \dot{x}^{\nu} \dot{x}^{\rho} \Gamma^{\mu}_{\nu\rho} \propto \dot{x}^{\mu}. \tag{20}$$

The term proportional to $\dot{x}^{\mu}$ in the latter equation can be absorbed by a change of parameter. The equation of geodesics then reduces to the form (6) it takes in Riemannian geometry when the parameter is the curvilinear abscissa. Curvature is defined through the parallel transport of a vector on an infinitesimal loop or parallelogram. Let the vector a^{μ} be parallely transported from x^{μ} to $x^{\mu} + \varepsilon^{\mu} + \eta^{\mu}$ through the alternative paths

---

[34] Weyl 1919a, pp. 100–1. Weyl also justified the general form of parallel displacement by requesting linearity with respect to the displaced vector and the possibility of forming infinitesimal parallelograms by displacing two infinitesimal segments; the latter condition implies linearity with respect to dx^{μ} and symmetry by permutation of the lower indices: see Weyl 1918b, p. 7.

[35] Weyl 1918a, chap. 1, §§1-2.

$x^\mu \rightarrow x^\mu + \varepsilon^\mu \rightarrow x^\mu + \varepsilon^\mu + \eta^\mu$, and $x^\mu \rightarrow x^\mu + \eta^\mu \rightarrow x^\mu + \varepsilon^\mu + \eta^\mu$. The resulting vectors are easily seen to differ by the amount

$$\Delta^\mu = R^\mu{}_{\nu\rho\sigma} a^\nu \varepsilon^\rho \eta^\sigma, \tag{21}$$

which characterizes the curvature and depends on the Riemann tensor[36]

$$R^\mu{}_{\nu\rho\sigma} = \partial_\rho \Gamma^\mu{}_{\nu\sigma} - \partial_\sigma \Gamma^\mu{}_{\nu\rho} + \Gamma^\mu{}_{\tau\rho}\Gamma^\tau{}_{\nu\sigma} - \Gamma^\mu{}_{\tau\sigma}\Gamma^\tau{}_{\nu\rho}. \tag{22}$$

Weyl regarded his construction of affinely connected manifolds as the paradigm of any true infinitesimal geometry. In this context, a geometry is first given in every tangent space; the geometrical properties in the tangent spaces at different points cannot be compared unless an additional structure, the connection, is introduced to allow comparison between the tangent spaces of two neighboring points. Weyl constructed his own metric geometry in this manner, assuming (pseudo-)Euclidean structure in each tangent plane, and then introducing an adequate connection. The local Euclidean structure is given by the results (3)

$$ds^2 = g_{\mu\nu} dx^\mu dx^\nu$$

of length measurements performed with a gauge (*Eiche, Maßstab*) l, that is, with a local standard of length $\sqrt{l}$. In Weyl's purely infinitesimal geometry, the choices of the gauge at different points should be a priori independent. In analogy with the case of affine connection, the comparison of the measurements done at different points is done by assuming the possibility of a uniform gauging in any infinitesimal neighborhood. Namely, for any given point x_0 there must exist a gauge transformation

$$l(x) \rightarrow \bar{l}(x) = \lambda(x)l(x) \tag{23}$$

such that the equality

$$\bar{l}(x_0 + dx) = l(x_0) \tag{24}$$

holds for any dx.

This condition implies that

$$dl/l = -d\lambda/\lambda = -(\partial_\mu \ln \lambda)dx^\mu \tag{25}$$

at point x_0. Since the choice of this point is arbitrary, there exists a field A_μ such that

$$dl/l = -A_\mu dx^\mu. \tag{26}$$

---

[36] Weyl 1918b, pp. 6–12; 1919, pp. 100–8. In modern terms, the tensor character of $R^\mu{}_{\nu\rho\sigma}$ results from its being the commutator of the covariant differential operators D_ρ and D_σ.

This is what Weyl calls the gauge connection. Under a gauge change λ, the metric field $g_{\mu\nu}$ changes to $\lambda g_{\mu\nu}$, and the gauge field A_μ changes to $A_\mu - \partial_\mu \ln \lambda$. The gauge field A_μ can be completely absorbed through a gauge change if an only if the "metric curl" (*metrischer Wirbel*) or "length curvature" (*Längenkrümmung, Streckenkrümmung*) $\partial_\mu A_\nu - \partial_\nu A_\mu$ vanishes. By Stokes' theorem, a gauge does not keep its length after travelling on a closed loop, unless the gauge curvature vanishes in the embraced region of space.[37]

Weyl next equips the manifold with an affine connection $\Gamma^\mu_{\nu\rho}$ and requires this connection to be compatible with the gauge connection. This means that the variation of the length of a vector a^μ during parallel transport should agree with the variation given by the gauge field

$$\delta \left(g_{\mu\nu} a^\mu a^\nu\right) = -(g_{\mu\nu} a^\mu a^\nu) A_\rho \mathrm{d}x^\rho, \quad \text{with} \quad \delta a^\mu = -\Gamma^\mu_{\nu\rho} a^\rho \mathrm{d}x^\nu. \tag{27}$$

The validity of this condition for any a^μ and the symmetry of the affine connection lead to

$$g_{\mu\sigma}\Gamma^\sigma_{\nu\rho} = \frac{1}{2}\left(\partial_\nu g_{\mu\rho} + \partial_\rho g_{\mu\nu} - \partial_\mu g_{\nu\rho}\right) + \frac{1}{2}\left(A_\nu g_{\mu\rho} + A_\rho g_{\mu\nu} - A_\mu g_{\nu\rho}\right). \tag{28}$$

This means that in a metric Weyl space, there is one and only one affine connection compatible with the metric structure defined by the metric tensor and the gauge field. In the Riemannian case for which the gauge field can be eliminated, the affine connection is a function of the metric tensor only.[38]

Weyl applied this new infinitesimal geometry to spacetime, interpreting the metric field $g_{\mu\nu}$ as the gravitational potential, the gauge field A_μ as the electromagnetic potential, and the gauge curvature $\partial_\mu A_\nu - \partial_\nu A_\mu$ as the electromagnetic field. Since electromagnetic and gravitational forces were then commonly believed to be the only fundamental forces, Weyl could claim the complete fusion of physics and geometry: "*Everything real* [Alles Wirkliche] *in the world is a manifestation of the world metric*; the physical concepts are none other than those of geometry."[39]

After a short-lived burst of enthusiasm, Einstein condemned Weyl's theory for the following reason: it implies that the rate of a clock depends on the path it has traveled in spacetime, to an extent incompatible with the accurate constancy of the frequency of spectral lines from astronomically distant sources. Weyl did not regard this objection as fatal, for he speculated that the times of concrete clocks could differ from the times given by the fundamental metric relations. Einstein and Pauli disliked this conventionalist escape, for it deprived general relativity from its main epistemological advantage: the fairly direct connection between theoretical quantities and actual physical processes. Weyl remained unshaken, in part because in his view clocks and rods were not primary geometrical entities

[37] Weyl 1918b, pp. 12–16; 1918c, pp. 41–2; 1919, pp. 108–11.

[38] Weyl 1918b, pp. 16–18; 1918c, pp. 32–3; 1919, pp. 111–12.

[39] Weyl 1918b, pp. 2 (citation, Weyl's emphasis), 24–5; 1918c, pp. 34–40; 1919, pp. 115–16, 242–53.

and therefore did not provide a direct assessment of metric relations. In addition, he had broader and higher reasons to believe in the necessary truth of his theory of spacetime.[40]

These reasons are found in his mathematics and in his philosophy. As we saw, his concept of infinitesimal geometry proscribed the direct comparison of distant geometrical objects. As Riemann had purged geometry from absolute distant parallelism, Weyl purged it from absolute distant congruence:

> Riemann's geometry preserves a last element of geometry at a distance [*Ferngeometrie*]—without any factual ground, as far as I can see; the contingent emergence of this theory from the theory of surfaces seems to be the only reason for this state of affairs. Indeed [Riemann's] metric form allows not only comparing the lengths of two vectors at the same point, but also at two arbitrarily distant points. *A true field geometry* [Nahegeometrie], *however, can only admit the principle of the transport of a length from one point to an infinitesimally close point,* and there is therefore as little ground to assume that length transport is integrable as there is to assume that direction transport is integrable.

As we saw, Weyl arrived at this view through a clear mathematical strategy: first impoverish the received mathematical structures of Euclidean space and Riemannian manifold, and then reintroduce structural furniture in a stepwise manner. The most rudimentary structure is that of a continuous manifold. The first furniture is differentiability, which implies local affine structure. Next comes the affine connection, which allows those local affine structures to communicate. Last comes the metric structure, first locally as a conformal structure, then globally with a metric connection. By analogy with the affine connection, the metric connection should not involve any direct comparison at a distance. This is why it involves a non-integrable gauge connection.[41]

Plausibly, Weyl found support to this view in Edmund Husserl's philosophy, in which he and his wife shared a vivid interest. His otherwise dryly mathematical *Raum Zeit Materie* contains, already in the first edition of Easter 1918, a largely Kantian introduction with Husserlian overtones. In particular, Weyl criticizes the naïve positing of an external reality and asserts the existence, in the form we give to any observation, of a non-empirical remnant that belongs to pure consciousness and resists any attempt to eliminate the ego:

> The actual world, each of its pieces and all their determinations, are and can only be given as intentional objects of consciousness . . . The given to consciousness is the starting point in which we must place ourselves in order to comprehend the sense and the justification of the positing of actuality . . . "Pure consciousness" is the seat of the philosophical a priori.

This Husserlian idea generalizes the Kantian necessity of the intuition of space in our apprehension of the external world. In order to reach the synthetic a priori essences of pure consciousness, Husserl and Weyl prescribe a distillation of the conceptual foundation of advanced physical theories. At the end of this process, every historical and contextual

---

[40] Einstein, comment to Weyl 1918c, p. 40; Weyl's reply to the latter, Weyl 1918c, pp. 41–2. Cf. Ryckman 1994; 2005, pp. 80–1, 86–9. Weyl renounced the electromagnetic interpretation of the gauge field when, in 1928, its relation to the phase of the Schrödinger wave function became clear (see Weyl 1968, vol. 2, p. 42).

[41] Weyl 1918c, p. 30.

contingency is eliminated, and the residual essence shines as self-evident: this essence is the "regional ontology" of the relevant domain of experience.[42]

There is textual evidence that in 1918 Weyl believed the coordinate systems of continuous manifolds to be just this residual essence. From later writings one gathers that he also included the metric structure in the infinitesimal neighborhood of an individual observer, as the form of our immediate intuition of space and time. This intuition, being necessarily homogenous (as it already was in Kant's transcendental philosophy), could not possibly extend to finite domains of spacetime:

> Only the spatiotemporally coinciding and the immediate spatiotemporal neighbor-hood have a directly clear meaning in intuition . . . The philosophers may have been correct that our space of intuition bears a Euclidean structure, regardless of what physical experience says. I only insist, though, that to this space of intuition belongs the ego-center and that the coincidences, the relations of the space of intuition to that of physics, becomes vaguer the further one distances oneself from the ego-center.

In order to extend our concept of spacetime to finite domains, Weyl needed an additional infinitesimal construct, the gauge connection. Weyl's impoverishing-refurnishing mathematical strategy thus had a Husserlian counterpart in the phenomenological method for extracting the essences of pure consciousness. In the fourth edition (1921) of *Raum Zeit Materie*, Weyl explained this philosophical dimension:

> My investigations [regarding the metric continuum] appear to me to be a good example of the essence-analysis [*Wesenanalyse*] striven for by phenomenological philosophy (Husserl); an example that is typical for such cases where a non-immanent essence is dealt with . . . Also, the example of space is most instructive for that question of phenomenology that seems to me particularly decisive: to what extent the delimitation of the essentialities [*Wesenheiten*] rising up to consciousness express a characteristic structure of the domain of the given itself, and to what extent mere convention participates to it.

Whether the Husserlian considerations motivated or followed Weyl's theory of gravitation and electricity, it plausibly contributed to his belief in the necessity of his own variety of infinitesimal geometry.[43]

As a consequence of his general rejection of congruence at a distance, Weyl rejected Helmholtz's approach to the foundations in geometry. In Weyl's opinion, Helmholtz assumed too much homogeneity by requiring the free mobility of rigid bodies throughout space; in addition, the resulting constant-curvature geometries were incompatible with Einstein's theory of gravitation. Weyl did not consider that Helmholtz's assumptions could be weakened to imply the local Euclidean character of space and yet allow for variable curvature. As a result, he found himself without any justification for the quadratic form of the metric in his own infinitesimal geometry. In the three first editions of *Raum Zeit*

---

[42]Weyl 1918a, pp. 3–4 (with a first footnote referring to Husserl's *Ideen*), cited in Ryckman 2005, pp. 115–16. Cf. Ryckman, 2005, Chap. 5.

[43]Weyl 1931, pp. 336, 339; 1921a, pp. 148–9; cited in Ryckman 2005, pp. 148, 157–8. Cf. Ryckman 1994, pp. 834–5; 2005, Chap. 6.

Materie, he simply posited this form. In the fourth, published in 1921, he sketched the following derivation of the necessity of the Pythagorean form.[44]

Weyl first defines a metric structure in tangent spaces through the Lie group G of congruence transformations acting transitively in each tangent space. He then introduces a "metric connection" $\Lambda^{\mu}_{\nu\rho}$ giving the "congruent transport" $\Lambda^{\mu}_{\nu\rho}a^{\nu}\mathrm{d}x^{\rho}$ of the vector a^{μ} from a point to a neighboring point. Intuitively, any such transport should be the combination of a parallel transport $-\Gamma^{\mu}_{\nu\rho}a^{\nu}\mathrm{d}x^{\rho}$ through an affine connection $\Gamma^{\mu}_{\nu\rho}$ and of an infinitesimal rotation $K^{\mu}_{\nu\rho}a^{\nu}\mathrm{d}x^{\rho}$, wherein $K^{\mu}_{\nu\rho}\mathrm{d}x^{\rho}$ belongs to the Lie algebra of the group G. That is to say, there should exist choices of the $K^{\mu}_{\nu\rho}$ and $\Gamma^{\mu}_{\nu\rho}$ coefficients such that

$$\Lambda^{\mu}_{\nu\rho} = K^{\mu}_{\nu\rho} - \Gamma^{\mu}_{\nu\rho}. \tag{29}$$

Weyl then makes two assumptions:

1. The choice of the metric connection is completely free.
2. This choice completely determines an affine connection.

The first assumption means that every set of values is possible for the coefficients $\Lambda^{\mu}_{\nu\rho}$, or else, any linear transformation of a vector system can be chosen to define a congruent transport. This should be so, Weyl tells us, because in the spirit of a purely infinitesimal geometry, the metric connection should not depend on the local metric structure; it should only be a function of the distribution of matter. The second assumption is a generalization of a fact demonstrated in the case of Riemannian and Weylian metric geometry, and promoted by Weyl to "the fundamental principle of infinitesimal geometry": there is a unique affine connection associated with given metric properties.[45]

Formally, the first assumption implies that there are n^3 independent choices of the $\Lambda^{\mu}_{\nu\rho}$ coefficients, if n is the dimension of the space manifold. If N denotes the dimension of the Lie algebra of the group G, there are nN independent choices for the $K^{\mu}_{\nu\rho}$ coefficients; and there are $n \cdot n(n + 1)2$ independent choices for the $\Gamma^{\mu}_{\nu\rho}$ coefficients because of their symmetry with respect to permutation of the lower indices. According to the second assumption, the decomposition (29) must be unique, so that the following relation holds:

$$n^3 = nN + n^2 \cdot (n + 1)/2, \text{ or } N = n(n - 1)/2. \tag{30}$$

The second assumption further requires that the Lie algebra of G should not contain any symmetric matrix (since such a matrix could as well contribute to the affine connection). Lastly, the trace of the generators of G must vanish in order that volume be conserved by congruence. The invariance group of a non-degenerate quadratic form clearly satisfies these three properties. In 1922, after much labor, Weyl succeeded in proving that no other kind of group did so. He thus believed to have proven the necessity of the locally

---

[44] On Weyl's rejection of Helmholtz's approach, cf. Weyl 1922a, pp. 337–40; 1923, p. 103.

[45] Weyl 1921a, pp. 127–8.

(pseudo-)Euclidean character of space by "pure conceptual analysis." In subsequent lectures on the problem of space, he presented this result as the crowning of his efforts in developing a purely infinitesimal geometry.[46]

As Weyl refused to regard Helmholtz's concept of rigidity as the proper foundation of the theory of space, he understandably sought a physical determination of the metric field $g_{\mu\nu}$ of general relativity that did not depend on rods and clock. In the first edition of *Raum Zeit Materie*, he used light signals to this effect. It is indeed evident that the metric tensor is determined up to a constant factor by knowledge of the light-like intervals dx^μ for which $g_{\mu\nu}dx^\mu dx^\nu = 0$. Weyl further believed that light propagation at a higher infinitesimal order determined the metric coefficients in an absolute manner. As Lorentz pointed out to him, this is not true because the null geodesics of light rays are invariant under the rescaling $g_{\mu\nu} \rightarrow \lambda g_{\mu\nu}$. This invariance is easily deduced from the definition of null geodesics as affine geodesics satisfying relation (6) and the nullness condition $\dot{x}^\mu \dot{x}_\mu = 0$. To compensate for his error, Weyl showed how the knowledge of particle geodesics, combined with the knowledge of null geodesics, fully determined the metric in a locally Minkowskian Weyl space (a fortiori, it does so for the pseudo-Riemannian space of general relativity). The demonstration goes as follows.[47]

According to the expression (28) of the affine connection for a given metric, the light geodesics depend neither on the scale of the quadratic tensor $g_{\mu\nu}$ nor on the linear tensor A_μ. In Weyl's words, they only determine the "conformal structure," that is, the properties that remain invariant by a local change of scale. Formally, two affine connections yield the same conformal structure if and only if they differ by an expression of the form

$$\Delta \Gamma^\mu_{\nu\rho} = \frac{1}{2}\left(\varphi_\nu \delta^\mu_\rho + \varphi_\rho \delta^\mu_\nu - \varphi^\mu g_{\nu\rho}\right), \tag{31}$$

wherein $g_{\mu\nu}$ is a metric tensor compatible with this structure and the φ_μ are arbitrary functions. Neither do the particle geodesics completely determine the metric connection; they determine only the "projective structure," which defines the change of the direction of a vector during infinitesimal transport (without constraining the change of its length). Formally, two metric connections yield the same projective structure if and only if the characteristic equation (6) of geodesics is preserved. This is so if and only if their difference $\Delta \Gamma^\mu_{\nu\rho}$ is such that for any vector $\dot{x}^\mu$, $\Delta \Gamma^\mu_{\nu\rho} \dot{x}^\nu \dot{x}^\rho$ is parallel to $\dot{x}^\mu$. Equivalently, this difference must have the form

$$\Delta \Gamma^\mu_{\nu\rho} = \delta^\mu_\nu \psi_\rho + \delta^\mu_\rho \psi_\nu, \tag{32}$$

wherein the ψ_μ are arbitrary functions. In order that two metric connections share both the same conformal and the same projective structure, their difference must be simultaneously of the forms (31) and (32). This can only happen if A_μ and ψ_μ vanish. Hence, the two connections must be equal.[48]

---

[46] Weyl 1921a, pp. 131–3; 1922a; 1922b; 1923, p. 140; Ryckman 2005, pp. 154–8.

[47] Weyl 1918a, pp. 182–4; 1921a, pp, 206–7, 285–6; 1921b; Lorentz 1923.   [48] Weyl 1921b.

To summarize, by some philosophico-mathematical distillation Weyl isolated the essence of infinitesimal geometry: a Lie group acting on every tangent space and a connection for comparing the group's actions in two neighboring tangent spaces. No direct comparison at a distance is ever possible in this philosophy; geodesic traveling is the only means to explore the structure of the manifold. In the least structured case, the Lie group is the group of translations, and the associated connection, called affine connection, is symmetric. In the metric case, the local Lie group is the group of isometries (including translations), and the associated connection, called metric connection, is arbitrary. In order that this connection uniquely determines the affine connection as its symmetric part, the group of isometries must be the (pseudo-)Euclidean group. In this case, the group's action can be defined through the quadratic form $g_{\mu\nu}$ that it leaves invariant, and a gauge connection is sufficient to determine the affine connection from the metric.[49]

Weyl's strategy of mathematical distillation amounts to a new kind of necessity arguments for physical theories. The cogency of these arguments is debatable. In the process of distillation, the contact with the empirical world is lost. For instance, a non-metric manifold and purely affine connection are abstractions that may be deemed irrelevant to a physical geometry, since the latter finds its original impulse in measuring practices. The idea of a non-integrable gauge connection contradicts the most basic requirement of any direct measurement, namely: that the size of the measuring unit should not depend on its history. As Weyl's geometry does not meet this requirement, it can hardly pretend to be a physical geometry concerning the structure of space and time. It remains true, however, that Weyl's reflections on space, time, and matter were enormously fertile, both from a mathematical and from a physical point of view.

Arthur Eddington

By allowing non-integrable gauge variations in spacetime, Weyl opened the Pandora box of increasingly exotic geometries in unified field theories. The first of his emulators was the Cambridge astronomer Arthur Eddington, leader of the 1919 solar eclipse expedition that spectacularly confirmed Einstein's theory, and most brilliant expositor of the new relativistic physics in Britain and elsewhere. His *Space, time and gravitation* of 1920 was a non-technical, philosophically oriented exposition of Einstein's theory, so well written and so full of ingenious dialogs, analogies, metaphors, and parables that it fascinated laymen and physicists alike. Out of a technical appendix to the French edition of this book grew *The mathematical theory of relativity* of 1923, which became the standard English text for learning general relativity. This book included an account of Weyl's theory as well as a further extension of his world geometry.[50]

---

[49] A given metric $g_{\mu\nu}$ and a given metric connection $\Lambda_{\nu\rho}^{\mu}$ do not determine a gauge connection. The metric connection determines an affine connection, which, together with the metric $g_{\mu\nu}$, enables one to compute the variation $\mathrm{d}l$ of the squared length of a vector during parallel displacement. A further restriction on the affine connection is needed in order that this variation takes the Weyl form $\mathrm{d}l/l = -A_\mu \mathrm{d}x^\mu$. Cf. Eddington 1923, pp. 217–18.

[50] Eddington 1920, 1923. On Eddington's involvement in relativity theory and philosophy, cf. the excellent Ryckman 2005, Chap. 7.

Eddington admired Weyl's theory for its tentative unification of gravity and electricity, for the attractiveness of its geometrical basis, and for ultimately divorcing concrete spacetime measurement from the abstract geometrical foundation of the world. As we saw, Weyl protected his theory from Einstein's clock-based objection by making concrete clocks beat at a rate different from that given by the ideal gauges of the basic spacetime manifold. While Einstein and Pauli condemned this move as opportunist conventionalism, Eddington applauded it as a healthy departure from the naïve assumption that ideal space and time were defined by the properties of material rods and clocks. In his opinion matter and its properties had to be derived from the geometrical basis provided by Weyl's theory, properly completed with atomistic assumptions: "In the game of world-building, we lose a point whenever we have to ask for extraordinary material specially prepared for the end in view."[51]

Yet Eddington was not fully satisfied with Weyl's theory. In his opinion, Weyl's gauges somehow compromised between ideal and physical geometry, because they agreed with concrete spacetime measurement in small neighborhoods, but ceased to do so for large-scale measurement:

> Weyl's theory of gauge-transformations occupies a position intermediate between pure mathematics and physics. He admits the physical comparison of length by optical methods . . . but he does not recognize physical comparison of length by material transfer . . . There is thus both a physical and a conventional element in his "length."

Eddington avoided this ambivalence by giving to the ideal metric a purely conventional character. In the end, he completely removed the metric from the mathematical foundation of the theory. The basic concept of the resulting theory is the affine connection $\Gamma^{\mu}_{\nu\rho}$, posited independently of any metric. Next comes the curvature tensor (Eq. (22))

$$R^{\mu}{}_{\nu\rho\sigma} = \partial_\rho \Gamma^{\mu}_{\nu\sigma} - \partial_\sigma \Gamma^{\mu}_{\nu\rho} + \Gamma^{\mu}_{\tau\rho}\Gamma^{\tau}_{\nu\sigma} - \Gamma^{\mu}_{\tau\sigma}\Gamma^{\tau}_{\nu\rho},$$

and its first contraction $R_{\mu\nu}$. There being no stricture on the choice of the $\Gamma^{\mu}_{\nu\rho}$ coefficients (besides symmetry with respect to the lower indices), the tensor $R_{\mu\nu}$ no longer has the symmetry of the usual Ricci tensor. Eddington separates the symmetric and antisymmetric parts:[52]

$$R_{\mu\nu} = G_{\mu\nu} + F_{\mu\nu}, \text{ with } G_{\mu\nu} = \frac{1}{2}\left(R_{\mu\nu} + R_{\nu\mu}\right) \text{ and } F_{\mu\nu} = \frac{1}{2}\left(R_{\mu\nu} - R_{\nu\mu}\right). \quad (33)$$

Eddington then identifies the tensor $G_{\mu\nu}$ with the metric $g_{\mu\nu}$ of the "natural geometry" that material rods measure, namely: he sets

$$G_{\mu\nu} = \lambda g_{\mu\nu}, \quad (34)$$

[51] Eddington 1923, p. 237.

[52] Eddington 1923, pp. 220–1 (citation). Cf. Goenner 2004, section 4.3; Ryckman 2005, Chap. 8; Smadja 2010, 139–42. I do not follow Eddington's notation. For mathematical convenience and for an easier comparison with Einstein's and Weyl's theories, Eddington introduced a purely "graphical" metric in his formulas and then selected the "in-tensors" that do not depend on the choice of this metric. His action principle, an extension of Weyl's, involves this graphical metric.

wherein λ is a universal constant, with the following bootstrap justification:[53]

> Any apparatus used to measure the world is itself part of the world, so that the natural
> gauge represents the world as self-gauging. This can only mean that the tensor $g_{\mu\nu}$
> which defines the natural gauge is not extraneous, but is a tensor already contained in
> the world-geometry.

Since $G_{\mu\nu}$ is the only symmetric second-rank tensor of the intrinsic world geometry,[54] it
is the only candidate for a $g_{\mu\nu}$. As Eddington remarks, this choice is not incompatible
with Einstein's general relativity, because Eq. (34) is equivalent to Einstein's fundamental
equation for the gravitational field in the absence of matter as long as the cosmological
constant has the finite value λ. Eddington believed this finite λ was needed to provide a
natural gauge and a well-defined size for elementary particles.

Eddington further proved the existence of a vector A_μ such that

$$F_{\mu\nu} = \partial_\mu A_\nu - \partial_\nu A_\mu, \tag{35}$$

which prompted him to identify this tensor with the electromagnetic field. He vaguely completed his theory with an action principle inspired from Weyl, the action being an invariant
built from $g_{\mu\nu}$ and A_μ. He praised the principle of least action for automatically yielding
the energy-momentum tensor and tensors attached with other conserved quantities (the
electric current), simply by taking the functional derivative (Eddington's "Hamiltonian
derivative") of the action with respect to the independent functions of which it is a functional. His theory of course contained Weyl's and Einstein's (with a finite cosmological
constant) as particular cases, by particularizing the choice of the affine connection. Its
alleged superiority to its forerunners lay in the cleaner separation between ideal and natural geometry and a well-defined strategy for moving from the ideal geometry to natural
geometry and dynamics. The ideal geometry was the "World Geometry which is the pure
geometry applicable to a conceptual graphical representation of all the quantities concerned in physics," whereas the Natural Geometry was "the single true geometry in the
sense understood by the physicist."[55]

In Eddington's opinion, both the basic relational structure of the world geometry and
the construction of physical quantities on this basis had rational necessity. The coordinate
system of the basic manifold and the attached tensors were little more than the recognition
that physical quantities are the results of measuring operations that involve arbitrary codes:

> To grasp a condition of the world as completely as it is in our power to grasp it, we
> must have in our minds a symbol which comprehends at the same time its influence
> on the results of all possible kinds of operations. Or, what comes to the same thing,
> we must contemplate its measures according to all possible measure-codes. It might
> well seem impossible to realise so comprehensive an outlook; but we shall find that

---

[53] Eddington 1923, p. 219.

[54] Without a formerly given metric, there is no way to form a variable scalar factor of $G_{\mu\nu}$ by contracting $G_{\mu\nu}$.

[55] Eddington 1923, p. 198. Eddington's take on Hamiltonian derivatives was more subtle and more evolving
than alluded here: cf. Smadja 2010.

the mathematical calculus of tensors does represent and deal with world-conditions just in this manner.

Then the affine connection or the Weylian metric must be introduced in order to compare the (affine) structural relations defined at various locations:[56]

> When resolved to ultimate terms, [structure] appears to resolve itself into a complex of relations. And further these relations cannot be entirely devoid of comparability; for if nothing in the world is comparable with anything else, all parts of it are alike in their unlikeness, and there cannot be even the rudiments of a structure.—The axiom of parallel displacement is the expression of this comparability, and the comparability postulated seems to be almost the minimum conceivable.

The affine connection being now given, the world builder must focus on the tensors that can be formed from it, since any potential representation of the world condition must have the tensor form. Among these tensors, there is only one, $G_{\mu\nu}$, that can represent the metric $g_{\mu\nu}$ of physical space. The other physical tensors are obtained by functional derivation of the action with respect to $g_{\mu\nu}$ and A_μ. Their conservation property (the vanishing of their 4-divergence) make them the obvious candidates for representing the energy-momentum tensor and the electromagnetic current, which are the central physical quantities in Einstein's theory and in electromagnetic theory. Eddington called this strategy for finding the physical tensors the "principle of identification":[57]

> We must proceed by inquiring first what experimental properties the physical tensor possesses, and then seeking a geometrical tensor which possesses these properties *by virtue of mathematical identities.*

As Eddington acknowledged, in his theory (unlike Weyl's) there are different classes of choices of the affine connection that are compatible with the observed physical tensors. In other words, there are many more degrees of freedom than there truly are in nature. Eddington turned this redundancy into an advantage of his theory:

> In order to discuss why the structure of the world is such that the observed phenomena appear, we must necessarily compare it with other structures of a more general type; that involves the consideration of "non-physical" quantities which exist in the hypothetical comparison-worlds, but are not of a physical nature because they do not exist in the actual world. If we refuse to consider any condition which is conceivable but not actual, we cannot account for the actual; we can only prescribe it dogmatically.

Eddington compared this situation with the kinetic theory of gases, which yields the same thermodynamic laws for a huge variety of molecular models:[58]

> This suggests striving for an ideal—to show, not that the laws of nature come from a special construction of the ultimate basis of everything, but that the same laws of nature would prevail for the widest possible variety of that basis.

---

[56] Eddington 1923, pp. 3, 224.   [57] Eddington 1923, p. 222.   [58] Eddington 1923, pp. 227, 106.

The principle of identification, Eddington tells us in *Space, time and gravitation*, is mainly the expression of the mind's predilection for permanence:

> When the eye surveys the tossing waters of the ocean, the eddying particles of water leave little impression; it is the waves that strike the attention, because they have a certain degree of permanence. The motion particularly noticed is the motion of the wave-form, which is not a motion of the water at all. So the mind surveying the world of point-events looks for the permanent things. The simpler relations, the intervals and potentials, are transient, and are not the stuff out of which mind can build a habitation for itself. But the thing that has been identified with matter is permanent, and because of its permanence it must be for mind the substance of the world. Practically no other choice was possible . . . Is it too much to say that mind's search for permanence has created the world of physics? So that the world we perceive around us could scarcely have been other than it is?

Eddington acknowledged a few gaps in his ideal deduction of the laws of nature, for instance the arbitrariness of the continuity assumption of the spacetime manifold, or the ignorance of atomicity and quanta in the description of matter. He nonetheless asserted the possibility of such a deduction:

> "Give me matter and motion," said Descartes, "and I will construct the universe." The mind reverses this. "Give me a world—a world in which there are relations—and I will construct matter and motion."

Eddington's idealism found its most poetic expression in the concluding lines of his *Space, time and gravitation*:[59]

> We have found a strange foot-print on the shores of the unknown. We have devised profound theories, one after another, to account for its origin. At last, we have succeeded in reconstructing the creature that made the foot-print. And Lo! It is our own.

Eddington's idealism went much further than Einstein's, and a little further than Weyl's. Although Einstein anticipated the emphasis on tensor structure and invariants, his trust in the truth of general relativity largely depended on its integrating the empirically corroborated components of earlier theories. Weyl, despite his Husserlian belief in the self-evidence of his extension of Riemannian geometry, acknowledged the mind's inability to grasp the essence of spacetime by mere introspection and our debt to the egregious critiques of earlier illusory evidences. Humans like him could access the self-evident essence only by applying the phenomenological method to the ultimate product of a long intellectual history, Einstein's theory of gravitation. Eddington seems to have had greater confidence in the mind's power to deduce the basic "relation-structure" of the world geometry. Yet his idealism was not a subjective idealism à la Berkeley. More like Kant and Weyl, he advocated a transcendental idealism in which the mind imposes necessary forms of understanding to the world. In order to fill these forms with actual content, Eddington needed the internal identification of invariants on the one hand, and the extraction of empirical invariants from

---

[59]Eddington 1920, pp. 178–80, 180–1, 182. Weyl's apocryphal citation of Descartes ignores his reduction of matter to pure extension, which makes his philosophy a closer anticipation of Eddington's.

highly corroborated theories of physics on the other hand. The principle of identification thus combined transcendental deduction and empirical induction.[60]

In Eddington's dichotomy between world geometry and natural geometry, the world geometry is uniquely determined by the mind, whereas the natural geometry and its Riemannian character depend on the contingent existence of invariable measuring rods or clocks. It may be easier to defend a nearly opposite thesis: there is much freedom in choosing the purely formal geometric basis of a theory, whereas pseudo-Riemannian geometry must be the natural geometry of the world if the world is measurable at all. In conformity with this thesis, the fundamental field theories of physics rely on increasingly broad kind of geometries, including non-Abelian fiber bundles and spaces of dimensions higher than four; whereas to this day the (pseudo)Riemannian geometry of general relativity still applies to natural geometry. By divorcing ideal geometry from natural geometry, Weyl and Eddington unwittingly opened the realm of ideal geometries. They failed to free natural geometry from the commonsensical demand of invariable measuring standards.

The EPS argument

In their criticism of the operational basis of general relativity, Weyl and Eddington both rejected the primacy of measuring rods and clocks. Whereas Eddington advocated a construction of these concrete gauges from unified field theory with a zest of atomism, Weyl recommended surveys based on free-falling particles and light rays. The latter approach has had some influence on the later history of general relativity, as John Synge and others used light-based geodesy to improve and clarify the observational interpretation of the theory. It invites a foundation of spacetime in which the geometrical structure would be derived from the necessities of geodetic measurement, in analogy with Helmholtz's derivation of the structure of ordinary space from the necessities of congruence-based measurement. In a remarkable contribution to a volume in honor of John Synge, the three theorists Jürgen Ehlers, Felix Pirani, and Alfred Schild (EPS) succeeded in doing just that.[61]

The references to Weyl and Helmholtz are explicit in EPS:

> In making these distinctions [between topological, differential, conformal, projective, affine, and metric structures] we follow essentially H. Weyl . . . The ideas that light propagation determines a conformal structure and free fall determines a projective structure on space-time have been clearly spelled out by Weyl.
>
> Our method has some similarity to Helmholtz's derivation of the metrics of spaces of constant curvature. According to Helmholtz and Lie, the existence and form of these metrics can be deduced from the qualitative assumption of the free mobility of rigid bodies. Similarly, we attempt to derive the conformal, projective, affine, and metric structures of space-time from some qualitative (incidence and differential-topological) properties of the phenomena of light propagation and free fall that are strongly suggested by experience. Not only the measurement of length but also that of time then appears as a derived operation.

---

[60] Cf. Ryckman 2005, p. 234, who rightly condemns the subjectivist-idealist reading of Eddington.

[61] Ehlers, Pirani, and Schild 1972.

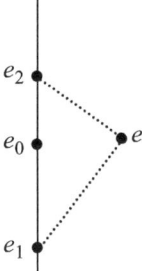

Fig. 5.2.   Synge's procedure for measuring the space-like interval e_0e. The light signals e_1e and ee_2 connect the event e to the world line of a clock-carrying particle through the event e_0.

Another important background to the EPS argument is Synge's method for measuring the interval between two events through clock-carrying particles and light signals. For a particle equipped with a standard clock, Synge measures the proper time ds of this particle by the number of ticks of its clock. The interval between two neighboring events of time-like separation can then be measured through the proper time of a particle connecting these two events. In order to measure the interval e_0e between two events e_0 and e of spatial separation, Synge uses the construction of Fig. 5.2, which implies a clock-carrying particle traveling through e_0, and two light signals connecting this particle with the event e.[62]

In Minkowski space, the vector relations

$$ee_1 = ee_0 + e_0e_1, ee_2 = ee_0 + e_0e_2 \tag{36}$$

and the self-perpendicularity

$$(ee_2)^2 = 0, \quad (ee_1)^2 = 0 \tag{37}$$

of the light vectors ee_1 and ee_2 imply that

$$(ee_0)^2 + 2(ee_0) \cdot (e_0e_1) + (e_0e_1)^2 = 0, \quad (ee_0)^2 + 2(ee_0) \cdot (e_0e_2) + (e_0e_2)^2 = 0. \tag{38}$$

The middle terms can be eliminated from this system by subtracting the product of the second equation by e_0e_1 from the product of the first equation by e_0e_2, with the result

$$(ee_0)^2 (e_0e_2 - e_0e_1) + (e_0e_1)^2e_0e_2 - (e_0e_2)^2e_0e_1 = 0, \tag{39}$$

and, after eliminating the common factor e_1e_2,

$$(ee_0)^2 = (e_0e_1) \cdot (e_0e_2) = - |e_0e_1||e_0e_2| \,. \tag{40}$$

[62]Ehlers, Pirani, and Schild 1972, pp. 63, 68, 65; Synge 1960, pp. 112–13.

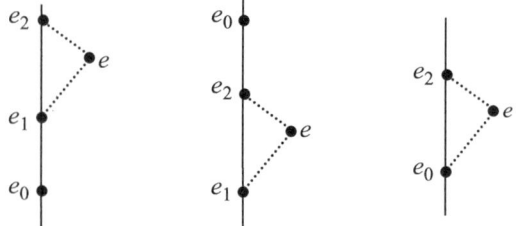

Fig. 5.3. Extension of Synge's procedure to positive time-like, negative time-like, and light-like intervals e_0e.

This means that the Minkowskian square of the space-like interval e_0e is the algebraic product of the times separating the event e_0 from the emission and the reception of the light signals by the particle.

This construction can be extended to intervals of time-like and light-like kinds between the events e and e_0. In the positive time-like case, the two light signals are emitted and received by the particle after its passing through e_0; in the negative time-like case, the two light signals are emitted and received by the particle before its passing through e_0; and in the light-like case, one of the signals passes through the event e_0 (see Fig. 5.3). This means that the Minkowskian distance from a given event to a neighboring event can be measured by the same clock-carrying particle and light signals emitted and received by this particle.

The given proof of the validity of this procedure presupposes the Minkowskian structure of small domains of spacetime. EPS had the ingenious idea of reversing the argument. Namely, they showed that Synge's operational procedure, once cleared from its pre-assumed metric aspects, implies the local Minkowskian structure. In their reasoning, the clock carried by the particle through e_0 can no longer be a metrical clock, since no measure of space or time is pre-assumed. EPS only assume that an arbitrary continuous parameter, the pre-time t, orders the successions of events through which the particle travels. They further assume that the set of all events form a 4-dimensional differentiable manifold. Concretely, they define this manifold through local "radar coordinates." Namely, they assume that in sufficiently small domains of spacetime an event can be smoothly characterized by the times t_1, t_2, t_1', t_2' at which two given particles P and P' emit and receive two light signals that connect it to the event (see Fig. 5.4).

Next, EPS assume that for any given event e_0 and for any particle through this event, there is a neighborhood of e_0 such that any event e in this neighborhood can be connected

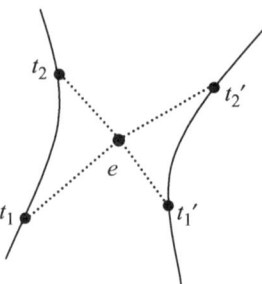

Fig. 5.4. Radar coordinates of an event e.

to this particle by a light roundtrip in the manner of Fig. 5.2. On the basis of this process, they define the scalar quantity g by

$$g = (t_1 - t_0)(t_2 - t_0), \tag{41}$$

where t_0, t_1, t_2 are the pre-times of the particle for the events e_0, e_1, e_2. This quantity vanishes when e and e_0 are connected by a light signal; it is positive when e_1 and e_2 are on the same side of e_0 on the world line of the particle, and negative when they are on different sides. In the special case when the event e belongs to the world line of the particle, it is equal to the square of the pre-time difference between e and e_0.[63]

The continuity of the spacetime manifold implies or at least suggests that for small values of the coordinate differences x^μ for the events e and e_0, the events e_1 and e_2 should be close to e. Consequently, in its first approximation the quantity g should be of second order with respect to x:[64]

$$g(x) \approx g_{\mu\nu} x^\mu x^\nu. \tag{42}$$

Since there are vectors for which this quadratic form vanishes, there are only three possibilities: degeneracy, $(+, +, -, -)$ signature, $(+, -, -, -)$ signature. Degeneracy is not an option, for it would reduce the effective number of dimensions of spacetime. Nor is the $(+, +, -, -)$ signature, because the domains of positive and negative g would then be two similar simply connected domains and there would be no distinction between past and present. We are left with the $(+, -, -, -)$ signature, which makes $g_{\mu\nu}$ a Minkowskian quadratic form.

The definition of this form depends on the choice of a particle through e_0 and from the choice of a pre-time on this particle (any differentiable, growing function of a pre-time being also a pre-time). However, two forms associated to different choices must vanish for the same set of time-like intervals $e_0 e$. Therefore, they can only differ by a scaling coefficient. The manifold is thus endowed with a conformal structure in every of its tangent spaces. From a physical point of view, this means that spacetime is locally Minkowskian, since the form $g_{\mu\nu}$, defined up to a scaling factor, is sufficient to compare the "length" of any vector with the length of a conventionally selected unit vector. Remarkably, EPS obtain this structure by the mere consideration of light echoes registered by an arbitrarily moving particle. Rigid rods, standard clocks, and inertial motion play no role in their derivation.[65]

Conformal geodesics can be intuitively defined on the manifold as null lines along which the tangent preserves its direction. Formally, these lines are the null geodesics of the fictitious metric associated with any $g_{\mu\nu}$ compatible with the local conformal structure. Indeed the latter geodesics preserve the direction of the tangent, and they are independent of the arbitrary scaling of $g_{\mu\nu}$, as we saw in the Weyl section of this chapter. There is, however, a

---

[63] Ehlers, Pirani, and Schild 1972, pp. 70–3.

[64] EPS give a different justification: if $\partial_\mu g$ did not vanish, the hypersurface $g = 0$ would be smooth around e_0, and there would be no room for space-like intervals $e_0 e$.

[65] Ehlers, Pirani, and Schild 1972, pp. 173–4.

more geometrical construction of the conformal geodesics: they are null lines imbedded in a light-like hypersurface. This is seen as follows.[66]

By definition, a light-like hypersurface is a surface $f(x) = 0$ such that its tangent planes $(\partial_\mu f) dx^\mu = 0$ have a null normal, that is,

$$g^{\mu\nu} \partial_\mu f \, \partial_\nu f = 0. \tag{43}$$

A null line $x^\mu(\tau)$ on such a surface satisfies

$$\dot{x}^\mu \dot{x}_\mu = 0 \text{ and } \dot{x}^\mu \partial_\mu f = 0, \tag{44}$$

which can only happen if $\dot{x}^\mu$ and $g^{\mu\nu} \partial_\nu f$ are proportional (because two orthogonal null vectors are necessarily parallel). Consequently, there exists a parameter τ for which

$$\dot{x}^\mu = g^{\mu\nu} \partial_\nu f. \tag{45}$$

If $Da^\mu / D\tau$ denotes the "absolute derivative" $\dot{a}^\mu + \{^{\ \mu}_{\nu\rho}\} \dot{x}^\nu a^\rho$ of the vector a^μ along the path $x^\mu(\tau)$, we have

$$\begin{aligned}
\frac{D \dot{x}^\mu}{D\tau} &= g^{\mu\nu} \frac{D \partial_\nu f}{D\tau} = g^{\mu\nu} \dot{x}^\rho D_\nu D_\rho f \\
&= g^{\mu\nu} g^{\rho\sigma} (D_\sigma f) (D_\nu D_\rho f) = g^{\mu\nu} D_\nu \left(\frac{1}{2} g^{\rho\sigma} D_\rho f D_\sigma f \right) = 0.
\end{aligned} \tag{46}$$

Therefore, the line $x^\mu(\tau)$ is a geodesic for the metric $g_{\mu\nu}$; and it is a conformal geodesic for the associated conformal structure.

As a consequence of the definition of the local g functions, we know that light lines are null lines with respect to the conformal structure defined by the $g_{\mu\nu}$ form. It remains to be seen whether they are conformal geodesics. EPS prove this by showing that the set ν_e of light lines from a given event e, characterized by $g(x) = 0$ in a not too large neighborhood, is a light-like surface. Call $n = (n_0, \mathbf{n})$ the normal at a given point p of this surface and $u = (u_0, \mathbf{u})$ an arbitrary null vector with respect to the form $g_{\mu\nu}$ at this point. A vector belonging to the intersection of the tangent plane with the double cone ν_p of the u vectors satisfies

$$n \cdot u = n_0 u_0 - \mathbf{n} \cdot \mathbf{u} = 0, \quad \text{with} \quad u_0 = |\mathbf{u}| \tag{47}$$

in a coordinate system for which $g_{01} = g_{02} = g_{03} = 0$. Ignoring the trivial solution $u = 0$, this condition is equivalent to

$$\cos(\mathbf{n}, \mathbf{u}) = n_0 / |\mathbf{n}| . \tag{48}$$

---

[66]Cf. Synge 1960, pp. 32–3.

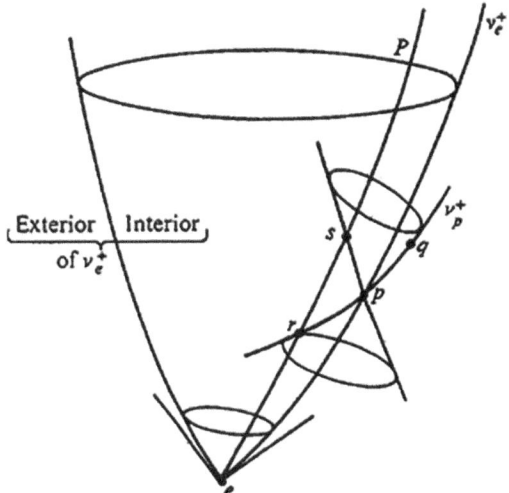

Fig. 5.5.   The EPS stacking of light cones. From
Ehlers, Pirani, and Schild 1972, p. 75.

For a space-like tangent plane, $n_0/|\mathbf{n}| > 1$, and the surface v_e^+ and the cone v_p intersect only at the origin of the cone. This is incompatible with v_e being made of null lines. For a time-like tangent space, $n_0/|\mathbf{n}| < 1$ and the intersection is a manifold of two dimensions. In this case (see Fig. 5.5), the world line of a particle P traveling nearly at the velocity of light from the event e could cross the double cone v_p twice, so that there would be three line signals (ep, rp, and ps) connecting the event x with this particle; and this is contrary to the assumption earlier made (in the definition of g) that in a not too large domain an event is related to a particle's world-line by two light signals only.[67] The only leftover possibility is that the normal n is a null-vector, in which case the surface v_e^+ and the cone v_p are tangent. Since the light rays are inscribed on the light-like surface v_e^+ and since they are null-lines, they must be conformal geodesics.[68]

We now come to the particle geodesics. EPS assume that there is a class of free falling particles for which the second derivative of the motion $\ddot{x}^\mu$ vanishes with respect to a specific choice of coordinates and parameter. Equivalently, these particles are projective geodesics in Weyl's sense: they are the geodesics of an affine connection $\mathrm{P}^\mu_{\nu\rho}$ defined up to an additive term of the form

$$\Delta \mathrm{P}^\mu_{\nu\rho} = \delta^\mu_\nu \psi_\rho + \delta^\mu_\rho \psi_\nu, \tag{49}$$

wherein the ψ_μ are arbitrary functions as in Eq. (32). In addition, we know that light rays are the geodesics of an affine connection $\mathrm{C}^\mu_{\nu\rho}$ defined up to

$$\Delta \mathrm{C}^\mu_{\nu\rho} = \frac{1}{2} \left(\varphi_\nu \delta^\mu_\rho + \varphi_\rho \delta^\mu_\nu - \varphi_\sigma g_{\nu\rho} g^{\mu\sigma} \right), \tag{50}$$

[67] Moreover, the interior of the cone v_e^+ would no longer define the domain of events causally related to e, because the event q, which is outside this cone, would be causally related to e through the portion er of the world line of particle P and the light ray rq.

[68] Ehlers, Pirani, and Schild 1972, pp. 74–6.

wherein the φ_μ are arbitrary functions as in Eq. (31) and $g_{\mu\nu}$ is any of the metrics compatible with the conformal structure. EPS further assume that light geodesics are the limits of particle geodesics for particles of infinite energy. This implies that the set of $C_{\nu\rho}^\mu$ connections must be included in the set of $P_{\nu\rho}^\mu$ connections. Since by Weyl's reasoning the intersection of the two sets is reduced to a single connection, there is one and only one affine connection $\Gamma_{\nu\rho}^\mu$ compatible with the conformal properties of light rays and the projective properties of free falling particles.[69]

For the sake of definiteness, let us select the normalized tensor $\bar{g}_{\mu\nu}$ such that $\det \bar{g}_{\mu\nu} = -1$ among the possible conformal metric tensors $g_{\mu\nu}$. The conformal connection is then given by

$$\bar{g}_{\mu\sigma} C_{\nu\rho}^\sigma = \frac{1}{2}\left(\partial_\nu \bar{g}_{\mu\rho} + \partial_\rho \bar{g}_{\mu\nu} - \partial_\mu \bar{g}_{\nu\rho}\right) + \frac{1}{2}\left(\varphi_\nu \bar{g}_{\mu\rho} + \varphi_\rho \bar{g}_{\mu\nu} - \varphi_\mu \bar{g}_{\nu\rho}\right), \qquad (51)$$

with an arbitrary φ_μ. There is one and only one choice of φ_μ for which this connection is compatible with the projective structure. Calling $\bar{\varphi}_\mu$ this choice and $\Gamma_{\nu\rho}^\mu$ the resulting connection, we have

$$\bar{g}_{\mu\sigma} \Gamma_{\nu\rho}^\sigma = \frac{1}{2}\left(\partial_\nu \bar{g}_{\mu\rho} + \partial_\rho \bar{g}_{\mu\nu} - \partial_\mu \bar{g}_{\nu\rho}\right) + \frac{1}{2}\left(\bar{\varphi}_\nu \bar{g}_{\mu\rho} + \bar{\varphi}_\rho \bar{g}_{\mu\nu} - \bar{\varphi}_\mu \bar{g}_{\nu\rho}\right). \qquad (52)$$

This connection is the connection associated with the Weyl metric defined by the forms $\bar{g}_{\mu\nu}$ and $\bar{\varphi}_\mu$. It is invariant under the gauge transformation

$$\bar{g}_{\mu\nu} \to g_{\mu\nu} = \lambda g_{\mu\nu}, \bar{\varphi}_\mu \to \varphi_\mu = \bar{\varphi}_\mu - \partial_\mu \lambda / \lambda. \qquad (53)$$

Thus we see that the EPS axioms of light propagation and free fall imply that spacetime is a locally Minkowskian Weyl space. An extraneous assumption is needed in order to retrieve the Riemannian spacetime of general relativity. For this purpose, EPS consider the parallel transport of a time-like vector by means of the affine connection $\Gamma_{\nu\rho}^\mu$ from a given point of spacetime to another through two different time-like paths. At the end of this operation the lengths of these two vectors differ by an amount given by the gauge curvature $\partial_\mu \bar{\varphi}_\nu - \partial_\nu \bar{\varphi}_\mu$. If this difference vanishes, the form $\bar{\varphi}_\mu$ can be eliminated by a gauge transformation and we are left with the Riemannian metric $g_{\mu\nu}$ only. Concretely, the time-like vectors may refer to the periods of two clocks originally beating at the same rate and transported elsewhere through two different paths. Then spacetime will be Riemannian if and only if the rates of the two clocks agree when they are brought together again. This condition is usually called the absence of "second clock effect," the first effect being time dilation in the sense of special relativity.[70]

The EPS derivation of the Weyl structure from the properties of light rays and free falling particles being highly formal, it will be useful to describe semi-concrete operations through which spacetime measurement can be devised in this context. This is not an easy

---

[69] The EPS reasoning is here more complex because they use intrinsic characterizations of the conformal and projective structures. Weyl's original reasoning is easier though perhaps less rigorous.

[70] Ehlers, Pirani, and Schild 1972, pp. 81–2.

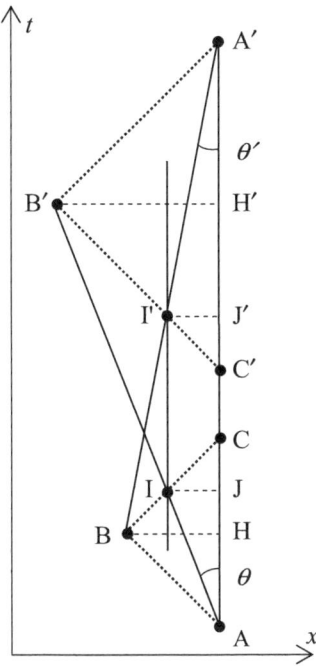

Fig. 5.6.   Construction of a parallel line II′ parallel to a given free-particle line AA′ through the light signals AB, BC, C′B′, B′A′, and the intersecting free-particle lines AB′ and A′B.

task, because reasoning is easily contaminated by preconceptions bound to the admission of a standard length or clock. Yet this can be done through Robert Marzke's ingenious construction of parallel lines through coincidences of particle and light geodesics. Let us first assume that spacetime is flat and Minkowskian, which is approximately the case in a sufficiently small neighborhood, and let us realize the coincidences of Fig. 5.6 through free falling particles and light signals. It will now be proved that the line II′ on this figure is parallel to the given time-like line AA′ by construction.[71]

Since the coincidences define a plane figure and since parallelism is invariant by any linear change of coordinates, we may choose rectangular coordinates (x, t) so that AA′ is parallel to the t-axis and light lines make the angles $\pm\pi/4$ with it. We therefore have, for the (ordinary Euclidean) lengths on the diagram:

$$\text{IJ} = \text{CJ}, \text{IJ} = \text{AJ}\tan\theta, \text{AJ} = \text{AC} - \text{CJ}, \text{AC} = 2\text{BH} = 2\text{A}'\text{B}\sin\theta', \qquad (54)$$

which together imply that

$$\text{IJ} = 2\text{A}'\text{B}\frac{\sin\theta\sin\theta'}{\cos\theta + \sin\theta}. \qquad (55)$$

[71] Marzke 1959; Marzke and Wheeler 1964, pp. 48–60. See also Castagnino 1971. Ehlers (1973, pp. 33–5), offers an operational construction of a comb made of a geodesic and a succession of perpendicular and mutually parallel teeth (space-like segments); it is not clear how a world-line parallel to the geodesic can be operationally generated from this comb.

Permuting primed and unprimed letters, we also have

$$I'J' = 2AB' \frac{\sin\theta \sin\theta'}{\cos\theta' + \sin\theta'}. \tag{56}$$

Therefore, $I'J'$ and IJ are equal if and only if

$$A'B(\cos\theta' + \sin\theta') = AB'(\cos\theta + \sin\theta). \tag{57}$$

This is indeed the case because

$$A'B\cos\theta' = A'H, \ A'B\sin\theta' = BH, \ BH = AH, \ AH + A'H = AA', \tag{58}$$

as well as the relations obtained by permuting primed and unprimed letters. The equality of IJ and $I'J'$ implies the parallelism of AA' and II', which remains true in any other system of linear coordinates.

Now suppose that the curvature of the Weyl space no longer vanishes and suppose that AA' is a small portion of the world-line of a free-falling particle (corresponding to a small increment of the pre-time of this particle). The geodesic character of the lines of the diagram (II' excepted) implies that the interval II' is parallel to AA' up to an error of second order with respect to AA'. Therefore, for any free-falling particle the world line of a neighboring particle can be adjusted so that it remains constantly parallel to the geodesic defined by the first particle (owing to geodetic deviation, the world line of this second particle cannot be a geodesic). The two particles define an invariable rod. In addition, the time along this particle can be redefined by requiring that the blips of a light signal bouncing between the two particles occur at equal time intervals (see Fig. 5.7).[72]

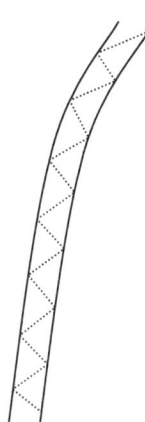

Fig. 5.7.   Bouncing-light clock.

[72] A more practicable test for a clock to give the correct time, whether or not its motion is geodetic, is found in Perlick 1987, 2007: at any given time, the experimenter let two particles fall freely from the clock-carrying particle with different initial velocities; the time of the clock is the true Weyl time (or the proper time in the pseudo-Riemannian case) if and only if a certain mathematical condition is met between the radar coordinates of these particles with respect to the clock-particle.

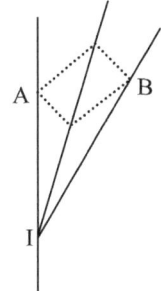

Fig. 5.8.   Construction of two congruent intervals IA and IB on two portions of geodesics that make an angle with each other. The construction requires a third free-falling particle passing through I, sending a flash of light to the other particles and detecting simultaneous echoes from them. The null character of the light lines implies the equality of the Minkowkian measures of IA and IB.

We have thus constructed a rod and a clock that keep their length and pace along the geodesic of a free falling particle. The construction is easily extended to time-like world-lines that are made of a succession of geodesics, because exact congruence is possible at each angular point of such broken lines (see Fig. 5.8). Further extension to an arbitrary time-like world line obtains by taking the limit of infinitely small geodesic portions. In this construction, the lengths of rods or the pace of clocks transported from one event to another through different paths need not remain in agreement with each other. This is what we should expect in a Weyl spacetime: it is possible to transform a given gauge $\bar{l}(x)$ to a gauge $l(x) = \lambda(x)\bar{l}(x)$ that does not vary on a given path, by taking

$$\ln \lambda[x(\tau)] = \int_0^\tau \bar{\varphi}_\mu[x(\tau')]\dot{x}^\mu(\tau')\mathrm{d}\tau' \qquad (59)$$

on this path; but it is impossible to obtain a uniform, path-independent l unless

$$\partial_\mu \bar{\varphi}_\nu - \partial_\nu \bar{\varphi}_\mu = 0. \qquad (60)$$

On a path for which the gauge has been made invariable, the integral $\int \sqrt{g_{\mu\nu}\mathrm{d}x^\mu \mathrm{d}x^\nu}$ gives the time of our inertial optical clock.

Although the depth and beauty of the EPS argument should not escape anyone who understands it, it can be criticized on various grounds. Firstly, EPS do not give complete mathematical demonstrations although they are much more careful than I have been in defining relevant neighborhoods and mappings for each of the properties they introduce and in spelling out the conditions of continuity and differentiability. They note: "A fully rigorous formalization has not yet been achieved, but we nevertheless hope that the main line of reasoning will be intelligible and convincing to the sympathetic reader." This hope has been so far justified, since no one seems to have challenged the mathematics of the argument. On the contrary, several commentators have emphasized the mathematical fertility of the EPS reasoning.[73]

Secondly, EPS fill up spacetime with events and processes that can only be virtual since not much of the world is the object of human observation and operation. Although the

---

[73] Ehlers, Pirani, and Schild 1972, pp. 69–70.

remark applies to any argument that relies on conditions of measurability, the difficulty may be more acute in the case of spacetime because we are never given a second chance to track an event: there is only one spacetime and one history. Thirdly, EPS's reliance on light signals is liable to the familiar objection that it makes their construction depend on electromagnetic theory, whose empirical truth may only be contingent. This is not a serious objection, because high-energy particles could replace light and because for the purpose of the argument light could be redefined as any entity that satisfies the EPS axioms of incidence between light rays and particle paths. It should be noted, however, that the EPS reliance on radar coordinates is less innocent than it seems: it implicitly assumes the universality of light propagation (independence on the velocity of the source, wavelength, and polarization), which is only natural to someone weaned in relativistic physics. Moreover, it only works in the absence of a material medium in the mapped region of spacetime. Strictly speaking, the EPS derivation of the Weylian structure of spacetime only applies to empty spacetime. The validity of this structure in regions occupied by bulk matter of high density is an unwarranted extrapolation.[74]

Fourthly, there is a difficulty in defining the subclass of free-falling particles that define the projective structure. Not every free-falling particle will do because there could be non-gravitational interactions, or a mass-heterogeneity of the particle could lead to multipolar effects. At first glance, it would seem that the relevant particles can only be defined as those who follow geodesics, and therefore cannot be used to define geodesics. As Ehlers, Pirani, and others have indicated, the logical circle can be avoided by an operational definition of the relevant particles. For instance, they can be defined as those who fall together irrespective of their mass and orientation (this excludes electric charge, which would make the fall mass-dependent; and this excludes multipolar structure, which would make the fall depend on orientation).[75]

Fifthly, the EPS axioms seem to be selected in an ad hoc manner, so that the known behavior of light signals, echoes, and particles is retrieved, and so that the causal structure of Minkowskian spacetime is retrieved. For most of the axioms, this objection carries little weight because the assumed properties can be tested in a direct manner by the mere observation of coincidences, or because they reflect the basic necessity of ordering events according to signaling possibilities. The geodesic axiom for free falling particles is more problematic. Its concrete verification requires knowledge that is not given in the EPS framework since it would imply the construction of clocks and free-falling reference frames. Moreover, nothing in the EPS argument tells us why we should expect the axiom to be true. It indeed amounts to a combination of the principle of equivalence and the principle of inertia, whose validity presupposes an odd pre-established harmony between the motions of all free-falling particles.[76]

The last axiom, the one excluding the second clock effect is even more problematic. As EPS emphasize, its operational meaning is much more complex than for the other axioms. It necessarily involves something like the earlier given construction of an inertial light

---

[74]These two last points are made in Ehlers 1973, pp. 23–4.

[75]Cf. Ehlers 1973, p. 30; Sklar 1977, pp. 259–60.

[76]On this point, cf. Brown 2005, pp. 14–18; Brown and Pooley 2006.

clock and the path-dependence of these clocks. In these circumstances, EPS judge it wise to reopen the possibility that spacetime might be Weylian instead of Riemannian. Although Weyl's interpretation of gauge curvature as the electromagnetic field is no longer viable, "one may, however, ask whether other interpretations ... relating the gravitational field to another universally conserved current, might contain some physical truth."[77]

There is another way to interpret EPS's failure to derive the Riemannian structure of spacetime: as an indication that light and particle geodesics cannot be a sufficient basis for the construction of spacetime because a natural axiomatization of their behavior does not comply with the measurability of spacetime. It is indeed a basic precondition of any measurement that the unit of measurement should not depend on its history. To accept the second clock effect is to renounce or at least to relax the measurability of spacetime. As there is no empirical indication that such a radical step is needed and as the very definition of physical geometry implies measurement, it may be wise to regard the existence of an invariable unit of time or length as a basic axiom of any spacetime theory. In this case, light rays and free falling particles need not be the only ingredients of a construction of spacetime; a standard clock or length may be assumed from the start.[78]

A last inconvenience of the EPS approach is its anti-pedagogical character. It can only be understood by someone who masters fairly advanced differential geometry. In it, the simpler aspects of space and time are subordinated to the difficult and highly abstract notion of a Weyl space. Namely, the Weyl structure successively leads to the Riemannian structure as a particular case, to the Minkowskian structure as a local approximation, and to the Euclidean space as a section of the former. In a more pedagogical and more historical approach, one would wish to proceed from the simplest to the most general structure.

5.3 The Helmholtzian approach, with morals

From rigid bodies to pseudo-Riemannian spacetime

Helmholtz's foundation of geometry is usually regarded as obsolete for at least two reasons: his concept of freely mobile rigid body is only compatible with a constant curvature, and rigid bodies are not compatible with relativity theory. We have already seen that the first objection is annihilated by the remark that free mobility can be restricted to small bodies, in which case it still implies the locally Euclidean character of space but becomes compatible with variable curvature. The second objection seems more lethal, despite Einstein's own reliance on rigid bodies and frames in his relativity theory. Yet it is not. Special relativity only excludes bodies that would instantaneously transmit a pressure from one point to another; it is perfectly compatible with the congruence criteria given in Chapter 4 as long as congruence is only meant between bodies that do not move with respect to each other. General relativity only excludes extended rigid bodies, it does not forbid the existence of approximately rigid bodies of small dimensions, which is all we need in the deduction of the locally Minkowskian character of the metric. In fact, modern general

---

[77] Ehlers, Pirani, and Schild 1972, pp. 69, 83 (citation).

[78] Marzke and Wheeler 1974, pp. 60–1 argue that Weyl's spacetime is incompatible with the exclusion principle (for it implies that particles with different histories are distinguishable).

relativity theory abundantly relies on the notion of rigid frames of references or tetrads attached to an observer's world-line. Of course, the implied rigidity is only local, and it is a kinematical sort of rigidity.[79]

More precisely, what general relativity allows is the concept of an approximately rigid body as a small system of particles that are maintained at very nearly constant mutual distances by optical means.[80] If we assume, as Synge does, the existence of standard clocks that measure time along the trajectory of any particle, the distance between two neighboring particles is said to be constant if and only if any flash of light emitted by the first particle and echoed by the second returns to the first particle after a constant lapse of its time.

Empirically, we know that it is possible to set the distances between four particles to four arbitrarily given values and to keep them constant, if these values are not too large; for a fifth particle, however, its distances from the first three particles determine its distance from the fourth. The first three particles, say A, B, C, thus define a neighborhood of a three-dimensional manifold, because the relative distance PQ of any two additional particles P and Q in the same neighborhood is a well-defined function f of their coordinates AP, BP, CP, AQ, BQ, CQ. Continuity and differentiability can be defined in the manner indicated in the earlier given account of Helmholtz's theory of space. As was shown in the same chapter, the free mobility of rigid bodies amounts to the assumption that the function f is completely determined by the mutual distances AB, AC, BC of the reference points.[81]

The concept of rigid body we thus reach is sufficiently restricted to be compatible with general relativity; it can be tested in a realistic manner; and it satisfies the axioms of Helmholtz's theory of space as long as the validity of these axioms is confined to small rigid bodies. Therefore, Helmholtz's deduction of the locally Euclidean character of space retains its validity in post-relativistic physics. For the sake of this deduction, we only need to assume that there exist clocks and signals for which the above-given definition of rigidity yields rigid bodies satisfying Helmholtz's axioms. I leave the precise nature of the clocks and signals open at this stage, because I want to introduce the more elementary space structure before the spacetime structure.

The present construction of locally Euclidean space amounts to the construction of a moderately extended rigid body that can be regarded as a reference frame. In a given frame, any event can be precisely located by measuring the coordinates of the point of the frame at which this event occurs. We now want to define time by means of motions referred to this reference frame. The best one can do, without prior knowledge of the laws of mechanics, is to select a frame such that an isolated particle can be in a state of permanent rest in every point of the frame (a posteriori, this frame will be a free falling non-rotating frame, and the particle will have to be charge-free and spherically symmetric). In this frame, it turns out that an isolated particle is either at rest or moves on a straight line. We may now build two identical catapults, such that the bullets they project from the same spot in the

---

[79] See, e.g., Synge 1960, pp. 114–18. The difficulties with the relativistic concept of rigidity vanish for infinitesimal domains because they have to do with the non-integrability of the rigidity conditions.

[80] More precisely, the variation of the mutual distances should be negligible compared to the inverse of the curvatures.

[81] See Chapter 4 above, p. 125.

same direction travel together, and use these catapults to define simultaneity. Or we may use light signals to the same effect, since light signals travel rectilinearly in any inertial frame and travel together when they are emitted at the same time from the same spot in the same direction. Two events will be said to be simultaneous if and only if bullets catapulted on the line joining their places of occurrence meet at the middle of this line.[82] Lastly, the time of an event can be defined as the distance traveled by a bullet catapulted from a fixed origin when this event is simultaneous with the passage of the bullet at this distance. The consistency of these definitions and their independence from the choice of the catapult (up to a change of unit) requires that all free-particle motions be uniform with respect to the defined time.

We thus see that the notion of inertial frame is bound to the necessity of defining a measurable time by means of free moving particles and that it requires a few presuppositions: the existence of extended rigid bodies, the free mobility of rigid bodies, and the existence of frames in which the motion of free particles has a special simplicity or symmetry. The latter presupposition is no more obvious than the geodetic axiom of EPS: it requires the same sort of pre-established harmony. However, we now understand that this harmony is necessary to a pre-mechanical definition of time.

My attentive reader may be worrying about circularity in the former chain of reasoning: rigid frames are defined by means of clocks, and the proper choice of these clocks is dictated by their agreement with an inertial time whose definition presupposes rigid frames. This circularity only affects the concrete possibility of building clocks by the mere consideration of particles and light signals (a difficulty already encountered in the EPS context). For the rest of our construction of spacetime all we need to assume is the existence of clocks such that the derived concept of inertial rigid frame leads to an inertial time that agrees with the time of these clocks. The means by which we might construct such clocks would be given in a further stage of reasoning, by identifying the dynamics compatible with the derived spacetime structure.

Once a first inertial frame and the associated time are defined, it is possible to define the uniform translation of another frame with respect to this frame. The new frame is easily seen to be inertial. However, the times defined in the two frames need not be identical since the procedures through which they are defined are evidently frame-dependent. All we can say is that to an event defined by sharp values of the space and time coordinates in one inertial system correspond well-defined values of these coordinates in another inertial system. We will now see that these transformations can be deduced in two different manners, one similar to the synthetic derivation of Pythagoras' theorem, the other similar to the Helmholtz–Lie reliance on infinitesimal group structure.

---

[82] Other conventions are possible for which the velocity of the bullets or the velocity of light would depend on the direction of propagation. For instance, it is possible, as Poincaré did, to define the "true" synchronicity of two events in any reference system through their synchronicity in a given reference frame (the ether frame). With respect to the corresponding "true time," the propagation of light is not isotropic in a moving frame, and clocks synchronized by ignoring this anisotropy only give the "apparent time." Reichenbach (1928) imagined still other asymmetric conventions: cf. Janis 2002; Brown 2005, pp. 95–8, 102–5. All these nonstandard conventions contradict Einstein's principle that a theoretical representation should not exhibit asymmetries that do not have an empirical counterpart.

The space coordinates in two inertial frames F and F′ can be chosen so that

$$x' = \gamma(x - vt), \quad y' = y, \quad z' = z, \tag{61}$$

wherein v is the velocity of any point of F′ with respect to F, and γ is a constant. With respect to the light propagation and inertial motion on which the definition of coordinates is based, the two frames F and F′ play completely symmetric roles and they are both isotropic. Therefore the relation of F′ to F should be the same as the relation of F to F′, except that the direction of the relative motion is inverted.[83] Granted that the units are the same in both frames, this symmetry implies that

$$x = \gamma(x' + vt'). \tag{62}$$

Moreover, the speed of light should be the same in both frames. Therefore, $x' = ct'$ must hold whenever $x = ct$. Multiplying $x' = \gamma(x - vt)$ by $x = \gamma(x' + vt')$ and dividing the result by tt', we get

$$c^2 = \gamma^2(c - v)(c + v), \text{ or } \gamma = (1 - v^2/c^2)^{-1/2}. \tag{63}$$

A little more algebra leads to the Lorentz transformation

$$x' = \gamma(x - vt), \quad t' = \gamma(t - vx/c^2), \tag{64}$$

which contains the Galilean transformation

$$x' = x - vt, \quad t' = t \tag{65}$$

as the limit $c = \infty$.

In the Helmholtzian deduction of the Lorentz transformations, we exploit the fact that the transformations for which the x' axis slides over the x axis form a one-parameter continuous group of linear transformations. Setting the origins of space and time in both frames at the same event, the transformation is reducible to a 2×2 matrix acting on (x, t). Owing to the isotropy of the frames, the transformation obtained by changing simultaneously the signs of x and x' in a given infinitesimal transformation should also be an infinitesimal transformation of the group. Consequently, the diagonal elements of the generator of the group must be zero.[84] In the degenerate case for which the two eigenvalues of this generator vanish, it can be represented by the matrix[85] $\begin{pmatrix} 0 & 1 \\ 0 & 0 \end{pmatrix}$. By exponentiation, this gives the Galilean transformation $x' = x - vt$, $t' = t$.

---

[83] This reciprocity is not as obvious as it seems: cf. Berzi and Gorini 1969.

[84] Diagonal matrices, which also satisfy the isotropy condition, are excluded because they correspond to a mere change of the units of time and length.

[85] The symmetric choice leads to the absurd $x' = x$.

In the non-degenerate case, the vanishing of the trace of the generator and the real character of its determinant imply that the two eigenvalues are either real or purely imaginary, with opposite signs. In the latter case, the transformations would be rotations, so that the iteration of a small boost of the frame of reference would absurdly lead back to the original frame. We are left with the real case for which the eigenvalues of the generator can be taken to be 1 and -1. The vanishing of the diagonal elements then implies the form $\begin{pmatrix} 0 & c \\ c^{-1} & 0 \end{pmatrix}$ of the generator, which generates the transformations

$$x' = x \cosh \varphi - ct \sinh \varphi, \quad t' = -xc^{-1} \sinh \varphi + t \cosh \varphi \qquad (66)$$

by exponentiation. This is just another way of writing the Lorentz transformations, since the condition $x = vt$ for $x' = 0$ implies $\tanh \varphi = v/c$.[86]

The similarity of this reasoning with Helmholtz's determination of the Euclidean character of space is striking. The counterpart of the Euclidean metric is the invariant form $c^2 t^2 - x^2$, more generally $c^2 t^2 - x^2 - y^2 - z^2$. There are a couple of differences, however. Firstly, in the case of space the existence of an invariant distance is assumed, whereas in the case of spacetime it is derived from the group structure. This difference is spurious, because in the former case we could very well start with the existence of bodies whose displacements define a Lie group of linear transformations on $\mathbf{R}^3$, with proper values of the dimensions of the subgroups that leave one or two points invariant. As Lie showed, this sole requirement implies the existence of a quadratic invariant (degenerate cases being excluded). To require that the displacements form a group is evidently the same as requiring the congruence of displaced bodies to be transitive (and symmetric).[87] As we saw in the case of rods, this transitivity implies rigidity.

Another difference between the space and spacetime cases comes from the indefinite sign of the metric in the latter case. If the invariant interval between two events is positive, there exists a reference frame such that the two events occur at the same point of space, and the invariant measures the time difference of the events in this frame. If the interval between the two events is negative, there exists a reference frame for which the two events are simultaneous, and the invariant measures their spatial distance. This is the Minkowskian distinction between intervals of space-like and time-like kinds. The constant c represents the maximum velocity that a reference frame can have with respect to another. In order not to violate causality, signals can only relate events separated by a time-like interval. Therefore, c must also be the maximum signaling speed between two stations. As this speed is isotropic and does not depend on the reference frame, it can be used to control rigidity and to synchronize clocks in the abovementioned manners.

---

[86] Ignatowski (1910) gave the first derivation of this kind (see above, p. 139); he used finite transformations only. For simplified versions of his demonstration, see Lévy-Leblond 1976; Pal 2003; Brown 2005, pp. 105–10. The present derivation is more automatic and more similar to Helmholtz's derivation of the Euclidean structure.

[87] Call **g** the (unique) transformation that brings the body B to congruence with B′, and **h** the transformation that brings the body B′ to congruence with B″. The symmetry of the congruence of B and B′ is equivalent to the invertibility of **g**; the transitivity of congruence between B, B′, and B″ is equivalent to the existence of the product **gh**.

In the Helmholtzian derivation of the Lorentz transformations, we could have done without the constancy of the velocity of light and without the concepts of inertial frame and time. All that is needed is the local existence of rigid Euclidean frames and the availability of a continuous pre-time en each frame. Assuming that the coordinate transformation from the frame F to the frame F′ is continuous and differentiable, it can be represented by a linear transformation in a small enough neighborhood of a given event, so that the points of F′ all have the same constant velocity with respect to F. Again, these transformations form a one-parameter continuous group for a given choice of the direction of the relative velocity. Symmetry considerations thus suffice to derive the local validity of the Lorentz group, in conformity with the EPS derivation of the conformal structure of spacetime without appeal to the projective structure induced by inertial motion. In contrast, the Einsteinian derivation of the Lorentz group appeals to inertial frames and to inertial time defined by inertial motion in rigid frames.

In the case of geometry, we saw that a choice of spatial coordinates for which the distance is given by $\sqrt{x^2 + y^2 + z^2}$ only needed to exist at small scale. Similarly, the former discussion of space and time measurement only requires the Minkowskian spacetime metric $c^2 t^2 - x^2 - y^2 - z^2$ in the vicinity of a given event. For given conventions of measurement, the extent to which time and space coordinates can still be chosen in accordance with this simple measure at a larger scale is an open empirical question. By analogy with the Riemannian structure of space, the most we can demand is that spacetime be a four-dimensional differentiable manifold with a metric $ds^2 = g_{\mu\nu} dx^\mu dx^\nu$ that locally takes the Minkowskian form for a proper class of coordinates. General relativity takes advantage of this more general spacetime geometry.

In the theory of space, it is a priori possible to give a metric definition of the coordinates, for instance as the distances from three rigidly connected reference points measured by means of rigid rods of a given class. In contrast, general relativity does not admit the extended rigid frames that a metric definition of coordinates would require.[88] In order to determine the interval between distant events in the context of this theory, physicists rely on the behavior of light rays and free-falling objects. The *local* determination of spacetime coordinates through a (locally) rigid frame and attached clocks nonetheless remains possible.

Operations and idealizations

We have now reached the end of a construction of the concepts of time and spacetime that has much analogy with the Helmholtzian construction of space. In both cases, we started with vague, private notions of distance and time such as counting steps or ordering events in our consciousness. In order to sharpen these notions, we introduced operational definitions that were partially conventional, partially restricted by empirically testable criteria. Thanks to these definitions, space and time are measured by numbers. For mathematical convenience and for the idealization of rigid bodies to be compatible with the idealization of infinitely precise measurement, the results of measurement are regarded as real numbers

---

[88] Cf., e.g., Landau and Lifchitz 1959, § 84.

for a given unit. Historically, the intuitive perception of the continuity of time led to its representation through proportional lengths; Euclidean geometric reasoning, based on figure and proportion, was used instead of algebraic reasoning over real numbers. Both kinds of reasoning require the idealization of the continuum, which itself derives from the idea of indefinitely sharpened measurement.[89]

In this Helmholtzian approach, space measurement and clock synchronization only make sense with respect to an extended rigid body. Space and time are intimately interrelated through this common reference. As soon as the ancestral intuition of instantaneous signaling is given up, the relativity principle implies that space and time coordinates intermix when passing from one reference frame to another. The intermixing is locally given by a Lorentz transformation. This remarkable result can be obtained in a way quite similar to Helmholtz's derivation of the (locally) Euclidean character of space. In fact, both results could be obtained simultaneously by including the choice of an orthonormal vector triad in the definition of a reference frame (this can be done without previous knowledge of the Euclidean metric, because right angles can be defined directly from the assumption of freely mobile rigid bodies). The spacetime coordinate transformations from one frame to another locally form a Lie group with ten generators (four for translations, three for rotations, and three for boosts). This group is the invariance group of the Minkowskian metric, called the Poincaré group.

Surveying and timing operations play a significant role in this Helmholtzian genesis of the concepts of space and time. They prompt the idealizations that lead to the mathematical structures of Riemannian and pseudo-Riemannian manifold. These idealizations have a life or their own, not only qua mathematical theories but also qua physical theories. Although the original operational definitions may be impractical for increased precision of measurement or for extreme scales, the idealizations are still useful when new, theory-dependent conventions of measurement and indirect methods of measurement are introduced. What can be learned from the operational genesis of the concepts of space and time is not a concrete operational foundation of them à la William Bridgman. The true lesson is that the pseudo-Riemannian structure of spacetime is deeply related to the possibility of rigid bodies and periodic processes or inertial motion in a restricted, local sense. This possibility itself derives from the assumed measurability of space and time, as long as measurement is meant to imply the existence of invariable standards.

Large and small scales

The idealization of primitive measures of space and time naturally leads to the locally Minkowskian character of spacetime. Conversely, we expect pseudo-Riemannian manifolds to be relevant to the physical world, at least at scales for which measurement remains conceivable. Space surveying by rigid rods and time measurement by concrete clocks only work at moderate scales. The exploration of smaller and larger scales requires optical means. Medium-scale experiments establish the rectilinear propagation of light, the laws

---

[89] I will return to this point in Chapter 6.

of geometrical optics, and the constancy of the velocity of light. In turn these results allow for goniometry, optical triangulation, microscopes, telescopes, and the inferential dating of events.[90]

The conformity of these indirect determinations of space and time with the direct ones can only be verified at the medium scale for which the latter are available. In order to extend the limits of perception, physicists have simply substituted the latter with the former. That is to say, they have conventionally defined rigidity and synchronicity so that light propagates rectilinearly with a constant velocity in a vacuum. The adequacy of this convention, as of any other convention of measurement, depends on the simplicity of the physics to which it leads. At large scale, modern physicists prefer to give it up and let the theory of general relativity determine the propagation of light in relation with spacetime geometry.

Whether or not spacetime concepts apply at a given scale depends on the mutual consistency of space and time determinations done by various direct or indirect means and on the integration of these determinations in a consistent physical theory. That the pseudo-Riemannian concept of spacetime should hold at arbitrarily large scales follows naturally from the validity of the pseudo-Euclidean concept at small scale. Thus, the reliance of our best cosmological theories on this concept should not be a surprise. More intriguing is the success of spacetime concepts at very small scales. Classical physics, which assumes perfectly localized events, works well at scales much smaller than that of ordinary perception. Although quantum mechanics and quantum field theory (in their standard formulations) no longer assume well-defined histories in space and time, they do allow for the indefinitely precise determination of a point-like event. This possibility somewhat justifies the use that the theory makes of sharply defined spacetime variables, be it in the classical basis for quantization, in the definition of a subclass of quantum states, or in the characterization of certain observables. Since the 1930s physicists have repeatedly speculated about a breakdown of spacetime concepts in relation with the infinities of quantum field theory or with the difficulties of a quantum theory of gravitation. Whether or not spacetime only is an emergent macroscopic structure still is an open question.

Conventions, structure, and measure

In the previous chapter we saw that the Helmholtzian construction of space implied a number of conventional choices for units and reference points and for the class of rigid bodies. The end product of this construction, a Riemannian manifold, bears two traces of this conventionalities: the freedom in the choice of coordinates, and the dependence of the metric on the class of rigid bodies. The only convention-free result is the locally Euclidean character of the metric. As we will now see, the situation is different for the Helmholtzian construction of spacetime.

Let us begin with a small region of spacetime. In this case, we have defined inertial frames, rods, and clocks in such a manner that the measured motion of a free particle is rectilinear and uniform and light propagates rectilinearly (in the approximation of geometrical

---

[90]Use of the latter instruments does not necessarily involve vision, since the virtual image they give can be optically transformed in a real image.

optics) at the universal, frame-independent velocity c in a vacuum (for a given choice of the length and time units). In effect, we have severely restricted the choice of rods and clocks by requiring that the laws of free motion and free propagation should be the simplest possible. The flat-spacetime subcase of Marzke's construction of bouncing-light clock shows that all Helmholtzian clocks are equivalent. The resulting Minkowskian metric is uniquely determined, save for the choice of length and time units.

We then assumed that the global spacetime was a pseudo-Riemannian manifold obtained by patching up small Minkowskian domains. In principle, the metric of this manifold can be determined by measuring relativist intervals of neighboring pairs of events with similar rods and clocks in a shared initial frame. The similarity of the rods and clocks used in different regions is here essential. It implies that two rods and clocks that gave identical results when they were once used in the same local frame will still give the same result if they are used again in another shared local frame in a different domain. This excludes the Weylian indetermination of the metric.

All events of the manifold may be labeled by arbitrary (non-metric) coordinates x^μ for instance by filling space with a dense network of pre-clocks. Call $\bar{x}^\mu$ a choice of coordinates that coincides with the coordinates of an inertial frame in the vicinity of a given event. In these coordinates and in this vicinity, we have

$$\mathrm{d}s^2 = \bar{g}_{\mu\nu}\mathrm{d}\bar{x}^\mu\mathrm{d}\bar{x}^\nu \approx \eta_{\mu\nu}\mathrm{d}\bar{x}^\mu\mathrm{d}\bar{x}^\nu = c^2\left(\mathrm{d}\bar{x}^0\right)^2 - \left(\mathrm{d}\bar{x}^1\right)^2 - \left(\mathrm{d}\bar{x}^2\right)^2 - \left(\mathrm{d}\bar{x}^3\right)^2. \quad (67)$$

As this expression should remain valid around a neighboring event of coordinates $\bar{x}^\mu + \delta\bar{x}^\mu$, we must also have $\partial\bar{g}_{\mu\nu}/\partial\bar{x}^\rho = 0$ and $\bar{\Gamma}^\mu_{\nu\rho} = 0$ for the associated connection. The metric at point x in the original coordinate system is therefore given by

$$g_{\mu\nu} = \partial_\mu\bar{x}^\rho\partial_\nu\bar{x}^\sigma\eta_{\rho\sigma}, \quad (68)$$

and the associated connection is given by

$$\Gamma^\mu_{\nu\rho} = \frac{\partial^2\bar{x}^\sigma}{\partial x^\nu\partial x^\rho}\frac{\partial x^\mu}{\partial\bar{x}^\sigma} \quad (69)$$

As we may repeat the same consideration in the vicinity of any event, the metric $g_{\mu\nu}$ is fully determined by local surveys with clocks and rods meeting the prescriptions of the Helmholtzian construction. These prescriptions imply rectilinear and uniform propagation for free particles and light rays in vacuum in the vicinity of a given event. Consequently, there exists a parameter τ such that the trajectory $\bar{x}^\mu(\tau)$ of the particles or rays satisfies $\ddot{\bar{x}}^\mu = 0$. The rays further satisfy $\dot{\bar{x}}^2 = 0$. Since the $\bar{\Gamma}^\mu_{\nu\rho}$ symbols vanish at the given event, these conditions imply that the equation of geodesic lines is satisfied at the given event. Since this is true for any event crossed by the trajectory, free particles must follow geodesics of the metric manifold, and light rays must follow null geodesics. We thus see that in the Helmholtzian approach the desired behavior of light and free particles in relativistic spacetime follows from the definition of rods and clocks.

This implication leads us to another way of determining the metric empirically: by identifying the trajectories of free particles and light rays, as Weyl recommended. Be it determined in this simple manner or through local clock-based surveys, the metric structure of spacetime seems perfectly defined. Does that mean that any conventional element has been eliminated from the theory? Surely not, the assumption that clocks and rods behave in ways locally compatible with the principle of inertia and the principle of relativity, or the assumption that free particles follow geodesic of spacetime are conventions or disguised definitions as Poincaré would put it. However, these conventions are designed so as to make the simplest objects behave in the simplest theoretical manner. Broadly speaking, we may impose two reasonable constraints on a good spacetime geometry: its metric structure should be determinable by concrete operations of measurement, and the global theory of matter and fields in this geometrical framework should be as simple as possible. A common mistake is to apply one of these two constraints independently of the other. The first constraint, used separately, would rigidify theoretically disadvantageous conventions. The second constraint, used separately, could lead to a theory in which space and time would lose most of their empirical significance. The most reasonable course, the one Einstein recommended, is to maintain the measurability of spacetime by concrete rigid rods or standard clocks but to let the global theory dictate their precise realization.[91]

The implementation of this agenda is not so easy, however. In harmony with the Helmholtzian approach, we may try to base our construction of clocks on the condition that the principle of inertia should hold and the velocity of light should be a constant in a certain class of reference frames in small domains of spacetime. This condition, however familiar it might sound to the ears of modern physicists, is not so easy to justify. As was already mentioned and as was denounced by Harvey Brown, the principle of inertial implies a strange pre-established harmony for the motions of free particles in certain frames.[92] We will return to this puzzle in section 5 of Chapter 7. More annoyingly for our present concern, the Helmholtzian conditions do not tell us how to build a workable clock. Marzke's construction of a bouncing-light clock, which I described in the EPS section, is purely ideal. At best one might imagine practical tests for the agreement of a given clock with this ideal clock, as Volker Perlick did.[93] For effective clock construction, we need more experimental and theoretical baggage. For instance, from the validity of quantum mechanics in the absence of gravitation, we may infer that excited atoms are natural clocks with well-defined frequency. By the strong principle of equivalence, the inference extends to free-falling atomic clocks. However, a more complete answer should include (moderately) accelerated clocks and would require a consistent theory of combined gravitation and quantum effects. The measurability conditions thus call for a deeper unification of our existing theories.[94]

---

[91] On Einstein's views, cf. Stachel 1977. On the need to take the global structure of the theory into account, cf. Friedman 1983, Chap. 7.

[92] Brown 2005, pp. 14–18; Brown and Pooley 2006.   [93] See n. 72 above.

[94] As Harvey Brown notes, a dynamical understanding of the Lorentz contraction similarly calls for a quantum-relativistic theory of the forces responsible for the cohesion of a rigid rod. Cf. Brown 2005, Chap. 7.

Relationism versus absolutism

The Helmholtzian approach leads to a relationist concept of space, since spatial relations are only defined with respect to concrete reference frames. In other words, there is no intrinsic definition of a point of space, no definition of it without reference to an extended rigid body. The choice of the reference frames may nevertheless be restricted in ways that allow some sort of absolutism. For instance, some nineteenth-century physicists admitted the ether as a natural frame.[95] As this ether frame turned out to be empirically indistinguishable from other frames moving uniformly with respect to it, this way of making space absolute must now be regarded as obsolete. Another way is to exploit the fact that the laws of mechanics, especially the law of inertia, only hold in the Galilean class of reference frames. In partial agreement with this fact, Newton regarded space as some sort of pervading substance with respect to which absolute position, velocity, and acceleration were defined. He thus unwittingly introduced theoretical asymmetries with no empirical counterparts, since absolute position and absolute velocity are not empirically detectable.

In the context of Newtonian and relativistic mechanics, a better kind of absolutism is to define a spacetime structure within which absolute acceleration makes sense, whereas absolute velocity and position do not.[96] This structure cannot be purely spatial, for it implies reference bodies moving with respect to one another. It is defined by the class of chronogeometric relations among (point-like) events that are invariant under a change of inertial frame; and it is characterized by the Galileo group or the Lorentz group. Events themselves are objective, for the occurrence of a (nearly) point-like event such as the birth of Isaac Newton is something all observers can agree about. By some stretch of the imagination, we may figure spacetime to be densely filled with events, and define the points of spacetime through equivalence classes of coinciding events.[97] The set of these points may be regarded as some sort of substance. The Minkowskian metric or its degenerate Galilean form define intrinsic structures of this substance and thus feed the absolutist conception of spacetime.

The opposite attitude, recommended by Ernst Mach in the context of classical mechanics and mentioned early in this chapter, is to stick to the relational view and to seek a material reference for the allegedly absolute accelerations. A possible reference is the distribution of remote matter. A full implementation of this program should treat nearby matter and remote matter on the same footing, and retrieve the confirmed predictions of Newton's theory on the basis of concrete spatial relations only. Julian Barbour came closest to this ideal in recent years.[98]

In the context of the theory of general relativity, the spacetime substance can be provided with an intrinsic structure defined by the chronogeometric relations that are invariant

---

[95] For example, Drude (1994, p. 9), Lorentz, and Larmor did so. Even Mach flirted with this idea (see above, p. 136).

[96] For a discussion of such constructs, cf. Earman 1989, Chap. 2.

[97] Although this definition of a continuum of events agrees with ordinary experience, it leaves the notions of individuality, identity, and proximity of events in the dark. Cf. Geroch 1978, pp. 3–4; Brown 2005, pp. 10–14.

[98] See n. 6 above.

under any change of coordinates: geodesic connection, light-connection, space-like or time-like intervals, etc. This structure differs from the spacetime structure of Newtonian mechanics, even though to a certain approximation it can be used to describe the same gravitational phenomena. As is well known, the discrepancy results from the conventional character of the distinction between inertial and dynamical effects.

Even though the spacetime structure of general relativity can be characterized without appeal to reference systems, general relativity is often regarded as a vindication of the relational view of space and spacetime. The reason is that this structure is to a large extent determined by the distribution of matter, in seeming conformity with Mach's endeavor to give a material reference to absolute acceleration. Has Machian relationism finally won? This is not so sure because, as Einstein himself came to realize, general relativity does not comply with Mach's principle: the distribution of matter does not completely determine the metric properties. General relativity rather seems to blur the very distinction between the absolute and relational concepts, for it treats spacetime and matter as mutually interacting dynamical objects.

The hole argument

In the Helmholtzian approach to space, spatial determinations are always done with respect to a reference frame that can be materialized through an extended rigid body. For any two points of this body, a distance can be defined on the basis of rigid-rod surveying. Any point of this body is determined by its distance from three fixed points (singular cases can be locally avoided). For this physical choice of coordinates, local surveys determine the coefficients of the metric $ds^2 = g_{\alpha\beta}dx_\alpha dx_\beta$. The result is not an arbitrary regular function of the coordinates x_α. It must be compatible with the choice of coordinates, namely: the distances between the point of coordinates x_α and the three reference points, as computed from the metric ds^2, must be equal to the coordinates x_α.

Such constraints on the $g_{\alpha\beta}$ can be avoided by leaving the physical interpretation of the coordinates x_α open. From this point of view, which is the one usually adopted, a given value of the triplet of coordinates corresponds to a well-defined point of the rigid reference body, but no metric interpretation of the coordinates is a priori given. Such an interpretation is found a posteriori in the following manner. Take three reference points on the reference frame. The metric $ds^2 = g_{\alpha\beta}dx_\alpha dx_\beta$ can be used to compute the distance between these three points and a fourth arbitrary point as a function of the coordinates of this point. This gives us three relations, from which we can determine the coordinates x_α as a function of concrete distances. Thus, the $g_{\alpha\beta}$ coefficients completely determine the intrinsic metric properties of space, all of which can be reduced to the knowledge of the distance of any two points as a function of their distances from three reference points.

The converse is not true. Two different sets of $g_{\alpha\beta}$ may lead to the same intrinsic geometry. Indeed for any diffeomorphism $x \to x'$ (short for $x_1, x_2, x_3 \to x'_1, x'_2, x'_3$), from the function $g_{\alpha\beta}(x)$ we can construct the new function

$$g'_{\alpha\beta}(x) = g_{\gamma\delta}(x')\frac{\partial x'_\gamma}{\partial x_\alpha}\frac{\partial x'_\delta}{\partial x_\beta} \tag{70}$$

such that

$$g'_{\alpha\beta}(x)\,\mathrm{d}x_\alpha\,\mathrm{d}x_\beta = g_{\alpha\beta}(x')\,\mathrm{d}x'_\alpha\,\mathrm{d}x'_\beta. \tag{71}$$

As long as the diffeomorphism $x \to x'$ leaves the coordinates of the reference points invariant, the metric $g'_{\alpha\beta}$ leads to the same intrinsic geometry as the metric $g_{\alpha\beta}$, because the distances calculated from the new metric are the same function of the x' coordinates as the distances calculated from the old metric were of the x coordinates.

To illustrate this point, consider the trivial case of a one-dimensional manifold. In terms of the arbitrary coordinate x, the metric is $\mathrm{d}s = \sqrt{g(x)}\mathrm{d}x$. A point M of the manifold is determined by its distance $s(x) = \int_0^x \sqrt{g(\xi)}\mathrm{d}\xi$ from the origin O (for which $x = 0$). The distance between two points M' and M'' is trivially given by the difference $s' - s''$ of their distances from the origin. Thus, any choice of g leads to the same intrinsic geometry. In harmony with this fact, any g can be obtained by diffeomorphic deformation of the uniform metric $\mathrm{d}s = \mathrm{d}x$: sending the point of coordinate x of the manifold to the point of coordinate $x' = \int \sqrt{g(x)}\mathrm{d}x$ leads to the desired metric.

A similar underdetermination occurs for the spacetime metric tensor of general relativity. Failure to appreciate this point leads to Einstein's hole argument of 1913, and to seeming violations of determinism. Suppose spacetime to be divided into a past and a future by a space-like hypersurface (for which the interval between any two points is space-like). There obviously are diffeomorphisms that reduce to the identity in the past of this hypersurface and differ from the identity in its future. As Einstein's equations for the gravitational field $g_{\mu\nu}(x)$ and other coupled fields are invariant through any diffeomorphism, different future values of the fields are compatible with the same past value.

In recent years, this seeming violation of determinism has been used as an objection against a peculiar substantivalist interpretation of spacetime in which the field function $g_{\mu\nu}(x)$ is regarded as an objective property of a spacetime substance. This is about the same as saying that the thermal dilation coefficient is an objective property of a material substance before knowing what the temperature scale is. Indeed, when expressed in arbitrary coordinates the metric function $g_{\mu\nu}(x)$ does not directly express empirically meaningful relations because the metric interpretation of the coordinates is only known through the function $g_{\mu\nu}(x)$ itself. What evolves deterministically in general relativity is not this function, but the diffeomorphically invariant relations among intervals between pairs of events.[99]

Thus falls the hole argument against spacetime substantivalism. Nothing prevents us from defining a spacetime substance through equivalence classes of coinciding events.[100] The underdetermination of the metric tensor does not interfere with this definition, because the concept of event precedes the concept of the pseudo-Riemannian manifold in which events are inscribed. Nor does this underdetermination threaten the objectivity of the metric structure of this substance, since it only reflects the lack of metric significance of the coordinates. Had Einstein and later revivers of the hole argument kept in mind the

---

[99] On the hole argument and its philosophical consequences, cf. Earman 1989, Chap. 9; Norton 2004 and bibliography there given.

[100] This statement should be nuanced by the remark of n. 97 above.

Helmholtzian genesis of space or spacetime, they would not have fallen into the error of defining spacetime before its measurement.

Conclusions

Let us return to our main concern, which is the viability of rationalist arguments in favor of a well-defined structure of time and space. Kant's original variety of transcendental philosophy no longer is a workable option, unless the Kantian table of categories is seriously revised to allow for the relativity of time and space, or unless the empire of categories is limited to the intuition of events occurring in the immediate neighborhood of a given percipient. Weyl, who suggested the latter way out, and Eddington, who did not bother to refer to Kant, invented their own varieties of transcendental idealism. In agreement with Husserl's notion of regional ontology, Weyl believed in the shining self-evidence of the mental structure implied in our experience of a given domain of the outer world. He regarded his pure infinitesimal geometry and his unified theory of gravity and electricity as the necessary and final expression of the psycho-mathematical structures adapted to our physical experience. Eddington similarly believed that the world geometry and its physical interpretation derived from basic-preconditions of our experience of the world, namely: the multiplicity of observations involving arbitrary codes, the translatability of observations performed in different codes, the existence of permanent features and conserved quantities in a comprehensible world.

Yet Weyl and Eddington did not derive their world structure by mere introspection. In order to ascertain the transcendental core of any future physics, they consulted history, contemporary physics, and contemporary strategies of mathematical construction. On the one hand, they were well informed of the historical evolution of space and time concepts from Newton to Einstein and they appreciated the power of Einstein's heuristics. On the other, they were guided by the inner constructive logic they perceived in the mathematical apparatus of infinitesimal geometry.

This logic, more or less impregnated with Husserlian phenomenology, brought Weyl to divorce the transcendentally certified world-geometry from the natural geometry explored with concrete rods and clocks. Eddington accentuated this move by removing the metric from the premises of the world geometry. In retrospect, we may judge that these two world builders overlooked two important consequences of the separation of fundamental and physical geometries. Firstly, this separation opens a much wider realm of possibilities for the fundamental geometry than ever dreamt in their philosophy: higher dimensions, torsion, *Fernparallelismus*, fiber bundles, non-commutative geometries, discrete geometries, etc. In raising the Weyl manifold or the affinely connected manifold to transcendental status, Weyl and Eddington updated Kant's error of raising Euclidean geometry to transcendental status. Secondly, by making physical geometry only a byproduct of a more fundamental world geometry, they allowed more freedom in the choice of the physical geometry than is empirically needed for an impressive range of scales.

The more sober, Helmholtzian variety of rationalism avoids these two pitfalls of Weyl's and Eddington's versions of transcendental idealism. For Helmholtz, the comprehensibility of the world implies the possibility of certain concrete operations (rather: ideal models of such operations). This possibility partly determines the structure of physical theory.

However, the comprehensibility of the world is hypothetical; and its precise content is regional, scale-dependent, and open to revision. In contrast, for the transcendental idealist the a priori conditions of experience are non-hypothetical and uniquely determined, and the associated operations need not be concrete. Among the deductions of spacetime structures encountered in this chapter, not only what I called the Helmholtzian deduction but also the deductions given by Ignatowski, Robb, and EPS fit into the Helmholtzian category of rationalism. For Ignatowski, comprehensibility implies the relativity of motion and the Lie-group structure of changes of reference frame. For Robb, comprehensibility implies the possibility of a partial, causal ordering of events. For EPS, comprehensibility implies the possibility of a geodesy based on light rays and free-falling particles. In the Helmholtzian deduction, comprehensibility implies the measurability of space and time through rigid rods and light-clocks.

Ignatowski's deduction of the Minkowskian structure suffers from the arbitrary selection of a limited class of reference frames. Robb's suffers from his artificial choice of the axioms of conical order. EPS is perhaps most satisfactory for the conceptual simplicity of its basis and for the mathematical elegance and rigor of its developments. However, this approach hardly deserves to be called a physical chronogeometry as long as it is not supplemented with a specific axiom for the constancy of measuring standards. If this axiom is admitted as part of the comprehensibility of nature, the pseudo-Riemannian structure of spacetime can be deduced in the manner suggested in section 3. Whatever be the weaknesses of these various Helmholtzian-rational deductions of spacetime structures, whichever be our favorite, there is the reassuring fact that their conclusions agree with one another. This convergence strongly indicates the local necessity of the Minkowskian structure and the global necessity of the pseudo-Riemannian or Weylian structure of spacetime as long as space and time remain measurable. Even if this structure turned out to fail at extremely small scales, it would make better sense to regard this failure as a dissolution of the concepts of space and time than to artificially redefine these concepts.

To say that the measurability of space and time leads to the pseudo-Riemannian structure of spacetime is not the same as saying that this measurability is obvious. As was just mentioned, it might fail at very small scales. Even at our scale, this measurability turns out to require a strange pre-established harmony of the motion of free particles, and its implementation in concrete, workable clocks such as atomic clocks raises additional difficulties. Fortunately, the implications of measurability for the structure of spacetime can be worked out without addressing these difficulties. This explains the success of Einstein's heuristics, and permits the moderate spacetime rationalism defended in this chapter.

6

NUMBERS AND MATH

> One may describe the situation by saying that the mathematician plays a game in which he himself invents the rules while the physicist plays a game in which the rules are provided by Nature, but as time goes on it becomes increasingly evident that the rules which the mathematician finds interesting are the same as those which Nature has chosen. (Paul Dirac, 1939)

> It might be asked, why in physical science generalization so readily takes the mathematical form. The reason is now easy to see. It is not only because we have to express numerical laws; it is because the observable phenomenon is due to the superposition of a large number of elementary phenomena that are *all similar to each other*.[1] (Henri Poincaré, 1902)

In a comparison of Plato's and Aristotle's philosophies published in 1597, Galileo's colleague at Pisa, Jacopo Mazzoni, wrote:

> There is no other question that has given place to more noble and beautiful speculation . . . than the question whether the use of mathematics in physical science as an instrument of proof and a middle term of demonstration, is opportune or not; in other words, whether it brings us some profit, or on the contrary is dangerous and harmful . . . It is well known that Plato believed that mathematics was quite particularly appropriate for physical investigations, which was the reason why he himself had many times recourse to it for the explanation of physical mysteries. But Aristotle held a quite different view and he explained the errors of Plato by his too great attachment to mathematics.

Modern physicists have decidedly opted for what Mazzoni believed to be the Platonist alternative. The most conspicuous feature of modern physical theory indeed is its being expressed mathematically, usually by means of functions, vector spaces, and functional spaces all based on the prior construct of real numbers. Is there any necessity for this pregnancy of mathematical analysis? This question haunts physical theory since the early modern times.[2]

In *Il saggiatore* (1623), Galileo famously declared:

> Philosophy is written in that great book which ever lies before our eyes—I mean the universe—but we cannot understand it if we do not first learn the language and grasp the symbols, in which it is written. This book is written in the mathematical language, and the symbols are triangles, circles and other geometrical figures, without whose

---

[1]Poincaré 1902, p. 87.   [2]Mazzoni 1597, p. 187, cited in Koyré 1943, pp. 420–1.

Physics and Necessity. First Edition. Olivier Darrigol.

help it is impossible to comprehend a single word of it; without which one wanders in vain through a dark labyrinth.

Closer to us, in 1939, the theoretical physicist Paul Dirac deplored that some features of the physical world, such as initial conditions and coupling constants, still resisted mathematical determination:

> This feature is so unsatisfactory that I think it safe to predict it will disappear in the future, in spite of the startling changes in our ordinary ideas to which we should then be led. It would mean the existence of a scheme in which the whole of the description of the universe has its mathematical counterpart, and we must suppose that a person with a complete knowledge of mathematics could deduce, not only astronomical data, but also all the historical events that take place in the world, even the most trivial ones. Of course, it must be beyond human power actually to make these deductions, since life as we know it would be impossible if one could calculate future events, but the methods of making them would have to be well defined. The scheme could not be subject to the principle of simplicity since it would have to be extremely complicated, but it may well be subject to the principle of mathematical beauty.

Reciprocally, Dirac propounded that every beautiful piece of mathematics (beauty having to do with symmetry and group-theoretical structure) would some day find an application in physics:[3]

> Pure mathematics and physics are becoming ever more closely connected, though their methods remain different. One may describe the situation by saying that the mathematician plays a game in which he himself invents the rules while the physicist plays a game in which the rules are provided by Nature, but as time goes on it becomes increasingly evident that the rules which the mathematician finds interesting are the same as those which Nature has chosen. It is difficult to predict what the result of all this will be. Possibly, the two subjects will ultimately unify, every branch of pure mathematics then having its physical application, its importance in physics being proportional to its interest in mathematics.

Galileo did not express the source of his mathematical conviction. Historians have offered many explanations, ranging from neo-Platonism to casual awareness of the power of mathematics in early modern technology. Dirac was probably extrapolating his own success in inventing new physical theories by exploring the most beautiful mathematical possibilities. He accepted the mathematical character of physics as a self-evident truth. In contrast, his brother-in-law Eugen Wigner discoursed upon "The unreasonable effectiveness of mathematics in the natural sciences" with the following moral:[4]

> The miracle of the appropriateness of the language of mathematics for the formulation of the laws of physics is a wonderful gift which we neither understand nor deserve. We should be grateful for it and hope that it will remain valid in future research and that it will extend, for better or for worse, to our pleasure, even though perhaps also to our bafflement, to wide branches of learning.

---

[3] Galileo 1623, cited in Burtt 1925, p. 64; Dirac 1939, p. 124. On Dirac's views, cf. Kragh 1980, Chap. 14.

[4] Wigner 1960, p. 14.

Should the book of nature be written in mathematical language, as Galileo claimed? Should every aspect of the physical world be amenable to mathematical analysis, as Dirac hoped? Should every domain of mathematics ultimately find application in physics, as Dirac predicted? Should the mathematical quality of nature forever remain a mystery, as Wigner suggested? This chapter is an attempt to answer these questions in a historically grounded manner. Its first section is a sketch of concerns with the mathematization of physics from Descartes to the nineteenth century; the second is a brief history of quantity from the ancient Greeks to the nineteenth century; the third deals with Helmholtz's essay on counting and measuring; the fourth with Poincaré's further elaboration of a philosophy of quantity in which the physical success of mathematical analysis receives a natural explanation.

6.1   From Descartes to the nineteenth century

Descartes offered an early modern answer to the question of the necessity of mathematics by arguing that pure extension, being the sole attribute of physical objects that resisted systematic doubt, should be the foundation of the entire system of the world. His opinion remained influential, especially in France, at least until the mid-eighteenth century. Newton and Leibniz also brought their share of mathematical reductionism, by making a geometric or algebraic theory of quantities one of the preconditions of their philosophies of nature. In practice, however, eighteenth-century natural philosophy was divided into a more qualitative, empirical branch called physics and a quantitative branch called mixed mathematics. Opinions varied on whether the entire domain of natural philosophy should be subjected to mathematics. Cartesians were most optimistic; Neo-Newtonians, the French encyclopedists for instance, believed that some aspects of nature would forever escape mathematical analysis; some chemists and early romantics regarded mathematics as a threat to a genuine philosophy of nature.[5]

Toward the close of the century, there was a growing frenzy at quantifying every domain of knowledge. Quantities of light, heat, and electricity were defined and measured. In Laplace's influential *Système du monde*, Newtonian astronomy became the norm for the physical sciences. Laplace identified physical principles that were counterparts of mathematical properties: for instance, the principle of superposition of small oscillations, which is a consequence of the linearity of the equations of motion. He delighted in this concrete power of mathematical analysis:[6]

> It is interesting thus to retrieve in natural phenomena the intellectual truths of analysis. This correspondence, of which the System of the World will offer us numerous examples, is one of the greatest charms attached to mathematical speculation.

On the side of philosophy, Kant's system offered a new explanation of the success of mathematics in physics. This explanation involves a reputedly obscure component of his doctrine, which is called schematism. For Kant, the pure forms of intuition do not in

---

[5]Descartes and Leibniz believed in *mathesis universalis*, a universal science modeled on mathematics. Cf. Rabouin 2009.

[6]Laplace 1797, p. 171. On the quantifying frenzy, cf. Heilbron 1993.

themselves carry any mathematical structure. They acquire such structure when objects presented to the intuition are subjected to the pure concepts of the understanding, or categories. The two first categories are quantity and quality, quantity referring to the synthesis of a multiplicity and quality to the reality of an object of perception. Categories of understanding and intuitions, Kant tells us, are heterogeneous, so that the former cannot be directly applied to the latter. Mediating devices are needed, which Kant calls *schemata*. A schema is a procedural rule for associating a mental image to a category. Time, being common to all representations, is sufficient to implement the schemata of the various categories.[7]

The schema of quantity is number, regarded as the ordering of a discrete series of events. This schema generates time as the pure image of quantity in internal intuition and space as the pure image of quantity in external intuition. The schema of quality is the degree of "filling" of a given instant of time with a sensation induced by an object. Kant calls this degree an "intensive quantity," in contrast with the "extensive quantities" for which the representation of the parts precedes the representation of the whole. In plain words, space and time are measurable, sensations are not. The doctrines of space and time, properly structured by the category of quantity via the schema of number, define Kant's mathematics. Therefore, for Kant arithmetic, geometry, and all mathematics are synthetic a priori. Since all our knowledge of the world occurs through representations in space and time, the world is inherently mathematical.

The basic idea behind the Kantian jargon is that all our knowledge of the world must be formulated in accordance with a priori conditions of the mind's logical and representational faculties. In the light of later criticism, we may judge that Kant overestimated his capacity to reach these conditions by a priori means and unconsciously injected empirical material in his derivation and application of the category of quantity. For example, his definition of extensive quantity implicitly draws on the practice of measurement by replication and addition of units. As we will later learn from Helmholtz, the compatibility of this practice with arithmetic can only be decided empirically. Another weakness of Kant's analysis is that it is limited to the mathematical character of space and time in our representation of physical phenomena. It does not address the question of the adequacy of mathematics for physical properties that do not have a geometric interpretation. On the contrary, Kant's definition of the degree of sensations as intensive, non-measurable, quantities suggests the non-mathematical character of much or our experience. Even density, in his opinion, should be treated as an intensive quantity and not, as was usually the case in early Newtonian physics, as the proportion of pores in a primordial homogeneous matter. This limitation is partially circumvented, in his *Metaphysical foundations of natural science*, by a general theory of matter and force in which space and time relations are the main ingredients. Even so, Kant judges that some alleged sciences of his time, such as chemistry, psychology, and the more qualitative domains of physics do not yet permit a reduction of this kind and therefore are not true sciences.[8]

---

[7] Kant 1781, *Analytik der Grundsätze*, Chaps. 1–2.

[8] Kant 1781, Chap. 2, section 3.2 (*Anticipationen der Wahrnehmung*); Kant 1786.

In the nineteenth century, the mathematization of physics gained considerable ground, though not necessarily in the neo-Newtonian form imagined by Laplace and his disciples. This evolution went along with increasingly precise experimentation in every domain of physics. Cumulative success, rather than philosophy, motivated these efforts. The measurability of physical properties and their mathematical representability were usually regarded as given and above critical scrutiny. The few natural philosophers who desired a transcendent justification of the power of mathematics could refer to Kant's system, at least before it became the target of empiricist criticism. As we will see in a moment, the first important nineteenth-century investigation of measurability in relation with mathematics was the one given by Helmholtz in 1887, in parallel with his earlier discussion of the foundations of geometry. Before we discuss the motivations and contents of this investigation, it will be useful to take a closer look at the ways in which physical quantities and their relations were represented from Greek antiquity to Helmholtz's times.

6.2 A historical sketch of quantity

The Euclidean heritage

That objects can be counted and that a length can be measured by comparing it with a multiple of a unit length are primitive notions that led the ancient Egyptians and the Babylonians to develop a theory of ratios and proportions. This theory was originally confined to commensurable quantities: the ratio of two quantities was simply given by the numbers of (equal) parts of the second quantity necessary to reproduce the first. The possibility of such comparison was taken for granted until Pythagoras of Samos, in the fifth century BC, proved that the diagonal of a square was incommensurable with its side. This discovery presumably prompted Plato's disciple Eudoxus of Cnidus to redefine ratios in a non-numerical manner. Euclid of Alexandria exploited this definition in the axiomatic theory of magnitude, ratio, and proportion that forms the fifth book of his *Elements*. This powerful theory is worth special attention, for it long remained the basis for the formulation of quantitative laws.[9]

The six first definitions of Euclid's fifth book read:

1. A magnitude is a part of a magnitude, the less of the greater, when it measures the greater.
2. The greater is a multiple of the less when it is measured by the less.
3. A ratio is a sort of relation in respect of size between two magnitudes of the same kind.
4. Magnitudes are said to have a ratio to one another which are capable, when multiplied, of exceeding one another.
5. Magnitudes are said to be in the same ratio, the first to the second and the third to the fourth, when, if any equimultiples whatever be taken of the first and third, and any equimultiples whatever of the second and fourth, the former equimultiples alike exceed, are alike equal to, or alike fall short of, the latter equimultiples respectively taken in corresponding order.
6. Let magnitudes which have the same ratio be called proportional.

---

[9]Cf., e.g., Vitrac 1994.

Definitions 1 and 2 involve the notion of a magnitude "measuring" another, which means that the latter magnitude can be obtained by juxtaposition of identical copies of the former, in analogy with a multiple length obtained by repeated translation of a given length. These definitions define multiples and submultiples of a given magnitude. According to definition 3, the ratio is a binary relation between two magnitudes of the same kind, and the more precise definition of this relation should correspond to our intuition of how much one magnitude is larger than the other. Definition 4 adds another precondition for the comparability of two magnitudes: that there should be multiples of the smaller quantity that exceed the larger. This condition corresponds to what we would now call the Archimedean property of a set of homogenous magnitudes, granted that an order relation has been previously defined on this set. Definition 5, traditionally attributed to Eudoxus, ingeniously defines the equality of ratios through the comparison of "equimultiples." Although equimultiplicity is a pre-numerical notion, for the convenience of the modern reader a multiple of a magnitude may be written as na, where n is an integer. In this notation, two magnitudes a and b are said to be in the same ratio as two other magnitudes a' and b' when for any two integers n and m the order of ma and nb is the same as the order of ma' and nb'.[10]

Euclid could have defined (whole) numbers as special kinds of ratios, for which the first quantity is a multiple of the second. This would have spared him the writing of the seventh book of his elements, in which he defines a number as "a multitude composed of units," a unit being "that by virtue of which each of the things that exist is called one" (meaning, presumably, that a number is defined by a set of objects once every property of each object is forgotten except its mere being). The usual reason evoked for the existence of this book of the elements is that Euclid followed the tradition of the older number-based theory of proportions. Another might have been that the Greeks and most mathematicians until the seventeenth century refused to regard ratios as numbers in general. Regarding fractions and "surd" (irrational) ratios as numbers is an algebraic step in which composition rules and their properties become more important than the combined objects. Newton notoriously took this step in his *Arithmetica universalis* of 1707:[11]

> By *number*, we understand not so much a multitude of units as the abstracted *ratio* of any quantity, to another of the same kind, which we take for unity. Number is threefold: integer, fracted, and surd, to which last unity is incommensurable.

The theory of proportions is all that early modern natural philosophers needed to express physical laws. For instance, the law of inertia implies the proportionality of traveled distance and elapsed time for a freely moving body. Galileo's law of fall states the proportionality of the descent of a free falling body with the square of the elapsed time. Newton's second law reads "The alteration of motion is ever proportional to the motive force impress'd." Note that by that time natural philosophers freely considered the product of two quantities (the square of a time for Galileo and the product of mass and velocity for Newton), although they did not write it in symbols. In essence, they always meant

---

[10] Euclid, in Heath and Heiberg 1908, vol. 2, pp. 113–14.   [11] Newton 1707, 1720, p. 2. Cf. Roche 1998.

relations between ratios of similar magnitudes. For instance, Newton's second law means $I/I' = (m/m')(u/u')$, if I is the impressed force (impulse), m the mass, and u the velocity increment.[12]

Equations and dimensions

In their mechanics, Galileo and Newton did not need equations as they reasoned geometrically and represented times by proportional lengths. The later development of algebraic methods under Leibniz's impulse prompted the Swiss and French Newtonians to reformulate Newtonian mechanics and its generic problems in the symbolism that is more familiar to us. In order to turn the proportionality relations into equations, a system of units had to be defined for the various quantities. Euler did just that in his *Mechanica*, by setting the units of length and time to Rhine foot and second, and the unit of velocity to a foot per second. In later writings, he favored natural units in which a velocity was represented by the square root of its equivalent height (we would now say that he made $g = 1$). Systematic reflection on units, dimensions, and homogeneity did not occur before Fourier's theory of heat, presumably because the consistency of a unit system and the effect of changes of units became more important questions when a non-mechanical phenomenon was mathematized.[13]

Fourier introduced basic units for length, time, mass, temperature, and heat and investigated how the measure of various combined quantities changed during a change of unit. Consistency requires that the various terms of an equation relating various quantities should change by the same factor during a given change of the base units. Equations meeting these conditions are intrinsically valid, and they do not depend on the arbitrary choice of units as long as the units of derived quantities, such as velocity, are defined through the units of the combined quantities. In his *Treatise on electromagnetism* of 1873, Maxwell introduced symbols for the unit of a quantity, for its measure, and for its intrinsic value, respectively, [L], l, and l [L] in the case of length. In this view, the equations of physics can be written in an intrinsic manner, as relations between the intrinsic values of quantities. In modern parlance, Maxwellian quantities belong to one-dimensional vector spaces, a unit is a basis in such a space, and the product of two quantities is a tensor product.[14]

As physicists gradually learned, the choice of a system of units has various conventional aspects: value of the base units, kind of the base units, laws used for defining the derived units, etc. Whatever be the chosen conventions, the laws of physics are usually expressed as equations involving quantities measured by real numbers. Further mathematical structure is imposed by considerations of homogeneity, and also by covariance with respect to changes of systems of reference. The vector or tensor character of some quantities results from the latter consideration; and further structuring results from the symmetries of modern field theories. We are left with the most basic question: why are there physical quantities at all?

---

[12]Newton 1687; 1729, p. 19.   [13]Euler 1736, vol. 1, pp. 11–12.

[14]Fourier 1822; Maxwell 1873. Cf. de Montagu 2010.

6.3   Helmholtz's *Counting and measuring*

Measurability before Helmholtz

The concept of physical quantity clearly presupposes measurement, as idealized in the Greek theory of ratios and as concretely performed with material devices. Physicists before Helmholtz almost never asked whether and why a given property could be measured. The probable reason for this silence is the commonsensical character of space and time measurement, from which every other physical measurement was believed to derive. The measurability of superficially non-mechanical entities, such as heat or electricity, was suggested by mechanical or substantialist analogy or by causal connection with mechanical properties (in the case of temperature for instance). During the nineteenth-century outburst of quantitative experimentation, the central question was not whether to measure but how to measure. The quantitative nature of physics was uncritically accepted.

The first questioning of measurability occurred in sciences in which the possibility of a quantitative approach was not so evident, namely in psychophysics and psychology. After Gustav Fechner, a physicist by training, had given his logarithmic relation between sensation and excitation, there was much debate among philosophers, physiologists, psychologists, and even mathematicians on whether sensations could be measured or not. Kantian orthodoxy suggested a negative answer, mechanistic analogies a positive one. In this context, preconditions of measurement were identified such as the stability of units, the transitivity of concrete comparison, and the additivity of the investigated property. In a physical context, Maxwell and Mach pointed out that temperature measurement presupposed the transitivity of thermal equilibrium, and Maxwell discussed the concrete addition of electric charges.[15]

Arithmetic empiricized

Helmholtz was first, however, to systematically address the issue of measurability. One reason for this priority may have been his awareness of the measurability debates in experimental psychology. Another plausible motivation is the Kantian parallel between internal and external intuition. As Helmholtz regarded his earlier philosophy of geometry as a revision of Kant's concept of space as a form of external intuition, he was naturally led to discuss the internal intuition of time and the associated schema of arithmetic.[16]

Helmholtz indeed introduces his essay of 1887 "On counting and measuring" through an analogy with the foundations of geometry. Kant, he recalls, regarded the axioms of arithmetic and geometry as "a priori given truths that further determine the transcendental intuition of time and space." In the case of geometry, Helmholtz had already shown against Kant that the axioms were of empirical nature, even though he agreed with Kant than the general notion of space remained a transcendental form of intuition. His essay of 1887 continues with the words:[17]

---

[15]Cf. Heidelberger 1986 (psychology); Hasok 2004 (temperature). Cf. Darrigol 2003a, pp. 535–45.

[16]Cf. Hertz 1921, Michell 1993, 1999; DiSalle 1993; Díez 1997. On the sources, contents, and reception of Helmholtz's essay, cf. Darrigol 2003a.

[17]Helmholtz 1887, pp. 356, 357.

Now it is clear that my empiricist theory, if it no longer admits that the axioms of geometry cannot and must not be proved, must also apply to the origin of the arithmetic axioms, which have a comparable relation to the temporal form of intuition.

Among the arithmetic axioms, Helmholtz includes the transitivity of equality, the associativity and commutativity of addition, and its compatibility with equality. His empiricist revision of arithmetic takes two steps. In the second and more straightforward step, he borrows from Hermann and Robert Grassmann the proof that all axioms of arithmetic result from the axiom $(a + b) + 1 = a + (b + 1)$. In the first step, he derives the latter axiom from a concept of number based on a psychological fact and on an adequate definition of addition. The psychological fact is our capacity to order a succession of mental states: "Counting is a procedure which rests on our ability to memorize the order in which acts of consciousness have occurred." Numbers (*Zahlen*) are an unlimited sequence of arbitrarily chosen signs in a given, conventionally fixed order. Their purpose is to fix in our memory the temporal order of past series of acts of consciousness. The numbers thus defined are ordinal numbers (*Ordnungszahlen*). The decimal notation is a convenient way to generate an unlimited system of signs with a definite order and no repetition and thus to define a communicable number system.[18]

Call Sa (successor of a) the number immediately following the number a. Helmholtz defines the sum $a+b$ of two numbers as the number $S^b a$ obtained by repeating b times the operation S. For example, $a+3 = SSSa$. This definition implies $S(a+b) = SS^b a = S^{b+1} a = a+Sb$. Since $Sb = b+1$, this is the same as Grassmann's axiom $(a+b)+1 = a+(b+1)$. Other arithmetic axioms follow from this one by mathematical induction. Multiplication is defined inductively by $1 \cdot a = a$ and $(Sb) \cdot a = b \cdot a + a$. Its properties of associativity, commutativity, and distributivity also follow by induction.[19]

Counting and measuring

Having thus reconstructed arithmetic, Helmholtz considers its application to concrete objects. The most immediate application is the determination of the number (*Anzahl*) of objects in a "group" of objects. Helmholtz proves that this number is independent of the order in which the number is counted. He emphasizes that objects of a given class are not necessarily countable: "No object should be permitted to disappear, or to fuse with others, or to split, or to be brought into existence." Whether this condition is met can only be decided empirically. Granted that objects of a given class can be counted, the number of objects of the reunion of two groups obviously equals the sum of the numbers of objects in each group. Helmholtz concludes that there is a perfect agreement between ordinal and cardinal numbers (*Zahl* and *Anzahl*). Ordinal numbers have, however, "the advantage of being accessible without recourse to external experience."[20]

Helmholtz next considers the counting of objects that are alike (*gleich*) in a given respect. Such objects are called units. The result of this counting is a concrete number

---

[18]Helmholtz 1887, pp. 360, 372.   [19]Helmholtz 1887, pp. 363–71. The Sa notation is mine.

[20]Helmholtz 1887, p. 372.

(*benannte Zahl*), defined by a pure number (*Zahl*) and the corresponding unit. Next comes the definition of a quantity (*Grösse*) and its measure:[21]

> We call quantities objects or attributes of objects which, when compared with other similar objects allow for the distinction of larger, equal, or smaller. When we can express a quantity by a concrete number, we call this number the value of the quantity, and we call the process through which we find this number the measurement of the quantity.

Helmholtz's basic question then is whether a given attribute of objects can be regarded as a quantity, and whether it can be measured. A specific attribute is defined by "a method of comparison" that enables us to decide whether two objects are alike in some respect. In order to be compatible with arithmetic axioms of equality, this method must meet the condition: *When two objects are found to be alike to a third, they must be alike*. This is transitivity, stated in a manner that includes symmetry.[22]

Helmholtz next reviews the methods of comparison for weight, distance, length, time, brightness, pitch, and electric intensity; and argues in each case that only experiment can confirm the validity of the condition. He emphasizes that transitivity is "not an objective law [*ein Gesetz von objectiver Bedeutung*], but a way to decide which physical relations could be recognized as equality." In other words, for Helmholtz transitivity does not refer to pre-existing quantities. On the contrary, it is a test for deciding what can be held to be a quantity.[23]

Helmholtz next introduces a physical connection (*Verknüpfung*) between two quantities of the same kind. In order to satisfy the arithmetic axioms of addition, the connection must be compatible with the physical equality ($a + b = a' + b'$ if $a = a'$ and $b = b'$), it must be commutative and it must be associative. Again, Helmholtz regards these conditions as an experimental test for identifying possible quantities. That the condition holds for all connections encountered in physics only means that physicists have already done the selection.[24]

Helmholtz then defines order among quantities: a first quantity is said to be larger than a second when it can be obtained by composition of the latter quantity with a third quantity. He implicitly assumes that the composition law is such that it generates a non-trivial order in this manner (this is not necessarily the case: the usual rule of composition of forces would otherwise lead to the conclusion that any force is larger than any given force). Helmholtz next proceeds to (uniform) division: "Quantities which can be added are also to be divided in general." He not only means the possibility to regard a given quantity as the sum of a number of equal quantities, but also the possibility of approximately expressing a given quantity as a multiple of a fixed unit, and to indefinitely improve the approximation by

---

[21] Helmholtz 1887, p. 375.

[22] Helmholtz 1887, pp. 375–6). Helmholtz sometimes speaks of *Gleichheit* of two quantities when he means of two objects. He uses the same word, quantity, to denote both a property defined by a method of comparison, and an object considered in regard to this property.

[23] Helmholtz 1887, pp. 377–80. Similarly, in Helmholtz's theory of perception the experienced correlations between sensations and voluntary impulses determine what can be regarded as external object.

[24] Helmholtz 1887, pp. 381–3.

introducing a series of sub-units. He implicitly assumes the Archimedean property, which, together with the existence of difference, allows the arbitrarily precise approximation of ratios by rational numbers.[25]

Helmholtz next discusses physical quantities for which no additive composition is known. These occur as "coefficients" in empirical laws or definitions relating different additive quantities. For instance, the optical index, specific weight, and conductivity of a substance are coefficients. Such quantities are known indirectly by measuring the additive quantities that they relate. The distinction between coefficients and additive quantities is however inessential, because a coefficient can become an additive quantity upon the discovery of a relevant additive connection.

The value of a coefficient generally depends on the choice of the units in which the related quantities are measured (unless the coefficient is dimensionless). For this reason, Helmholtz introduces the multiplication and division of concrete numbers as Maxwell had already done in his *Treatise*. In the Gaussian view according to which all quantities can be generated from length, time, and mass, all units have the form $[L]^{\alpha}[T]^{\beta}[M]^{\gamma}$, wherein $[L]$, $[T]$, and $[M]$ are the units of space, time, and mass and α, β, γ are three positive or negative integers. Helmholtz is very brief on vector calculus in Grassmann's and Hamilton's guises, his main purpose being "to show the meaning and justification of calculation with pure numbers and the possibility of their application to physical quantities."[26]

A measurable world

We may now summarize the essential components of Helmholtz's analysis. Helmholtz first introduces ordinal numbers, in a manner independent of external experience and related to our ability to order acts of consciousness in time. He then derives the usual arithmetic axioms by mathematical induction from natural definitions of addition and multiplication. Next, he characterizes quantity through the equality of objects with respect to a given method of comparison. To make quantity measurable, he further requires a concrete operation of addition (and difference). Measurement means the division of a given quantity into equal units, with a rest to be divided into sub-units.

The consistency of Helmholtz's definition of measurable quantity requires concrete equality and concrete addition to obey the corresponding arithmetic laws. Helmholtz insists that only experience can tell whether these properties are met. If they are, the method of comparison and the procedure of addition define a measurable quantity. Helmholtz's concept of measurement thus combines conventional and objective elements: it depends on a choice of concrete physical operations that is partly free and partly constrained by experiment.

There are obvious differences between Helmholtz's and Kant's conceptions of quantity. The first difference regards the status of arithmetic. Whereas for Kant number is the pure schema of quantity and is therefore prior to any experience, for Helmholtz (ordinal) number is based on a psychological fact of experience: our ability to order events. In practice, this first difference is tenuous since the two thinkers agree on our innate ability

---

[25]Helmholtz 1887, pp. 383–5.   [26]Helmholtz 1887, p. 390.

to order events and agree on making it the basis of arithmetic. A more consequential difference regards the relation between number and measure. Whereas for Kant the arithmetic schema automatically applies to any scientific representation, for Helmholtz the concrete realization of arithmetic operations requires the satisfaction of certain empirical criteria of measurability. The degree to which nature lends itself to a description through measurable quantities is an open question. This is true for any kind of measurement, including those of space and time.

We may now give Helmholtz's answer to the questions raised at the beginning of this chapter. To the extent that nature is measurable, it is amenable to mathematical representation. The relevant mathematics is first and foremost arithmetic, because measurement hinges on the concrete realization of arithmetic relations and operations. Helmholtz's stance on the measurability of nature reminds us of his stances on causality and on the possibility of geometry. In every case, experience is the only judge of the existence of some types of regularities in the perceived world. However, our past experience encourages us to assume such regularities, and a world without such regularities could hardly be comprehended.

Some criticism

Helmholtz's essay is decidedly empiricist, for it bases numbers and their applicability to nature on experience. Some aspects of this empiricism are open to criticism. Firstly, the measurability criteria (arithmetic properties of the concrete equality and the concrete addition), which Helmholtz regards as his main insight, are never used by physicists. Yet they are not pure figments of Helmholtz's mind. Their validity is a posteriori verified, for they result from the theories in which the relevant quantities are imbedded (including theories of the apparatus). This is no coincidence. In the quantitative exploration of new phenomena, physicists tentatively regard some property as a quantity for some theoretical, metaphorical, experimental, or sensorial reason, and then, they try to measure this property by whatever means this reason suggests. If the measured numbers are found to be related to other measured quantities in a regular manner, the measurability criteria for equality are implicitly met, because these numbers would otherwise be ill-defined. If the resulting laws are simple and general, some other criteria such as the associativity of concrete addition are likely to be met. In other words, the global testing of a theory by quantitative experiments entails an implicit test of the measurability of relevant quantities. Helmholtz's criteria remain useful as conditions that any quantitative theory must harbor a theory of measurement in which models of concrete equality and addition are conceivable. This stricture has far-reaching structural consequences, to which we will return in a moment.

It is not clear whether Helmholtz, by emphasizing measurability, also meant that every quantity occurring in the equations of physics should correspond to a principally measurable quantity. The physics of his time did not have complex wave functions, vectors and operators in Hilbert space, spinor fields, or other highly abstract entities whose connection to measurable quantities is partial and indirect. But it had the thermometric (non-absolute) temperature, which Maxwell, Mach, and Duhem refused to regard as a genuine quantity because of the high arbitrariness of the convention of measurement defining a temperature scale. Helmholtz did not follow them, because in his view every convention of measurement was acceptable as long as it remained compatible with the arithmetic criteria. Whatever position one adopts in this regard, it remains true that every temperature scale can be turned into another admissible scale by applying to it a strictly monotonous and regular function, and that the equations of thermodynamics should retain their form under this change of scale. As Mach puts it, thermometric temperature is just a numerical labeling of the degrees of heat. In modern language, it is a purely ordinal measurement, without intrinsic additive structure. Additive structure only exists for the thermodynamic temperature, which is defined though the ratio of the heats exchanged in a bithermal reversible engine.[27]

This relaxation of Helmholtzian measurability can be pushed to the extreme by ceasing to require both the cardinal and the ordinal structure, and defining measurement by "the assignment of numerals to objects or events according to rule," as Stanley Smith Stevens did in 1946. This liberal concept of measurement is widespread in psychology, and it has recently been the target of Joel Michell's criticism. According to Michell, much of quantitative psychology is illusory, because the simple relations propounded between the liberally defined quantities do not reflect any genuine quantitative structure. Before measuring some property, psychologists should test the compatibility of this property with Helmholtz's additivity criteria or with extensions of these criteria to indirect measurement. On the one hand, this demand may seem to strong, since not even physicists perform such tests. On the other, it seems reasonable in a field in which simple regularities are not likely to be frequent.[28]

On the mathematical side, Helmholtz's empiricism induces him to base arithmetic on a universal feature of internal experience, our awareness of temporal successions. As Helmholtz knew, in 1882 the mathematician Paul Du Bois Reymond had gone further in this direction by defining number through external experience, as "that which remains in our mind when everything that distinguishes the [sensed] objects vanishes and only their awareness of their being separated is retained." Du Bois similarly derived rational numbers from the necessities of experience, and he even conceived an empiricist view of analysis in which every theorem was rephrased as a statement involving only rational numbers at a given, indefinitely improvable precision. Helmholtz seems to have approved this sort of finitism, for he declared that rational numbers were all physicists needed as long as the monstrous functions imagined by contemporary mathematicians did not occur in physics.[29]

Du Bois's and Helmholtz's empirical constructivism run against the contemporary arithmetization of analysis. In the latter trend, numbers had to be defined through set-theoretical constructs, for instance as classes of equipotent classes, independently of any empirical intuition. The rigor of analysis required an arithmetic definition of real numbers, for instance as cuts in the infinite set of rational numbers; a cut being a partition of this set into two subsets such that every element of the first subset is inferior to every element of the second (for example, the rationals whose square is smaller, respectively larger

[27]Cf. Darrigol 2003a, pp. 542–4, 555.   [28]Stevens 1946; Michell 1993, 1999.

[29]Du Bois-Reymond 1882, p. 32; Helmholtz 1887, p. 385. The empiricism of Helmholtz, Du Bois, and his brother Emil (Helmholtz's closest friend), had a common origin in the empirical physiology of Johannes Müller.

than two). The most influential proponents of this and other set-theoretical approaches, Richard Dedekind and Georg Cantor, rejected Du Bois's and Helmholtz's empiricism as a return to naïve intuitions. Here is Dedekind's credo:

> The whole of analysis is a necessary consequence of *thought* per *se*; only after the throughout *pure* development of thought (without recourse to the quantity-representations) are we in a position to conceive the concept of quantity with full exactness.

On the one hand, Cantor's and Dedekind's distrust of the intuitive, geometric conception of analysis was well founded, for this conception had brought many inconsistencies and paradoxes. On the other, these mathematicians overlooked the fact that their own rigorous constructions of the concepts of number, continuity, and limits drew on old intuitions of empirical origin. For instance, Dedekind concept of cut looks like a set-theoretical, arithmetic reconstruction of the Eudoxian definition of ratios. In the notation earlier introduced, the ratios n/m of any multiplicities n and m for which $na > mb$ define a cut in the set of rationals. As we will see in a moment, this kinship did not escape Poincaré's attention.[30]

The somewhat loose and old-fashioned character of Helmholtz's definition of number does not affect his concept of measurement. The only thing that matters in this context is the possibility of a definition of numbers prior to that of measurement. Helmholtz's purpose would have been served just as well by a set-theoretical definition or by earlier formalist definitions à la Grassmann. One may still regret another mathematical consequence of Helmholtz's empiricism. He seems to have underestimated the fact that the equations of mathematical physics, by their very form, refer to quantities defined with indefinite precision through an infinite range. As Poincaré soon argued, this extrapolation of empirical measurability explains why real number analysis pervades physical theory.

6.4   Poincaré on number and quantity

Helmholtz, as a physiologist and physicist, was evidently more interested in the measurability of nature than on the kind of mathematics needed in physical theory. In contrast, Henri Poincaré, as a mathematician with an eye on physics, wanted to understand why mathematical analysis had been so successful as a tool for formulating physical laws and theories. His *Science and hypothesis* of 1902 begins with a first chapter on "Number and quantity" drawn from essays written in the early 1890s. This chapter begins with the following dilemma:

> The very possibility of mathematical science seems an insoluble contradiction. If this science is deductive only in appearance, wherefrom does it draw this perfect rigor that no one thinks of questioning? If, on the contrary, all its propositions can be derived from each other by the rules of formal logic, how is it that mathematics do not reduce themselves to a huge tautology?

[30] Dedekind 1872 (cuts); [c. 1882], p. 199 (citation). On Dedekind, Cantor, and Frege, cf. Dugac 1976; Belna 1996. On their attacks on Du Bois, cf. Darrigol 2003a, pp. 533–534.

To answer this question, Poincaré examines "the purest part of mathematical thought," arithmetic. His analysis has enough similarity with the arithmetic section of Helmholtz's "Zählen und Messen" to suggest a historical connection.[31]

Arithmetic and induction

Poincaré assumes the operation $a + 1$ to be already given for every number a, by whatever means. Then he defines the addition $a + b$ inductively through $a + (b + 1) = (a + b) + 1$ and derives its commutativity and associativity also by induction. Mathematical induction, Poincaré comments, offers a way to condense in a single formula an infinity of syllogisms. It is the very origin of the mathematician's ability to induce general truths from particular ones. It is neither reducible to pure logic nor to experience, because of its infinite character. It is "the prototype of the synthetic a priori judgment," or "the affirmation of the power of the mind which knows itself capable of conceiving the indefinite repetition of the same act as soon as this act is once possible."[32]

    Poincaré thus shares with Helmholtz (and Grassmann) the emphasis on inductive definitions and derivations, as well as the need of a prior definition of succession. There are, however, some differences. Helmholtz is more concerned with the latter definition, which he finds in psychological experience, than with the nature of mathematical induction, which he takes for granted. In general, Poincaré emphasizes the mathematician's need to transcend experience, whereas Helmholtz anchors all knowledge on idealized experience. Helmholtz and Poincaré both betray Kant in their own ways, Helmholtz by denying the a priori character of arithmetic, Poincaré by evacuating or bracketing Kantian intuition. Whereas for Kant number is the pure schema of quantity and therefore appeals to internal intuition, for Poincaré numbering is more vaguely defined as a mental faculty, of which we become aware through experience:

> We have the faculty of conceiving that a unit can be added to a collection of units; it is through experience that we have the occasion to exert this faculty and that we become aware of it.

Poincaré nevertheless agrees with Kant about the synthetic a priori character of arithmetic judgments.[33]

The physical and mathematical continua

Poincaré next discusses "mathematical quantity and experience" in order to show the origins and consistency of the mathematical continuum. To this end he introduces Helmholtz's distinction between arithmetic equality and concrete equality:

> The properties of equality are not an experimental truth exposed to refutation through future, refined experiments. I prefer to believe with Helmholtz that we give the name of equality to everything in the external world that meets the preconceived idea we have of mathematical equality.

---

[31] Poincaré 1902, p. 1. On Poincaré's mathematics in relation to physics, cf. Ly 2007.

[32] Poincaré 1902, pp. 12–13.   [33] Poincaré 1902, p. 37.

Poincaré then gives a new twist to the transitivity criterion: whereas Helmholtz requires physical quantity to comply with this criterion, Poincaré argues that the concrete appreciation of the intensity of a property violates the transitivity of equality. According to Ernst Heinrich Weber, the sensations produced by two weights A and B can be mutually discernable without being discernable from the sensation produced by a third, intermediate weight C. Even if a more precise method is used to compare the weights, Poincaré goes on, the precision of this comparison is always finite, so that there are always choices of the weights A, B, C for which

$$A = C, B = C, A < B.$$

This paradoxical triplet of relations defines the "physical continuum." The mathematical continuum then emerges from our attempts to solve the contradictions inherent in this empirical notion.[34]

In Poincaré's opinion, Dedekind's cuts provide the desired construct of the mathematical continuum. They solve the paradox of the physical continuum by indefinitely increasing the precision of the distinction between two numbers and requiring that at the limit of infinite precision the following property of the physical continuum still holds: when the continuum is parted in two successive parts (two half-lines) these two parts have a common boundary. Poincaré thus agrees with the arithmetizers of mathematics that a rigorous concept of the continuum has to rest on integral numbers and adequate definitions. But he insists that experience has motivated and inspired this construction: "The mathematical continuum ... has been created from bits and pieces [*de toutes pièces*] by our mind, but it is experience that has provided the occasion."[35]

By continuum, Poincaré means any linear quantity in which equality and order are defined, not only real numbers. As the indefinite divisibility and continuity of a quantity does not necessarily imply measurability, Poincaré further introduces the equality of intervals through a special convention of measurement. This convention, once given, allows a definition of addition for intervals starting from the same point: the interval AD is said to be the sum of the intervals AB and AC when the interval BD is equal to the interval AC. Poincaré of course requires this addition to be commutative and associative. He refers to "von Helmholtz's magisterial work" (1887) for a fuller discussion of this issue.[36]

Poincaré's conception of measurement has indeed much similarity with Helmholtz's. Poincaré believes that any measurement requires conventions of equality and addition and that the arbitrariness of these conventions is only limited by arithmetic properties such as transitivity, commutativity, and associativity. For example, in his lectures on heat theory, he follows Maxwell and Mach in defining the equality of the temperatures of two bodies through the absence of expansion or contraction during thermal contact; he notes that only experiment can tell whether this convention leads to a transitive equality; and he temporarily adopts the further convention that equal temperature intervals correspond to equal

[34] Poincaré 1892b, p. 75; 1893; 1902, pp. 29–41.

[35] Poincaré 1902, p. 35. Poincaré may here have been inspired by Du Bois's empiricism.

[36] Poincaré 1902, pp. 41–8; 1893, p. 33 (citation).

scale intervals on the mercury thermometer. In his article of 1898 on the measurement of time, he insists that simultaneity as well as the equality of time intervals are conventional, the best conventions being those for which the laws of mechanics and optics have their simplest expression. In a later discussion of the same topic, in 1908, he verifies that the conventional definition of simultaneity through the exchange of light signals meets the transitivity criterion for moving observers.[37]

To summarize, for Poincaré experience gives us the *occasion* to develop notions of number and quantity. However, the rigor of mathematical constructs is incompatible with the intrinsic vagueness of experience. The constructive tools of the mathematicians must rest on a priori faculties of the mind:

> To sum up, the mind has the faculty of creating symbols, and thus it has constructed the mathematical continuum, which is nothing but a particular system of symbols. The power of the mind is only limited by the necessity of avoiding any contradiction. However, the mind only uses this faculty when experience gives it a reason to do so.

In this view, one need not wonder why all modern physical theories rely on the mathematical continuum and real numbers: being motivated by experience, mathematical constructs must naturally apply to experience. However, the extent to which they apply remains open.[38]

Elementary phenomena

Poincaré also explains why infinitesimal calculus and differential equations are so pervasive in physical theories. He argues that the physicist's understanding of phenomena requires their analysis into "elementary phenomena" obeying the same law. This is why principles of superposition and composition are frequently used in physical theories. For example, the evolution of a physical system during a finite time interval is derived from a uniform law for the evolution in infinitesimal intervals of time; the state of a spatially extended system is built by assuming that a given point of the system is only affected by its neighboring points; small motions are superposed; arbitrary signals are analyzed into harmonic components; etc. The definition of the elementary phenomena and the reasons for their simplicity may of course vary as physics evolve. However, this definition must be possible if nature is amenable to mathematical analysis:[39]

> It might be asked, why in physical science generalization so readily takes the mathematical form. The reason is now easy to see. It is not only because we have to express numerical laws; it is because the observable phenomenon is due to the superposition of a large number of elementary phenomena that are *all similar to each other*; and in this way differential equations are quite naturally introduced. It is not enough that

---

[37] Poincaré 1892a, pp. 16–17; 1898; 1908. On time, cf. Darrigol 1995, p. 37.

[38] Poincaré 1902, p. 40. Cf. Heinzmann 2001b and Ly 2007.

[39] Poincaré 1902, p. 187. Poincaré's argument resembles Kant's demand of the synthetic unity of apperception, from which he derives his categories. It anticipates Frank Wilczek's recent remark that symmetry and locality are the reason why many laws of physics take the form of differential equations (Wilczek 2006).

each elementary phenomenon should obey simple laws: all those that we have to combine must obey the same law; then only is the intervention of mathematics of any use. Mathematics teaches us, in fact, to combine like with like. Its object is to divine the result of a combination without having to reconstruct that combination element by element. If we have to repeat the same operation several times, mathematics enables us to avoid this repetition by telling the result beforehand by a kind of induction.

A similar kind of homogeneity presides over the concept of space according to Poincaré: we are able to combine and mutually compensate two sorts of displacements, those of our body and those of objects; the displacements have the structure of a group; they can be as small as we wish; and we can imagine their indefinite repetition. Poincaré thus arrives at the Lie Group structure and at its synthetic a priori character: "The general concept of group preexists in our minds, at least potentially. It is imposed on us not as a form of our sensibility, but as a form of our understanding." This concept is not specifically geometric; it is the mathematical embodiment of the general idea of the synthesis of phenomena by indefinitely repeating the application of the same operation to an elementary phenomenon. Poincaré thus foresaw and explained the pervasiveness of Lie groups in modern physical theory.[40]

Poincaré nonetheless understood that not every field of inquiry could benefit from the group-theoretical combination of operations:

> For that purpose all these operations must be similar; in the contrary case we must evidently make up our minds to working them out in full one after the other, and mathematics will be useless. It is therefore thanks to the approximate homogeneity of the matter studied by physicists that mathematical physics came into existence. In the life sciences the following conditions are no longer to be found: homogeneity, relative independence of remote parts, simplicity of the elementary fact; and that is why biologists are compelled to have recourse to other modes of generalization.

Then, why is it that the world of inanimate of objects, or at least some aspects of it, has the sort of homogeneity that makes them amenable to mathematical analysis? Poincaré shuns this question.[41]

After Poincaré

Although Poincaré's philosophy of quantity went further than Helmholtz's in identifying the kind of mathematics needed in physical theory, it was expressed in ordinary language and fell short of explicating the formal constructs to which it was alluding. A memoir of 1901 by the German mathematician Otto Hölder provides such explication, although Hölder himself rather saw his contribution as an axiomatic implementation of Helmholtz's doctrine of quantity.[42]

Hölder's interest in the theory of quantity presumably derived from his earlier study of David Hilbert's *Foundations of geometry* of 1899. Hilbert based synthetic geometry on a

---

[40]Poincaré 1902, p. 90. Like Helmholtz, Poincaré only considers homogenous geometries for which the curvature is constant.

[41]Poincaré 1902, pp. 187–8.   [42]Hölder 1901, p. ln.

small number of demonstrably independent and non-contradictory axioms. In addition, he showed that metric properties could be derived from purely geometric axioms for the equality of segments (*Strecken*) and the disposition of points on a line. To Hölder this meant that an axiomatic theory of quantity could be given in the special case of segments. His goal was to extend the axiomatic approach to more general quantities. This extension could serve as "a preparation for the case when a physical state can be measured." Hölder knew Maxwell's additivity argument for the measurability of charge, as well as the existence of properties, such as hardness, that had degree but no quantity.[43]

Hölder had seven axioms of quantity, the independence of which he could prove. Call Q the set of quantities of a given kind. In modern notation, the axioms read:[44]

    I. $\forall(a, b)\in Q^2, (a = b)$ or $(a < b)$ or $(b < a)$
   II. $\forall a\in Q, \exists b\in Q : b < a$
  III. $\forall(a, b)\in Q^2, \exists c\in Q : c = a{+}b$
  IV. $\forall(a, b)\in Q^2, a < a{+}b$ and $b < a{+}b$
   V. $\forall(a, b)\in Q^2, a < b \Rightarrow \exists\, x\in Q : a{+}x = b, \exists y\in Q : y{+}a = b$
  VI. $\forall(a, b, c)\in Q^3, (a{+}b){+}c = a{+}(b{+}c)$
 VII. $\forall(A, B)\in[\mathcal{P}(Q){-}\varnothing]$ such that $[Q = A\cup B, A\cap B=\varnothing, (\forall\, (a, b)\in A\times B : a < b)],$
       $\exists \xi \in Q : [\{a\in Q\,|\,a < \xi\} \subset A$ and $\{b\in Q\,|\,\xi < b\}\subset B]$

This list does not include axioms of equality such as symmetry and transitivity, because Hölder regards equal quantities as identical. He is nonetheless aware that in concrete cases this simplification requires empirical testing.[45] Also missing are the commutativity axiom and the Archimedean axiom: this is because Hölder derives both from his seven axioms. The last axiom is an axiom of continuity clearly obtained by transposition of Dedekind's cuts. Hölder easily proves that every couple (a, b) of quantities defines a Dedekind cut among rational numbers, that is, a real number to be interpreted as the ratio a/b.[46]

To some extent, Hölder was right to regard his theory of quantity as a refinement of Helmholtz's. They both defined quantity so as to fit structural properties of numbers. Hölder's refinement consisted in enlarging the arithmetic basis. There were important differences, however. Whereas Helmholtz continued the Euclidean tradition of whole number or rational measure, Hölder admitted the newly arithmetized irrationals. Whereas Helmholtz regarded his axioms of quantity as criteria for selecting measurable properties of concrete objects, Hölder regarded them as the foundation of a mathematical theory of quantity on the same footing as Hilbert's abstract geometry. To him the non-contradiction and independence of the axioms mattered more than applications to physics. Consequently, his axioms were not well adapted to empirical tests of measurability. Hölder had neither the easily testable transitivity of equality nor the commutativity of addition, and the continuity expressed in his last axiom of course eluded any test.

---

[43] Hilbert 1899; Hölder 1900, p. 54.   [44] Hölder 1901, pp. 5–6.   [45] Hölder 1901, p. 4n.

[46] In a last section, Hölder shows that Hilbert's axioms for segments imply that distances obey the axioms of quantity. Cf. Michell 1999, pp. 74–5, who regards this remark as an anticipation of the theory of conjoint measurement.

In 1931, the Czech-American philosopher of science Ernest Nagel modified Hölder's axioms so as to make them verifiable, in Helmholtz's spirit. For Nagel, quantities are particular objects like for Helmholtz; for Hölder, they are universals applying to a class of objects. In 1951, Patrick Suppes imitated Nagel, with weaker axioms corresponding to a broader concept of measurement. The modern "theory of measurement" derives from these efforts. It is a formal attempt to express the general conditions for the existence of measurable quantities. These quantities satisfy a number of axioms that allow their correspondence with real numbers. The imbedded concept of measurement is said to be representational, because it requires a separate definition of quantity and number. The extent to which this concept can be realized empirically depends on the choice of the axioms of quantity. Although these developments satisfy the contemporary taste for formalism, and although they bring refined conditions of direct and indirect measurability, from a broader viewpoint they add little to Helmholtz's and Poincaré's earlier insights into the relation between measurement and mathematization.[47]

Conclusions

In this chapter we have encountered several answers to the question of the necessity of mathematics in physics. The first kind of answer is ontological: it rests on the conviction that the physical world is reducible to a form of being that is inherently mathematical. This form was numbers for the Pythagoreans, geometry for Descartes, the play of matter and forces for the Newtonians, clockwise mechanism for neo-Cartesians and British natural philosophers of the nineteenth century. In this approach, the necessity of mathematics boils down to the necessity of these ontologies, which has been addressed in earlier chapters.

Another argument for the necessity of mathematics is transcendentalism, as defined by Kant and pursued by many of his disciples. In this view, scientific knowledge presupposes the representation of phenomena in space and time and their subsumption under the category of quantity. The result is inherently mathematical, because arithmetic and geometry are the mental dispositions that permit this subsumption. Yet Kant was not a complete mathematical reductionist, for he admitted empirical, sensation-based investigations that resisted quantification. Among those he included the chemistry and psychology of his time. Although he did not deny the legitimacy of such disciplines, he refused to regard them as sciences *stricto sensu*. Kant's view suffers from a usual defect of his doctrine: it overestimates the purity, the necessity, and the applicability of the categories of understanding.

The ontological and transcendental arguments for the necessity of mathematics in physics have affinities with the Greek concept of quantity according to which ratios and proportions are defined abstractly, without reference to experience. In this concept as in Kant's, quantity precedes number. Numbers can even be defined as ratios, a move that enabled eighteenth-century natural philosophers to express the laws of physics under the form

---

[47]Nagel 1931; Suppes 1951; Kranz, Suppes, Luce, and Tversky 1971; Luce and Suppes 1981. Cf. Michell 1993, pp. 198–9. On the origins of the modern theory of measurement, cf. Diez 1997; Michell 1999. Other influential discussions of measurement and quantity after Poincaré were Campbell's and Carnap's: cf. de Courtenay 2008.

of numerical equations. The applicability of the Greek doctrine of ratios to the physical world is natural for a Cartesian who believes in an essentially geometrical world, since Euclid's main application of this doctrine is to geometry. The abstract character of this doctrine and its precedence over arithmetic long sheltered it from questions about its relation to the practice of measurement.

The situation changed in the second half of the nineteenth century, as a consequence of the rise of empiricist views on the foundation of mathematics and other sciences. An extreme variety of empiricism is that of Paul Du Bois-Reymond, who believed whole numbers, rationals, and even a finitist version of infinitesimal calculus to be abstractions drawn from experience. In this view, the success of mathematics in physics is simply explained by its empirical origin and nature. Helmholtz's empiricism is more nuanced, for it does not presuppose the quantitative nature of the external world. In his view, the evidence of ordinal numbering comes from internal experience, and its applicability to the external world is not a priori warranted. In order that objects should be countable or measurable, they must obey certain empirical criteria, such as the stability of the counted objects or the transitivity of the concrete comparison of the measured objects.

The last and most profound view is what could be dubbed Poincaré's *occasionism*. In this philosophy, experience gives us the motivation to develop and the opportunity to apply the mathematical concepts of whole numbers, mathematical continuum, real number, differential equations, groups, etc. However, experience only is the occasion; it cannot be the foundation because the inherent vagueness of sensorial experience is incompatible with mathematical rigor. The true foundation is the mind's ability to conceive symbolic operations and to imagine their indefinite repetition. We could, in principle, imagine some mathematics without any empirical motivation; but this would be a vain activity; for Poincaré the only mathematics worth pursuing are those motivated by experience (if only remotely). In this case, the success of mathematics in physics is no wonder: mathematics was created for the very purpose it is serving.[48]

One need not follow Poincaré in his evacuation of strictly pure, internally driven mathematics. It is sufficient to agree with him that a large portion of mathematics was created with (partly) empirical motivations. This portion includes arithmetic, analysis, and group theory, which are the universal language of modern physical theory. Of course, Poincaré's thesis justifies only a partial success of mathematics in its application for physics: the tools designed by mathematicians are infinitely sharper than the definition of the concrete procedures that motivated their introduction. Can this definition be sharpened indefinitely? Is the mathematical tool still adequate when this definition is sharpened? Perhaps these questions are better left unanswered.

With the help of Helmholtz's insights into measurement and Poincaré's concept of elementary phenomenon, we may, however, more precisely define what is at stake in the application of mathematics to physics. Measurement requires counting units and defining them into subunits in order to enhance the precision. These operations necessitate whole numbers and generate rational numbers. The idealization of indefinite precision yields real numbers or a similar construct, because indefinitely increasing the number of subunits

---

[48] Cf. Heinzmann 2001b, who uses "occasionalism" (I prefer to reserve this term to Malebranche's doctrine).

yields something like a decimal number with an unlimited number of decimals. For the sake of mathematical convenience, it is better to adopt this idealization, even though the precision of an actual measurement is always finite. Real-number analysis indeed tends to be simpler than its finitist counterparts, and it is required by the geometric idealizations used in our apprehension of spatial relations. We thus see that a fundamental characteristic of modern physical theory, its reliance on real-number analysis, corresponds to the ideal of indefinite measurability. As long as the world is measurable, its states can be represented by a set of real numbers representing the potential results of measurements (including measurements of space and time). Physical phenomena can then be expressed as relations between these numbers.

According to Poincaré, the physicist's comprehension of nature requires a further property of these relations: they should be expressible through the repeated application of the same law to "elementary phenomena" into which the investigated phenomenon can be decomposed. The existence and homogeneity of the elementary phenomena may have various origins, and it depends on the scale at which the phenomena are considered. As we saw in the case of mechanics (secular principle) and as is expressed by the renormalization group of modern quantum field theories, the desired regularity may depend on the approximate independence of the phenomena of a given scale from the phenomena on a lower scale;[49] or, as imagined by Poincaré and by Ludwig Boltzmann, it may reflect a statistical regularity deriving from the law of large numbers. Whatever be its justification, decomposition into elementary phenomena is a precondition for physics as we know it. Otherwise the physical world would be more akin to a biological system and would elude the mathematical approach.

In brief, measurability and a certain kind of homogeneity bring phenomena into the dominion of mathematical analysis. Although this explanation of the success of mathematics in physical theory may be the best we can have, it is necessarily incomplete. As was already mentioned, modern physical theory assumes a wealth of non-measurable quantities, and it extrapolates some measurable quantities to scales at which measurement is practically, even sometimes theoretically impossible. One could still argue that the non-measurable quantities must be simply related to measurable quantities in order to acquire physical meaning and must therefore share their mathematical essence. Also, the extrapolation of some measurable quantities beyond the possibilities of actual measurement could be seen as an aspect of Poincarean homogeneity. With such arguments, however, there would be no limit to the empire of mathematics.

In reality, many aspects of nature elude measurement and lack the kind of homogeneity assumed in mathematical physics. Physicists sometimes claim reductions of these aspects to regular processes at a lower scale. For instance, meteorology is traced to a combination of fluid dynamics and thermodynamics, or chemistry is regarded as applied quantum mechanics. These reductions, however, are only useful to the extent that they furnish concepts directly applicable at the scale of the investigated phenomena, such as the concept of meteorological front, or the concept of hybrid orbitals in chemistry. Different fields of investigation call for different notions of comprehensibility. In physics, comprehensibility

---

[49] On the autonomy of scales and the renormalization group, cf. Cao and Schweber 1993.

often implies measurability and homogeneity. In other fields, vaguer structural analogies, analysis into organs and functions, and other tools of qualitative understanding are more likely to be fruitful.[50]

We might conclude that physicists have deliberately selected those aspects of the world that can be subjected to mathematical analysis, and we might even define physics as the study of the mathematical aspects of the world. This would make the success of mathematics in physics a tautology. Fortunately, there is more to this success than the artificial imposition of mathematics on an inherently amorphous substratum. Good physicists differ from Eddington's fisherman, who does not realize that the law of minimal fish size is a consequence of the mesh size of his net. They may sometimes fall into this error, for instance when they push too hard the extraction of regularities in complex systems. There is no doubt, however, that nature, when properly carved at its joints, obeys mathematical laws of extremely general validity. Behind the apparent diversity and heterogeneity of the world, there are implacable numerical regularities that escape human control. Some of us wish the world would have been patched together in a looser manner. Others prefer to see the world as a huge theorem. No one knows, ultimately, how much mathematical character nature must have so that we may live and think.[51]

---

[50] On the differences between biological and physical explanation, cf. Fox Keller 2002.

[51] Eddington 1939, pp. 17–31 (who thereby preaches a kind of selective subjectivism). For a witty discussion of the selectionist explanation of the success of mathematics, see Wilczek 2006. On adequate idealization, cf. McMullin 1985.

7

CLASSICAL FIELD THEORIES

> The general type of stress is not suitable as a representation of a magnetic force, because a line of magnetic force has direction and intensity, but has no third quality indicating any difference between the *sides* of the line, which would be analogous to that observed in the case of polarized light. We must therefore represent the magnetic force at a point by a stress having a single axis of greatest or least pressure, and all the pressures at right angles to this axis equal.[1] (James Clerk Maxwell, 1861)

The most fundamental theories of today's physics are all based on the field concept, that is, on physical quantities specified at every point of space and time. Why do we need field theories? Are there general constraints to which any field theory should be subjected? Are the known field theories the only possible ones? These are the questions addressed in this chapter, in the classical case in which the fields have a fairly direct operational significance.

In Chapter 2 we encountered various mechanical concepts of interaction: through connected systems, by direct action at a distance, through collisions, and through stressed continuous media. The two last concepts are the only ones compatible with relativity theory. We already examined the question of the necessity of relativistic particle dynamics in the collision approach. The same question may be now raised for field dynamics in the stress approach. In a relativistic context, the mechanistic reduction of stresses to molecular forces or to hidden motion is excluded. All we may require is that the stress-based dynamics satisfies the energy principle and the principle of least action. As argued in Chapter 3, energy conservation has some necessity independent of mechanical reduction. This is unfortunately not true for the principle of least action, at least as long as one refrains from quantum-theoretical considerations. All we can get by combining energy conservation, relativity, and the concept of stress is the vanishing divergence of the energy-momentum tensor. This property of the field dynamics no more implies the existence of a stationary action than the energy principle implies the principle of least action for ordinary mechanics.

In this chapter, we will nonetheless presuppose that the basic field equations of relativistic field theories can be derived from an invariant action. The gain in simplicity is considerable, because the invariance of a scalar is much easier to discuss and because the superiority of the energy-momentum tensor as a concept more physical than action becomes illusory in the context of general relativity (we will return to this point). In the

---

[1] Maxwell 1861–2, p. 164,

Physics and Necessity. First Edition. Olivier Darrigol.
© Olivier Darrigol 2014. Published in 2014 by Oxford University Press.

following selection of field theories, stress considerations are used only in the elementary case of static fields, for which relativity does not play a role.

The main purpose of this chapter is to show that in a relativistic, action-based framework, the choice of field theories is severely limited by a principle that may be called the *Faradayan principle* because it implicitly appears in Michael Faraday's old description of field stresses. In a preliminary vague statement, this principle states that field dynamics should depend only on field properties that can be tested through the motion of point-like test particles. As is shown in the first, historical section of this chapter, in their classical volume on field theory Lev Davidovich Landau and Evgeny Mikhailovich Lifshitz used this principle in an impressive argument for the superiority of the received electromagnetic theory over other vector field theories. They were not the first to offer an argument of this kind. In 1861, James Clerk Maxwell proved that Faraday's assumption of a magnetic stress system that shares the axial symmetry of the lines of force implied empirically valid formulas for magnetic and electromagnetic forces.

The rest of this chapter is devoted to sharper formulations of the Faradayan principle and to their consequences on the possible field theories, first in the case of static field, then for relativistic fields. In the static case for which all matter is at rest, the following principles are assumed:

FP1a: *The force density at a given point of a material body derives from field stresses that depend only on the force that an isotropic test particle would experience at this point.*

This principle only makes sense if isotropic particles can actually be used to explore the field.[2] Accordingly, we must also assume the principle

FP1b: *The forces experienced by two different isotropic particles in the same field are proportional.*

As is shown in the second section of this chapter, the FP1 principles imply that in a linear approximation (for the relation between source and field) the forces between two isotropic particles must vary as the inverse square of the distance.

The rest of the second section offers a generalization of these principles in the general case of moving matter and varying field. A classical field theory is said to be Faradayan if it obeys the principles:

FP2 a: *The equations of motion for the field and the material particles derive from a Minkowski-invariant action whose field term must have the same symmetry as the matter-dependent part of the action.*

FP2 b: *For given 4-velocities, the 4-accelerations of two different isotropic particles in the same field are proportional.*

The principle FP2a is an adaptation of FP1a to a Lagrangian and Minkowskian framework; it warrants that the field dynamics in a given neighborhood depend only on the testable effects of this field on a particle in this neighborhood. The principle FP2b is a

---

[2]The test particles must have the symmetry of a geometrical point (isotropy) in order to exclude dipoles and higher multipoles.

reformulation of FP1b so as to include a possible velocity-dependence of forces; it warrants the testability of the field by isotropic particles.

In the third section of this chapter, it is shown that the simplest theories that satisfy FP2 are:

- electromagnetism in the (linear) vector case.
- general relativity in the tensor case.
- Nordström's theory of gravitation in the scalar case.

Remarkably, the equivalence principle is not needed in the derivations of Einstein's and Nordström's theories of gravitation. The latter theory can be excluded by extending FP2b to compound particles with moving parts. In the tensor case, for which the equations of general relativity are retrieved, the Minkowskian framework loses its concrete metrical significance and the testable geometry becomes pseudo-Riemannian.

The fourth section is devoted to the relativistic generalization of Faraday's stresses into an energy-momentum tensor. It is explained why this tensor permits a natural generalization of principle FP1a in the vector case only. It is also shown that the Gupta–Feynman derivation of Einstein's gravitational field equations, which is based on the idea that the energy-momentum tensor of the gravitational field should act as a source in the field equations, implicitly requires the Faradayan principles FP2. Although this derivation shares the present approach's feature of not assuming the pseudo-Riemannian geometry of spacetime from the start, it is less economical for it assumes two forms of the equivalence principle and implies more difficult calculations.[3]

The fifth and last section is an attempt to correct a shared defect of the Faradayan and Gupta–Feynman approaches, the reliance on a Minkowski metric that ends up having no concrete metric significance. This attempt is based on the *super-Faradayan principle* according to which every aspect of the field dynamics, *including the geometrical aspects*, should be defined by the motion of point-like test particles. It is shown that this principle, the principle of least action, and a causality principle together lead to Einstein's theory of gravitation and electromagnetism when applied to a pre-metric differentiable, 4-dimensional manifold of events.

7.1 Landau, Faraday, and Maxwell

The Maxwell–Lorentz equations

In vector form and in Gaussian units, the Maxwell–Lorentz equations of electromagnetism read

$$\nabla \times \mathbf{E} = -\frac{1}{c}\frac{\partial \mathbf{H}}{\partial t}, \ \nabla \times \mathbf{H} = \frac{1}{c}\left(\frac{\partial \mathbf{E}}{\partial t} + \rho \mathbf{v}\right) \tag{1}$$

---

[3]An interesting approach, proposed by Robert Wald, consists in deriving Einstein's free-field equations as the only non-linear extension of the Pauli–Fierz equations of a spin-two field that meets a requirement of perturbative consistency. This approach shares with the present one the advantage of not relying on the equivalence principle. However, its formal implementation is difficult. Cf. Wald 1986; Straumann 2000, pp. 17–19.

$$\nabla \cdot \mathbf{E} = \rho, \ \nabla \cdot \mathbf{H} = 0 \tag{2}$$

$$\mathbf{f} = \rho \left(\mathbf{E} + \frac{\mathbf{v}}{c} \times \mathbf{H} \right), \tag{3}$$

where $\mathbf{E}$ and $\mathbf{H}$ are the electric and magnetic fields, c is the velocity of light, ρ is the microscopic density of electric charge, $\mathbf{v}$ is its velocity, and $\mathbf{f}$ is the electromagnetic force density acting on the charge carriers (electrons and other charged particles). The historical genesis of these equations is complex: it implied empirical studies of electricity and magnetism, belief in contiguous action, mechanical models, elimination of auxiliary quantities, and reduction of macroscopic electrodynamics to a microphysical theory in which electrons or ions are held responsible for the interaction between an immaterial ether and the imbedded matter. Despite some nice symmetry and the simplification brought by the vector notation, the mathematical form of these equations is still complex, and so too was the model with which Maxwell arrived at their macroscopic version. Ludwig Boltzmann famously wondered: "Was it a god who wrote these signs?"[4]

In 1908, Hermann Minkowski gave the relativistic invariant form of these equations:

$$\partial_\mu F^{\mu\nu} = j^\nu, \ \text{with} \ F_{\mu\nu} = \partial_\mu A_\nu - \partial_\nu A_\mu \tag{4}$$

$$f^\mu = j_\nu F^{\mu\nu}, \tag{5}$$

where A_μ is the 4-potential, $F_{\mu\nu}$ the derived electromagnetic field tensor, and j_μ the 4-current. The correspondence with the non-relativistic description is given by

$$\mathbf{E} = -\frac{1}{c} \frac{\partial \mathbf{A}}{\partial t} - \nabla A_0, \ \ \mathbf{H} = \nabla \times \mathbf{A}, \ \ j_0 = \rho, \ \mathbf{j} = \rho \mathbf{v}/c. \tag{6}$$

In the case of a single point-like particle of charge q, the 4-current may be written as

$$j_\mu(x) = q \int \delta^4[x - \bar{x}(\tau)] \dot{x}_\mu(\tau) \mathrm{d}\tau, \tag{7}$$

wherein $\bar{x}_\mu(\tau)$ denotes the trajectory of the particle in terms of the arbitrary parameter τ and $\dot{x}_\mu$ its derivative with respect to this parameter. The equation of motion of the point charge reads

$$m\ddot{x}^\mu = q\dot{x}_\nu F^{\mu\nu}. \tag{8}$$

This equation and the field equation (4) can be obtained by varying the action[5]

$$S = S_M + S_I + S_F, \tag{9}$$

---

[4]Boltzmann 1891–3, vol. 1, p. 96, from the introductory monologue of Goethe's Faust.

[5]The total action in the one-particle case of course leads to a divergent self-interaction. In order to avoid this divergence, one must think of a continuous dust in which each particle of the dust is only subjected to the (field-mediated) action of the other particles.

with

$$S_M = -m \int_{\tau_1}^{\tau_2} \frac{1}{2} \eta_{\mu\nu} \dot{x}^\mu \dot{x}^\nu \mathrm{d}\tau, \; S_I = -q \int_{\tau_1}^{\tau_2} A_\mu[\bar{x}(\tau)] \dot{x}^\mu \mathrm{d}\tau, \; \text{and} \; S_F = -\frac{1}{4} \int F_{\mu\nu} F^{\mu\nu} \mathrm{d}^4 x,$$

(10)

if $\eta_{\mu\nu}$ denotes the Minkowskian metric.

Landau's argument

In this relativistic formulation, the Maxwell–Lorentz equations become so simple that one is tempted to deduce them by a priori means, as Landau and Lifshitz did in their lectures on field theory. The form of the action S_M of a free particle, they argue, is the evident relativist generalization of the ordinary action $\int \frac{1}{2} m v^2 \mathrm{d}t$. The action S_I for the coupling of the particle is the simplest action that can be formed by combining linearly a vector field and the 4-velocity of the particle (Landau here concedes that only experience can tell us whether nature agrees with this simplest choice). This action is invariant under the gauge transformation

$$A_\mu \to A_\mu + \partial_\mu \xi.$$

(11)

From this invariance, Landau infers that the tensor $F_{\mu\nu} = \partial_\mu A_\nu - \partial_\nu A_\mu$ is the only linear combination of the field derivatives that has physical meaning. He further requires that the field action S_F should only be a function of physically meaningful quantities, and concludes that the simplest field action (leading to second-order, linear field equations) must be

$$S_F = -\alpha \int F_{\mu\nu} F^{\mu\nu} \mathrm{d}^4 x,$$

(12)

the coefficient α depending on the system of units.[6]

In this last step, Landau requires that the field dynamics should involve only field variables that affect the motion of particles in the field. In mathematical terms, this implies that the field action should have the same symmetries as the particle action. We will now see that Faraday made a similar assumption in his description of electric and magnetic interactions.

Faraday's lines of force

As is well known, Faraday represented the interaction between electrified or magnetized bodies by "lines of force" in the intervening space. He originally defined these lines in an operational manner, for instance as the lines giving the orientation of little magnetic needles in the magnetic case. There is no doubt, however, that he meant these lines to have some physical reality. In particular, he explained the forces between electrified or magnetized bodies as a consequence of a tension along the lines of force and a mutual repulsion

---

[6]Landau and Lifshitz 1951, §§ 8, 16, 27. These authors use the form $-m \int \mathrm{d}s$ for the action of a particle.

of these lines. In modern terms, this means that he assumed a system of stresses sharing the axial symmetry of a small portion of a line of force. Or else, he assumed that the field dynamics, represented by the stresses, had the same symmetry as the force experienced by a test particle in the field. Having little mathematical capability, Faraday did not try to express this intuition in a formal manner.[7]

Maxwell's stresses

Maxwell did it for him in the first part of his memoir of 1861 "On physical lines of force," in which the Maxwell equations appeared for the first time. This essay begins with a model of the magnetic field in which the magnetic lines of force represent vortices in a fluid of pressure p modified by centrifugal pressure. The resulting stress system has the same symmetry as the system assumed by Faraday. If $\mathbf{H}$ denotes a vector directed along the lines of force and measuring the rotation in the corresponding vortices, this stress system has the form

$$\sigma_{ij} = -p\delta_{ij} + \mu H_i H_j, \tag{13}$$

where $p - \mu H^2$ represents the tension along the lines of force, and μH^2 represents the mutual repulsion of the lines of force. From the net effect of these stresses on the sides of an infinitesimal cube, Maxwell derived the force density

$$f_i = \partial_j \sigma_{ij} = -\partial_i p + H_i \partial_j(\mu H_j) + \mu H_j(\partial_j H_i - \partial_i H_j) + \mu \partial_i \left(\frac{1}{2}H^2\right), \tag{14}$$

or

$$\mathbf{f} = (\nabla \cdot \mu\,\mathbf{H})\mathbf{H} + (\nabla \times \mathbf{H}) \times \mu\,\mathbf{H} + \mu\nabla\left(\frac{1}{2}H^2\right) - \nabla p. \tag{15}$$

In the three first terms of the latter formula, Maxwell recognized the exact expressions of the magnetic force acting on the distribution $\nabla \cdot \mu\mathbf{H}$ of magnetic masses (in the old theory of magnets), the electromagnetic force on the current $\mathbf{j} = \nabla \times \mathbf{H}$, and the force responsible for the tendency of diamagnetic or paramagnetic bodies to move along a gradient of field intensity (with a sign depending on the polarizability μ). This stunning coincidence persuaded Maxwell that there was some truth in the vortex representation of magnetism. For us who no longer believe in mechanical models of the field, there remains the striking ability of Faraday's simple stress system to represent electromagnetic phenomena.[8]

There is a common key to Faraday's and Landau's successes in deriving the basic structure of classical electrodynamics: they both assume that field dynamics, represented by

---

[7] Faraday 1839, §§1231, 1297, 1369, 1373. Cf. Gooding 1978; Darrigol 2000, Chaps. 1, 3.

[8] Maxwell 1861, part I. Cf. Siegel 1991, pp. 56–65; Darrigol 2000, pp. 147–149.

stresses in Faraday's case and by a field action in Landau's case, should have the same symmetry as the effect of the field on point-like test particles. We will now investigate the relation between these two arguments more closely, and extend them to other kinds of fields.

7.2   Toward a definition of Faradayan theories

The static case

According to principle FP1b, the force acting on an isotropic particle at point $\mathbf{r}$ of the field has the form $q\mathbf{X}(\mathbf{r})$, where q is a constant depending on the particle (its charge) and $\mathbf{X}$ is a vector field that is the same for all particles. For a dust made of such particles, the force density has the form $\rho\mathbf{X}$, where ρ denotes the charge density of the dust. Following FP1a, this force must derive from a stress system that depends only on the vector $\mathbf{X}$. The only homogenous tensors that can be built from this vector are linear combinations of $X_i X_j$ and $X^2\delta_{ij}$. Therefore, the stress system must have the form

$$\sigma_{ij} = \alpha\, X_i X_j + \beta\, X^2 \delta_{ij}, \tag{16}$$

wherein α and β are two constants. Equating the resulting force density to $\rho\mathbf{X}$, we get

$$\partial_j \sigma_{ij} = \rho\, X_i, \text{ or } \alpha\, \mathbf{X}(\nabla \cdot \mathbf{X}) + \alpha(\nabla \times \mathbf{X}) \times \mathbf{X} + (\beta + \alpha/2)\nabla X^2 = \rho \mathbf{X}. \tag{17}$$

Taking the scalar product of this equation with $\mathbf{X}$, we get

$$\rho = \alpha\, \nabla \cdot \mathbf{X} + (\beta + \alpha/2)\frac{\mathbf{X} \cdot \nabla X^2}{X^2}. \tag{18}$$

If we further assume the relation between the density ρ and the field $\mathbf{X}$ to be linear, the second term must vanish identically. The constant β must therefore be equal to $-\alpha/2$, and the relation

$$\rho = \alpha\, \nabla \cdot \mathbf{X} \tag{19}$$

must hold.

Equation (17) then implies that

$$(\nabla \times \mathbf{X}) \times \mathbf{X} = \mathbf{0}. \tag{20}$$

To the field $\mathbf{X}$, we may superpose the field $\mathbf{Y}$ created by any single point charge. Through relation (19), the isotropy of this field with respect to the source charge implies that it is radial and that its intensity varies as the inverse square of its distance from the charge. We therefore have

$$\nabla \times \mathbf{Y} = \mathbf{0}. \tag{21}$$

In addition, the assumed linearity for the relation between field and source implies that

$$[\nabla \times (\mathbf{X} + \mathbf{Y})] \times (\mathbf{X} + \mathbf{Y}) = \mathbf{0}. \tag{22}$$

The three preceding equations together imply that

$$(\nabla \times \mathbf{X}) \times \mathbf{Y} = \mathbf{0}. \tag{23}$$

As the vector $\mathbf{Y}$ can have any direction we wish at any given point, we must have

$$\nabla \times \mathbf{X} = \mathbf{0}. \tag{24}$$

Therefore, there exists a potential ψ such that

$$\mathbf{X} = -\nabla\psi. \tag{25}$$

Combining this equation with Eq. (19), we get the Poisson equation

$$\Delta\psi = -\alpha^{-1}\rho. \tag{26}$$

Consequently, the forces are Newtonian. They correspond to electrostatic forces if α is positive, and to gravitational forces if α is negative and the charge q is identified with the gravitational mass.[9]

General definition

The particles are now allowed to move slowly, so that the field's dependence on the sources remains approximately the same as in the static case. As the forces derive from a potential, the equation of motion of a set of particles of charges q_i can be obtained by varying the action

$$S = \sum_i \left[-q_i \int_{t_1}^{t_2} \psi[\mathbf{r}_i(t)]\mathrm{d}t + \int_{t_1}^{t_2} \frac{1}{2}m_i\dot{\mathbf{r}}_i^2\mathrm{d}t \right] \tag{27}$$

with respect to the function $\mathbf{r}_i(t)$ for fixed values of $\mathbf{r}_i(t_1)$ and $\mathbf{r}_i(t_2)$. In this procedure, the only degrees of freedom are the coordinates of the particles. In the contiguous-action philosophy, the values of the potential ψ should be regarded as additional degrees of freedom. Thus, there should be a total action whose variation with respect to the trajectory of a particle yields the equation of motion of this particle and whose variation with respect to ψ gives Poisson's equation

$$\Delta\psi + \sum_i q_i\delta[\mathbf{r} - \mathbf{r}_i(t)] = 0. \tag{28}$$

---

[9]The mere axial symmetry of the stress tensor would allow for Yukawa forces. The latter are excluded by further requiring that the stress tensor should be built from $\mathbf{X}$.

The action

$$S = \int_{t_1}^{t_2} \int \frac{1}{2}(\nabla \psi)^2 \mathrm{d}^3 x \mathrm{d}t - \sum_i \int_{t_1}^{t_2} q_i \psi[\mathbf{r}_i(t)] \mathrm{d}t + \sum_i \int_{t_1}^{t_2} \frac{1}{2} m_i \dot{\mathbf{r}}_i^2 \mathrm{d}t \qquad (29)$$

fulfils this purpose.[10]

In the general case of arbitrarily fast motion, it will be assumed that the equations of motion derive from an action whose form is a relativistic generalization of the former expression. That is to say, the action must have three terms: a field term which, at least in a first approximation, is an invariant quadratic form of the derivatives of a potential, an interaction term that involves this potential (linearly) and the trajectory of the particles in an invariant manner, and a matter term which is the usual action for free relativistic particles.[11] This assumption preserves two aspects of Faraday's notion of contiguous action. Firstly, it makes the evolution of the potential at a given point depend on its value at infinitesimally close points only. Secondly, it implies the existence of a conserved energy-momentum tensor. As long as this tensor can be decomposed into a field term and a matter term, the vanishing of the divergence of this tensor implies that the force acting on matter derives from field stresses.

We now look for a counterpart to Faraday's assumption that the field stresses should depend on locally testable forces only. More broadly, this means that the field dynamics should not depend on field variables that do not affect the motion of particles in the field. This will be the case if principle FP2a holds, namely if the field term of the action shares any symmetry of the interaction term (together with the matter term). Indeed the irrelevance of some field variables in the equation of motion of the particles implies that the interaction term of the action should be invariant under any change of these variables; and the invariance of the field action under such changes implies that they do not affect the field dynamics.

Lastly, Faraday's idea that a field is characterized by the motion of test particles requires principle FP2b, which states that for a given velocity of the test particle and for a given field, its acceleration is determined up to a proportionality coefficient. In other words, the inertia of a particle should not depend on the external field in which it is immersed. As we will see in a moment, the previous restrictions on the form of the action do not automatically warrant this property. It must be assumed separately.

To summarize, it is assumed that the field action should have the same symmetry as the rest of the action, and that the acceleration of a particle in the field is proportional to a definite function of field and velocity. This is the gist of the Faradayan principles FP2. In the sequel, the consequences of these principles are investigated in the cases for which the field potential is a vector, a tensor, or a scalar.

---

[10] Again, the divergent self-interaction can be eliminated by taking the limit of a continuous dust (n. 5).

[11] Evidently, this form of the matter-dependent part of the action is not a realistic representation of macroscopic matter in general. It is nonetheless sufficient for our purpose, which is the analysis of the consequences of ideal measurability through test particles that interact only with the investigated field.

7.3   The vector, tensor, and scalar cases

Vector fields

From now on, Minkowskian notation is used with the metric

$$\mathrm{d}s^2 = \eta_{\mu\nu}\mathrm{d}x^\mu \mathrm{d}x^\nu = (\mathrm{d}x^0)^2 - (\mathrm{d}x^1)^2 - (\mathrm{d}x^2)^2 - (\mathrm{d}x^3)^2. \tag{30}$$

If the field potential is a 4-vector A_μ, the most general field action that is quadratic in the derivatives of the potential is[12]

$$S_F = \int \left[\alpha\, \partial_\mu A_\nu \partial^\mu A^\nu + \beta\, \partial_\mu A_\nu \partial^\nu A^\mu + \gamma (\partial_\mu A^\mu)^2 \right] \mathrm{d}^4 x. \tag{31}$$

For one particle only, the interaction term of the action is

$$S_I = -q \int_{\tau_1}^{\tau_2} A_\mu \left[\bar{x}(\tau) \right] \dot{x}^\mu \mathrm{d}\tau, \tag{32}$$

and the mass term is

$$S_M = -m \int_{\tau_1}^{\tau_2} \frac{1}{2} \eta_{\mu\nu} \dot{x}^\mu \dot{x}^\nu \mathrm{d}\tau, \tag{33}$$

wherein $\bar{x}_\mu(\tau)$denotes a trajectory of the particle in terms of the arbitrary parameter τ and $\dot{x}_\mu$ its derivative with respect to this parameter.[13]

Variation with respect to this trajectory (with fixed extremities) yields the equation of motion of a particle:

$$m\ddot{x}^\mu = q\dot{x}_\nu F^{\mu\nu}, \tag{34}$$

with

$$F_{\mu\nu} = \partial_\mu A_\nu - \partial_\nu A_\mu. \tag{35}$$

A posteriori, the parameter τ must be proportional to the curvilinear abscissa s along the trajectory because of the relations

$$\mathrm{d}(m\dot{s}^2)/\mathrm{d}\tau = 2m\dot{x}_\mu \ddot{x}^\mu = 2m\dot{x}_\mu \dot{x}_\nu F^{\mu\nu} = 0. \tag{36}$$

The equation of motion (34) evidently satisfies the principle FP2b.

---

[12]One could allow terms involving directly the potential (for the mass term of a massive vector field). However, the gauge invariance implied by FP2a would later exclude such terms.

[13]Another possible form of the mass term is $S_M = -m \int_{\tau_1}^{\tau_2} \sqrt{\eta_{\mu\nu} \dot{x}^\mu \dot{x}^\nu}\mathrm{d}\tau$. It will be used in the scalar case, in which it is slightly more convenient.

The interaction term S_I of the action and the resulting equation of motion are invariant under the gauge transformation[14]

$$A_\mu \rightarrow A_\mu + \partial_\mu \xi \tag{37}$$

(more exactly, the transformed action differs from the original action by a constant only). According to FP2a, the field action S_F should have the same symmetry. This condition implies that $\beta = -\alpha$ and $\gamma = 0$, so that the field action becomes:

$$S_F = \int \alpha F_{\mu\nu} F^{\mu\nu} d^4 x. \tag{38}$$

The resulting field equation is

$$-4\alpha \partial^\mu F_{\mu\nu} = j_\nu, \tag{39}$$

with[15]

$$j_\mu = q \int \delta^4(x - \bar{x}) \dot{x}_\mu d\tau. \tag{40}$$

Maxwell's equations correspond to the case $\alpha = -1/4$. Other negative values of this constant would lead to the same theory up to a change of units. A positive value is not acceptable, for it would lead to negative field energy. Consequently, the usual electromagnetic theory is the only Faradayan vector field theory for which the field equations are linear.[16]

Tensor fields

If the field potential is a tensor $h_{\mu\nu}$ of second order, the interaction term of the action takes the form[17]

$$S_I = -q \int_{\tau_1}^{\tau_2} \frac{1}{2} h_{\mu\nu}(\bar{x}) \dot{x}^\mu \dot{x}^\nu d\tau. \tag{41}$$

The tensor $h_{\mu\nu}$ can be restricted to be symmetric, because an antisymmetric component would disappear by contraction with a symmetric tensor. Again, the mass term is given by Eq. (33):

---

[14] If the Minkowskian metric were replaced by a pseudo-Riemannian metric, the action $S_I + S_M$ would further enjoy general covariance and conformal invariance. These symmetries would not bring anything new in the vector case, because the gauge invariant field action $-(1/4) \int F^{\mu\nu} F_{\mu\nu} d^4 x$ shares them.

[15] The limits of integration τ_1 and τ_2 must be infinitely large in order to include the whole trajectory of the particle.

[16] The gauge invariance is intimately related to the conservation of current $\partial_\mu j^\mu = 0$ since the interaction term can be rewritten as $- \int j_\mu A^\mu d^4 x$. Accordingly, the conditions $\beta = -\alpha$ and $\gamma = 0$ express the compatibility of the field equations with the conservation of current.

[17] More complex forms of this action would not be linear with respect to the tensor field.

$$S_M = -m \int_{\tau_1}^{\tau_2} \frac{1}{2} \eta_{\mu\nu} \dot{x}^\mu \dot{x}^\nu \, d\tau.$$

By analogy with the equation of geodesics on a Riemannian manifold, the resulting equation of motion is

$$\left(m \eta_{\mu\nu} + q h_{\mu\nu}\right) \ddot{x}^\nu = -q \dot{x}^\nu \dot{x}^\rho \, \Gamma_{\mu\nu\rho}, \tag{42}$$

with

$$\Gamma_{\mu\nu\rho} = \frac{1}{2} \left(\partial_\nu h_{\mu\rho} + \partial_\rho h_{\mu\nu} - \partial_\mu h_{\nu\rho} \right). \tag{43}$$

With the notation

$$g_{\mu\nu} = (m/q) \eta_{\mu\nu} + h_{\mu\nu}, \tag{44}$$

this equation of motion implies that

$$d \left(g_{\mu\nu} \dot{x}^\mu \dot{x}^\nu \right) / d\tau = \partial_\rho g_{\mu\nu} \dot{x}^\mu \dot{x}^\nu \dot{x}^\rho - 2\Gamma_{\mu\nu\rho} \dot{x}^\mu \dot{x}^\nu \dot{x}^\rho = 0, \tag{45}$$

so that the parameter τ must a posteriori be proportional to the parameter s' obtained by integrating

$$ds' = \sqrt{g_{\mu\nu} dx^\mu dx^\nu} \tag{46}$$

along the trajectory of the particle.

The equation of motion (42) does not comply with the principle FP2b, for it yields non-proportional values of the acceleration for different values of the ratio q/m. In order to avoid this defect, we must assume that the charge q and the mass m are proportional. A proper choice of units makes them identical. Consequently, every particle has the same equation of motion, in conformity with the weak equivalence principle. A further consequence of the identity of mass and charge is that the equation of motion of a particle as well as the action

$$S_P = -m \int_{\tau_1}^{\tau_2} \frac{1}{2} g_{\mu\nu} \dot{x}^\mu \dot{x}^\nu \, d\tau \tag{47}$$

from which it derives are invariant under the invertible differentiable transformation

$$x_\mu \rightarrow X_\mu(x), \quad g_{\mu\nu} \rightarrow g_{\rho\sigma}(\partial x^\rho / \partial X^\mu)(\partial x^\sigma / \partial X^\nu), \tag{48}$$

which formally correspond to a change of coordinates on a Riemannian manifold. According to a well-known theorem, the only invariant combinations of the derivatives of the tensor $g_{\mu\nu}$ are obtained by taking traces of the Riemann curvature tensor $R_{\mu\nu\rho\sigma}$ or of tensor powers of it. The simplest of these traces is

$$R = g^{\mu\rho} g^{\nu\sigma} R_{\mu\nu\rho\sigma}, \tag{49}$$

which leads to the field action[18]

$$S_F = \frac{1}{2k^2} \int R \sqrt{-\det g}\, d^4x. \tag{50}$$

As is well known, the total action $S_F + S_P$ leads to Einstein's field equation[19]

$$R_{\mu\nu} - \frac{1}{2} g_{\mu\nu} R = -k^2 \theta_{\mu\nu}, \tag{51}$$

wherein $R_{\mu\nu}$ is the Ricci tensor and $\theta_{\mu\nu}$ is the energy-momentum tensor such that

$$\theta_{\mu\nu} \sqrt{-\det g} = m \int \delta^4(x - \bar{x}) \frac{d\bar{x}_\mu}{ds'} \frac{d\bar{x}_\nu}{ds'} ds' \tag{52}$$

in the case of a single particle. The non-linear character of this field equation results from the requested invariance.

So far the metric change $\eta_{\mu\nu} \to g_{\mu\nu}$ has been used in a purely formal manner. In order to appreciate its physical implications, consider the periodic bouncing of a ball between two walls in a region of space in which a uniform, static gravitational potential reigns and suppose the temporal component h_{00} to be dominant. The action for the free motion of the ball is

$$S_P = \int \sqrt{(1 + h_{00}) c^2 dt^2 - dr^2}. \tag{53}$$

It differs from the gravity-free action of a ball by the substitution

$$t \to t\sqrt{1 + h_{00}}. \tag{54}$$

Consequently, this primitive clock is slowed down by the gravity potential (if $h_{00} > 0$). This simple consideration strongly suggests that the original Minkowskian metric loses its chronogeometric significance in a gravitation field. The true metric, which is the one that can be explored through simple physical devices involving free falling particles and light signals, is the pseudo-Riemannian metric $g_{\mu\nu}$. The Minkowskian metric corresponds

---

[18] This action implies the second derivatives of the potential $h_{\mu\nu}$; but it does so linearly, so that the second-order terms disappear by integration (cf. Landau and Lifshitz 1951, §93).

[19] Cf. Landau and Lifshitz 1951, Chap. 11. Adding a constant Λ to the curvature R in Eq. (50) would lead to the cosmological term $g_{\mu\nu}\Lambda$ on the left-hand side of Einstein's equation. With this modification, Newton's theory would no longer be the weak-field, low-velocity limit of Einstein's theory.

to a purely fictitious situation in which the gravitational field would be cancelled without removing its sources.[20]

Scalar fields

In the case of a scalar potential, the only field action that is a quadratic form of the derivatives of the potential is:

$$S_F = \frac{1}{2k^2} \int \partial_\mu \varphi \, \partial^\mu \varphi \, \mathrm{d}^4 x. \tag{55}$$

The action of a particle immersed in this field has the form

$$S_P = -q \int \varphi(\bar{x}) \mathrm{d}s - m \int \mathrm{d}s, \quad \text{with } \mathrm{d}s = \sqrt{\eta_{\mu\nu} \mathrm{d}\bar{x}^\mu \mathrm{d}\bar{x}^\nu}. \tag{56}$$

Varying the total action leads to the field equation

$$\partial^\mu \partial_\mu \varphi = -qk^2 \int \delta^4(x - \bar{x}) \mathrm{d}s \tag{57}$$

and to the equation of motion of a particle

$$(m + q\varphi) \frac{\mathrm{d}^2 x_\mu}{\mathrm{d}s^2} = q \left(\partial_\mu \varphi - \frac{\mathrm{d}x_\mu}{\mathrm{d}s} \frac{\mathrm{d}x^\nu}{\mathrm{d}s} \partial_\nu \varphi \right). \tag{58}$$

As in the tensor case, compatibility with FP2b requires the proportionality of q and m. These parameters become equal in proper units. The resulting theory satisfies the weak equivalence principle and leads to attractive forces (as results from the signs in Eqs. (57) and (58)). It was first proposed by the Finnish theorist Gunnar Nordström in 1913 as a theory of gravitation.[21]

The particle action of this theory can be rewritten as

$$S_P = -m \int \Phi(\bar{x}) \mathrm{d}s, \quad \text{with } \Phi = 1 + \varphi. \tag{59}$$

This action is evidently invariant under the conformal transformation

$$\Phi \to \Phi', \, x \to x' \tag{60}$$

---

[20] Feynman (1995, pp. 66–9) gives a more rigorous argument of this kind. The reasoning is somewhat analogous to the way in which H. A. Lorentz derived the contraction of length from the Lorentz invariance of the Maxwell–Lorentz equations before relativity theory. Norbert Straumann (2000, pp. 24–5) astutely relies on the computable behavior of a hydrogen atom.

[21] Nordström 1913a, b. On this theory and its role in the genesis of general relativity, cf. Pais 1982, pp. 232–7.

such that

$$\Phi'(x') = \lambda^{-1}(x)\Phi(x) \tag{61}$$

and

$$\eta_{\mu\nu}dx'^{\mu}dx'^{\nu} = \lambda^2(x)\eta_{\mu\nu}dx^{\mu}dx^{\nu}. \tag{62}$$

The scaling parameter λ is not arbitrary. It must satisfy the equation $\partial^2 \ln \lambda = 0$. The proof of this condition goes as follows.

For the infinitesimal conformal transformation defined by

$$x'_{\mu} = x_{\mu} + \xi_{\mu} \text{ and } \lambda = 1 + \varepsilon, \tag{63}$$

the identity (62) implies that

$$\lambda^2 \eta_{\rho\sigma} = \partial_{\rho} x'_{\mu} \partial_{\sigma} x'^{\mu}, \tag{64}$$

or, to first order,

$$2\varepsilon \eta_{\rho\sigma} = \partial_{\sigma}\xi_{\rho} + \partial_{\rho}\xi_{\sigma}. \tag{65}$$

The trace of this equation gives

$$8\varepsilon = 2\partial_{\mu}\xi^{\mu}. \tag{66}$$

Its contraction with $\partial^{\rho}\partial^{\sigma}$ gives

$$2\partial^2\varepsilon = 2\partial^2(\partial_{\mu}\xi^{\mu}) = 8\partial^2\varepsilon, \tag{67}$$

so that $\partial^2\varepsilon$ must vanish. Any finite transformation can be obtained as the limit of the product of n small transformations when the size ε_i of each of the latter transformations goes to zero and the number n becomes infinite. The scale parameter of this product is

$$\lambda = (1 + \varepsilon_1)(1 + \varepsilon_2)\ldots(1 + \varepsilon_i)\ldots(1 + \varepsilon_n). \tag{68}$$

Its logarithm is

$$\ln \lambda = \sum_{i=1}^{n} \ln(1 + \varepsilon_i) \sim \sum_{i=1}^{n} \varepsilon_i. \tag{69}$$

As the d'Alembertian of each ε_i differs from zero only by second-order terms, the d'Alembertian of $\ln \lambda$ must also vanish.

According to FP2a, the field action should share the conformal symmetry of the particle term. We will now see that the action given by Eq. (55) spontaneously enjoys this property! The effect of the infinitesimal conformal transformation (63) on partial derivatives is given by

$$\partial'_\mu = \partial_\mu - (\partial_\mu \xi^\rho)\partial_\rho. \tag{70}$$

Consequently, we have

$$
\begin{aligned}
\partial'^2 \Phi' &= \left[\partial^\mu - (\partial^\mu \xi^\rho)\partial_\rho\right]\left[\partial_\mu - (\partial_\mu \xi^\sigma)\partial_\sigma\right](\Phi - \varepsilon\,\Phi) \\
&= \partial^2\Phi - \varepsilon\partial^2\Phi - \Phi\partial^2\varepsilon - \left(\partial_\mu \xi_\nu + \partial_\nu \xi_\mu\right)\partial^\mu\partial^\nu\Phi - \left(\partial_\mu\Phi\right)\left(2\partial^\mu\varepsilon + \partial^2\xi^\mu\right).
\end{aligned}
\tag{71}
$$

Using the vanishing of $\partial^2\varepsilon$, relation (65), and the derived relation

$$2\partial^\mu\varepsilon + \partial^2\xi^\mu = 0, \tag{72}$$

we have

$$\partial'^2 \Phi' = \partial^2\Phi - 3\varepsilon\,\partial^2\Phi. \tag{73}$$

Consequently, the product $\Phi^{-3}\partial^2\Phi$ is invariant. The 4-volume element being itself multiplied by $1 + 4\varepsilon$, the integral

$$2k^2 S_F = \int \partial_\mu\Phi\partial^\mu\Phi\mathrm{d}^4 x = -\int \Phi\partial^2\Phi\mathrm{d}^4 x \tag{74}$$

is invariant.[22]

Therefore, Nordström's theory is a Faradayan theory. The full Nordström–Einstein theory, in which the action for matter is not restricted to that of a dust, is obtained by requiring that this action should share the conformal invariance of the action of a particle. As Einstein and Adriaan Fokker showed in 1914, this theory admits a geometrical interpretation. The particle action (59) may indeed be rewritten as

$$S_P = -m \int \mathrm{d}\sigma, \tag{75}$$

with the new metric

$$\mathrm{d}\sigma^2 = g_{\mu\nu}\mathrm{d}x^\mu\mathrm{d}x^\nu = \Phi^2\eta_{\mu\nu}\mathrm{d}x^\mu\mathrm{d}x^\nu. \tag{76}$$

The equation of motion (58) of a particle is the same as the equation of the geodesics of the corresponding Riemannian manifold. As long as the choice of

---

[22]The integral $\alpha \int \Phi^4\mathrm{d}^4 x$, being also invariant, could also contribute to the field action.

coordinates is restricted to preserve the diagonal form of the metric, the scalar curvature (trace of the Ricci tensor) is

$$R = -6\Phi^{-3}\eta_{\mu\nu}\partial^{\mu}\partial^{\nu}\Phi. \tag{77}$$

Considering that the determinant of the metric $g_{\mu\nu}$ equals $-\Phi^8$, the invariant integral of this curvature is

$$\int R\sqrt{-\det g}\,d^4x = -6\int \Phi\eta_{\mu\nu}\partial^{\mu}\partial^{\nu}\Phi d^4x = 6\int \eta^{\mu\nu}\partial_{\mu}\Phi\partial_{\nu}\Phi d^4x, \tag{78}$$

which is proportional to the expression (55) of the field action.[23]

Just as in the tensor case, the analysis of space and time measurement processes in this theory leads to the conclusion that the new metric $g_{\mu\nu}$ is the observed one. The original Minkowskian metric corresponds to a fiction in which clocks and rods would not be affected by the gravitational field in which they are immersed. In order to decide between this theory and Einstein's, we may extend the principle FP2b to include compound bodies. This extension implies a form of the equivalence principle according to which the time of fall of the center of mass of a compound body does not depend on the velocity of its components. The tensor theory meets this criterion and the scalar one does not. This is proved as follows.

In the scalar case, the equation of motion of a particle in a static field takes the form

$$(1+\varphi)\gamma\frac{d(\gamma\,\mathbf{v})}{dt} = -c^2\,\nabla\varphi + \gamma^2\,\mathbf{v}\cdot(\mathbf{v}\cdot\nabla\varphi). \tag{79}$$

Taking $\varphi = -gz/c^2$ and projecting over the z-axis, this gives

$$\left(1-\frac{gz}{c^2}\right)\gamma\frac{d(\gamma\dot{z})}{dt} = g\left(1-\gamma^2\frac{\dot{z}^2}{c^2}\right), \quad \text{with } \gamma = 1/\sqrt{1-v^2/c^2}, \tag{80}$$

so that the particle falls slower when it is released with a horizontal velocity. Consequently, the center of mass of a system made of two identical particles originally moving toward each other at horizontal, equal, and opposite velocities would fall with acceleration depending on the internal kinetic energy of the system.

We will now see that the tensor theory does not have this defect. In this case, the resulting equation of motion of a particle in a static, uniform field limited to the Γ^{μ}_{00} components is

$$d(\gamma\,\mathbf{v})/dt = \gamma\mathbf{g}. \tag{81}$$

Projection of this equation on a vertical axis yields

$$\dot{\gamma}\dot{z} + \gamma\ddot{z} = \gamma g. \tag{82}$$

[23] Einstein and Fokker 1914.

Its scalar product with **v** leads to

$$c^2 \dot{\gamma} = \gamma g \dot{z}. \tag{83}$$

Consequently, the fall of the particle obeys the equation

$$\ddot{z} = g\left(1 - \dot{z}^2/c^2\right), \tag{84}$$

which does not depend on the initial horizontal velocity.

7.4   Energy-momentum considerations

The energy-momentum tensor of a given field theory is most conveniently derived from its action by the following method, which can be traced to Hermann Weyl. When written in arbitrary curvilinear coordinates, the action S depends on the metric tensor g. Call $\Theta_{\mu\nu}$ the symmetric tensor such that

$$S(g + \delta g) = \frac{1}{2} \int \Theta_{\mu\nu} \delta g^{\mu\nu} \sqrt{-\det g} \, d^4 x. \tag{85}$$

The invariance of the action under a change of arbitrary coordinates and the vanishing of its variation with respect to the non-metric fields (or particle trajectories) imply the relation

$$\Theta^\nu_{\mu\,;\nu} = 0, \tag{86}$$

wherein the semi-colon denotes covariant derivation with respect to the metric g. In a Minkowskian system of coordinates (if there is any), this relation becomes

$$\partial^\mu \Theta_{\mu\nu} = 0, \tag{87}$$

which has the same form as the conservation of energy-momentum. As can be shown in simple cases, the tensor $\Theta_{\mu\nu}$ differs from the canonical energy-momentum tensor only by physically irrelevant terms.[24]

The Faradayan vector theory is the electromagnetic theory, for which the total action reads

$$S = \int -\frac{1}{4} g^{\mu\rho} g^{\nu\rho} F_{\mu\nu} F_{\rho\sigma} \sqrt{-\det g} \, d^4 x - q \int A_\mu(\bar{x}) d\bar{x}^\mu - m \int_{\tau_1}^{\tau_2} \frac{1}{2} g_{\mu\nu} \dot{x}^\mu \dot{x}^\nu d\tau \tag{88}$$

in curvilinear coordinates. Varying with respect to the metric tensor, and setting $g_{\mu\nu} = \eta_{\mu\nu}$ and $\tau = s$ in the end, we get

$$\Theta^{\mu\nu} = T^{\mu\nu} + \vartheta^{\mu\nu}, \tag{89}$$

---

[24]Weyl 1917, §2. Cf. Landau and Lifchitz 1951, §94. This procedure is rooted in earlier efforts by Klein, Hilbert, and Einstein to derive the equations of gravitation from a variational principle.

wherein

$$T^{\mu\nu} = -F^{\mu\rho}F^\nu{}_\rho + \frac{1}{4}\eta^{\mu\nu}F^2 \tag{90}$$

is the energy-momentum tensor of the electromagnetic field and

$$\vartheta_{\mu\nu} = m \int \delta^4(x-\bar{x})\frac{\mathrm{d}\bar{x}_\mu}{\mathrm{d}s}\frac{\mathrm{d}\bar{x}_\nu}{\mathrm{d}s}\mathrm{d}s \tag{91}$$

is the energy-momentum tensor of the particle of mass m. This result and the derived relation

$$\partial^\mu T_{\mu\nu} = -\partial^\mu \vartheta_{\mu\nu} = -m \int \delta^4(x-\bar{x})(\mathrm{d}^2\bar{x}_\nu/\mathrm{d}s^2)\mathrm{d}s \tag{92}$$

is what we would expect from a relativistic generalization of the relation (14) between the force applied on a particle and the stress in the surrounding space. Moreover, the form (90) of the energy-momentum tensor of the electromagnetic field generalizes the form (13) of the stress tensor as expected from the Faradayan principle FP1.

No such simple generalization of static Faradayan considerations should be expected in the scalar and tensor cases, because the Minkowskian metric here loses its physical meaning. In the tensor case, the Weyl procedure leads to the relation

$$\theta^{\mu\nu}{}_{;\nu} = 0 \tag{93}$$

for the material energy-momentum tensor

$$\theta_{\mu\nu} = \frac{m}{\sqrt{-\det g}} \int \delta^4(x-\bar{x})\frac{\mathrm{d}\bar{x}_\mu}{\mathrm{d}s'}\frac{\mathrm{d}\bar{x}_\nu}{\mathrm{d}s'}\mathrm{d}s' \tag{94}$$

with $\mathrm{d}s' = \sqrt{g_{\mu\nu}\mathrm{d}x^\mu\mathrm{d}x^\nu}$ and $g_{\mu\nu} = \eta_{\mu\nu} + h_{\mu\nu}$. This relation of course does not imply the conservation of the energy-momentum of matter alone, for it has the form

$$\partial_\mu\theta^{\mu\nu} = -\Gamma^\nu_{\mu\rho}\theta^{\mu\rho} - \Gamma^\mu_{\rho\mu}\theta^{\rho\nu}. \tag{95}$$

The right-hand side of this equation somehow accounts for the energy-momentum of the gravitational field. Although energy-momentum tensors can be formally defined for this field, they are only covariant with respect to a fictitious Minkowskian metric, and they depend on arbitrary conventions. As is well known, precise localization of the energy in a gravitational field does not make any physical sense.[25]

In the scalar case, functional derivation of the action

$$S = \frac{1}{2k^2} \int \eta^{\mu\nu}\partial_\mu\Phi\partial_\nu\Phi\mathrm{d}^4x - m \int \Phi(\bar{x})\sqrt{\eta^{\mu\nu}\mathrm{d}\bar{x}_\mu\mathrm{d}\bar{x}_\nu} \tag{96}$$

---

[25]Cf., e.g., Weinberg 1972, §7.6.

with respect to the metric $\eta_{\mu\nu}$ leads to the conserved tensor

$$\Theta_{\mu\nu} = \partial_\mu \Phi \partial_\nu \Phi - \frac{1}{2}\eta_{\mu\nu}\partial_\rho \Phi \partial^\rho \Phi + \Phi \vartheta_{\mu\nu}. \tag{97}$$

The last term involves the field Φ besides the "bare" energy-momentum tensor $\vartheta_{\mu\nu}$ of a particle given by Eq. (91), whereas the relativistic generalization of the Faradayan static theory would suggest dependence on the particle variables only. Again, this anomaly derives from the purely formal character of the Minkowskian metric. In the curved spacetime induced by the true, observable metric, precise energy localization does not make any more sense than in the tensor theory.

Despite the unavoidable ambiguity of the energy-momentum tensor of the gravitational field in Einstein's theory of gravitation, Suraj Gupta and Richard Feynman managed to derive Einstein's equations by regarding this tensor as an additional source in an otherwise linear gravitational field equation in Minkowski spacetime. The physical justification for this procedure is a form of the equivalence principle according to which any energy should behave like a gravitational mass. In symbols, the Gupta–Feynman assumption reads

$$\mathrm{D}^{(2)}h = k^2(\vartheta + T), \tag{98}$$

wherein $\mathrm{D}^{(2)}$ is a differential operator of second order, h is the gravitational field tensor, ϑ the energy-momentum tensor of matter, and T the energy-momentum tensor of the gravitational field. Feynman obtains the precise form of the operator $\mathrm{D}^{(2)}$ by requiring that it should automatically imply the conservation law

$$\partial \cdot (\vartheta + T) = 0 \tag{99}$$

for the energy-momentum of matter and field. In symbols this condition reads

$$\partial \cdot \mathrm{D}^{(2)} = 0. \tag{100}$$

The resulting field action,

$$S_F^0(\eta, h) = \frac{1}{2k^2}\int \left(\eta^{\mu\nu}\eta^{\nu\tau}\eta^{\rho\sigma} - \eta^{\mu\tau}\eta^{\nu\rho}\eta^{\sigma\upsilon}\right)\Gamma_{\mu\nu\rho}(h)\Gamma_{\sigma\tau\upsilon}(h)\sqrt{-\det \eta}\,\mathrm{d}^4x \tag{101}$$

is that of the Pauli–Fierz massless field of spin 2.[26]

If matter is represented by a single particle, the equation of motion of this particle reads

$$\frac{\mathrm{d}^2 \bar{x}^\mu}{\mathrm{d}s^2} + \Gamma_{\nu\rho}^\mu \frac{\mathrm{d}\bar{x}^\nu}{\mathrm{d}s}\frac{\mathrm{d}\bar{x}^\rho}{\mathrm{d}s} = 0, \tag{102}$$

---

[26] Gupta 1954; Feynman 1995 [1962–3]. For the derivation of $\mathrm{D}^{(2)}$, cf. Feynman 1995, pp. 43–4. The factor $\sqrt{-\det \eta}$, which equals *one* for the Minkowskian metric η, is here only for the sake of a forthcoming argument. A more usual form of the Pauli–Fierz Lagrangian is $-\frac{1}{4}(\partial_\rho h_{\mu\nu})(\partial^\rho h^{\mu\nu}) + \frac{1}{2}(\partial_\rho h_{\mu\nu})(\partial^\nu h^{\rho\mu}) - \frac{1}{2}(\partial^\nu h_{\mu\nu})(\partial_\mu h^\rho_\rho) + \frac{1}{4}(\partial_\mu h^\rho_\rho)(\partial^\mu h^\rho_\rho)$.

and the Minkowskian expression of its energy-momentum tensor is

$$\vartheta_{\mu\nu} = m \int \delta^4(x - \bar{x}) \frac{\mathrm{d}\bar{x}_\mu}{\mathrm{d}s} \frac{\mathrm{d}\bar{x}_\nu}{\mathrm{d}s} \mathrm{d}s. \tag{103}$$

Consequently, the relation

$$(\partial + \Gamma) \cdot \vartheta \equiv \partial_\mu \vartheta^{\mu\nu} + \Gamma^\nu_{\mu\rho} \vartheta^{\mu\rho} = 0 \tag{104}$$

holds. In order to determine the energy-momentum T of the gravitational field, Feynman requires that this equation should be an automatic consequence of the field equation (98). In symbols, this condition reads

$$0 = (\partial + \Gamma) \cdot (k^{-2}\mathrm{D}^{(2)}h - T) = -\partial \cdot T + \Gamma \cdot k^{-2}\mathrm{D}^{(2)}h - \Gamma \cdot T. \tag{105}$$

As Γ and $\mathrm{D}^{(2)}h$ depend linearly on h, the tensor T can be developed as a sum

$$T = \sum_{n=2}^{\infty} T_n \tag{106}$$

of terms of increasing degree with respect to h, such that

$$\partial \cdot T_2 = \Gamma \cdot k^{-2}\mathrm{D}^{(2)}h \text{ and } \partial \cdot T_{n+1} = -\Gamma \cdot T_n \text{ for } n \geq 2. \tag{107}$$

In addition, Feynman requires the existence of an action from which the field equations derives at every order. He expects to obtain the exact field equations at the end of this complex iterative procedure.[27]

The problem can be simplified by noting that in terms of the tensor G defined by

$$G = \mathrm{D}^{(2)}h - k^2 T, \tag{108}$$

the Gupta–Feynman problem amounts to finding a second-rank tensor G that is a function of h and of its derivatives up to second order and that satisfies the three conditions

(i) $(\partial + \Gamma) \cdot G = 0$,
(ii) G is the functional derivative of a Minkowski-invariant action S_F with respect to h.
(iii) G reduces to $\mathrm{D}^{(2)}h$ at the lowest order in h.

We now introduce

$$\tilde{G} = G/\sqrt{-\det g}, \quad \text{with} \quad g = \eta + h. \tag{109}$$

---

[27] Through a clever modification of the original action, Stanley Deser managed to obtain Einstein's equations in only one step of this kind (Deser 1970). Cf. Alvarez 1989, pp. 562–4.

Using the well-known relation

$$\Gamma^{\mu}_{\mu\nu} = \partial_{\nu} \ln \sqrt{-\det g}, \tag{110}$$

condition (i) is equivalent to

$$\partial_{\mu} \tilde{G}^{\mu\nu} + \Gamma^{\nu}_{\mu\rho} \tilde{G}^{\mu\rho} + \Gamma^{\mu}_{\rho\mu} \tilde{G}^{\rho\nu} = 0. \tag{111}$$

If we formally regard $\eta + h$ as a metric g, this means that the covariant derivative $\tilde{G}^{\mu\nu}{}_{;\mu}$ vanishes with respect to this metric. During an infinitesimal change of coordinates

$$x_{\mu} \rightarrow x_{\mu} + \xi_{\mu}(x) \tag{112}$$

the variation of the metric can be written as

$$\delta g^{\mu\nu} = \xi^{\mu\;;\nu} + \xi^{\nu\;;\mu}. \tag{113}$$

Using conditions (i) and (ii), the resulting variation of the action S_F is[28]

$$\delta S_F = \int \tilde{G}_{\mu\nu}(\xi^{\mu\;;\nu} + \xi^{\nu\;;\mu})\sqrt{-\det g}\,\mathrm{d}^4 x = -2 \int \xi^{\mu} \tilde{G}^{\nu}_{\mu\;;\nu}\sqrt{-\det g}\,\mathrm{d}^4 x = 0. \tag{114}$$

This means that the action S_F is invariant under arbitrary changes of coordinates. From the theory of invariants on Riemannian manifolds, we know that the simplest invariant of the requested kind has the form

$$S_F = \frac{1}{2k^2} \int R\sqrt{-\det g}\,\mathrm{d}^4 x. \tag{115}$$

The compatibility of this action with condition (iii) is easily seen by noting that it is equivalent to the action $S_F^0(\eta + h, h)$ obtained in replacing the metric η with the metric $\eta + h$ in the expression (101) of the action for the Pauli–Fierz massless field.[29] To summarize, the Gupta–Feynman strategy leads to Einstein's equations for any tensor theory of gravitation in which gravitational energy acts a source of the gravitational field itself.[30]

Key implicit assumptions of this derivation are that the 4-divergence relations $\partial \cdot (\vartheta + T) = 0$ and $(\partial + \Gamma) \cdot \vartheta = 0$ should automatically follow from the gravitational field equations. Feynman offers no justification for these assumptions. It will now be seen that they directly result from the Faradayan principle FP2a. In the linear approximation, the coupling term of the action (for a fixed $\vartheta_{\mu\nu}$) has the form

$$S_I = -\frac{1}{2} \int h_{\mu\nu} \vartheta^{\mu\nu}\,\mathrm{d}^4 x \tag{116}$$

---

[28] Cf. Landau and Lifshitz 1951, §94.   [29] For a proof of this point, cf. Landau and Lifchitz 1951, §93.

[30] A more complex reasoning of this sort (without the benefit of the Weyl formula) is found in Feynman 1995, pp. 82–7.

and is therefore invariant under the gauge transformation

$$h_{\mu\nu} \rightarrow h_{\mu\nu} + \partial_\mu \alpha_\nu + \partial_\nu \alpha_\mu \tag{117}$$

for any $\alpha_\mu(x)$. According to FP2a, the field action S_F at the same approximation should be invariant under the same transformations. Hence comes the identity

$$\partial_\mu \frac{\delta S_F}{\delta h_{\mu\nu}} = 0, \tag{118}$$

which is the same as the constraint

$$\partial \cdot \mathbf{D}^{(2)} = 0, \tag{119}$$

leading to the Pauli–Fierz theory.

Beyond the linear approximation, the action S_P of a material particle is invariant under the infinitesimal transformation (112–13) for any infinitesimal $\xi_\mu(x)$. According to the identity (114), the invariance of the field action under these transformations is equivalent to the identity $\tilde{G}^\nu_{\mu\ ;\nu} = 0$, which is the condition for $(\partial + \Gamma) \cdot \vartheta = 0$ to follow automatically from the field equations.

Thus, the Gupta–Feynman approach implicitly appeals to the principle FP2a, and also to FP2b through the weak-equivalence principle. As we saw, the Faradayan principles are themselves sufficient to select Einstein's theory of gravitation as the simplest (lowest-order) tensor theory of gravitation. The Faradayan derivation of Einstein's equations is easier than the Gupta–Feynman derivation. It is also more economical, since it does not resort to the equivalence principle.

7.5   The super-Faradayan approach

The success of the Faradayan approach suggests some hidden connection of the Faradayan principles with the revisions of space and time concepts implied in relativity theory. In the static case, the Faradayan principle FP1a implies field stresses of a form incompatible with their concrete realization through an elastic medium.[31] This result agrees with Faraday's preference for a pure, non-mechanical force concept over any mechanical ether theory. It also suggests the lack of a privileged, ether-bound reference system and thus leads to the premises of Minkowskian covariance. The Minkowskian structure of spacetime still involves a privileged class of reference systems in which inertial motion is rectilinear and uniform. In the Faradayan approach, the motion of particles is controlled by fields that are themselves defined through the motion of test particles. According to the principles FP2, no third physical entity can intervene in the relation between particle motion and field. This suggests that the spacetime structure should itself correspond to a dynamical field.

---

[31] Indeed the stress tensor of a single-constant elastic body is $\sigma_{ij} = K(\partial_i u_j + \partial_j u_i + \delta_{ij}\partial_k u_k)$, where $\mathbf{u}(\mathbf{r})$ denotes the elastic deformation.

The formal deployment of the Faradayan approach indeed implies that the effective, observable metric depends on a dynamical tensor field. But the implication is only indirect: the principle FP2 assumes a flat Minkowskian structure that ends up being purely fictitious. In order to mend this defect, it is tempting to push the Faradayan approach to the extreme by assuming that all aspects of field dynamics, *including the geometrical aspects*, are defined by point-like test particles. This super-Faradayan principle agrees with Harvey Brown's idea that the metrical properties of spacetime should be regarded as deriving from a more primitive, pre-metrical dynamics.[32]

Let us begin with a continuous, differentiable, 4-dimensional manifold of (classes of equivalence of) potential events. This topological framework may be regarded as a natural idealization of our intuitive knowledge that (point-like) events are locally mapped by three space coordinates and a time coordinate, no matter how these coordinates are concretely defined. A most basic observation in this manifold of events is the motion of a point-like particle, defined by the value $\bar{x}^\mu(\tau)$ of its coordinates for every value of the arbitrary continuous parameter τ. Action at a distance being excluded, this motion should be controlled by fields. The equation of motion must be covariant since the choice of coordinates is very free at this pre-metric stage. The easiest way to meet this condition is to assume an invariant action constructed from the field and from the derivatives of the motion. As this action must also be independent of the parameter τ, it can only depend on the velocity $\dot{x}^\mu = d\bar{x}^\mu/d\tau$. No such action can be built for a scalar field. For a vector field $A_\mu(x)$, we have the possibility

$$S_P = -q \int A_\mu[\bar{x}(\tau)]\dot{x}^\mu d\tau. \tag{120}$$

By Faradayan principle FP2a, the field action should only depend on $F_{\mu\nu} = \partial_\mu A_\nu - \partial_\nu A_\mu$. No invariant field action can be formed from the latter field only.

For a symmetric tensor field $g_{\mu\nu}(x)$, we have the unique possibility

$$S_P = -m \int \sqrt{g_{\mu\nu}[\bar{x}(\tau)]\dot{x}^\mu \dot{x}^\nu} d\tau. \tag{121}$$

In this case, it is possible to define an invariant field action. The simplest non-trivial choice is

$$S_F = \frac{1}{2k^2} \int (R + \Lambda)\sqrt{-\det g} d^4x, \tag{122}$$

wherein R is the scalar Riemann curvature associated with the metric $g_{\mu\nu}$, and k and Λ are two constants. At this stage, the signature and the metric interpretation of the quadratic

---

[32] See Brown 2005. Brown's own suggestion (pp. 161-163) is to exploit the severe restriction that general covariance imposes on the dynamics. If the conditions of Lovelock's theorem are met, the field equation can only be Einstein's equation, and the divergence of the energy-momentum tensor of singularities representing the immersed particles must vanish. Consequently, the particles must move along the geodesics of the metric defined by the tensor field (because the equations (95), (102), and (104) are equivalent). In the approach suggested below, the particles come first, and the invariance of the field action results from the invariance of the particle action.

form $g_{\mu\nu}$ are unknown. The signature $(+, -, -, -)$ can easily be derived from the compatibility of the free-particle motions with a concept of causality, in the EPS manner.[33] The metric interpretation is obtained through the geodetic construction of an invariable rod and bouncing-light clock along any time-like trajectory, as was done in the EPS framework.[34] In the present case, the equation of geodesics depends on the $g_{\mu\nu}$ tensor only (there is nothing like the Weyl gauge field),[35] and the operational time agrees with the measure $\int \sqrt{g_{\mu\nu}\mathrm{d}x^{\mu}\mathrm{d}x^{\nu}}$.

The metric $g_{\mu\nu}$ being given, it becomes possible to form an invariant action for the vector field A^{μ}. Its simplest expression is

$$S_F = \alpha \int g^{\mu\rho}g^{\nu\sigma}F_{\mu\nu}F_{\rho\sigma}. \qquad (123)$$

Similarly, we may form the invariant actions

$$S_P = -m \int \Phi[\bar{x}(\tau)]\sqrt{g_{\mu\nu}\dot{x}^{\mu}\dot{x}^{\nu}}\mathrm{d}\tau \text{ and } S_F = \frac{1}{2k^2} \int g^{\mu\nu}\partial_{\mu}\Phi\,\partial_{\nu}\Phi\mathrm{d}^4x \qquad (124)$$

for the scalar field $\Phi(x)$. The restriction of the resulting theory to the case of a flat $g_{\mu\nu}$ is Nordström's theory, which we earlier excluded by extending principle FP2b to compound bodies. Thus, for tensor rank inferior or equal to two, the super-Faradayan principle privileges Einstein's theory for gravitation and the concomitant generalization of Maxwell's electromagnetic theory. There could be more complex expressions of the action: of higher degree, or combining scalar, vector, and tensor fields. However, it remains true that the simplest possible choices are those of the Einstein–Maxwell theory of gravitation and electromagnetism.

The super-Faradayan principle leads to the locally Minkowskian structure of spacetime when applied to a pre-metric manifold of events, if only the manifold has dimension four and if the particles' motion is compatible with a partial ordering of events. The principle of inertia thereby receives a remarkably simple justification, whereas it was brutally assumed in the EPS approach (through the projective structure) and it derived from counter-intuitive conditions for the measurability of time in the Helmholtzian approach.[36] As Harvey Brown emphasized, inertial motion presupposes a mysterious harmony between the free motions of a priori independent particles. Minkowskian geometry captures this harmony without explaining it. The Faradayan view reveals the secret of inertia as a field-based covariant guiding process for the test particles.

The super-Faradayan approach has its own imperfections. In particular, its starting point, the manifold of events, is not as pre-metric as it would seem. In order to have operational significance, the possibility of mapping an event by four coordinates implies a sort of measurement, for instance the radar detections imagined by EPS. In chapter 4, I have argued that from a physical point of view the concept of manifold necessarily presupposes

[33]See Chapter 5, p. 163.   [34]See Chapter 5, pp. 166–9.

[35]As far as I can see, there is no particle action that yields the equations of geodesics for a Weyl connection.

[36]See Chapter 5, pp. 170, 173, 180.

the concrete determination of coordinates. However, this determination depends on arbitrary conventions and it can be as indirect as one wishes. As a result, the metric structure that may be attached to a given concrete determination of the coordinates remains artificial without further theoretical justification (just as the metric structure of the temperature of a dilation thermometer lacks any deep physical meaning). This arbitrariness of coordinate determinations justifies the requirement of general covariance for processes described on the manifold.

Conclusions

General relativity can be reached without *initially* making the basic physical assumptions Einstein used on his path to this theory. These assumptions are the locally Minkowskian character of spacetime, the equality of inertial and gravitational mass; the local equivalence of a gravitational field with inertial forces in an accelerated frame; and the related interpretation of gravitation as an effect of the curvature of spacetime. In the Faradayan and super-Faradayan approaches, the basic assumption is that field dynamics exclusively depend on field properties that can be tested through the motion of point-like test particles; and Einstein's assumptions are recovered only at the end of the reasoning for tensor fields. In particular, the equality of gravitational and inertial mass turns out to be a consequence of the definiteness of the measured force field; the effective curvature of spacetime results from the consequences of the Faradayan field dynamics on spacetime surveys; and in the super-Faradayan approach the locally Minkowskian character of spacetime derives from a causal-ordering principle.

In the Faradayan approach, the simplest (lowest-order) Faradayan theories are Nordström's theory of gravitation in the scalar case, the Maxwell–Lorentz electromagnetic theory in the vector case, and Einstein's theory of gravitation in the tensor case. A touch of equivalence principle or some stretching of the principle FP2b excludes Nordström's theory. In the super-Faradayan approach, we directly reach Einstein's theory for combined gravitation and electromagnetism. This leaves us with a very limited spectrum of classical theories, compared with the wealth of field theories that today's physicists call classical. For instance, the Faradayan approach excludes Yukawa's theory, non-Abelian gauge theories, and any spinor field theory. This state of affairs is tolerable because the latter theories are only classical in a formal sense: they serve as the basis for quantized field theories and are not meant to apply directly to the classical, large-scale interactions of macroscopic bodies. For the latter kind of phenomena the only relevant fields belong to electromagnetism and gravitation, which are known to obey Maxwell's and Einstein's theories. One could not hope better agreement with the Faradayan selection of classical theories.

A common weakness of the super-Faradayan and Faradayan approaches is their relying on the principle of least action. If they are meant to provide a rational selection of classical field theories, they should only rely on principles that can themselves receive a rational justification. As was repeatedly mentioned, such justification is still lacking for the principle of least action. However, we know that every fundamental (classical) theory complies with this principle; we know that conservation laws are intimately related to the symmetries of the action;[37] and we know that the action has a physical interpretation as a phase in the quantized theories. When combined with the Faradayan field definition in an amorphous 4-dimensional manifold, the action principle leads, at the lowest algebraic and differential orders, to Einstein's theory of gravitation and electromagnetism. Whether we regard this result as further evidence of the power of the action principle or as an argument for the necessity of our best classical theories, it still demonstrates the power of the Faradayan focus on test particles.

This focus is necessary if we exclude direct action at a distance and if we replace it with action mediated by measurable fields. As in most cases of rational necessity defended in this book, this necessity is bound to a refutable idea of the comprehensibility of nature. In the Faradayan approach, this idea implies the measurability of spacetime, a field dynamics compatible with the resulting Minkowskian or pseudo-Riemannian structures, and the measurability of the fields by point-like test particles. In the super-Faradayan approach, it only implies the continuous manifold of events, a compatible field dynamics, and the measurability of the fields by point-like test particles.

These approaches involve both fields and particles. However, their success should not be regarded as an argument to exclude monistic theories, be they field-free formulations or particle-free. The test particles of the Faradayan reasoning could be field singularities, and the fields could be formal tools for expressing retarded action. What matters in this reasoning is that something can play the role of particles and something can play the role of a field in an operational sense. Then there is a well-defined though limited notion of the comprehensibility of nature that leads, together with the principle of least action, to the Einsteinian classical field theories.

[37] The relation is especially tight in generally covariant theories, since David Lovelock has proven that any divergenceless, second-order symmetric tensor built from a tensor field and from its first and second derivatives derives from an action principle (in dimension four, the only such tensor is the Einstein tensor with the cosmological term). Cf. Lovelock 1971, p. 500.

8

QUANTUM MECHANICS

We suggest that quantization be understood as a deformation of the structure of the algebra of classical observables, rather than as a radical change in the nature of the observables.[1] (François Bayen, Moshé Flato, Christian Fronsdal, André Lichnerowicz, and Daniel Sternheimer, 1978)

It is important to note that quantum mechanics can be obtained from classical mechanics merely by reducing the ontological premises without incorporating new empirical components. This will be demonstrated in detail within the framework of the quantum logic approach to quantum mechanics. Consequently, in quantum mechanics just those parts of classical mechanics are missing which are not intuitive and which do not correspond to plausible reasoning. This means that quantum mechanics is more intuitive than classical mechanics – a result which is paradoxical at first glance.[2] (Peter Mittelstaedt, 2003)

Quantum theory is simply a new type of probability theory. Like classical probability theory it can be applied to a wide range of phenomena. However, the rules of classical probability theory can be determined by pure thought alone without any particular appeal to experiment (though, of course, to develop classical probability theory, we do employ some basic intuitions about the nature of the world). Is the same true of quantum theory? Put another way, could a 19th century theorist have developed quantum theory without access to the empirical data that later became available to his 20th century descendants? In this paper it will be shown that quantum theory follows from five very reasonable axioms which might well have been posited without any particular access to empirical data.[3] (Lucien Hardy, 2001)

In earlier chapters, I have defended some rational necessity of much of classical physics, including classical mechanics (relativistic or not) and the field theories of electromagnetism and gravitation. This necessity implies the assumption of a causal, spacetime description of phenomena, regarded as a basic feature of the comprehensibility of phenomena at our scale. Until the late nineteenth century, most physicists judged this assumption so natural that they did not hesitate to extend it to the invisible micro-world. The situation changed radically with the advent of quantum mechanics, which denies the possibility of a causal, space-time description of atomic phenomena. Granted that quantum laws do not have the kind of rational necessity found in classical theories, do they nonetheless obey other kinds of necessity?

---

[1] Bayen et al. 1978, p. 62   [2] Mittelstaedt 2003, p. 282.   [3] Hardy 2001, p. 1.

Physics and Necessity. First Edition. Olivier Darrigol.
© Olivier Darrigol 2014. Published in 2014 by Oxford University Press.

This chapter examines various answers to this question. The first section is devoted to two kinds of historical necessity, one driven by the analogy between classical and quantum theory, the other by the analogy between matter and light. These analogies led to two different versions of the new mechanics: the Heisenberg–Born–Jordan matrix mechanics and the de Broglie–Schrödinger wave mechanics, which occurred in two different contexts: atomic constitution and the statistical properties of matter and light. The second section is devoted to a mathematical kind of necessity of quantum mechanics: the possibility of deriving quantum mechanics by deforming the Poisson algebra of classical mechanics. Around 1940, José Moyal and Hilbrand Groenewold discovered that quantum mechanics admitted a phase-space formulation that increased its formal similarity with Hamiltonian mechanics: one just had to replace the Poisson bracket by the more complicated "Moyal bracket." In the late 1970s, the mathematicians Jacques Vey, André Lichnerowicz, and Simone Gutt found that the latter bracket could be obtained by deforming the Poisson bracket and that this deformation was essentially unique. This means that in a deep mathematical sense classical mechanics already contains quantum mechanics.

The two remaining sections of this chapter deal with attempts to base quantum mechanics on axioms more natural than those of the standard textbook formulation. Since the invention of quantum mechanics, many different axiomatics have been proposed for quantum theory, with various motivations including mathematical rigor and completeness, conceptual and structural clarity, broader scope (for systems resisting standard quantization), deeper interpretation or ontology, physical transparency and necessity. I only retain the axiomatics that have the necessitarian motivation. This choice explains, for instance, why I do not discuss the C*-algebraic approach, despite its seductive generality and mathematical power.[4] Section 8.3 deals with the so-called "quantum logic," which purports to derive the basic structure of quantum mechanics from simple, natural requirements about the logic of Yes-No empirical questions. In this approach, which originated in John von Neumann's rigorous reformulation of quantum mechanics, the specificity of quantum phenomena is traced to the existence of incompatible questions. The last and fourth section is devoted to a more recent kind of axiomatics, based on natural constraints on a statistical-operational definition of physical states. In this approach, which originated in a seminal paper of 2001 by Lucien Hardy, discreteness, probability, and information are the most important notions. In the past few years, heavy weight has been placed on the information-theoretic meaning of the axioms, sometimes with reductionist intentions.

None of these necessity arguments have absolute value, they only show that the basic predictive apparatus of quantum mechanics follows from a few simple or natural assumptions, the truth of which ultimately depends on experiment. Nevertheless, the identification of such assumptions has two advantages: firstly, it tells us the price to be paid if quantum mechanics is to be replaced by a more general theory in some domains of experience (and there is no lack of attempts at such generalizations motivated by quantum field theory or by

---

[4] A C*-algebra is a Banach*-algebra over $\mathbf{C}$ (i.e., a complete normed vector space over $\mathbf{C}$, in which every element a has an adjoint a^*, with $(a^*)^* = a$, $(a+b)^* = a^* + b^*$, $(ab)^* = b^* a^*$, and $(\lambda a)^* = \lambda^* a^*$ for any other element b and for any complex number λ) for which the norm and the adjunction satisfy $\|a^* a\| = \|a\|^2$. Such algebras generalize the algebra of observables of standard quantum mechanics (bounded operators in Hilbert space). Pascual Jordan's (1932) and Irving Ezra Segal's (1947) early attempts to define quantum theory directly through an algebra of observables motivated the study of C*-algebras. Cf. Primas 1981, pp. 163–78.

gravitation theory); secondly, it may shed light on the specificity of quantum phenomena and contribute to the interpretation debate. Although the necessity arguments or natural axiomatics described in this chapter do not directly address these deeper questions, they may be as instructive as a more direct approach would be.[5]

8.1 Historical necessity

Quantum mechanics being universally regarded as a strikingly counter-intuitive theory, one immediately expects that its strangest features could only be introduced by some sort of unwanted necessity, be it of empirical or rational origin. The history of quantum theory indeed contains a rich stock of arguments for the necessity of features of the quantum world. The present historical sketch focuses on the steps that ultimately contributed to the construction of quantum mechanics. A fuller history would better explain when and why these steps were taken, and it would give a better idea of the distribution and competition of efforts in various directions of research. But it would not help much to understand why quantum mechanics was reached in the end.[6]

Failures of classical theory

Quantum theory began with arguments for the necessity to depart from the received theories of mechanics and electrodynamics. In the second-half of the nineteenth-century, electromagnetic radiation was known to reach a well-defined state of thermal equilibrium by interaction with matter. This state is the so-called *blackbody radiation*, which can be observed within a uniformly heated cavity with absorbing walls. As a consequence of a thermodynamic theorem established by Gustav Kirchhoff around 1860, the spectrum of this radiation is universal: namely, the energy per unit volume has a well-defined (continuous) distribution over frequencies. By the end of the century, this distribution was empirically known to decrease exponentially for high frequencies. In addition, the joint application of (macroscopic) thermodynamics and electrodynamics implied theoretical restrictions on the form of this distribution (Wien's displacement law, and the Stefan–Boltzmann law) which were well verified by experiments. It was therefore hoped that the theoretical study of the electromagnetic interaction between material radiators and radiation would yield the form of the universal spectrum. After long efforts, in 1900 Max Planck succeeded in this task, but only at the price of introducing foreign elements in the theory: the Planck constant h and the associated energy-elements $h\nu$ for the energy of a radiator of frequency ν.[7]

---

[5] For a lucid justification of the constructive, axiomatic approach to the foundations of quantum mechanics, see Grinbaum 2007.

[6] For a fuller description of the shortcomings of this historical sketch, cf. Darrigol 2009, §4. There are two major treatises covering the whole history of quantum theory: Jammer 1966 and Mehra and Rechenberg 1982–7, henceforth abbreviated as *CD* and *HD*. Much valuable work is now being published by a research group on the history of quantum physics at the Max Planck Institut für Wissenschaftsgeschichte in Berlin. For a brief history and a bibliography, cf. Darrigol 2003. A history of the constructive role of classical analogies is found in Darrigol 1992, henceforth abbreviated as *CQ*.

[7] Cf. Kuhn 1978; Needell 1980; *CQ*, part A; Gearhart 2002.

Planck did not interpret this innovation as a rupture with received theory. The true revolution occurred in the years 1905-08, with a series of proofs that the interaction of thermalized matter with electromagnetic radiation yielded an absurd spectral distribution: the Rayleigh-Jeans law for which the spectral energy density diverges quadratically for high frequencies. The first assumption of these proofs was that the interaction between matter and radiation obeyed the laws of the Maxwell–Lorentz theory of electrodynamics. The other assumption was some weak sort of ergodicity: the interaction had to be such that almost every initial condition of the global micro-system (radiation plus material entities) would lead to the same macroscopic behavior in the long run. The three main proofs differed substantially in the details. Albert Einstein's proof of 1905 relied on the random interaction between cavity radiation and thermalized resonators; James Jeans's proof of 1905 rested on Maxwellian statistical mechanics applied to a gas interacting with radiation; Hendrik Lorentz's proof of 1908 was based on the application of Josiah Willard Gibbs's statistical mechanics to a set of electrons interacting with cavity radiation. The convergence of these three proofs increased the plausibility of their baffling conclusion.[8]

The first and perhaps most convincing of these proofs, Einstein's, goes as follows. Consider a set of linear electric oscillators with a broad, quasi-continuous range of frequencies, and suppose that these oscillators interact with the radiation included in a cavity with mirroring walls and also with a gas that has the well-defined temperature T. According to Kirchhoff's radiation theorem, the final spectrum of the radiation does not depend on the thermalizing entities and should therefore be identical with the universal blackbody spectrum. From measurements of the specific heats of solids at moderate temperatures, the equipartition of energy predicted by Maxwell and Boltzmann was known to hold for the degrees of freedom of the interacting gas and oscillators. Consequently, each oscillator should have the (time-) average energy

$$U = kT, \qquad (1)$$

where k is Boltzmann's constant. Call $u_\nu d\nu$ the energy per unit volume of the radiation whose frequency is comprised between ν and $\nu + d\nu$. According to Planck, the interaction between one of the resonators and the surrounding radiation implies the relation

$$u_\nu = (8\pi\nu^2/c^3)U \qquad (2)$$

between the average energy U of the resonator and the spectral density u_ν of the radiation at the frequency ν of the resonator. Consequently, the equilibrium spectrum of the radiation should be given by

---

[8]Cf. Kuhn 1978. Earlier derivations of the Rayleigh–Jeans distribution by Lord Rayleigh (1900) and Lorentz (1903) were only meant to apply to low-frequency radiation. Although Jeans (1905) and Lorentz (1908) removed this limitation, they speculated that an exceedingly long time might be needed for the thermal energy transfer from matter to high-frequency radiation. Their critics noted that in this case the observed blackbody spectrum would no longer be a true state of equilibrium, so that its universality and its compliance with thermodynamic laws would be very difficult to understand. Lorentz soon recognized the helplessness of the situation. Jeans did the same a couple of years later.

$$u_\nu = (8\pi \nu^2 / c^3) kT, \tag{3}$$

in contradiction with the observed blackbody spectrum and with the obvious requirement that the total energy should be finite.[9]

As Planck's relation (2) plays an important reasoning in this and later reasoning, a concise derivation will now be given. This derivation may be criticized for assuming the definiteness, smoothness, and isotropy of the distribution of the radiation over various modes without proof.[10] However, the validity of these assumptions is strongly suggested by the empirical definiteness and universality of the blackbody spectrum.

We may assimilate the resonator with an electron of charge e and mass m constrained to move along the x axis and elastically attached to the origin of this axis. The equation of motion of this electron in the external field $\mathbf{E}$ (supposed to be uniform at the scale of the resonator) is

$$m(\ddot{x} + \omega_0^2 x) - (2e^2/3c^3)\dddot{x} = eE_x, \tag{4}$$

if ω_0 denotes the angular frequency of the free oscillations of the electron. The second term represents the damping force due to the emission of radiation by the accelerated electron. For the frequencies of interest, this term is very small compared to the elastic and inertial force, so that the resonator only interacts with the Fourier components of the radiation that have an angular frequency very close to ω_0. The exciting field E_x can be written as a sum of contributions from the proper modes of the mirroring cavity in which it is immersed, labeled by the discrete index s:

$$E_x = \sum_s a_s \cos(\omega_s t + \varphi_s). \tag{5}$$

Accordingly, the permanent part of the solution of the equation of motion (4) can be written as

$$x = \sum_s \frac{a_s}{|Z(\omega_s)|} \cos(\omega_s t + \psi_s), \tag{6}$$

where

$$Z(\omega) = (m/e)(\omega_0^2 - \omega^2) + \mathrm{i}(2e/3c^3)\omega^3. \tag{7}$$

On the one hand, the average energy U of the resonator reads

$$U = m\omega_0^2 \overline{x^2} = m\omega_0^2 \sum_s a_s^2 / 2|Z(\omega_s)|^2. \tag{8}$$

---

[9] Einstein 1905a, which also has the lightquantum. Cf. Klein 1963; CD, Chap. 1.3; HD, vol. 1; Büttner, Renn, and Schemmel 2001.

[10] Planck's electromagnetic H-theorem provides such a proof, but only at the price of the hypothesis of "natural radiation." Cf. CQ, Chap. 3.

On the other hand, the quadratic average of the exciting field reads

$$\overline{E_x^2} = \frac{1}{2}\sum_s a_s^2 = \int_0^\infty J(\omega)d\omega, \tag{9}$$

if $J(\omega)d\omega$ denotes the contribution of the modes s such that $\omega < \omega_s < \omega + d\omega$. The comparison of these two expressions leads to

$$U = m\omega_0^2 \int_0^\infty \frac{J(\omega)}{|Z(\omega)|^2}d\omega. \tag{10}$$

Granted that the distribution $J(\omega)$ is well-defined and smooth, the narrowness of the resonance leads to the simpler expression

$$U = m\omega_0^2 J(\omega_0) \int_0^\infty \frac{d\omega}{|Z|^2(\omega)}. \tag{11}$$

Using again the narrowness of the resonance, the latter integral can be computed under the approximation

$$Z(\omega) = (2m\omega_0/e)(\omega_0 - \omega) + i(2e/3c^3)\omega_0^3. \tag{12}$$

The resulting expression of the average resonator energy is

$$U = (3\pi c^3/4\omega_0^2)J(\omega_0). \tag{13}$$

Granted that the radiation is isotropic, the x-component of the electric field contributes one sixth of the energy density $(E^2 + B^2)/8\pi$ of the electromagnetic field. Therefore, the spectral energy distribution of the radiation is given by

$$u_\nu d\nu = (1/8\pi) \times 6 \times J(\omega_0)d\omega_0 \text{ (with } \omega_0 = 2\pi\nu), \text{ or } u_\nu = (8\pi\nu^2/c^3)U \tag{14}$$

in conformity with Eq. (2) that was to be demonstrated.[11]

Quantum derivations of the blackbody law

In 1900, Planck obtained by obscure theoretical means a blackbody law that fitted experiments excellently. In 1906, Einstein derived this law by assuming that blackbody radiation was in equilibrium with electric resonators whose energy was restricted to be an integral multiple of the quantum $h\nu$, wherein h is Planck's constant and ν the frequency of the resonator. The following is a slightly modified version of his considerations.[12]

---

[11] A somewhat similar proof is found in Pauli 1929, pp. 1522–5. I use Gaussian units for which the Coulomb force between two point charges q and q' is qq'/r^2 if their mutual distance is r.

[12] Einstein 1906. Cf. Klein 1965. Einstein originally discretized the microcanonical expression of the entropy of a resonator. He used a discretization of the canonical distribution (as done below) in Einstein 1907b.

Consider a harmonic oscillator with the frequency v, the mass m, and the quadratic Hamiltonian

$$H(q, p) = p^2/2m + 2\pi^2 m v^2 q^2, \tag{15}$$

and suppose that this resonator is in contact with a thermostat at the temperature T. The Gibbs–Boltzmann distribution law for the energy of this oscillator yields the average energy

$$U = \frac{\int H e^{-H/kT} \, dq dp}{\int e^{-H/kT} \, dq dp}. \tag{16}$$

Owing to the quadratic character of the Hamiltonian, the surface in the (q, p)-plane comprised between the ellipses $H(p, q) = E$ and $H(p, q) = E + dE$ does not depend on the value of E. Consequently, the average energy of the resonator may be rewritten as

$$U = \frac{\int_0^\infty E e^{-E/kT} \, dE}{\int_0^\infty e^{-E/kT} \, dE}, \tag{17}$$

with the result

$$U = kT. \tag{18}$$

Now assume with Einstein that the energy of the oscillator is restricted to the discrete values nhv. The natural discrete counterpart of Eq. (17) is

$$U = \frac{\sum_{n=0}^\infty nhv \, e^{-nhv/kT}}{\sum_{n=0}^\infty e^{-nhv/kT}} = \frac{hv}{e^{hv/kT} - 1}. \tag{19}$$

Together with relation (2), this yields

$$u_v = \frac{8\pi h v^3}{c^3} \frac{1}{e^{hv/kT} - 1}, \tag{20}$$

which is Planck's law. As Einstein did not fail to see, a weakness of this derivation is that the quantization of the resonator contradicts the classical derivation of relation (2).

This argument by Einstein was the first intimation of a sharp concept of quantization, according to which the states of a microphysical entity are restricted in a discrete manner depending on the quantum of action. Planck preferred to assume that the states of the resonator with an energy differing by less than a quantum counted only as one state in the combinatorial computation of the resonator's entropy. By the Solvay congress of

1911, most experts agreed that some discontinuity had to be introduced in the dynamics of microphysical entities in order to save the phenomena (blackbody radiation and also the low-temperature behavior of specific heats), although there was much variety of opinion about the manner in which this discontinuity should be introduced and on whether it should be reducible to some underlying mechanism of a more familiar kind.[13]

Einstein's discrete quantization was most daring, for it made it difficult to understand how electromagnetic radiation could interact with quantized resonators, unless radiation itself had a discontinuous structure. As we will see in a moment, Einstein had already imagined this possibility. Whereas the implied lightquanta long remained marginal, the discrete quantization of the states of material entities soon gained ground: a few theorists including Arthur Erich Haas, William Nicholson, Niels Bjerrum, and Niels Bohr began using the quantum of action for the purpose of discrete selection among the possible states of classical models of atoms or molecules. This selection was supposed to determine the normal state of atoms and also to explain the discrete character of atomic or molecular spectra. The remarkable success of Bohr's attempt in this direction largely contributed to establish discrete quantization for matter.[14]

The frequency rule

Scattering experiments performed in Ernest Rutherford's Manchester laboratory in the early 1910s suggested that atoms were made of a central positive nucleus with a few electrons orbiting around it. In 1912–13, Bohr tried to model atoms as series of concentric electronic rings the filling of which corresponded to the chemical periods. This model being inherently unstable (both mechanically and radiatively), he used Planck's quantum of action to select rings endowed with a special, classically unaccountable kind of stability. The precise way he did that in his famous trilogy of 1913 was inspired by Balmer's formula for the visible spectrum of the hydrogen atom, which is the $m = 2$ case of the more general Rydberg formula:[15]

$$\nu_{nm} = K \left(\frac{1}{m^2} - \frac{1}{n^2} \right),$$ (21)

where ν_{nm} is the frequency of the observed lines, m and n are two integers, and K is the so-called Rydberg constant.

In analogy with the quantization of a harmonic oscillator, Bohr assumed that the single electron he posited in the hydrogen atom could only exist in a series of "stationary states" determined by a rule of the form:

$$E_n = -\alpha \, nh\bar{\nu}_n,$$ (22)

---

[13] Cf. Kuhn 1978; Barkan 1993.   [14] Cf. Heilbron 1964, 1977.

[15] Bohr 1913. Cf. Heilbron and Kuhn 1969; Pais 1991, Chap. 8; *CD*, Chap. 2; *HD*, vol. 1, Chap. 2; *CQ*, Chap. 5; Kragh 2012.

where E is the binding energy of the electron, $\bar{\nu}$ its orbital frequency, α a numerical constant, and n a positive integer. Bohr further assumed that ordinary mechanics applied to the motion of the electron in the Coulomb field of the nucleus. The resulting Kepler motion satisfies the relation[16]

$$(-E/\bar{\nu})^3 = \pi^2 \mu\, e^4/2\bar{\nu}, \tag{23}$$

if μ denotes the mass of the electron and e its charge. Together with the quantum rule (22), this relation implies the quantized energy values

$$E_n = -Kh/n^2, \text{with } K = \pi^2 \mu e^4/4\alpha^3 h^3. \tag{24}$$

The Rydberg formula can then be rewritten as

$$h\nu_{nm} = E_n - E_m. \tag{25}$$

Bohr found that the choice $\alpha = 1/2$ yielded an accurate value of the Rydberg constant K.

From these formal considerations, Bohr inferred that the emission of a spectral line involved two stationary states, and that the frequency of this line depended on the energy of these two states according to the frequency rule (25). This was a very daring assumption as it contradicted the classical equality of radiation frequency and orbital frequency. Yet it is difficult to imagine any plausible alternative in the context of discrete quantization. One could try (as Fritz Hasenöhrl did in 1911)[17] to identify the spectral frequencies with the orbital frequencies of discrete states depending on two indices. This would be artificial, however, as there would be no natural way to relate radiation to change of state.

In favor of the frequency rule, Bohr could have argued that it resulted from energy conservation applied to the emission of a lightquantum. He did not, because he agreed with most of his contemporaries that Einstein's lightquantum was incompatible with the well-established wave properties of radiation. In Bohr's view, Maxwell's theory of free electromagnetic radiation was necessary to the very definition of the concepts necessary to define the properties of emitted and absorbed radiation. Lightquanta being too paradoxical, Bohr rather left the coupling between continuous radiation and quantum jumping in the dark.[18]

Despite this inherent incompleteness, Bohr's theory quickly achieved new successes that enhanced its credibility. It correctly predicted that some spectral lines originally attributed to hydrogen had to be ascribed to traces of ionized helium. It gave a handle on the x-ray spectra of higher elements. And it correctly explained the inelastic collisions between electrons and mercury atoms observed by Philipp Frank and Gustav Hertz. This last success was especially important as it indirectly confirmed the strange frequency rule: the energy communicated by the electrons to the mercury atoms turned out to be identical to the frequency of the resonance line of these atoms multiplied by Planck's constant.[19]

---

[16] Again, Gaussian units are used (see n. 11).   [17] Cf. Heilbron 1964, p. 211.

[18] On Einstein's lightquantum and its difficult reception, cf. Klein 1963, 1964; Wheaton 1983.

[19] Cf. CD, Chap. 2; HD, vol. 1, Chap. 2; Pais 1991, Chap. 8; Heilbron 1974.

Yet, until 1915 Bohr did not believe in the generality of his frequency rule. For instance, he ascribed the splitting of spectral lines in magnetic and electric fields to a correction to this rule. In 1916, two masterful contributions to his theory proved the generality of the frequency rule. Einstein showed that simple probabilistic assumptions on the relation between the quantum jumping of the atoms of a gas and the intensity of the surrounding radiation led to Planck's blackbody law if and only if the frequency rule applied to the radiation emitted or absorbed during the quantum jumps. Arnold Sommerfeld and his collaborators showed that relativistic, electric, and magnetic perturbations of the Kepler motion in the hydrogen atom could be quantized in such a manner that the frequency rule applied generally.[20]

In 1918, Bohr clearly formulated the two "fundamental assumptions" of his theory:

1. That an atomic system can exist permanently only in a certain series of states corresponding with a discontinuous series of values of its energy, and that any change of the energy of the system including absorption and emission of electromagnetic radiation must take place by a transition between two such states. These states are termed "the stationary states" of the system.

2. That the radiation absorbed or emitted during a transition between two stationary states is "unifrequentic" and possesses a frequency v, given by the relation $E' - E'' = hv$, where h is the Planck constant and where E' and E'' are the values of the energy of the two states under consideration.

The first assumption is the existence of stationary states, the second is the frequency rule. Bohr regarded them as the unshakable pillars of his theory. They were indeed more directly related to experiments than other assumptions of his theory. Until at least 1925, they remained the two basic postulates of the quantum theory, despite the vicissitudes of most other assumptions. From the beginning, Bohr was not sure to what extent ordinary mechanics should apply to the motion in stationary states. He was more confident that this motion could be represented by well-defined trajectories, whatever the appropriate mechanics might be. Note, however, that his postulates did not depend on this assumption. As we will see in a moment, they survived the failure of the orbital representation in the years 1924–5.[21]

The correspondence principle

In 1913, Bohr noted that his assumptions, despite their evident incompatibility with ordinary electrodynamics, were able to reproduce the predictions of this theory in the limiting case for which the quantum numbers were very high and the quantum jumps were very small. Specifically, the frequency of the light emitted during a transition from the n state to the $n - \tau$ state approaches the frequency of the τ harmonic component of the motion in these states when n is very large. This convergence only holds if the constant α in the quantum rule (22) is exactly one-half. Bohr originally used this argument to consolidate

[20]Einstein 1916b, 1917; Sommerfeld 1916; Schwarzschild 1916. Cf. Klein 1979; Eckert 2013, §7.4; *CD*, Chap. 3.1; *HD*, vol. 1, Chap. 5.1; *CQ*, Chap. 6.

[21]Bohr 1918, p. 5.

this choice of the value of α in his theoretical expression of the Rydberg constant. As he further noted in 1914, the asymptotic form of the energy of the stationary states can in fact be derived from the asymptotic agreement between classical and quantum-theoretical spectrum. The argument goes as follows.[22]

The asymptotic agreement between classical and quantum-theoretical spectrum requires that

$$E_n - E_{n-\tau} \sim \tau \, h\bar{\nu}_n. \tag{26}$$

Since the large number n can be regarded as a quasi-continuous variable, this is equivalent to

$$\frac{dE_n}{dn} \sim h\bar{\nu}_n. \tag{27}$$

Differentiation of relation (23) for the Kepler motion yields

$$d(E_n/\bar{\nu}_n) = -dE_n/2\bar{\nu}_n, \tag{28}$$

so that

$$d(E_n/\bar{\nu}_n) \sim -(h/2)dn, \tag{29}$$

which implies the asymptotic validity of the quantum rule

$$E_n = -nh\bar{\nu}_n/2 \tag{30}$$

and of the resulting energy spectrum. This sort of argument can be used to guess the quantum rule for other periodic systems, as Bohr soon did.

As was already mentioned, in 1916 Sommerfeld and his collaborators managed to quantize a wider class of systems, the so-called multiperiodic systems which comprise the relativistic Kepler problem, and the hydrogen model perturbed by a constant electric or magnetic field (Stark and Zeeman effects). In these cases, the blind application of the generalized quantum rules and of the frequency rule yields far many more spectral lines than observed experimentally. For instance, in the case of the (normal) Zeeman effect quantum jumps are allowed in which the magnetic quantum number varies by any amount, whereas the only observed lines correspond to the variations -1, 0, $+1$ of this quantum number (triple splitting of the unperturbed lines).

Bohr noticed that in such cases the classical spectrum was in better qualitative agreement with the true spectrum than the quantum-theoretical spectrum. Indeed, the effect of a constant uniform magnetic field on the Kepler motion is a uniform precession at the Larmor frequency $\bar{\nu}_L$, following which the Fourier spectrum of the orbital motion involves the frequency triplet $\bar{\nu}_0 - \bar{\nu}_L$, $\bar{\nu}_0$, $\bar{\nu}_0 + \bar{\nu}_L$ (wherein $\bar{\nu}_0$ denotes the frequency of the unperturbed

---

[22] Bohr 1913, pp. 8–9; 1914. Cf. *CQ*, pp. 87–9.

motion). In order to remedy this defect of the quantum theory, Bohr complemented it with the idea that the possibility and probability of a given quantum transition should be determined by the existence and intensity of the corresponding harmonic component of the (dipolar moment of) motion in the initial stationary state. The precise definition of this "correspondence" derived from the condition that the classical and quantum theoretical spectrum should asymptotically agree.[23]

In the simplest case of a single quantum number n, this condition implies that the frequency of the line emitted during a jump from n to $n - \tau$ should be asymptotically equal to the frequency of the τ harmonic of the orbital motion:

$$\nu_{n,n-\tau} \sim \tau \bar{\nu}_n. \tag{31}$$

It also requires the asymptotic proportionality of the probability $A^n_{n-\tau}$ of this quantum jump with the intensity $|a_\tau(n)|^2$ of the corresponding harmonic of motion:

$$\nu^{-3}_{n,n-\tau} A^n_{n-\tau} \propto |a_\tau(n)|^2. \tag{32}$$

For moderate values of the quantum number n, the first of these two relations is no longer valid; Bohr nevertheless assumed that the second relation approximately held and that it was exact whenever the classical intensity $|a_\tau(n)|^2$ vanished. In other words, he excluded any quantum jump for which the "corresponding" harmonic component of the classical electric moment vanished; and he generally estimated the probability of a quantum jump from the intensity of this harmonic component.

This assumption is what Bohr named "correspondence principle" in 1917. Contrary to a common belief, this principle was not the mere condition that classical and quantum-theoretical spectra should asymptotically agree. Rather, the principle required that some of the relations satisfied by the classical harmonics of motion should be preserved (exactly or approximately) for the "corresponding" quantum-theoretical intensities. The relevant "correspondence" associated the quantum jump from n to $n - \tau$ with the τ harmonic of the classical motion.

Thanks to this principle, Bohr could derive the much needed selection rules, according to which certain quantum numbers, such as the magnetic quantum number m or the azimuthal quantum number k, can only vary by certain amounts ($\Delta m = 0, \pm 1$; $\Delta k = \pm 1$). With Hendrik Kramers's help, he also derived the approximate values of the intensities of the lines of the hydrogen atom. Broadly speaking, the correspondence principle is a relation between the periodicity properties of a classical model associated with the atomic system on the one hand, and the true spectrum of this system on the other hand. Bohr sometimes used the principle deductively, as a way to deduce properties of the spectrum from computable properties of the classical motion. Some other times, he used it inductively as a way to infer properties of the orbital motion from the empirically known spectra. He for instance did so in the perturbation theory and in the theory of the helium atom that he developed with Kramers. Altogether, the principle was very useful, and Bohr took

[23] Bohr 1918. Cf. *CD*, Chap. 3.2; *HD*, vol. 1, Chaps. 2–5; *CQ*, Chap. 6; Meyer-Abich 1965; Petruccioli 1993.

it as a hint to a future, complete quantum theory that would be a rational generalization of classical electrodynamics.[24]

Quantum mechanics

Bohr originally believed that the motion of the electrons in stationary states should be represented by well-defined orbits, whose periodicity properties were the proper basis for the application of the correspondence principle. This is why in 1924 he resisted Wolfgang Pauli's pressure to reject the orbital picture. By that time, difficulties had accumulated in applying this picture to the anomalous Zeeman effect, to the helium atom, and to optical dispersion. At the beginning of 1925, the difficulties became so severe that Bohr capitulated. Yet the correspondence principle did not follow the orbits to the grave. Bohr now believed that the classical orbital model could still have a "symbolic" relation to the true motion in stationary states. In the preceding months, Kramers, Max Born, and Werner Heisenberg had shown that certain classical relations between harmonic components of motion in the classical model could be translated into exact quantum-theoretical relations between intensities. The key to this translation was Bohr's "correspondence" between τ harmonic and $n \rightarrow n - \tau$ jump. In his contribution to this development, Born saw the dawn of a new *Quantenmechanik*.[25]

The correspondence principle thus became a tool for the symbolic translation of classical relations into quantum-theoretical relations, regardless of any descriptive value of the classical motion. In the spring of 1925, Heisenberg had the brilliant idea of applying this symbolic translation directly to the classical equations of motion. The following is a simplified version of his reasoning.[26]

Consider a mechanical system whose configuration depends on the single coordinate q and whose every motion is periodic (anharmonic oscillator). The equation of motion can be written under the form

$$\ddot{q} = \alpha_1 q + \alpha_2 q^2 + \dots . \tag{33}$$

Any given solution of this equation can be developed into a Fourier series

$$q = \sum_{\tau=-\infty}^{+\infty} q_\tau, \text{with } q_\tau = a_\tau e^{2\pi i \tau \bar{\nu} t}. \tag{34}$$

---

[24]Cf. *CQ*, Chaps. 6–7; Darrigol 1997. The correspondence principle was coolly received in Munich: cf. Hendry 1984; Heilbron 1983; *CQ*, pp. 137–45, Chap. 8; Seth 2010.

[25]Kramers 1924; Born 1924; Kramers and Heisenberg 1925. Cf. *CD*, Chap. 5.1; *HD*, vol. 2, Chap. 3.5; Dresden 1987, Chap. 8; *CQ*, Chap. 9. On the contemporary crisis, cf. Hendry 1984; *CQ*, Chap. 8. On John van Vleck's contemporary use of the correspondence principle, cf. Duncan and Janssen 2008.

[26]Heisenberg, 1925. Cf. *CD*, Chap. 5.1; *HD*, vol. 2, Chap. 3.5; *CQ*, Chap. 10; Rüdinger 1985. The idea of symbolic translation was originally associated with the BKS idea of virtual radiation in stationary states (Bohr, Kramers, and Slater 1924): cf. Klein 1970c; Dresden 1987, Chaps. 6, 8; *CD*, Chap. 4.3; Hendry 1984, Chap. 5; *CQ*, Chap. 9.

In terms of the harmonic components q_τ, the equation of motion can be rewritten as

$$-(2\pi\tau\bar{v})^2 q_\tau = \alpha_1 q_\tau + \alpha_2 \sum_{\tau'+\tau''=\tau} q_{\tau'} q_{\tau''} + \dots . \tag{35}$$

The correspondence principle yields the translation rules:

$$\tau\bar{v} \to v_{n,n-\tau}, \quad |q_\tau|^2 \to A^n_{n-\tau} v^{-3}_{n,n-\tau} \tag{36}$$

inspired by the asymptotic relations (31) and (32). The translation of the preceding form of the equation of motion requires a slight extension of this correspondence as

$$a_\tau e^{2\pi i\tau\bar{v}t} \to q_{n,n-\tau} = a_{n,n-\tau} e^{2\pi i v_{n,n-\tau} t}, \tag{37}$$

in which $a_{n,n-\tau}$ is a complex "quantum amplitude" involving a phase factor. The choice of n is immaterial in this translation rule since any jump from n to $n - \tau$ corresponds to a τ harmonic as long as n is larger than τ. We may therefore translate the form (35) of the equation of motion as

$$-4\pi^2 v^2_{n,n-\tau} q_{n,n-\tau} = \alpha_1 q_{n,n-\tau} + \alpha_2 \sum_{\tau'+\tau''=\tau} q_{n,n-\tau'} q_{n-\tau', n-\tau'-\tau''} + \dots \tag{38}$$

The translation choice $q_{\tau''} \to q_{n-\tau', n-\tau'-\tau''}$ ensures that the oscillation frequency of every term in this equation be the same. Indeed Bohr's frequency rule implies that

$$v_{n,n-\tau} = v_{n,n-\tau'} + v_{n-\tau', n-\tau}. \tag{39}$$

In terms of the matrix $\mathbf{q}$ whose elements are $q_{m,n}$, Eq. (38) reads

$$\ddot{\mathbf{q}} = \alpha_1 \mathbf{q} + \alpha_2 \mathbf{q}^2 + \dots . \tag{40}$$

Hence, the quantum-mechanical equation of motion is simply obtained by substituting the matrix $\mathbf{q}$ for the coordinate q in the equation of motion and replacing ordinary products with matrix products.[27]

In the same spirit, to any dynamical variable $g(t)$ of the system we may associate the Hermitian matrix $\mathbf{g}(t)$ whose elements have the form $g_{mn}(0)e^{2\pi i v_{m,n} t}$. In particular, to the invariable energy H, we associate a diagonal matrix $\mathbf{H}$ whose diagonal elements are the energies of the various stationary states. As a consequence of the frequency rule (25)

$$E_m - E_n = h v_{m,n},$$

the equation

$$\dot{g}_{mn} = 2\pi i v_{m,n} g_{mn} \tag{41}$$

---

[27]Matrices do not explicitly occur in Heisenberg's paper.

can be rewritten as[28]

$$\dot{\mathbf{g}} = (i/\hbar)[\mathbf{H}, \mathbf{g}].$$ (42)

Introducing the momentum matrix $\mathbf{p} = m\dot{\mathbf{q}}$, the Hamiltonian matrix $\mathbf{H}$ has the form

$$\mathbf{H} = \mathbf{p}^2/2m + V(\mathbf{q}).$$ (43)

We therefore have

$$\mathbf{p}/m = \dot{\mathbf{q}} = (i/\hbar)[\mathbf{H}, \mathbf{q}] = (i/\hbar)[\mathbf{p}^2/2m, \ \mathbf{q}],$$ (44)

or

$$\mathbf{p}[\mathbf{p}, \mathbf{q}] + [\mathbf{p}, \mathbf{q}]\mathbf{p} = -2i\hbar\mathbf{p}.$$ (45)

The identity

$$\frac{\mathrm{d}}{\mathrm{d}t}[\mathbf{p}, \mathbf{q}] = [\dot{\mathbf{p}}, \mathbf{q}] + [\mathbf{p}, \dot{\mathbf{q}}] = 0$$ (46)

further implies that the commutator $[\mathbf{p}, \mathbf{q}]$ is diagonal. Calling k_n the diagonal elements of this commutator, condition (45) implies that

$$p_{mn}k_m + p_{mn}k_n = -2i\hbar p_{mn}.$$ (47)

Granted that every quantum transition is allowed ($p_{mn} \neq 0$), this reduces to the condition:

$$k_m + k_n = -2i\hbar$$ (48)

for every choice of m and n such that $m \neq n$. This can be true only if $k_n = -i\hbar$ for any n, namely:[29]

$$[\mathbf{q}, \mathbf{p}]_{nn} = i\hbar.$$ (49)

Remembering that the commutator is diagonal, we have the quantum rule[30]

$$[\mathbf{q}, \mathbf{p}] = i\hbar.$$ (50)

---

[28] Cf. Born and Jordan 1925 for a different derivation of this equation, and Hund 1974, pp. 227–9.

[29] Heisenberg obtained this relation in a different way and in a different form, by symbolic translation of the expression $J = \oint p\,dq$ of the action variable. Cf. Darrigol 2009, pp. 160–1.

[30] Hund 1974, pp. 227–9 shows that the choice of this quantum rule implies the equivalence between the quantum version of Hamilton's equations and Eq. (42) applied to $\mathbf{p}$ and $\mathbf{q}$. But he does not deal with the reciprocal implication.

Taking

$$V(\mathbf{q}) = -\frac{1}{2}m\alpha_1\mathbf{q}^2 - \frac{1}{3}m\alpha_2\mathbf{q}^3 - \ldots, \tag{51}$$

the equation of motion (40) is equivalent to Hamilton's equations:

$$\dot{\mathbf{q}} = \frac{\partial \mathbf{H}}{\partial \mathbf{p}}, \quad \dot{\mathbf{p}} = -\frac{\partial \mathbf{H}}{\partial \mathbf{q}}. \tag{52}$$

These two equations are easily seen to derive from the equations

$$\dot{\mathbf{q}} = (\mathrm{i}/\hbar)[\mathbf{H}, \mathbf{q}], \quad \dot{\mathbf{p}} = (\mathrm{i}/\hbar)[\mathbf{H}, \mathbf{p}] \tag{53}$$

combined with the quantum rule (50). Therefore, the fundamental equations of Heisenberg's quantum mechanics (in the Born–Jordan form) can be written as Eqs. (42) and (50):[31]

$$\dot{\mathbf{g}} = (\mathrm{i}/\hbar)[\mathbf{H}, \mathbf{g}] \text{ for any } \mathbf{g}(\mathbf{q}, \mathbf{p}), \text{ and } [\mathbf{q}, \mathbf{p}] = \mathrm{i}\hbar.$$

Remembering that the matrix $\mathbf{H}$ is a diagonal matrix whose diagonal elements represent the energies of the successive stationary states and that the squared modulus of the element q_{mn} of the matrix $\mathbf{q}$ is proportional to the probability of a transition between the states m and n, the basic problem of Heisenberg's quantum mechanics (in the Born–Jordan form) is to find two infinite matrices $\mathbf{q}$ and $\mathbf{p}$ such that $[\mathbf{q}, \mathbf{p}] = \mathrm{i}\hbar$ and the matrix $\mathbf{H}(\mathbf{q}, \mathbf{p})$ is diagonal.

As Schrödinger remarked in 1926, this problem can be described in a more abstract manner as the following three-step problem:[32]

(i) Find two Hermitian operators $\mathbf{q}$ and $\mathbf{p}$ such that $[\mathbf{q}, \mathbf{p}] = \mathrm{i}\hbar$ in an infinite-dimensional vector space,

(ii) Diagonalize the Hermitian operator $\mathbf{H}(\mathbf{q}, \mathbf{p})$,

(iii) Find the matrix elements of the operator $\mathbf{q}$ in the basis that diagonalizes $\mathbf{H}$.

A simple way to accomplish the first step is to pick the operators

$$\mathbf{q} : \psi(q) \to q\psi(q) \text{ and } \mathbf{p} : \psi(q) \to -\mathrm{i}\hbar\partial\psi/\partial q \tag{54}$$

that act in the vector space of complex functions $\psi(q)$. The second step amounts to solving the (time-independent) Schrödinger equation

$$H(q, -\mathrm{i}\hbar\partial/\partial q)\psi(q) = E\psi(q). \tag{55}$$

---

[31] Cf. Born and Jordan 1925.

[32] Schrödinger 1926c. In modern English and in all rigor, what is here called a Hermitian operator is only a self-adjoint operator.

Granted that the spectrum of possible values of the energy E is discrete, the Hermitian character of the operator $\mathbf{H}$ warrants that these values are real and that the eigenfunctions are orthogonal with respect to the Hermitian scalar product. The latter property reads:

$$\int \psi_m^*(q)\psi_n(q)\mathrm{d}q = 0 \ \text{ if } \ m \neq n. \tag{56}$$

Lastly, the matrix elements of the operator $\mathbf{q}$ can be obtained through the formula

$$q_{mn} = \int \psi_m^*(q)q\psi_n(q)\mathrm{d}q, \tag{57}$$

provided that the eigenfunctions have been normalized.

So far, we have followed Heisenberg in regarding the quantum operators as time-dependent and the vectors on which they act as time-independent. In this picture, the equation of motion (42)

$$\dot{\mathbf{g}} = (\mathrm{i}/\hbar)[\mathbf{H}, \mathbf{g}]$$

can be integrated as

$$\mathbf{g}(t) = \mathrm{e}^{\mathrm{i}\mathbf{H}t/\hbar}\mathbf{g}(0)\mathrm{e}^{-\mathrm{i}\mathbf{H}t/\hbar}. \tag{58}$$

Consequently, the unitary transformation $\mathrm{e}^{-\mathrm{i}\mathbf{H}t/\hbar}$ absorbs the time-dependence of the operators. The resulting time-dependence of the vectors is

$$|\psi(t)\rangle = \mathrm{e}^{-\mathrm{i}\mathbf{H}t/\hbar} |\psi(0)\rangle, \tag{59}$$

or

$$\mathrm{i}\hbar\frac{\partial |\psi\rangle}{\partial t} = \mathbf{H}|\psi\rangle. \tag{60}$$

In the wave representation of the vectors, this gives the (time-dependent) Schrödinger equation[33]

$$\mathrm{i}\hbar\frac{\partial \psi(q, t)}{\partial t} = H\left(q, -\mathrm{i}\hbar\frac{\partial}{\partial q}\right)\psi(q, t). \tag{61}$$

To summarize, Bohr's two quantum postulates and the correspondence principle suggest a symbolic translation of the classical equation of motion (for a bound system with one degree of freedom) in which quantum amplitudes oscillating at the Bohr frequencies correspond to the harmonic components of the classical motion. There is only one such translation that agrees with the two postulates. The result of this translation is the Heisenberg–Born–Jordan form of quantum mechanics, including the canonical

---

[33] A similar derivation is found in Dirac 1927.

commutation rule. Heisenberg's quantum mechanics implicitly contains the fundamental equations of Schrödinger's wave mechanics.[34]

Interpretation

In Heisenberg's quantum mechanics, the time average of the matrix **g** is a diagonal matrix whose nth diagonal element g_{nn} represents the time average of the dynamical variable g in the nth stationary state. This interpretation derives from the correspondence principle, which makes the large n limit of g_{nn} the zero-frequency component of the Fourier development of $g(t)$ in the nth stationary state (in the Bohr–Sommerfeld theory). It was part of Paul Dirac's genius to understand that the transformation properties of quantum mechanics generate a complete interpretation of quantum mechanics from this tiny bit of interpretation. In the simplest case of one degree of freedom, his reasoning went as follows.[35]

Consider the diagonal elements

$$g_{\alpha'\alpha'} = \langle \alpha' | \, \mathbf{g} \, | \alpha' \rangle \tag{62}$$

of the operator **g** in a scheme for which the matrix $\boldsymbol{\alpha}$ corresponding to the dynamical variable $\alpha(q, p)$ is diagonal:[36]

$$\boldsymbol{\alpha} \, | \alpha' \rangle = \alpha' \, | \alpha' \rangle . \tag{63}$$

Evidently, the elements $g_{\alpha'\alpha'}$ do not depend on the choice of the Hamiltonian. We may therefore consider $\boldsymbol{\alpha}$ as a fictitious Hamiltonian. In this case, the elements $g_{\alpha'\alpha'}$ represent the time average of the variable g, and the canonically conjugate variable β varies linearly in time (in the Bohr–Sommerfeld theory). Therefore, these elements also represent the average value of g when the variable α takes a given value α' and the value β' of the variable β is uniformly spread. The latter interpretation no longer refers to the fictitious Hamiltonian and therefore remains valid when the true evolution is restored.

Accordingly, the element $\delta(\mathbf{g} - g')_{\alpha'\alpha'}$ represents the β-average of $\delta(g - g')$ when $\alpha = \alpha'$. By definition of Dirac's δ function, the only values of β that contribute to this average are those for which g is close to g', and these contributions have equal weight. Therefore, the result should be the relative probability that $g = g'$ when $\alpha = \alpha'$ (and β is uniformly spread). Using the completeness relation

$$\int | g' \rangle \langle g' | \, \mathrm{d} g' = \mathbf{1} \tag{64}$$

for the vectors $| g' \rangle$ that diagonalize **g**, we have

---

[34]This simplified history ignores important contributions to the new quantum mechanics by Pauli, Norbert Wiener, Dirac, and others; and it does not tell the story of spin and statistics: cf. *CD*, Chaps. 3.4, 5.2; *HD*, vol. 4, part 1; Kragh 1990; *CQ*, Chap. 12; Waerden 1960.

[35]Dirac 1927. Cf. Kragh 1990, Chap. 2; *CD*, Chap. 6.2; *HD*, vol. 4, part 1; *CQ*, Chap. 12.

[36]Dirac 1927 did not introduce state vectors. His entire reasoning was based on the transformation matrices that left the fundamental equations (42), (50) invariant.

$$\langle \alpha' | \, \delta(\mathbf{g} - g') \, | \alpha' \rangle = \int \langle \alpha' | g'' \rangle \delta(g'' - g') \langle g'' | \alpha' \rangle \, dg'' = \langle \alpha' | g' \rangle \langle g' | \alpha' \rangle = |\langle \alpha' | g' \rangle|^2. \qquad (65)$$

Therefore, the expression $|\langle \alpha' | g' \rangle|^2$ represents the probability that the variable g takes the value g' when the variable α takes the value α'. Dirac regarded this result as the full interpretation of the quantum formalism. To this day, it remains the basis for concrete applications of quantum mechanics. It also inspired Heisenberg's and Bohr's subsequent considerations on the intuitiveness and completeness of quantum mechanics, which need not be discussed here.

Wave–particle duality

Let us now consider the historical origins of the other form of quantum mechanics, wave mechanics. As is well known, Erwin Schrödinger obtained his wave mechanics without any (initial) recourse to the quantum mechanics of Heisenberg, Born, and Jordan. The story begins with Einstein's suggestion, expressed in 1905, that in some respects light of a given frequency v behaved as if it were made of discrete quanta hv. The inference was based on the expression

$$S(v) - S(V) = k(E/hv)\ln(v/V) \qquad (66)$$

of the entropy variation of the low-density blackbody radiation (obeying Wien's law) of energy E and of frequency comprised between v and $v + dv$, when the volume varies from V to v. Einstein then used Boltzmann's relation

$$S = k \ln W \qquad (67)$$

to derive

$$W = (v/V)^{E/hv} \qquad (68)$$

for the probability W of a fluctuation in which the radiation is confined within the fraction v/V of the volume V of the cavity. Comparing this result with the probability $(v/V)^N$ that the N molecules of a gas be found within a fraction v/V of the available volume, Einstein hypothesized that the energy E of the radiation was made of distinct quanta hv. He used this "heuristic assumption" to explain the existence of a frequency threshold in the photoelectric effect and predict the relation

$$hv = P + eV \qquad (69)$$

between the frequency v of the incoming radiation, the work P needed to extract an electron from the surface of the metal, the charge e of the electron, and the electric potential V needed to stop the photoelectrons.[37]

Most of Einstein's contemporaries rejected the lightquantum hypothesis, and Einstein himself came to admit his inability to conciliate it with the well-established wave properties

---

[37] Einstein 1905a. Cf. Klein 1963; Kuhn 1978, Chap. 7; *CD*, Chap. 1.3; *HD*, vol. 1.

of light. The situation changed in the early 1920s when new experiments on the interactions between x rays and matter strongly supported the hypothesis. Some of these experiments, concerning the photoelectric effect induced by x rays, were done by Maurice de Broglie in Paris. In 1923, his younger brother Louis de Broglie speculated that the wave–particle duality of light extended to matter.[38]

Louis imagined that both matter and light involved corpuscles sliding on waves, the relativistic 4-momentum p of the (free) corpuscles being related to the 4-wavevector k of the (plane monochromatic) wave through the covariant relation

$$p = \hbar k, \tag{70}$$

whose time- and space-components yield

$$v = E/h \text{ and } \lambda = h/|\mathbf{p}| \tag{71}$$

for the frequency v and the wavelength λ of the wave as functions of the energy E and the 3-momentum $\mathbf{p}$ of the corpuscles.[39] Emboldened by this relativistic way of associating waves to particles of matter, de Broglie suggested that the quantum conditions of Bohr and Sommerfeld resulted from the synchronicity of the motion of electrons orbiting around a nucleus with the associated wave motion.

De Broglie also noted that Maupertuis's principle of least action and Fermat's principle of least time were equivalent when the former was applied to the motion of a corpuscle (in a potential) and the latter to the motion of the associated wave. Formally,

$$\delta \int p_\mu \mathrm{d}x^\mu = 0 \text{ is equivalent to } \delta \int k_\mu \mathrm{d}x^\mu = 0, \tag{72}$$

as long as the relation $p = \hbar k$ can be generalized to particles moving in a conservative field of force. De Broglie expected this correspondence to hold only when the associated wave could locally be approximated by plane waves, as is the case in the ray-optics limit of wave optics. In the general case, he expected the matter waves to undergo diffraction. With foresight, he wrote: "The new dynamics of the material point is to the old dynamics (including Einstein's) what undulatory optics is to geometrical optics."[40]

Wave mechanics

In 1926, Erwin Schrödinger took this suggestion seriously and translated it into the task of finding the wave equation that would yield the classical trajectories in the approximation of locally plane waves. Schrödinger's first derivation is found in notebook entries written at the turn of the years 1926–27. The first assumption is that matter waves of a given frequency should obey a wave equation of the form

$$\Delta \psi + \frac{\omega^2}{C^2(\mathbf{r})} \psi = 0, \tag{73}$$

[38] Broglie 1923a, b, c, 1924. Cf. Wheaton 1983, part 5; *HD*, vol. 1, Chap. 5.4; Darrigol 1993.

[39] Einstein had already used these relations in the limited case of massless light quanta, e. g. in Einstein 1909a, b, 1917.

[40] Broglie 1923b, p. 83.

which is for instance known to apply to monochromatic sound waves in an isotropic heterogeneous elastic medium. In order to determine the function $C(\mathbf{r})$, one may consider large values of the angular frequency ω for which there exist solutions that can be locally approximated by plane waves. The local value of the phase velocity of these waves is $C(\mathbf{r})$. According to de Broglie, it is also given by the ratio E/p of the energy and the momentum of the associated particle (since $E = \hbar\omega$ and $p = \hbar k$). Taking into account the relativistic relation

$$(E - V)^2 = p^2 c^2 + m^2 c^4 \tag{74}$$

between the energy and the momentum of a particle of mass m immersed in the potential $V(\mathbf{r})$, we have

$$\frac{\omega^2}{C^2} = \frac{p^2}{\hbar^2} = \frac{1}{c^2 \hbar^2}[(E - V)^2 - m^2 c^4]. \tag{75}$$

Injecting this expression into the wave equation (73), we get the time-independent form of the Klein–Gordon equation for a particle in the potential V. Schrödinger managed to solve this equation in the case of the hydrogen atom, and found an energy spectrum that intolerably departed from the one given by Sommerfeld's successful theory of the fine structure of the hydrogen atom. He then took the non-relativistic limit of expression (75), which leads to the wave equation

$$\Delta\psi + \frac{2m}{\hbar^2}(E - mc^2 - V)\psi = 0. \tag{76}$$

This is the time-independent Schrödinger equation (save for the inclusion of mc^2 in the energy E).[41]

Schrödinger never published this derivation. In print he favored the following derivation, based on comparing the Hamilton–Jacobi equation of classical mechanics, which yields the classical trajectories, with the eikonal approximation of wave optics, which yields the trajectory of rays as predicted by geometrical optics.[42]

Ignoring polarization, the optical wave equation in a medium of variable index $n(\mathbf{r})$ reads

$$\Delta\varphi - \frac{n^2}{c^2}\frac{\partial^2\varphi}{\partial t^2} = 0, \tag{77}$$

where c is the velocity of light in a vacuum. For a monochromatic wave of angular frequency ω, this reduces to

$$\Delta\varphi + \frac{n^2\omega^2}{c^2}\varphi = 0. \tag{78}$$

In the approximation of optical geometry, the undulation can be locally approximated by a plane monochromatic wave. The wave φ can then be written under the form

---

[41] Schrödinger's relevant notebook is lucidly analyzed in Kragh 1982.

[42] Cf. Hanle 1977; Wessels 1979; Kragh 1982; *HD*, vol. 5; Joas and Lehner 2009; Renn 2013.

$$\varphi(\mathbf{r}, t) = a(\mathbf{r}) \cos [\omega t + \xi(\mathbf{r})],\tag{79}$$

$\xi(\mathbf{r})$ being a quickly varying phase and $a(\mathbf{r})$ a slowly varying amplitude. And the wave equation can be replaced with the "eikonal equation"

$$(\nabla \xi)^2 = n^2 \omega^2 / c^2.\tag{80}$$

For a given index function $n(\mathbf{r})$, the rays are obtained by solving this equation and drawing the lines orthogonal to the surfaces $\xi = $ constant.[43]

For a non-relativistic particle of mass m moving in a potential $V(\mathbf{r})$ and for a given value of the energy E (which is conserved), the possible trajectories can be similarly determined by solving the Hamilton–Jacobi equation

$$(\nabla S)^2 = 2m(E - V)\tag{81}$$

and drawing the lines orthogonal to the surfaces $S = $ constant. It is therefore tempting to regard the dimensionless ratio $S/\hbar$ as a phase and to seek the (time-independent) wave equation of which the Hamilton–Jacobi equation is the eikonal approximation. The correspondence

$$2m(E - V)/\hbar^2 \leftrightarrow n^2 \omega^2 / c^2\tag{82}$$

gives this equation as

$$\Delta \psi + \frac{2m(E - V)}{\hbar^2} \psi = 0, \quad \text{or} \left(-\frac{\hbar^2}{2m} \Delta + V \right) \psi = E\psi.\tag{83}$$

This is the time-independent form of the Schrödinger equation.[44]

In the case for which V is the Coulomb potential of an electron in the field of the nucleus, Schrödinger found that this equation admitted a solution if and only if the energy E was positive or else belonged to a discrete series of negative values identical with those obtained in Bohr's non-relativistic theory of the hydrogen atom. He thus knew he was on the right track.[45]

In conformity with the Einstein–de Broglie relation between energy and frequency, Schrödinger assumed that to a solution of energy E corresponded an oscillation of frequency E/h. As a simple ansatz for the most general oscillation he took

$$\psi = \sum_{\alpha} u_{\alpha} e^{-iE_{\alpha} t / \hbar},\tag{84}$$

[43]Cf. Landau and Lifshitz 1951, Chap. 7, § 43. On the history of this approach, cf. Kragh 1982; Joas and Lehner 2009.

[44]A similar derivation is found in Schrödinger 1926b.

[45]Schrödinger 1926a.

where u_α is a solution of the time-independent Schrödinger equation with the energy E_α. This oscillation is the general integral of the equation

$$i\hbar\frac{\partial\psi}{\partial t} = \left(-\frac{\hbar^2}{2m}\Delta + V\right)\psi. \tag{85}$$

Schrödinger assumed the validity of this equation for non-conservative systems in which the potential V depends on time. He showed that Kramers's dispersion formula resulted from the first-order solution of this equation when V described the dipolar interaction with an incoming monochromatic electromagnetic wave. In this calculation, he took the original ψ to be the wave function of the fundamental state of the hydrogen atom, and he generated the outgoing electromagnetic field through the dipolar moment

$$\mathbf{M} = e\int \mathbf{r}\psi(\mathbf{r}, t)\psi^*(\mathbf{r}, t)\,\mathrm{d}^3r. \tag{86}$$

The latter expression derived from the interpretation of the product $e\psi\psi^*$ as some electric density within the atom, about which more will be said in a moment.[46]

As Schrödinger promptly realized, in a system of several interacting particles the matter waves no longer belong to three-dimensional space. The analogy between the action function and the phase of the wave indeed makes the relevant space identical with the $3n$-dimensional configuration space, if n is the number of particles (when spin is ignored). The Schrödinger equation then reads[47]

$$i\hbar\frac{\partial\psi(\mathbf{r}_1, \mathbf{r}_2, \ldots \mathbf{r}_n; t)}{\partial t} = \left(-\frac{\hbar^2}{2m}\sum_{k=1}^{n}\Delta_k + V(\mathbf{r}_1, \mathbf{r}_2, \ldots \mathbf{r}_n)\right)\psi(\mathbf{r}_1, \mathbf{r}_2, \ldots \mathbf{r}_n; t). \tag{87}$$

Interpretation

Being guided by the analogy between matter and light, Schrödinger originally assumed that some real wave process occurred within the atom. This naïve interpretation did not square with the $3n$-dimensional character of the relevant space. In the end, Schrödinger assumed that $|\psi|^2$ was "some sort of weight-function" in configuration space, as suggested by the conservation of its integral (owing to the Hermitian character of the Hamiltonian). More concretely, he assumed that the true electric density and current (in ordinary space) derived from this weight-function. This assumption yields the correct dispersion formula; but it fails to explain spontaneous emission because the resulting electric distribution is stationary in any stationary state. In order to derive the frequencies and intensities of the emitted lines, Schrödinger had to import extraneous elements such as Bohr's frequency rule or

[46]Schrödinger 1926e. Although Schrödinger assumed the frequency E/h in his earlier papers, Eq. (85) only appears in the last installment of his theory. He earlier preferred a second-order equation that only applied to stationary states and still contained the energy E.

[47]Schrödinger 1926b for the time-independent version; Schrödinger 1926e for the time-dependent version.

Heisenberg's relation between the matrix-elements of the polarization and the intensity of spectral lines.[48]

Max Born's scattering theory of 1926 offered another handle on the interpretation of the wave function. The object of this theory was the scattering of electrons by a potential or by an atom. In the potential case, Born developed the scattered wave as a sum of plane monochromatic waves:

$$\psi = e^{-iEt/\hbar} \sum_{k} a_{k} e^{ik \cdot r}, \tag{88}$$

the sum being extended to all the vectors k compatible with the asymptotic energy E. In conformity with de Broglie's idea that a plane wave represents the motion of a free particle, Born decided that $|a_{k}|^2$ should be proportional to the probability that the particle be scattered in the direction of the vector k. In the case of electron-atom scattering, Born developed the global outgoing wave function as a sum of (tensor) products of plane electron waves and stationary atomic waves, and he interpreted the squared modulus of a given coefficient in this development as the probability of a scattering event in which the electron and the atom end up in the states described by the corresponding wave functions.[49]

Dirac offered a similar bit of interpretation of the wave function in his own version of the time-dependent perturbation theory. While in his dispersion theory Schrödinger interpreted the perturbed wave function

$$\psi(r, t) = \sum_{n} c_{n}(t) u_{n}(r) \tag{89}$$

as giving the relative weight $|\psi(r, t)|^2$ of the various positions r of the electron, Dirac identified $|c_{n}(t)|^2$ with the probability that the system be found in the n stationary state. Again, the Hermitian character of the total Hamiltonian warrants the conservation of the sum of these probabilities. In the case of an atom perturbed by a plane electromagnetic wave, Dirac used this interpretation to derive the value of Einstein's B_{n}^{m} coefficients for the probability of induced quantum jumps.[50]

Born and Dirac thus inaugurated the interpretation of the wave function as a means to derive the probability of certain classically defined variables, namely a scattering angle or the final energy of the atom. Schrödinger formally did the same for the position of the electron, although by "weight-function" he probably meant a spread of the substance of the electron and not a statistical outcome of position measurements. The question now is: Can we extend the statistical interpretation to any classically defined variable of the system?

[48] Schrödinger 1926e, p. 135. The frequency rule occurs in Schrödinger 1926a; Heisenberg's expression of intensities occurs in Schrödinger 1926d, p. 465.

[49] Born 1926a, b. In the first paper, Born succinctly treated the problem of the scattering of an electron wave by an atom originally in a stationary state. In the second paper, he also treated the simpler case of scattering by a fixed center of force. Cf. *CD*, Chap. 6.1; Konno 1978; *HD*, vol. 3, Chap. 5.6; Gyeong Soon 1996; Beller 1990.

[50] Dirac 1926. The same paper contains the description of indistinguishable particles by symmetric or anti-symmetric waves.

A simple (though somewhat formal) way to answer this question is to imagine that the system is subjected to a very brief and very small interaction whose potential takes significant values only when the variable takes a given value. For the sake of simplicity, suppose there is only one degree of freedom. A given dynamical variable is a function $g(q,p)$ of the coordinate q and the momentum p. The potential of the imagined interaction is proportional to $\delta(t-\tau)\delta[g(q,p)-g_0]$. If $\psi(q,\tau)$ is the normalized wave function of the system immediately before the time τ of the perturbation, first-order perturbation theory à la Schrödinger–Dirac yields

$$P_\psi(g_0) = \int \psi^*(q,\tau)\delta[g(q,-i\hbar\partial/\partial q)-g_0]\,\psi(q,\tau)\mathrm{d}q \qquad (90)$$

for the probability that the system leaves the state $\psi(q,\tau)$ (which can be regarded as a stationary state at the time scale of the perturbation). Owing to the definition of the perturbation, this integral also represents the probability that the variable g takes the value g_0.

Call $u_g(q)$ the normalized eigenfunctions of the Hermitian operator $g(q,-i\hbar\partial/\partial q)$ (the index g is sufficient in the non-degenerate case), and assume that they form a basis in the space of wave functions. Then the unique decomposition

$$\psi(q,\tau) = \int c_g(\tau)u_g(q)\mathrm{d}g \qquad (91)$$

leads to

$$P_\psi(g_0) = \int c_g{}^* \delta(g-g_0)c_g\mathrm{d}g = |c_{g_0}|^2. \qquad (92)$$

In Hilbert-space language, this means that the probability (density) that the dynamical variable g takes the value g_0 when the system is in the state ψ is the square of the modulus of the projection of this state over the eigenstate of the operator $g(q,-i\hbar\partial/\partial q)$ that has the eigenvalue g_0. This rule is a slight generalization of the rule obtained by Dirac through algebraic quantum mechanics (for Dirac, the ψ state would be characterized by a specific value of another dynamical variable). There are of course some mathematical difficulties bound with the discrete, continuous, or mixed nature of the spectrum of the $g(q,-i\hbar\partial/\partial q)$ operators. In general, part of the above-written integrals over g would have to be replaced by a discrete sum; and, as is well known, the eigenfunctions can be normalized only in the discrete case.[51]

We have thus arrived at a fully interpreted quantum mechanics through the wave approach. The result is entirely equivalent to that obtained through the quantum jumping approach. As was already mentioned, Heisenberg and Bohr soon argued that the statistical interpretation yielded the maximal answer to any conceivable experimental question about

[51] This account of the wave-based interpretation of quantum mechanics is purely fictitious. In reality Schrödinger, Dirac, and Jordan combined intuitions gained in both the wave and matrix approaches. Cf. Beller 1999.

quantum systems. Heisenberg's arguments were more dependent on the quantum jumping approach, and Bohr's on the wave approach. Both physicists nevertheless admitted the formal and predictive equivalence of the two forms of quantum mechanics.[52]

Historical necessity

Having identified the main innovative steps in the history of quantum theory, one may wonder whether these steps were in some sense necessary. The question is not easy to answer because the implied necessity may belong to different categories. In a first category, the novel elements are deduced from well-defined physico-mathematical principles, in a quasi-rational way. In a second category, the novel elements are induced from well-established empirical data (together with well-confirmed, lower-level theories). In a third category, they have a pragmatic necessity resulting from the intractability or unavailability of alternative approaches. In a fourth and last category, they may be the resultant of psychological or social factors, implying for instance the authority of a leader. In a philosophical dreamworld, the two first categories would be dominant. Needless to say that in the real world the first and second kinds of necessity are often contaminated by the third and fourth. Also, the distinction between deductive and inductive necessity can only be a loose one, because induction usually requires established principles, and because principles often have a partly empirical origin.

An additional difficulty is the variability of the kind of necessity of a given innovation when a fine time scale is used. Most frequently, the innovation is initiated by a single actor for reasons that have to do with his personal itinerary, his cultural immersion, and his psychological character. At this early stage, he may be the only one to regard his move as inductively or deductively necessary, while other actors may be skeptical and regard the move as arbitrary. At a later stage a critical debate usually occurs, at the end of which the majority of experts agree that the step must be taken for reasons that may vary from case to case: confirmation of the move by new empirical data, theoretical consolidation of the original deduction, availability of independent deductions that lead to the same result, or compatibility with independent, fruitful developments.[53]

In most of the following discussion, I will judge the necessity of the innovative steps at the end of this second stage. The above-given simplified history does not provide enough information on the first stage. Whether, for instance, Bohr's familiarity with Harald Høffding's philosophy or Born's awareness of the anti-causal philosophies of the Weimar period inspired their most daring moves could only be judged from a much more detailed and diversified history. At any rate, the closest approximations to inductive or deductive necessity are more likely to be found in the justification stage.[54]

---

[52] Cf. Beller 1999.

[53] These two stages are vaguely similar to Hans Reichenbach's distinction between context of discovery and context of justification.

[54] Social constructivists would agree with me that this second stage is essential in stabilizing the basic constructs of science. However, their analysis of the stabilizing process tends to underestimate rational constrains or to reduce them to socially defined systems of beliefs.

The early-twentieth-century conclusion that ordinary electrodynamics could not yield equilibrium for thermal radiation comes close to the ideal of deductive necessity. As was mentioned, the lack of rigor in the implied deductions was compensated by the multiplicity and variety of derivations of the same result. Moreover, the status of one of these derivations, the Gibbsian one given by Lorentz, rose with the conviction that Gibbs's ensembles correctly represented thermodynamic equilibrium despite the lack of a firm foundation.

The introduction of quantum discontinuity obeyed a weaker necessity of the inductive kind. Einstein's and Bohr's discrete quantization was the simplest way to account for Planck's blackbody law and for the spectrum of the hydrogen atom. Yet it was highly problematic for two reasons: it implied a non-classical selection among classically defined states, and it made it very difficult to imagine a plausible mechanism for the interaction between atoms and radiation. For the latter reason, Planck long preferred a division of phase space into cells of equal a priori probability.

As is well known, in 1911 Paul Ehrenfest and Henri Poincaré proved that the canonical distribution of energy over resonators could not yield a finite energy for cavity radiation unless there was a finite energy threshold for the excitation of the resonators.[55] This proof is largely illusory, because it depends on an unwarranted extension of Gibbs's canonical distribution law to systems that no longer obey the laws of classical dynamics.[56] The true reason why Einstein's idea of a sharp quantization came to dominate over more timid attempts was the multiple, successful applications it had in the context of the Bohr–Sommerfeld theory.

Similar comments can be made about Bohr's frequency rule. It is tempting to say that Bohr read the rule in the Balmer–Rydberg formula. In reality, this inference has the typical underdetermination of any inductive reasoning: the hydrogen spectrum can be derived from the frequency rule only when this rule is combined with a few other assumptions including the existence of stationary states and the truth of the laws of wave optics for the emitted radiation. We saw that Bohr himself did not believe in the generality of the frequency rule until he became aware of Sommerfeld's and Einstein's contributions to his theory in 1916. The assumption of stationary states and the frequency rule gained credibility and became Bohr's two "postulates" after their simultaneous application yielded correct results in an increasing variety of situations involving spectra, atomic structure, and atomic collisions. This happened despite the evident incompleteness of the theory (it left the radiation mechanism in the dark) and despite its reliance on classical concepts belonging to an incompatible electrodynamics.

The very definition of stationary states and the statement of the frequency rule required classical concepts: energy and frequency. Bohr struggled to show that these concepts could be defined in the quantum realm through a limited use of classical theory that did not contradict the quantum postulates. Most important, his correspondence principle pointed to a deep formal analogy between classical electrodynamics and the evolving quantum

---

[55] Cf. Klein 1970a; *CD*, p. 53.

[56] In his statistical mechanics, Gibbs assumed the validity of Hamiltonian dynamics. Although Einstein did not in his own statistical mechanics, he still assumed continuous evolution, invariance of the volume element in phase space, and some weak ergodicity. In the 1910s, there already were reasons to doubt the validity of any of these requirements in the case of quantum systems.

theory. He hoped that in the long run this analogy would project the consistency of the former theory over the latter. The quantum postulates would remain intact in this process.

Although Bohr reached the correspondence principle by analogy with classical electrodynamics, he insisted on the formal character of this analogy and emphasized the contrast between the quantum postulates and the continuity of classical radiation processes. In order to judge the necessity of this principle, one must first be aware that in Bohr's original view this principle was a relation between the periodicity properties of the motion in stationary states (whether or not this motion obeyed classical mechanics) and the properties of the emitted radiation. There were three arguments in favor of the necessity of this principle: It warranted the asymptotic agreement between the empirical predictions of classical and quantum theory; in the deductive mode, it provided the selection rules and good estimates of the intensities of some spectral lines; through Bohr's more obscure appeal to the inductive mode, it led to a plausible classification of the elements.

The magic of the correspondence principle did not catch on outside Copenhagen. By 1924, the idea of well-defined orbits in the atom, which the principle seemed to require, was under much criticism. Even Bohr ended up rejecting this idea in early 1925. Yet in Bohr's circle the confidence never died that correct quantum-theoretical relations could be extracted by analogy with classical multiperiodic systems, whether or not the motion of such systems truly represented the motion in stationary states. This confidence even increased in 1923–24 when Kramers, Born, and Heisenberg managed to translate some classical relations into what they (correctly!) believed to be exact quantum-mechanical relations. One reason for this belief was the empirical relevance of these relations. Another was the automatic agreement between the large-quantum-number limit of these relations and the corresponding classical relations. Still another was the fact, first emphasized by Kramers, that these relations involved only the basic quantities entering Bohr's postulates and no longer referred to the suspicious orbits. In the spring of 1925, Heisenberg's conviction that he had discovered quantum mechanics resulted from these three qualities of the symbolic translation, together with the consistency and completeness of the resulting computational scheme.

Heisenberg's quantum mechanics may be regarded as a necessary consequence of Bohr's two postulates (discrete stationary states, and frequency rule) and of a rule for translating the equations of motion of a classical periodic system (expressed in Fourier form) into relations between "quantum amplitudes" directly related to the observable quantities that enter the two quantum postulates. This rule itself derived from the correspondence principle, whose plausibility rested on the asymptotic validity of classical electrodynamics and on successful applications (of a different kind) in the earlier quantum theory. One might then wonder why quantum mechanics was not discovered earlier, say in 1917, when Bohr already had the two postulates as the pillars of his theory and the correspondence principle as a constructive tool. One reason is that before 1924 no one wished to banish orbital parameters from quantum theory. Another is that before Heisenberg's breakthrough of June 1925, no one imagined that the "correspondence" counterparts of the Fourier components of a periodic classical motion would completely characterize the quantum-mechanical motion just as these components sufficed to define the classical motion.

On the side of wave mechanics, the story began with de Broglie's extension of the wave-particle duality to particles of finite mass. Although the extension was natural from a formal, relativistic point of view, it could easily pass for crazy speculation. The receptivity of Paul Langevin, Einstein, and Schrödinger depended on a few favorable circumstances. Firstly, the lightquantum, which provided the basis for de Broglie's extension, was gaining momentum (literally and metaphorically). Secondly, de Broglie's successfully applied his theory to a wide spectrum of problems including the derivation of the Bohr–Sommerfeld rule, the analogy between Fermat's and Maupertuis's principles, and the derivation of Planck's quantum cells (for the statistics of gas molecules). Thirdly, in 1925 Einstein retrieved the de Broglie waves through a different route: he designed a quantum theory of gas degeneracy by analogy with Satyendra Nath Bose's corpuscular derivation of Planck's law, and found that the theoretical fluctuation of his quantum gas implied wave behavior in conformity with de Broglie's relations. Being also involved in quantum-gas theory, Schrödinger appreciated the force of Einstein's reasoning.[57]

It would nonetheless be excessive to speak of a deductive or inductive necessity of de Broglie's waves. In 1925, they still were a bold assumption without direct experimental counterpart.[58] De Broglie was himself shy in his suggestion of electron diffraction.[59] A stronger necessity can be seen in the deduction of the Schrödinger equation. De Broglie's idea that the classical dynamics of a particle should be to wave mechanics what geometrical optics is to wave optics automatically leads to the time-independent Schrödinger equation in the non-relativistic limit. In addition, the success of this equation in determining the stationary states of the hydrogen atom could hardly be regarded as pure chance. One may still be perplexed by the coincidence that made the Schrödinger equation appear just a few months after Heisenberg's quantum mechanics. There is little to justify this timing besides the contemporary willingness to renounce electronic orbits in atoms.[60]

A last question of special philosophical interest is the necessity of the now standard probabilistic interpretation of the formalism of quantum mechanics or wave mechanics. In Dirac's quantum-mechanical approach, the starting point is the allegation that for a sharply defined value of the energy (corresponding to a stationary state) the conjugate phase is uniformly spread. The ensuing deduction of the whole interpretation only requires the transformation properties of the fundamental equations of quantum mechanics (invariance by unitary transformations). The necessity of this interpretation should therefore be measured by the necessity of the starting point. Dirac justified his starting point through the correspondence principle, arguing that in the large quantum-number limit a stationary state may be represented by a revolving electron whose phase varies uniformly in time. Thus, Dirac was willing to admit that energy and phase retained a meaning in quantum mechanics. More generally, he assumed that any dynamical variable and its canonical conjugate retained a meaning in quantum mechanics although it was impossible to have initial conditions in which both variables were exactly determined. There is no evident necessity

---

[57] Einstein 1924, 1925a, b. Cf. *CD*, pp. 248–9; *HD*, vol. 1, Chap. 5.3; Forman 1969; Hanle 1977.

[58] Walther Elsasser tried to relate de Broglie waves to anomalies observed in Göttingen for the scattering of low-energy electrons. Cf. Born 1926b; *CD*, pp. 249–51; Russo 1981.

[59] The suggestion only appears in Broglie 1923b, p. 549, not in the *Thèse* (Broglie 1924).

[60] Jürgen Renn judiciously notes that both approaches were severely constrained by asymptotic correspondence with classical theory (with classical electrodynamics in Heisenberg's case, and with classical mechanics in the Hamilton–Jacobi form in Schrödinger's case). Cf. Renn 2013.

for this persisting relevance of classical concepts in the quantum context. Nevertheless, the harmony of Dirac's statistical interpretation with the transformation properties of quantum mechanics pleaded for the uniqueness of this interpretation.

In the matter-wave approach, Born's probabilistic interpretation of scattered electron waves seems unavoidable. Indeed the naïve interpretation of the wave as dilute matter would imply that only a fraction of an electron is detected at a given angle. Any attempt to save this interpretation by building wave packets of very small size would fail because of the spreading of the wave packets. Similarly but less stringently, Dirac's statistical interpretation of the perturbed Schrödinger wave of an irradiated atom seems to result from the nature of the problem, granted that long after the interaction the atom can only be found in a stationary state. By itself Born's probabilistic interpretation of scattered waves leads to the full statistical interpretation of wave mechanics through the earlier given idealization of the measuring process. Although this idealization is far remote from any concrete measurement device, it seems legitimate if only the concept of external potential is admitted in the theory.

As long as one is concerned only with precise operational rules for applying the quantum formalism and not with deeper interpretive issues, one thus comes to feel that the standard statistical interpretation of wave or matrix mechanics is unavoidable. The ability of this interpretation to correctly represent the outcome of experiments in the quantum regime has rarely been contested. The apple of later discord rather was the possibility of defining or measuring physical quantities more than quantum mechanics allows.

To sum up, the historical genesis of quantum mechanics can be regarded as a series of bold, imaginative, but firmly supported steps. The first two constructive steps, the introduction of discrete stationary states and the frequency rule, were taken with full awareness of their problematic character and they were later consolidated by multiple successes of their combined application. These assumptions have counterparts in modern quantum mechanics, although stationary states are no longer regarded as the only possible states. In contrast, the auxiliary reliance on classical concepts posed more and more problems and led to the severe crisis of 1924–5. Amid this crisis, Heisenberg's invention of a first form of quantum mechanics strikingly confirmed Bohr's idea that the correspondence principle promised a "rational generalization" of classical electrodynamics. The contemporary but largely independent invention of a closely related wave mechanics strengthened the air of inevitability of quantum mechanics. The statistical interpretation of this theory largely derives from its mathematical structure, combined with a touch of correspondence.

As was already the case for the history of classical mechanics, rational arguments from the history of quantum mechanics lack rigor and purity. They rely on arbitrary idealizations and they are often contaminated by appeal to experimental knowledge. These arguments nevertheless suggest natural assumptions from which quantum mechanics might follow deductively.

8.2   The deformation of classical mechanics

One suggestion we can take from history is that quantum mechanics should be some sort of "rational generalization" (as Bohr put it) of classical mechanics. The standard recipe for building a quantum-mechanical Hamiltonian, namely, replacing canonical pairs with non-commutating operators, does not meet anyone's spontaneous idea of rationality. Around 1930, Hermann Weyl and Eugen Wigner independently discovered reciprocal ways of associating operators with functions in ordinary phase space. Some fifteen years later, Joseph Moyal and Hilbrand Groenewold used this correspondence to reformulate quantum mechanics in phase space. In particular, they translated the product of two operators into the "Moyal product" of the corresponding functions in phase space, and the commutator of two operators into a generalization of the Poisson bracket of two functions in phase space. In the 1970s, mathematical studies of deformations of the classical Poisson algebra by André Lichnerowicz's group, by Jacques Vey, and by Simone Gutt led to a proof that the Moyal bracket was the unique continuous deformation of the Poisson bracket. This section is a historico-critical analysis of this astonishing discovery, beginning with the earliest conceptions of quantization and ending with comments on the sort of necessity it implies for quantum mechanics.

Early connections between classical and quantum mechanics

The history of quantum mechanics, the correspondence principle, and the Bohrian interpretation of this theory in terms of classical observation all suggest an intimate connection between classical and quantum mechanics. In the simple case for which the classical Hamiltonian function has the form

$$H(p, q) = p^2/2m + V(q), \tag{93}$$

the quantum-mechanical Hamiltonian has the same form

$$\mathbf{H} = \mathbf{p}^2/2m + V(\mathbf{q}), \tag{94}$$

and the equations of motions in the Heisenberg picture still are Hamilton's equations

$$\dot{\mathbf{q}} = \frac{\partial \mathbf{H}}{\partial \mathbf{p}}, \quad \dot{\mathbf{p}} = -\frac{\partial \mathbf{H}}{\partial \mathbf{q}}. \tag{95}$$

The only formal difference hinges on the commutation rule

$$[\mathbf{q}, \mathbf{p}] = i\hbar. \tag{96}$$

In the Schrödinger picture, the Schrödinger equation

$$i\hbar \frac{\partial \psi(q, t)}{\partial t} = H\left(q, -i\hbar \frac{\partial}{\partial q}\right) \psi(q, t) \tag{97}$$

is the only first-order wave equation whose eikonal approximation leads to the Hamilton–Jacobi equation

$$\frac{\partial S}{\partial t} = H\left(q, \frac{\partial S}{\partial q}\right) \tag{98}$$

for the phase S of the waves.

There is much boldness in either way of generating quantum-mechanical formalism. One way introduces non-commuting quantities; the other turns particles into waves. No classical physicist would have taken the resulting equations seriously. Yet the formal-mathematical kinship between classical and quantum mechanics goes even deeper than suggested by their historical constructions. In 1925, Dirac noted that Heisenberg's equations, once rewritten under the form

$$[\mathbf{q}, \mathbf{p}] = i\hbar, \quad \dot{\mathbf{g}} = (i/\hbar)[\mathbf{H}, \mathbf{g}], \tag{99}$$

were the exact counterpart of the classical equations

$$\{q, p\} = 1, \quad \dot{g} = -\{H, g\}, \tag{100}$$

in which the Poisson bracket of two functions f and g of q and p is defined as

$$\{f, g\} = \frac{\partial f}{\partial q}\frac{\partial g}{\partial p} - \frac{\partial f}{\partial p}\frac{\partial g}{\partial q} \tag{101}$$

That is to say, the quantum-mechanical equations can be obtained through the simple correspondence

$$\{\ ,\ \} \to (i\hbar)^{-1}[\ ,\ ]. \tag{102}$$

In mathematical terms, the Poisson algebra of classical Hamiltonian infinitesimal evolutions seems to be mapped into the Poisson algebra of quantum Hamiltonian infinitesimal evolutions. In his earliest work on quantum mechanics, Dirac emphasized this correspondence and used it abundantly to adapt classical methods of resolution to the quantum domain.[61]

Unfortunately, the correspondence between the two theories is not as simple as Dirac wished. In December 1925, Heisenberg warned Dirac that the alleged correspondence between Poisson brackets and commutators could not hold for every pair of quantities. For instance, if $[\mathbf{q}, \mathbf{p}]$ corresponds to $i\hbar\{q, p\}$, then $[\mathbf{q}^2, \mathbf{p}^2]$ cannot correspond to $i\hbar\{q^2, p^2\}$ because

$$[\mathbf{q}^2, \mathbf{p}^2] = \mathbf{q}[\mathbf{q}, \mathbf{p}^2] + [\mathbf{q}, \mathbf{p}^2]\mathbf{q} = \mathbf{qp}[\mathbf{q}, \mathbf{p}] + \mathbf{q}[\mathbf{q}, \mathbf{p}]\mathbf{p} + \mathbf{p}[\mathbf{q}, \mathbf{p}]\mathbf{q} + [\mathbf{q}, \mathbf{p}]\mathbf{pq} = 2i\hbar(\mathbf{qp} + \mathbf{pq}) \tag{103}$$

whereas by definition (101) we have[62]

$$\{q^2, p^2\} = 4qp. \tag{104}$$

[61] Dirac 1925.

[62] Heisenberg to Dirac, 1 Dec. 1925, AHQP (photocopies of letters of Dirac's private collection, in Berkeley's OHST), discussed in *CQ*, p. 318. One might still hope to avoid trouble by properly symmetrizing the monomes $\mathbf{q}^m\mathbf{p}^n$ corresponding to $q^m p^n$. The impossibility of this escape is proved in Groenewold 1946, pp. 449–450: there does not exist any correspondence $g \to \mathbf{g} = \Omega(g)$ such that $[\Omega(f), \Omega(g)] = i\hbar\Omega\left(\{f, g\}\right)$.

Weyl and Wigner

This counterexample brings forth a more general difficulty in defining the quantum-mechanical operator **g** corresponding to a given classical quantity $g(q, p)$. Whenever the power series-development of the g function involves terms of the form $p^m q^n$, there is an ambiguity in the quantum translation. For instance, should pq translate into **pq**, **qp**, or $\frac{1}{2}$(**pq** + **qp**)?

In 1927, Hermann Weyl addressed this question from a group-theoretical point of view in which the Hermitian operators **q** and **p**, such that [**q**, **p**] = i$\hbar$, are regarded as generators of the Lie algebra of unitary evolutions of the quantum system. Namely, any such evolution can be obtained by taking the exponential of a linear combination i(α**q** + β**p**) with real coefficients α and β. Besides, any Hermitian operator **g** of the Lie algebra can be obtained as a superposition

$$\mathbf{g} = \int \tilde{g}(\alpha, \beta) \mathrm{e}^{\mathrm{i}(\alpha \mathbf{q} + \beta \mathbf{p})} \mathrm{d}\alpha \mathrm{d}\beta, \tag{105}$$

with

$$\tilde{g}^*(\alpha, \beta) = \tilde{g}(-\alpha, -\beta). \tag{106}$$

Weyl introduced this decomposition in analogy with Fourier analysis, and because he wanted to circumvent the unboundedness of the operators **p** and **q** of quantum mechanics (which is an obstacle to its Hilbert-space representation).[63]

The condition (106) implies that the ordinary Fourier transform of the coefficients $\tilde{g}(\alpha, \beta)$,

$$g(q, p) = \int \tilde{g}(\alpha, \beta) \mathrm{e}^{\mathrm{i}(\alpha q + \beta p)} \mathrm{d}\alpha \mathrm{d}\beta \tag{107}$$

is a real function of the real variables q and p. Weyl naturally interpreted this function as the classical quantity corresponding to the quantum quantity **g**. He thus had in hand a well-defined, group-theoretically sound way of associating a quantum operator to a classical quantity. In compact form, the recipe reads:

$$\mathbf{g} = \frac{1}{4\pi^2} \int g(q, p) \mathrm{e}^{\mathrm{i}[\alpha(\mathbf{q}-q) + \beta(\mathbf{p}-p)]} \mathrm{d}q \mathrm{d}p \mathrm{d}\alpha \mathrm{d}\beta, \tag{108}$$

which implies the following quantization of monomes:[64]

$$q^m p^n \rightarrow \frac{1}{2^n} \sum_{k=0}^{m} C_n^k \mathbf{p}^{n-k} \mathbf{q}^m \mathbf{p}^k. \tag{109}$$

---

[63] Weyl 1927, pp. 27–8.

[64] For rule (109), cf. Cf. Zachos, Fairlie, and Curtright 2005, p. 30. Weyl (1927, p. 17) gave the rules for constructing $\tilde{h}$ from $\tilde{f}$ and $\tilde{g}$ so that **h** = **fg**. So too did von Neumann (1931), who used the Weyl–Fourier analysis of self-adjoint operators in a proof of the uniqueness of the Schrödinger representation of the operators **p** and **q**.

Five years later, Eugen Wigner addressed the seemingly unrelated question of the representation of quantum states in phase space. His aim was to devise a quasi-classical approximation strategy for the quantum statistical averages

$$\langle \mathbf{g} \rangle = \mathrm{Tr}(\boldsymbol{\rho}\mathbf{g}), \text{ with } \boldsymbol{\rho} = e^{-\beta\mathbf{H}}/\mathrm{Tr}(e^{-\beta\mathbf{H}}). \tag{110}$$

In order to ease comparison with the classical statistical average[65]

$$\bar{g} = \int \rho(q,p)g(q,p)\mathrm{d}q\mathrm{d}p, \tag{111}$$

he associated a phase-space distribution $\rho(q,p)$ to any density operator $\boldsymbol{\rho}$ according to the formula

$$\rho(q,p) = 2\int \mathrm{d}q' e^{2ipq'/\hbar} \langle q - q'| \boldsymbol{\rho} |q + q' \rangle, \tag{112}$$

and showed that for any quantity $\mathbf{g}$ that is a sum of a function of $\mathbf{p}$ and a function of $\mathbf{q}$, the quantum average could be replaced with a phase-space average over this distribution:[66]

$$\mathrm{Tr}(\boldsymbol{\rho}\mathbf{g}) = \int \rho(q,p)g(q,p)\mathrm{d}q\mathrm{d}p. \tag{113}$$

In the case of a pure state $\boldsymbol{\rho} = |\psi\rangle \langle\psi|$, the associated phase-space distribution reads[67]

$$\rho(q,p) = 2\int \mathrm{d}q' e^{2ipq'/\hbar} \psi^*(q + q')\psi(q - q'). \tag{114}$$

Wigner did not tell how he had arrived at this miraculous formula. He only indicated that Leo Szilard and himself had obtained it a few years earlier in another context. Possibly, the two friends had been looking for a phase-space formulation of quantum mechanics, in an attempt to reduce quantum-mechanical probabilities to ordinary probabilities. Wigner's paper indeed contains the phase-space counterpart of the Schrödinger equation:

$$\frac{\partial \rho}{\partial t} = -\frac{p}{m}\frac{\partial \rho}{\partial q} + \sum_{n=0}^{+\infty} \frac{(-1)^n \hbar^{2n}}{(2n+1)!} \frac{\partial^{2n+1} V}{\partial q^{2n+1}} \frac{\partial^{2n+1} \rho}{\partial p^{2n+1}}, \tag{115}$$

---

[65] For the sake of simplicity, I only give the formulas in the case of one degree of freedom.

[66] As we will see in a moment, this property in fact holds for any Weyl-ordered operator g. No one—except Alexander Blum, who kindly communicated me this information—seems to have noticed that Dirac had already used the Wigner transform in a discussion of the relation between the Hartree–Fock approximation and the Thomas atom (Dirac 1930, pp. 382–3).

[67] Wigner 1932, pp. 750, 753.

for a Hamiltonian of the form $\mathbf{H} = \mathbf{p}^2/2m + V(\mathbf{q})$. This is the quantum-theoretical generalization of the classical equation of evolution of a density in phase-space:

$$\frac{\partial\rho}{\partial t} = -\frac{p}{m}\frac{\partial\rho}{\partial q} + \frac{\partial V}{\partial q}\frac{\partial\rho}{\partial p} \tag{116}$$

Wigner did not fail to note that the quantum-theoretical density ρ, unlike its classical approximation, could take negative values and therefore could not be interpreted as a true probability:[68]

> Of course [the density $\rho(q,p)$] cannot be really interpreted as the simultaneous probability for coordinates and momenta, as is clear from the fact that it may take negative values. But of course this must not hinder the use of it in calculations as an auxiliary function which obeys many relations we would expect from such a probability.

Moyal, Dirac, and Groenewold

On the one hand, Weyl ascribed an operator to any function in phase space. On the other, Wigner ascribed a function in phase-space to any operator. Neither of them realized that the implied correspondences were the inverse of each other, presumably because the context and the relevant kind of operator differed. Whereas Weyl wanted to construct the Hermitian operators of quantum mechanics from functions in phase-space, Wigner sought a phase-space representation of the density operators or wave functions of quantum mechanics. Some ten years elapsed before two marginal theorists, José Moyal and Hilbrand Groenewold, discovered the Weyl-Wigner reciprocity in systematic explorations of the phase-space representation of quantum mechanics.[69]

Around 1940, the electrical engineer Moyal discovered that the quantum-mechanical average of any operator could be represented as an ordinary average in phase space, if only the $\mathbf{q}$'s and $\mathbf{p}$'s were properly ordered in the expression of the operator. He took this as an indication that joint probability distributions in q and p could adequately represent quantum evolution, despite the non-positive character of these distributions. Dirac opposed this idea and blocked its publication. In contrast, the statistician Maurice Bartlett supported Moyal and helped him develop his theory. In 1944, Moyal approached Dirac a second time. The reaction was again negative. Dirac argued that phase-space averages generally failed to represent quantum averages, because for instance the phase-space average of pq must be the same as the average of qp, whereas the quantum averages of $\mathbf{pq}$ and $\mathbf{qp}$ differ by $i\hbar$. As Moyal explained in his reply, Dirac had misread him, because in Moyal's theory the two kinds of average only agree if the quantum operator is ordered according to Weyl's prescription. Thus, in Dirac's example the theory only requires that the phase-space average of pq should be equal to the quantum average of $\frac{1}{2}(\mathbf{pq} + \mathbf{qp})$.[70]

Dirac then raised a more serious objection. He noted that in the case of a harmonic oscillator characterized by the Hamiltonian $H = \frac{1}{2}(p^2 + q^2)$, Moyal's theory implied a

---

[68] Wigner 1932, pp. 750n, 751.   [69] On this history, cf. Curtwright and Zachos 2011.

[70] Moyal to Dirac, 18 Feb. 1944; Dirac to Moyal, 20 Apr. 1945; Moyal to Dirac, 29 Apr. 1945. Cf. Ana Moyal 2006, Chaps. 1–3 (biography), appendix 2 (Moyal–Dirac correspondence).

non-vanishing energy fluctuation in any energy eigenstate, against the standard view that the energy is sharply defined in such states. Indeed, the Weyl quantization of H^2 and the commutation rule $\mathbf{qp} - \mathbf{pq} = i\hbar$ together lead to[71]

$$(p^2 + q^2)^2 \to (\mathbf{p}^2 + \mathbf{q}^2)^2 + \hbar^2, \tag{117}$$

so that

$$\overline{H^2} - \bar{H}^2 = \langle \mathbf{H}^2 + \hbar^2/4 \rangle - \langle \mathbf{H} \rangle^2 = \hbar^2/4. \tag{118}$$

The source of this paradox is the naïve idea that the quantum-theoretical expectation value of any physical quantity is obtained by Weyl-quantizing its classical expression $g(q, p)$ and forming $\langle \mathbf{g} \rangle = \mathrm{Tr}(\boldsymbol{\rho}\mathbf{g})$. This prescription, or the equivalent identification of the Wigner phase average $\bar{g}$ with the expectation value of g, is generally incompatible with the postulate of sharply defined values of a quantity in its eigenstates. This incompatibility, of which Dirac's paradox is an illustration, can be avoided by interpreting the quantum expression $\langle \mathbf{g} \rangle$ as the expectation value of the classical quantity obtained in replacing the $\mathbf{q}$ and $\mathbf{p}$ in the expression of $\mathbf{g}$ by ordinary numbers.[72]

Moyal did not envision this subterfuge. He accepted Dirac's criticism, and published his theory in 1949 with a warning that its statistical predictions did not entirely agree with those of standard quantum mechanics. He even suggested experimental tests of the differences. In 1946, the Dutch theoretical physicist Hilbrand Groenewold, who did not have Dirac in his way, had already published a systematic study of hidden-variable theories which also contained a phase-space representation of quantum mechanics. Whereas Moyal interpreted this representation as a plausible, if imperfect, statistic-deterministic interpretation of this theory, Groenewold used it as a means to argue the impossibility of such an interpretation. Both theorists nevertheless obtained very nearly the same formal apparatus. They were both aware of Weyl quantization, and Groenewold knew about Wigner's introduction of the Wigner function. Bartlett and Moyal obtained this function by inverting the Weyl quantization.[73]

Bartlett became aware of Wigner's contribution in the summer of 1945. Moyal, informed by Bartlett, immediately told Dirac that his brother in law had anticipated the phase-space representation. Dirac remained unimpressed. To his earlier objections, he added that the phase-space representation of quantum mechanics, unlike that of classical mechanics depended on the choice of the canonical pair (q, p). He commented:

> If you depart so much from the usual classical ideas is there any point in trying to fit things into a classical framework? What advantages does your system have over the

---

[71] Dirac to Moyal, 11 and 18 May 1945.

[72] Under this prescription, the expectation value of the quantity g is the phase average of the quantity g', obtained by symmetrizing g (if necessary) and replacing all ordinary products by star products (defined below, pp. 273–4) in the polynomial expression of g. For instance, the expectation value of H^2 is the phase-space average of $H^{2\prime} = H * H$, so that the fluctuation in an energy eigenstate is $\overline{H * H} - \bar{H}^2 = \langle \mathbf{H}^2 \rangle - \langle \mathbf{H} \rangle^2 = 0$. Cf. Zachos, Fairlie, and Curtright 2005, pp. 17–18.

[73] Moyal to Dirac, 15 and 26 May 1945; Moyal 1949; Groenewold 1946.

usual statistical interpretation of quantum mechanics? Any results that you get from
your system must either conform to the usual quantum mechanics or else be incorrect.
I think your kind of work would be valuable only if you can put it in a very neat form.

In the eyes of modern adepts of the phase-space representation, the formal apparatus
of Moyal's published memoir looks clear and elegant, and its notation may be more
transparent than Groenewold's. This apparatus will now be given in modern notation.[74]

Quantum mechanics in phase space

If $\tilde{g}(\alpha, \beta)$ denotes the inverse Fourier transform of the classical quantity $g(q, p)$, the Weyl
quantization formula (108) implies that

$$\langle q' | \mathbf{g} | q'' \rangle = \int \tilde{g}(\alpha, \beta) \langle q' | e^{i(\alpha \mathbf{q} + \beta \mathbf{p})} | q'' \rangle \, d\alpha d\beta. \tag{119}$$

Using the Weyl identity[75]

$$e^{i(\alpha \mathbf{q} + \beta \mathbf{p})} = e^{i\alpha \mathbf{q}} e^{i\beta \mathbf{p}} e^{-(1/2)[i\alpha \mathbf{q}, i\beta \mathbf{p}]} = e^{i\alpha \mathbf{q}} e^{i\beta \mathbf{p}} e^{i(\hbar/2)\alpha\beta}, \tag{120}$$

we have

$$\langle q' | e^{i(\alpha \mathbf{q} + \beta \mathbf{p})} | q'' \rangle = e^{i(\hbar/2)\alpha\beta} e^{i\alpha q'} \delta(\hbar\beta + q' - q''), \tag{121}$$

and

$$\langle q' | \mathbf{g} | q'' \rangle = \frac{1}{\hbar} \int \tilde{g}(\alpha, q''/\hbar - q'/\hbar) e^{i(\alpha/2)(q'+q'')} d\alpha = \frac{1}{h} \int dp e^{ip(q'-q'')/\hbar} g\left(\frac{q'+q''}{2}, p\right). \tag{122}$$

Changing the variables to $q = (q' + q'')/2$ and $q_- = (q'' - q')/2$, and inverting the Fourier
transform yields the Wigner formula (112):

$$g(q, p) = 2 \int dq_- e^{2ipq_-/\hbar} \langle q - q_- | \mathbf{g} | q + q_- \rangle.$$

This means that the Wigner formula it the exact inverse of the Weyl quantization formula.
There is a one-to-one correspondence between distributions in phase-space and quantum
operators.

As Moyal and Groenewold both realized, the Weyl–Wigner correspondence is not
unique. However, it enjoys two important properties: it yields real (though non-positive)
values for the phase-space distributions g associated to Hermitian operators (physical quantities or density matrices); and it translates quantum averages into ordinary
phase-space averages as in Eq. (113):

---

[74]Moyal to Dirac, 21 Aug. 1945; Dirac to Moyal, 31 Oct. 1945.

[75]This identity is often called the Glauber identity, even though it plays an important role in Weyl 1927, p. 27.

$$\text{Tr}(\boldsymbol{\rho}\mathbf{g}) = \int \rho(q,p)g(q,p)\mathrm{d}q\mathrm{d}p.$$

More generally, for any two real functions $f(q,p)$ and $g(q,p)$ and for the associated Weyl operators $\mathbf{f}$ and $\mathbf{g}$,

$$\text{Tr}(\mathbf{fg}) = \int f(q,p)g(q,p)\mathrm{d}q\mathrm{d}p. \tag{123}$$

Indeed, Weyl's definition of $\mathbf{f}$ and $\mathbf{g}$ implies that

$$\text{Tr}(\mathbf{fg}) = \int \tilde{f}(\alpha,\beta)\tilde{g}(\alpha',\beta')T(\alpha,\alpha',\beta,\beta')\mathrm{d}\alpha\mathrm{d}\beta\mathrm{d}\alpha'\mathrm{d}\beta' \tag{124}$$

with

$$T(\alpha,\alpha',\beta,\beta') = \text{Tr}\left[e^{i(\alpha\mathbf{q}+\beta\mathbf{p})}e^{i(\alpha'\mathbf{q}+\beta'\mathbf{p})}\right] = e^{i(\hbar/2)(\alpha\beta-\alpha'\beta')}\text{Tr}\left[e^{i\alpha\mathbf{q}}e^{i(\beta+\beta')\mathbf{p}}e^{i\alpha'\mathbf{q}}\right]$$

$$= e^{i(\hbar/2)(\alpha\beta-\alpha'\beta')}\int \mathrm{d}q\mathrm{d}p e^{i(\alpha+\alpha')q}e^{i(\beta+\beta')p} = 4\pi^2\delta(\alpha+\alpha')\delta(\beta+\beta') \tag{125}$$

while

$$\int f(q,p)g(q,p)\mathrm{d}q\mathrm{d}p = 4\pi^2 \int \tilde{f}(\alpha,\beta)\tilde{g}(-\alpha,-\beta)\mathrm{d}\alpha\mathrm{d}\beta. \tag{126}$$

In order to translate quantum-mechanical equations into phase-space equations, it is convenient to introduce the star product $f * g$ whose associated operator is the operator $\mathbf{fg}$. On the one hand, after Eq. (122) this definition requires that

$$\langle q'|\,\mathbf{fg}\,|q''\rangle = \frac{1}{h}\int \mathrm{d}p e^{ip(q'-q'')/\hbar}[f * g](q_+,p), \text{ with } q_+ = \frac{q'+q''}{2}. \tag{127}$$

On the other hand, the definitions of $\mathbf{f}$ and $\mathbf{g}$ imply that

$$\langle q'|\,\mathbf{fg}\,|q''\rangle = \int \mathrm{d}q\,\langle q'|\,\mathbf{f}\,|q\rangle\,\langle q|\,\mathbf{g}\,|q''\rangle$$

$$= \frac{1}{h^2}\int \mathrm{d}q\mathrm{d}p'\mathrm{d}p'' e^{ip'(q'-q)/\hbar}e^{ip''(q-q'')/\hbar}\,f\left(\frac{q'+q}{2},p'\right)g\left(\frac{q+q''}{2},p''\right) \tag{128}$$

Using the Taylor developments

$$f\left(\frac{q'+q}{2},p'\right) = \sum_{n'=0}^{+\infty}\frac{1}{n'!}\left(\frac{q-q''}{2}\right)^{n'}\partial_q^{n'}f(q_+,p'),$$

$$g\left(\frac{q+q''}{2},p''\right) = \sum_{n''=0}^{+\infty}\frac{1}{n''!}\left(\frac{q-q'}{2}\right)^{n''}\partial_q^{n''}g(q_+,p''), \tag{129}$$

the identities

$$(q - q'')^{n'} e^{ip''(q-q'')/\hbar} = (-i\hbar)^{n'} \partial_{p''}^{n'} e^{ip''(q-q'')/\hbar}, (q - q')^{n''} e^{ip'(q'-q)/\hbar} = (i\hbar)^{n''} \partial_{p'}^{n''} e^{ip'(q'-q)/\hbar},$$
(130)

and integrating by parts so that all the derivatives act on f or g, we get

$$\langle q' | \mathbf{fg} | q'' \rangle = \frac{1}{h^2} \int dq dp' dp'' e^{iq(p''-p')/\hbar} e^{i(p'q'-p''q'')/\hbar}$$

$$\times \sum_{n'n''} \left(\frac{i\hbar}{2}\right)^{n'+n''} \frac{(-1)^{n''}}{n'! n''!} \partial_q^{n'} \partial_p^{n''} f(q_+, p') \partial_q^{n''} \partial_p^{n'} g(q_+, p'')$$

$$= \frac{1}{h} \int dp e^{ip(q'-q'')/\hbar} \sum_{n'n''} \left(\frac{i\hbar}{2}\right)^{n'+n''} \frac{(-1)^{n''}}{n'! n''!} \partial_q^{n'} \partial_p^{n''} f(q_+, p) \partial_q^{n''} \partial_p^{n'} g(q_+, p) \quad (131)$$

Comparing with Eq. (127), this gives

$$[f * g](q, p) = \sum_{n'n''} \left(\frac{i\hbar}{2}\right)^{n'+n''} \frac{(-1)^{n''}}{n'! n''!} \partial_q^{n'} \partial_p^{n''} f(q, p) \partial_q^{n''} \partial_p^{n'} g(q, p),$$
(132)

or, in the symbolic notation used by Groenewold and Moyal,

$$f * g = f e^{(i\hbar/2)(\overleftarrow{\partial}_q \overrightarrow{\partial}_p - \overleftarrow{\partial}_p \overrightarrow{\partial}_q)} g,$$
(133)

in which the arrows above the derivatives indicate on which side they are operating.[76]
    To first order in $\hbar$, this formula yields

$$f * g \approx fg + (i\hbar/2)(\partial_q f \, \partial_p g - \partial_p f \, \partial_q g) = fg + (i\hbar/2)\{f, g\}.$$
(134)

For the skew-symmetric part, which is the phase-space counterpart of the commutator $[\mathbf{f}, \mathbf{g}]$, we have

$$f * g - g * f \approx i\hbar \{f, g\}.$$
(135)

The quantum-mechanical equation of motion

$$i\hbar \dot{\boldsymbol{\rho}} = [\mathbf{H}, \boldsymbol{\rho}]$$
(136)

translates into

$$\dot{\rho} = \{\{H, \rho\}\},$$
(137)

---

[76] Groenewold 1946, pp. 451–2.

wherein the Moyal bracket is defined by

$$f * g - g * f = i\hbar\{\{f, g\}\}. \tag{138}$$

When $\hbar$ reaches zero, the Moyal bracket reduces to the Poisson bracket and the equation of motion (137) turns into the classical equation of motion

$$\dot{\rho} = \{H, \rho\}. \tag{139}$$

These considerations give a precise meaning to the relation between commutators and Poisson brackets. Quantum and classical mechanics now being both represented in phase space, it becomes clear that the Lie algebra of quantum evolutions is a continuous deformation of the Lie algebra of classical evolutions. As Wigner anticipated, this formulation helps us understand the semi-classical limit of quantum mechanics. The quantum nature of processes can be appreciated by identifying the departures of the phase-space density from its classical value. This is why the Wigner function is popular among physicists who try to understand the transition from quantum to classical behavior, for instance decoherence and Schrödinger cats.[77]

Equivalence of all deformations of the Poisson brackets

The existence of the Moyal bracket implies that the basic structure of quantum mechanics can be obtained by deforming the Lie algebra of Hamiltonian mechanics. This result confirms Dirac's intuition that the two theories are not so remote from each other, at least from a mathematical point of view. A few mathematicians established a much stronger connection in the 1970s.

The relevant concept of deformation is the one introduced by Murray Gerstenhaber in 1964. Accordingly, a deformed Poisson bracket is a C^∞ bilinear alternate function $\{f, g\}_\hbar$ of the phase-space functions f and g and of the parameter $\hbar$ defining a Lie Algebra (satisfying the Jacobi identity) in the space of phase-space functions, varying continuously with $\hbar$ and coinciding with the usual Poisson bracket for $\hbar = 0$. In 1974, André Lichnerowicz and his collaborators Moshé Flato and Daniel Sternheimer discovered non-trivial first-order differential deformations of the Poisson bracket on curved symplectic manifolds. All such deformations are trivial in the flat $\mathbf{R}^{2n}$ case which is the most commonly encountered in Physics.[78]

The following year, Jacques Vey proved the existence of non-trivial deformations of infinite differential order when the third Betti number[79] of the manifold vanishes. In the flat $\mathbf{R}^{2n}$ case, Vey's deformed bracket is identical to the Moyal bracket. The Lichnerowicz

---

[77] Cf. Haroche and Raimond 2006, appendix.

[78] Flato, Lichnerowicz, and Sternheimer 1975.

[79] The Betti numbers are topological indicators defined as the rank of the successive homology groups of a topological space. Intuitively, the zeroth correspond to the number of connected components, the first to the numbers of tunnels, the second to the number of voids, etc.

group soon noted this coincidence. Following a suggestion by Flato, they conceived a new concept of quantization based on deforming the Poisson algebra:

> These developments encourage attempts to view quantum mechanics as a theory of functions or distributions on phase space, with deformed products and brackets. *We suggest that quantization be understood as a deformation of the structure of the algebra of classical observables, rather than as a radical change in the nature of the observables.*

In particular, Flato, Lichnerowicz, and Sternheimer hoped to help theoretical physicists quantize the constrained Hamiltonian systems that occur in some quantum field theories. Most interestingly, in 1979 Lichnerowicz and Simone Gutt established that all deformations of the Poisson bracket were mutually equivalent on symplectic manifolds for which the second Betti number vanishes. In the flat $\mathbf{R}^{2n}$ case, all deformations are equivalent to the Moyal bracket.[80]

The equivalence of two deformed brackets $\{f, g\}_\hbar$ and $\{f, g\}'_\hbar$ is there defined as the existence of a differential operator

$$T = \mathrm{Id} + \sum_{s=1}^{\infty} \hbar^s T_s \tag{140}$$

such that

$$T\{f, g\}'_\hbar = \{Tf, Tg\}_\hbar \tag{141}$$

for any pair f, g of phase functions. This equivalence evidently preserves the equation of motion

$$\dot{\rho} = \{H, \rho\}_\hbar. \tag{142}$$

In addition, the equivalence

$$T(f *' g) = Tf * Tg \tag{143}$$

for the associated star products preserves the possibility of translating this equation in operator language. Indeed if we replace the Weyl transform (105)

$$\mathbf{g} = W(g) = \int \tilde{g}(\alpha, \beta) e^{i(\alpha \mathbf{q} + \beta \mathbf{p})} \mathrm{d}\alpha \mathrm{d}\beta$$

with the transform

$$W'(g) = W(Tg) = \int \tilde{g}(\alpha, \beta) e^{i(\alpha \mathbf{q} + \beta \mathbf{p})} \tilde{T}(\alpha, \beta) \mathrm{d}\alpha \mathrm{d}\beta, \tag{144}$$

we have

$$W'(f)W'(g) = W(Tf)W(Tg) = W(Tf * Tg) = W'[T^{-1}(Tf * Tg)] = W'(f *' g). \tag{145}$$

---

[80] Vey 1975; Bayen et al. 1978a, p. 62; Lichnerowicz 1979, 1982; Gutt 1979. In the general case, all deformed brackets are equivalent to a Vey bracket, namely a bracket that matches the Moyal bracket in infinitesimal neighborhoods of the manifold.

Thus, the T-equivalence amounts to an alternative ordering of the quantum operators.[81]

Lichnerowicz's and Gutt's proofs of the equivalence of all star products or of all deformations of the Poisson bracket when the second Betti number vanishes are based on a powerful theorem by Vey regarding the Hochschild cohomology induced by the star product or regarding the Chevalley cohomology induced by the bracket. This result having important bearings on the question of the necessity of quantum mechanics, I will give an elementary proof in the case of a single degree of freedom (one q and one p).[82]

A general star product is a one-parameter deformation of the ordinary product of two functions, defined by

$$f * g = fg + \sum_{n=1}^{\infty} \alpha^n C_n(f, g), \tag{146}$$

the coefficients C_n being bilinear differential operators of order at least one (with respect to each variable) acting on the smooth functions f and g, such that

$$C_0(f, g) = fg, \ C_1(f, g) - C_1(g, f) = \{f, g\}, \tag{147}$$

and such that the product is associative:

$$(f * g) * h = f * (g * h). \tag{148}$$

For $\alpha = i\hbar/2$, this is a broad generalization of the Groenewold-Moyal product.

Call $\partial C(f, g, h)$ the trilinear functional generated from the bilinear functional $C(f, g)$ by the rule

$$\partial C(f, g, h) = fC(g, h) - C(fg, h) + C(f, gh) - hC(f, g), \tag{149}$$

and call $\partial B(f, g)$ the bilinear function generated from the linear functional $B(f)$ by the rule[83]

$$\partial B(f, g) = fB(g) + gB(f) - B(fg). \tag{150}$$

On the one hand, the associativity condition implies that for any $n \geq 1$, ∂C_n is an expression involving only the coefficients C_m of index $m < n$ and vanishing when these coefficients vanish. On the other hand, the equivalence (143),

$$T(f *' g) = Tf * Tg,$$

---

[81] The modified Wigner transforms do not preserve the identity of phase-space and quantum averages since $\mathrm{Tr}[W'(f)W'(g)] = \mathrm{Tr}[W(Tf)W(Tg)] = \int TfTg\,dpdq \neq \int fg\,dpdq$. However, $\mathrm{Tr}[W'(f)W'(g)] = \int f *' g\,dpdq$ remains true, so that the interpretive prescription of n. 72 (above, p. 271) extends to every star product.

[82] This proof has some similarity with the general proof found in Gutt and Rawnsley 1998.

[83] ∂ is the coboundary operator of the Hochschild cohomology.

induced by the differential operator (140),

$$T = \mathrm{Id} + \sum_{n=1}^{\infty} \alpha^n T_n,$$

turns the coefficient C_n into C_n' such that $C_n' - C_n - \partial T_n$ is an expression involving only the coefficients C_m of index $m < n$, and vanishing when these coefficients vanish.

By definition, the products $*$ and $*'$ are equivalent up to the order $n = 1$. Suppose that that their equivalence has been proved up to the order $n - 1$. Then the coefficients C_m and C_m' can be brought to coincide up to this order, and associativity at order n implies that $\partial(C_n - C_n') = 0$. It will now be shown that this condition implies the possibility of a choice of the differential operator T_n such that $C_n' - C_n - \partial T_n$ is a skew-symmetric bilinear functional of differential order one with respect to q and with respect to p.

For this purpose, the following lemma will be used: any bidifferential form C of finite differential order such that $\partial C = 0$ can be brought to the form $c\{f, g\}$ (where c is a constant) by a finite sequence of transformations. Call k the maximal order of the partial derivations occurring in the expression of C. This form may be written as

$$C(f, g) = \sum_{l=0}^{k} (\partial_l^k f \, D_l g + \partial_l^k g \, E_l f) + \dots, \tag{151}$$

wherein $\partial_l^k = \dfrac{\partial^k}{\partial^l p \partial^{k-l} q}$, D_l and E_l are differential operators of order inferior to k, and the suspension marks indicate terms involving only partial derivatives of order inferior to k. As the ∂ operation enjoys the property

$$\partial(A \otimes B) = \partial A \otimes B - A \otimes \partial B \tag{152}$$

(the tensor products being defined by $A \otimes B(f, g) = A(f)B(g)$ etc.), and since the operation ∂ lowers the differential order by one unit, the condition $\partial C = 0$ implies that $\partial D_l = \partial E_l = 0$ for the differential operators D_l and E_l. Consequently, these operators are of first order and can be written as

$$D_l = a_l \frac{\partial}{\partial q} + b_l \frac{\partial}{\partial p} \text{ and } E_l = a_l' \frac{\partial}{\partial q} + b_l' \frac{\partial}{\partial p}, \tag{153}$$

so that

$$\partial C(f, g, h) = \sum_{l=0}^{k} [(\partial \partial_l^k (f, g)(a_l \partial_q h + b_l \partial_p h) - (a_l' \partial_q f + b_l' \partial_p f) \partial \partial_l^k (g, h)] + \dots \tag{154}$$

For $k > 1$, we have

$$-\partial \partial_l^k (f, g) = \partial_p f \, \partial_{l-1}^{k-1} g + \partial_q f \, \partial_l^{k-1} g + \dots, \tag{155}$$

Grouping similar terms of highest order in the expression of ∂C, its vanishing requires that

$$a_l = a_l', b_l = b_l', b_l = a_{l+1}', b_l' = a_{l+1}, \tag{156}$$

so that C has the restricted form

$$C(f,g) = \sum_{l=0}^{k} \partial_l^k f(a_l \partial_q g + a_{l+1} \partial_p g) + f \leftrightarrow g + \ldots \tag{157}$$

If T is defined so that

$$T(u) = u - \sum_{l=0}^{k+1} a_l \partial_l^{k+1} u, \tag{158}$$

we have

$$\partial T(f,g) = \sum_{l=0}^{k} \partial_l^k f(a_l \partial_q g + a_{l+1} \partial_p g) + f \leftrightarrow g + \ldots \tag{159}$$

Therefore, $C - \partial T$ only contains terms of differential order inferior to k. We may repeat this operation until the differential order is one. The result has the form

$$C(f,g) = a \partial_q f \, \partial_q g + b \partial_p f \, \partial_p g + c \partial_p f \, \partial_q g + d \partial_q f \, \partial_p g. \tag{160}$$

Through an equivalence based on a quadratic T, we may subtract the symmetric part of this form to get[84]

$$C(f,g) = c(\partial_q f \, \partial_p g - \partial_p f \, \partial_q g). \tag{161}$$

Applying this series of equivalences to $C_n' - C_n$, we arrive at

$$C_n' - C_n = c_n\{f,g\}. \tag{162}$$

Then the equivalence operator

$$T = \mathrm{Id} - \alpha^{n-1} c_n \tag{163}$$

turns C_n into

$$C_n + c_n C_1 = C_n + c_n\{f,g\} = C_n'. \tag{164}$$

---

[84] This is a particular case of Vey's theorem (for $m = 2$ and dim M $= 2$), which states that every differential m-cocycle C on a symplectic manifold M is the sum of the coboundary of a differential $(m-1)$-cochain B and a 1-differential skew-symmetric m-cocycle A. In symbols, $C = \partial B + A$ if $\partial C = 0$.

Therefore, the star products $*$ and $*'$ are equivalent at order n if they are equivalent at order $n - 1$. Their full equivalence follows by induction. The corresponding brackets $f * g - g * f$ and $f *' g - g *' f$ are also equivalent.[85]

Deformation and necessity

The equivalence of all deformations of the Poisson bracket is a highly remarkable result, for it conveys a deep mathematical necessity to quantum mechanics: the generating algebra of this theory is, up to an isomorphism, the only possible deformation of the Lie algebra of the infinitesimal evolutions of Hamiltonian mechanics. So to speak, quantum mechanics is implicitly contained in classical mechanics. Recent results by Maurice de Gosson confirm this genetic relationship. In particular, Heisenberg's uncertainty relation can be given a meaning in classical mechanics: there exists a phase-space distribution for which this relation is compatible with the Hamiltonian flow.[86] A more spectacular result, concerns the relation between the classical group of Hamiltonian evolutions in phase-space and the quantum-mechanical group of unitary evolutions. For quadratic Hamiltonians, mathematicians have known for some time that the covering group of the former group is identical with the latter. In 2011, Gosson and Basil Hiley proved that for an arbitrary Hamiltonian there still is a one-to-one correspondence between the two kinds of evolution:[87]

> In this way we have shown that the mathematical formalism of the theory of Schrödinger's equation is already present in classical mechanics, and is in fact a reformulation of Hamiltonian dynamics in terms of operators.

Gosson and Hiley warned their reader against over-interpreting their claims:

> We are *not* claiming that we are deriving quantum mechanics from classical mechanics; what we are doing is the following: knowing that quantum mechanics exists, we show that the mathematical formulation of quantum mechanics in its Schrödinger formulation lies within Hamiltonian mechanics. This does not imply that quantum mechanics—as a *physical* theory—can be reduced to classical mechanics.

Similarly, Sternheimer has warned against confusing mathematical with physical results:

> A word of caution may be needed here. It is possible to intellectually imagine new physical theories by deforming existing ones... Nevertheless such intellectual constructs, even if they are beautiful mathematical theories, need to be somehow confronted with physical reality in order to be taken seriously in physics. So some physical intuition is still needed when using deformation theory in physics.

Surely, proofs that classical mechanics, qua mathematical theory, implicitly contains the mathematical apparatus of quantum mechanics do not imply that this apparatus can be

---

[85] In the general case, the elimination of A (see the previous note) requires that the second Betti number of the manifold should be zero (in the de Rham cohomology engendered by the external derivation of differential forms).

[86] Gosson and Hiley 2009. In the two-dimensional case, this distribution is the uniform distribution within an ellipse of surface h; the $2n$ case requires Gosson's subtler notion of "symplectic capacity."

[87] Gosson and Hiley 2011, p. 1434.

interpreted in a physically meaningful manner. Only a Diracian belief that every beautiful mathematics should someday find an application in the world could prompt us to think so. It remains true, however, that quantum mechanics has a striking kind of mathematical necessity, or perhaps even a transcendental necessity if one shares Poincaré's belief that the Lie group structure implied in the mathematical definition of the deformed dynamics is a necessary form of understanding.[88]

8.3   Quantum logic

A time-honored way of showing the necessity of quantum mechanics is to seek operational reasons for the strange algebra of quantum-mechanical observables. The basic idea, introduced by John von Neumann in 1932, is to examine the consequences of the possible incompatibility of measurements performed on a physical system. As every measurement can be regarded as the answer to a series of (compatible) Yes-No questions, the problem boils down to characterizing formal extensions of ordinary (Boolean) logic to propositions whose conjunction and disjunction are not always defined. We will first see how von Neumann and Garrett Birkhoff developed this idea in a powerful memoir of 1936. Essentially, they proved that simple, natural assumptions for the lattice of propositions made it isomorphic with the lattice of subspaces of a generalized Hilbert space (in the case of finite dimension). We will then see how later contributors to quantum logic, especially Constantin Piron in Geneva and George Mackey at Harvard, improved on this pioneering study by consolidating its operational foundation, by extending it to infinite dimension, and by deriving the quantum-mechanical description of states and evolutions.

From von Neumann's spectral theorem to a new logic

John von Neumann invented the "logic of quantum mechanics" in the early 1930s, while working on an appropriate mathematical foundation for quantum mechanics. The two pillars of this foundation were the complex Hilbert space of infinite dimension and the spectral theorem for self-adjoint operators in this space. For finite dimension (N), it had long been known that every Hermitian matrix H admits a sequence of mutually orthogonal eigenvectors $\psi_1, \psi_2, \ldots, \psi_N$ and a sequence of eigenvalues $\lambda_1, \lambda_2, \ldots, \lambda_N$ such that $H\psi_n = \lambda_n \psi_n$. In the general (possibly degenerate) case, several eigenvectors may correspond to the same eigenvalue and a given eigenvalue λ_i defines a linear subspace E_i that can have any dimension from 1 to N. This is why David Hilbert preferred a more intrinsic version of the spectral theorem according to which Hermitian operators with a number I of distinct eigenvalues admit the spectral decomposition

$$H = \sum_{i=1}^{I} \lambda_i P_i, \qquad (165)$$

where P_i is the orthogonal projector onto the subspace E_i. These projectors satisfy $P_i^2 = P_i$; they are Hermitian operators with the eigenvalues 0 and 1; and the associated subspaces

---

[88] Gosson and Hiley 2011, p. 1418. Sternheimer 1998, p. 2. On Poincaré's belief, see Chapter 6, p. 202–3.

are mutually orthogonal. The theorem is easily generalized to bounded operators in an infinite-dimensional Hilbert space.[89]

In quantum mechanics, the operators representing some physical quantities are not bounded and they are not even defined in the whole Hilbert space. For instance the momentum operator, $-i\hbar\nabla$, when applied to a normalized wave function, may lead to a wave function of infinite norm, and its eigenfunctions are sine waves that do not belong to the Hilbert space. Moreover, the set of eigenvalues is not always discrete; it may be continuous or mixed. This is why von Neumann generalized the notion of Hermitian operator to that of *self-adjoint* ("hypermaximal") *operator*, for which the domain of definition is only required to be dense in the Hilbert space and for which the usual relation of conjugation $\langle\varphi|\,H\,|\psi\rangle = \langle\psi|\,H|\varphi\rangle^*$ holds. He then showed that Hilbert's form of the spectral theorem lent itself to a generalization in which such operators admit the spectral decomposition

$$H = \int_{-\infty}^{+\infty} \lambda P_\lambda \mathrm{d}\lambda, \tag{166}$$

the *spectral measure* $P_\lambda \mathrm{d}\lambda$ being defined so that for any measurable subset Λ of **R**, the integral $P_\Lambda = \int\limits_{\lambda\in\Lambda} P_\lambda \mathrm{d}\lambda$ is an orthogonal projector. This measure is such that the subspaces associated with the projectors P_Λ and $P_{\Lambda'}$ are orthogonal whenever the subsets Λ and Λ' are disjoint.[90]

The projectors P_Λ were the basis of von Neumann's statistical interpretation of the quantum formalism. He regarded them as observables with the possible values 0 and 1. The value 1 corresponds to states for which the value of the observable H belongs to the subset Λ, and the value 0 to states for which the value of H belongs to the complementary subset. In logical terms, the projector P_Λ is associated with the bipolar proposition: "The value of the observable H belongs to the subset Λ." In his influential *Mathematical foundations of quantum mechanics* of 1932, von Neumann commented:[91]

> The relation between the projectors and the properties of a physical system allows for a sort of logical calculus [*eine Art Logikkalkül*] with these projectors. However, in contrast with the calculus of ordinary logic, this calculus is enlarged by the concept of "simultaneous decidability" [*gleichzeitige Entscheidbarkeit*] that is characteristic of quantum mechanics.

Birkhoff and von Neumann's propositional calculus

Von Neumann and Garret Birkhoff developed this new "sort of logical calculus" or "propositional calculus" in a brilliant memoir of 1936. The basic idea is to associate every proposition a about the state of a physical system with the invariant subspace A of the

---

[89] See von Neumann 1930, pp. 58–9; 1932, pp. 56–8. On the history of von Neumann's foundations, cf. Lacki 2000.

[90] Von Neumann 1930, pp. 58–62.

[91] Von Neumann 1930, p. 253. On the history of quantum logic, cf. Jammer 1974, Chap. 8; Rédei 2007; Dalla Chiara, Giuntini, and Rédei 2007.

associated projector. The relation "*a* implies *b*" is identified with the inclusion of the associated subspaces, the generalized conjunction "meet of *a* and *b*" with the intersection of the associated subspaces, the general disjunction "join of *a* and *b*" with the closed linear sum of the associated subspaces, the negation of *a* with the orthogonal complement of the associated subspace, the always false proposition "0" with the trivial subspace $\{0\}$, and the always true proposition "1" with the entire Hilbert space $\mathscr{H}$. In symbols, we have

$a \leq b$	$a \wedge b$	$a \vee b$	$\bar{a}$	0	1
$A \subset B$	$A \cap B$	$A + B$	$A^{\perp}$	$\{0\}$	$\mathscr{H}$

The relation $a \leq b$ is a relation of partial order; the meet $a \wedge b$ is the greatest lower bound of *a* and *b* with respect to this relation; the join $a \vee b$ is the least upper bound of *a* and *b*. Thus, the set of propositions forms what mathematicians call a *lattice*. This lattice is *orthocomplemented*, namely: it has a minimal element 0 and a maximal element 1; and to every *a* corresponds an orthocomplement $\bar{a}$ satisfying $\bar{\bar{a}} = a$, $a \wedge \bar{a} = 0$, $a \vee \bar{a} = 1$; if $a \leq b$ then $\bar{b} \leq \bar{a}$.[92]

The usual operations of logic share the orthocomplemented lattice structure. In addition, they enjoy the property of distributivity:

$$a \wedge (b \vee c) = (a \wedge b) \vee (a \wedge c) \text{ and } a \vee (b \wedge c) = (a \vee b) \wedge (a \vee c), \qquad (167)$$

which makes the set of propositions a *Boolean lattice*. The new propositional calculus is not distributive, as is easily seen for bidimensional *A* and one-dimensional *B* and *C*. Distributivity only holds within subsets of *mutually compatible propositions* whose associated projectors commute. The condition for the compatibility of *a* and *b* is

$$a = (a \wedge b) \vee (a \wedge \bar{b}) \text{ and } b = (a \wedge b) \vee (\bar{a} \wedge b), \qquad (168)$$

which corresponds to the condition

$$A = (A \cap B) + (A \cap B^{\perp}) \text{ and } B = (A \cap B) + (A^{\perp} \cap B) \qquad (169)$$

for the commutativity of the orthogonal projectors onto the subspaces *A* and *B*. The compatibility condition is easily seen to be equivalent to the distributivity of the sublattice engendered by *a*, *b*, $\bar{a}$, and $\bar{b}$.[93]

In a Hilbert space of finite dimension, the new calculus enjoys the weaker *modular property*:

$$\text{If } a \leq c \text{ then } a \vee (b \wedge c) = (a \vee b) \wedge c. \qquad (170)$$

---

[92] Birkhoff and von Neumann 1936, pp. 827–30. On lattice theory, cf. Birkhoff 1940; Grätzer 2011.

[93] Birkhoff and von Neumann 1936, pp. 830–1, 833 (compatibility).

This is so because on the one hand, if $A \subset C$, then A and $B \cap C$ are both included in $(A + B) \cap C$, which implies that $A + (B \cap C) \subset (A + B) \cap C$; and because on the other hand by repeated use of the identity

$$\dim(X + Y) = \dim X + \dim Y - \dim X \cap Y, \qquad (171)$$

the condition $A \subset C$ implies that[94]

$$\dim[A + (B \cap C)] = \dim[(A + B) \cap C] = \dim(A + B) + \dim C - \dim(B + C). \qquad (172)$$

Birkhoff and von Neumann then raise the two following questions:

1. Is every orthocomplemented modular lattice isomorphic to the lattice of subspaces in a Hilbert space or similar construct?
2. Do the axioms of an orthocomplemented modular lattice have a natural physical interpretation?

If these two questions can be answered positively, then quantum mechanics acquires some sort of necessity as a consequence of the natural calculus of propositions concerning tests performed on the system.

From modular lattices to projective geometry

In their answer to the first question, Birkhoff and von Neumann rely on a theorem by Birkhoff, according to which *any irreducible complemented modular lattice of finite dimension defines a projective geometry of finite dimension*. By definition, *a complemented lattice* is a lattice for which there exist a 0 and a 1 and for which every element a admits a complement a' such that $a \vee a' = 1$ and $a \wedge a' = 0$ (this operation need not be an orthocomplementation). A lattice is said to be *irreducible* if there is a third minimal non-zero element smaller than the join of any two distinct minimal non-zero elements of the lattice (the rationale of this definition will be explained in a moment). The dimension of the lattice is said to be finite if there is a maximal value for the number of elements in a chain $0 < a_1 < a_2 < \ldots . < 1$. An abstract *projective geometry of finite dimension* is defined by a set of elements of increasing but bounded "dimension" called points, lines, planes, etc. and satisfying the four following axioms:

P$_1$: Two distinct points are contained in one and only one line.

P$_2$: If A, B, C are points not all on the same line, and D and E are two distinct points such that B, C, D are on a line and C, A, E are on a line, then there is a point F such that A, B, F are on a line and also D, E, F are on a line.

P$_3$: Every line contains at least three points.

P$_4$: The set of points on lines through any k-dimensional element and a fixed point not on the element is a $(k + 1)$-dimensional element, and every $(k + 1)$-dimensional element can be defined in this way.

[94] Birkhoff and von Neumann 1936, pp. 831–3. Richard Dedekind introduced the notion of modular lattice in 1897: cf. Corry 1996, pp. 121–9.

A convenient model of these axioms is the set of vector subspaces of a finite-dimensional vector space, a one-dimensional subspace being identified with a point, a bidimensional one with a plane, and so forth.[95]

It is easy to see that a projective geometry defines a complemented modular lattice if two elements a and b are ordered according to the relation "a is on b." Then the meet of two elements is their intersection, and their join is the smallest element that contains both (or the set of all points contained on lines joining points of these two elements). The modularity of this lattice is proved by reasoning similar to that used for the subspaces of a Hilbert space. The 0 of the lattice is the empty set, and the 1 is the maximal element. If a is a given element, by P_4 the 1 can be obtained by successive joining of points to this element: $1 = a \vee e_1 \vee e_2 \vee \ldots$. These points can be chosen outside a, since they would otherwise not contribute to the join. Therefore $e_1 \vee e_2 \vee \ldots$ defines a complement to a.

Reciprocally, any irreducible complemented modular lattice of finite dimension (superior to 2) defines a projective geometry. The first step of the proof is the definition of the *dimension*[96] of an element a of the lattice by the maximal value of the number n of elements of a chain $0 < a_1 < a_2 < \ldots < a_n = a$ connecting this element with 0. In a modular lattice, the dimension satisfies the identity

$$\dim a \vee b + \dim a \wedge b = \dim a + \dim b, \tag{173}$$

which Birkhoff borrowed from earlier lattice theory. For a simple proof, we will need the following definitions and lemmata.[97]

Definitions: b covers a if $a < b$ and if there exists no element c such that $a < c < b$; *b at most covers a* if b covers a or $a = b$. A *maximal chain* $a < x_0 < x_1 < \ldots < x_n = b$ between two elements a and b is a chain whose length cannot be increased by intercalating further elements (equivalently, x_i covers x_{i-1} for $1 \leq i \leq n$).

Lemma 1: In a modular lattice, if b covers a, then $b \vee c$ at most covers $a \vee c$.

For a proof *ad absurdum*, suppose that there exists d such that $a \vee c < d < b \vee c$. Then $a \leq d \wedge b \leq b$ because $a < b$ and $a < d$; $a \neq d \wedge b$ because otherwise $a \vee c = (d \wedge b) \vee c = d \wedge (b \vee c) = d$; and $b \neq d \wedge b$ because otherwise $b \leq d$, which implies that $b \vee c \leq d$ since $a \vee c < d$ implies that $c < d$. Therefore, $a < d \wedge b < b$, in contradiction with b covers a.

Lemma 1′: In a modular lattice, if b covers a, then $b \wedge c$ at most covers $a \wedge c$.

The proof is left to the reader.

Lemma 2: In a modular lattice, if $b \neq c$ and if b and c both cover a, then $b \vee c$ covers b and c.

The proof goes as follows.

By Lemma 1, if b and c cover a, then $b \vee c$ at most covers $a \vee c = c$ and $a \vee b = b$. The equality $b \vee c = c$ is excluded because, together with $b \neq c$ and b covers a, it would imply

[95] Birkhoff 1935.

[96] This dimension exceeds the usual geometric dimension by one unit; it is identical with the vector-dimension in the vector-subspace representation of the associated projective geometry.

[97] For a different proof, cf. Birkhoff 1940, p. 40.

that $a < b < c$, which is incompatible with c covers a. By symmetric reasoning, $b \vee c = b$ is also excluded, so that $b \vee c$ covers b and c.

Lemma 3 (Jordan-Dedekind property): All maximal chains between two elements a and b have the same length.

The proof goes as follows.

The property is obvious if b covers a. By induction, suppose that the property has been proved for any two elements between which there exists a maximal chain of length $n - 1$. Then consider a maximal chain C: $a < x_1 < x_2 < ... < x_n = b$, and suppose there exists another maximal chain C': $a < y_1 < y_2 < ... < y_m = b$. If $y_1 = x_1$, the chain $x_1 = y_1 < y_2 < ... < y_m < b$ must have the length $n - 1$ since there exists a maximal chain $x_1 < x_2 < ... < x_n = b$ of length $n - 1$ between x_1 and b; whence C' has length n. If $y_1 \neq x_1$, by Lemma 2 $x_1 \vee y_1$ covers x_1 and y_1 because x_1 and y_1 both cover a. Since any maximal chain of the kind $x_1 < x_1 \vee y_1 < ... < b$ has length $n - 1$, any maximal chain of the kind $y_1 < x_1 \vee y_1 < ... < b$ also has length $n - 1$. Hence any maximal chain between y_1 and b has length $n - 1$, and the chain C' must have length n.

We are now equipped to prove the dimensional identity (173). Consider the maximal chain $a \wedge b = a_0 < a_1 < ... < a_n = a$. By lemma 1', $a_i \vee b$ at most covers $a_{i-1} \vee b$. By Lemma 3, this implies that $\dim(a_i \vee b) \leq \dim(a_{i-1} \vee b) + 1$. Hence $\dim(a \vee b) \leq \dim b + n$. Lemma 3 further implies that $n = \dim a - \dim(a \wedge b)$, so that

$$\dim a \vee b + \dim a \wedge b \leq \dim a + \dim b. \tag{174}$$

Dual reasoning on the chain $b = b_0 < b_1 < ... < b_m < a \vee b$ by means of Lemmata 1 and 3 yields the opposite inequality.

Having properly defined lattice dimension, we may call point an element of the lattice of dimension one, plane an element of the lattice of dimension two, and so forth. The join of two distinct points has dimension two, and axiom P_1 holds trivially because this join cannot be included in a "line" of dimension two without being equal to this line. Axiom P_2 is verified because

$$\dim c = \dim[(a \vee b) \wedge (d \vee e)] = \dim a \vee b + \dim d \vee e - \dim[(a \vee b) \vee (d \vee e)] =$$
$$4 - \dim[(a \vee e) \vee (b \vee d)] = 4 - \dim a \vee e - \dim b \vee d + \dim[(a \vee e) \wedge (b \vee d)] = 1. \tag{175}$$

The first part of P_4 holds because if $\dim a = k$, $\dim e = 1$, and $a \wedge e = 0$, then $\dim a \vee e = \dim a + \dim e = k + 1$. The second part also holds, because for a complement e' of a one-dimensional element e of a $(k+1)$-dimensional element b, modularity implies that

$$e \vee (e' \wedge b) = (e \vee e') \wedge b = b \tag{176}$$

and because

$$\dim(e' \wedge b) = \dim[e \vee (e' \wedge b)] + \dim[e \wedge (e' \wedge b)] - \dim e = \dim b - 1 = k. \tag{177}$$

As for P_3, it holds by the definition of irreducibility. This ends the proof of Birkhoff's theorem for the equivalence of an irreducible complemented modular lattice with a projective geometry for finite dimension.

The qualification "irreducible" is used because an arbitrary lattice is the direct product of irreducible lattices. In order to see that, define the "conjointness" of two points by their being either equal or having a third point on their join. This relation is a relation of equivalence and it can therefore be used for partitioning the points of the lattice into disjoint sets of mutually conjoint points. Then it can be shown that the lattice is the direct product of the irreducible sublattices defined by taking the join of all points in each of the sets from this partition. In particular, a lattice is irreducible if and only it satisfies Birkhoff's criterion that any two of its points are conjoint.[98]

In later considerations, it will be useful to know that in the case of an orthocomplemented modular lattice Birkhoff's criterion of irreducibility is equivalent to the following criterion due to Piron: *The lattice is said to be irreducible if 0 and 1 are the only elements that are compatible with every element of the lattice.* Let us first prove that Birkhoff's criterion implies Piron's. If the latter criterion fails to apply, then there is an element a of the lattice differing from 0 and 1 and such that $(a \wedge x) \vee (a \wedge \bar{x}) = a$ for every other element x. Consequently, if a point x does not belong to a, then $a \wedge x = 0$, $a \wedge \bar{x} = a$, and $\bar{a} \vee x = \bar{a}$, so that x belongs to $\bar{a}$. Call e a point of a and f a point of $\bar{a}$. A point g on $e \vee f$ belongs either to a or to $\bar{a}$. In the first case, modularity gives $g \leq a \wedge (e \vee f) = e \vee (a \wedge f) = e$; in the second, it gives $g \leq f$. Hence Birkhoff's criterion is not met. In order to prove the converse proposition that Piron's criterion implies Birkhoff's criterion, first consider a line $e \vee f$ generated by two incompatible points e and f. Then $g = (e \vee f) \wedge \bar{f}$ or $h = (e \vee f) \wedge \bar{e}$ provides a third point on the line: otherwise g would coincide with e, h would coincide with f, and the ensuing $e \leq \bar{f}$ would make e and f compatible. For a line generated by two compatible points e and f, Piron's irreducibility implies the existence of elements of the lattice that are incompatible with both e and f. Among these elements, there must be at least one point, say x, because by axiom P$_4$ every element is the join of points and because compatibility with two elements implies compatibility with their joins. Since e and x are incompatible, the line $e \vee x$ includes a third point g; similarly the line $f \vee x$ includes a third point h. If x belongs to $e \vee f$, it provides a third element on this line. If it does not, the conditions for P$_2$ are met and a third element obtains by taking the intersection of $e \vee f$ with $g \vee h$. This ends the proof of the equivalence between Birkhoff's and Piron's criteria of irreducibility.[99]

The vector space interpretation of projective geometry

The main result reached so far is that every irreducible complemented modular lattice of finite dimension (higher than 2) is a projective geometry. This theorem was essential to Birkhoff and von Neumann because it enabled them to exploit familiar results of projective geometry. As was mentioned earlier, the vector subspaces of a finite-dimensional vector

---

[98] Cf. Birkhoff 1935, pp. 746–7. For the projective geometry associated with subspaces of the quantum-mechanical Hilbert space, a third point on the join of two points corresponds to a superposition of the (pure) quantum states associated with these two points. Therefore, axiom P$_3$ corresponds to the superposition principle of quantum theory. Cf. Jauch 1968, pp. 106–8.

[99] Piron 1976, p. 464, theorem XX. My proof differs from Piron's as it does not presuppose Birkhoff's reduction theorem.

space satisfy the axioms of projective geometry. As was known since the previous century, for dimension higher than three the only possible projective geometries are of this kind. For readers not familiar with projective geometry, the following is an outline of the proof.[100]

Firstly, it is easy to see that Desargues's theorem holds in every projective geometry of dimension higher than three: for two non-coplanar triangles ABC and A'B'C' such that the lines AA', BB', and CC' intersect in a common point, the points AB ∩ A'B', AC ∩ A'C', BC ∩ B'C' are on the same line. Indeed if we call α the plane BCB'C', β the plane ACA'C', γ the plane ABA'B', π the plane ABC, and π' the plane A'B'C', we have

$$AB \cap A'B' = (\gamma \cap \pi) \cap (\gamma \cap \pi') = \gamma \cap (\pi \cap \pi'), \tag{178}$$

$$AC \cap A'C' = \beta \cap (\pi \cap \pi'), \quad BC \cap B'C' = \alpha \cap (\pi \cap \pi'), \tag{179}$$

so that the three intersection points belong to the line $\pi \cap \pi'$. Secondly, for any plane imbedded in a Desarguesian space, it is possible to define the sum and the product of points on a line (for three given reference points O, I, U) in such a manner that the line acquires the structure of a division ring $\mathbf{K}$. The construction, given in Figs. 8.1 and 8.2, is inspired by perspective drawings in which parallel lines converge on an ideal horizon. The axioms of projective geometry, together with the Desarguesian axiom, imply that the operations defined by this construction are unique (up to an isomorphism depending on the choice of O, U, I) and that they satisfy the algebraic axioms of a division ring.

The construction of Fig. 8.3 then provides the coordinates of a point in a perspective plane. To the point of coordinates x and y we can associate the vector-line of $\mathbf{K}^3$ engendered

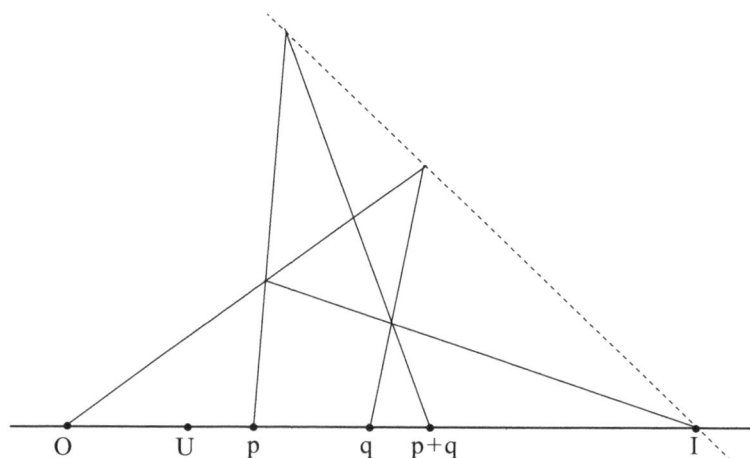

Fig. 8.1 Construction of the sum p+q of the points p and q on a line with an origin O and an ideal point I. This construction is intuitively justified by regarding the dotted line as the horizon of a perspective drawing.

[100]See, e.g., Garner 1981.

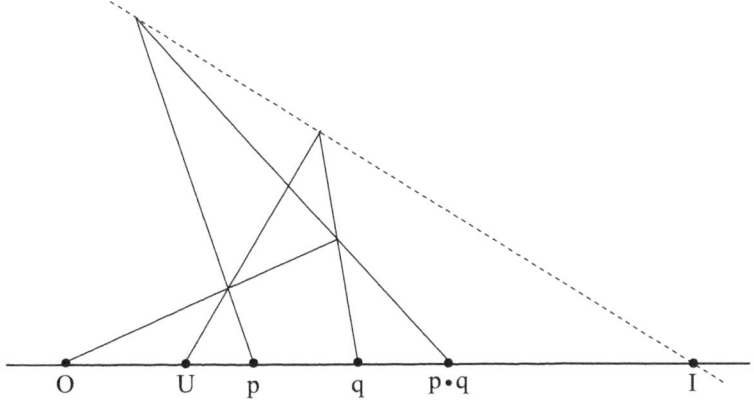

Fig. 8.2   Construction of the product p·q of two points on a line with an origin O and a unit U.

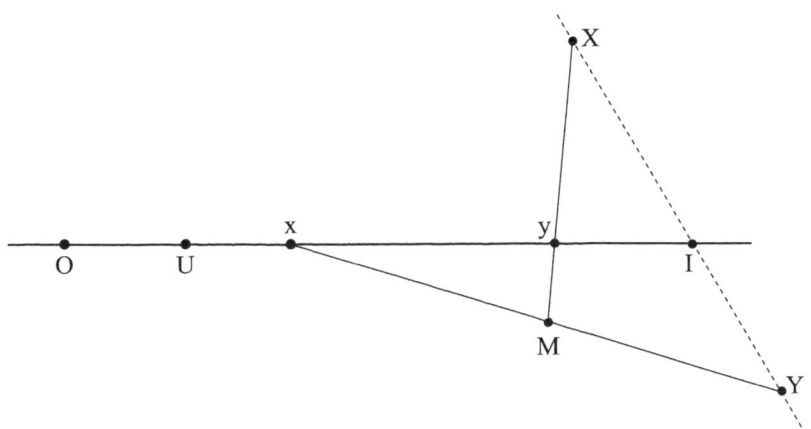

Fig. 8.3   Construction of the coordinates x and y of a point M with respect to the ideal point I and the ideal axes OX and OY.

by the vector $(1, x, y)$. Then a line of the projective plane is easily seen to correspond to a vector-plane of $\mathbf{K}^3$. More elaborate constructions yield similar results for projective geometries of dimension N higher than three. The objects of these geometries are thus interpreted as vector subspaces of $\mathbf{K}^N$.

From orthocomplemented modular lattices to generalized Hilbert spaces

The division ring on which the vector space is built is arbitrary. It may even be non-commutative (in which case Pappus's theorem does not hold). We will now see that the orthocomplementation of the lattice associated with the projective geometry brings a restriction on this division ring. From quantum mechanics we already know that the vector subspaces of a Hilbert space define a projective geometry whose associated lattice is orthocomplemented. The demonstration of this fact remains unchanged if we replace the Hilbert space by a more general kind of space which I shall call $\mathbf{K}^*$-space. Such a space is obtained by replacing the field of complex numbers with any division ring $\mathbf{K}$ that supports a "star conjugation" with the properties

$$(x + y)^* = x^* + y^*, (xy)^* = y^* x^*, \text{and } x^{**} = x \tag{180}$$

for any two elements x and y of the division ring, and if the Hermitian product of two vectors a and b is replaced with the definite form $< a, b >$ such that

$$< a, b + c > = < a, b > + < a, c >, < a + b, c > = < a, b > + < a, c >, \tag{181}$$

$$< b, a\lambda > = < b, a > \lambda, < b\lambda, a > = \lambda^* < b, a >, < a, b > = < b, a >^* \tag{182}$$

for any three vectors a, b, c and for any element λ of the division ring. Reciprocally, *every orthocomplemented projective geometry of dimension $N \geq 3$ is isomorphic to the set of vector subspaces of a $\mathbf{K}^*$-space of dimension N.* Birkhoff and von Neumann's proof of this remarkable theorem being somewhat opaque, I will provide a simple proof in the case $N = 3$.[101]

In the three-dimensional vector space associated with the projective geometry, let a vector be represented by a triplet (x, y, z) of coordinates, and a vector plane by the triplet $[X, Y, Z]$ of coefficients in the equation $Xx + Yy + Zz = 0$ (left-right order matters since the division ring may not be commutative). It is always possible to choose three basis vectors so that each of them is orthogonal to the plane engendered by the two others (orthogonality being the vector-space interpretation of orthocomplementation). In this basis, we have

$$(1, 0, 0)^{\perp} = [1, 0, 0], (1, \alpha, 0)^{\perp} = [1, f(\alpha), 0], (1, 0, \beta)^{\perp} = [1, 0, g(\beta)] \tag{183}$$

because the z-axis (resp. the y axis) belongs to any plane that is orthogonal to a vector belonging to the xy plane (resp. the xz plane). If $\Delta = (1, \alpha, \beta)$, $D = (1, \alpha, 0)$, and $\Delta^{\perp} = [1, \alpha', \beta']$, the identity $(\Delta + D)^{\perp} = \Delta^{\perp} \cap D^{\perp}$ leads to[102]

$$[1, -\alpha^{-1}, 0]^{\perp} = (1, f^{-1}(-\alpha^{-1}), 0) = (1, -f(\alpha)^{-1}, -\beta'^{-1}(1 - \alpha' f(\alpha)^{-1})). \tag{184}$$

This relation and the relation obtained by permuting the second and third axes imply that

$$\alpha' = f(\alpha), \beta' = g(\beta), \tag{185}$$

and

$$f^{-1}(-\alpha^{-1}) = -f(\alpha)^{-1}, g^{-1}(-\beta^{-1}) = -g(\beta)^{-1}. \tag{186}$$

---

[101] Birkhoff and von Neumann 1936, pp. 837–43 (appendix).

[102] In order to alleviate the notation, I write $f(\alpha)^{-1}$ instead of the more correct $[f(\alpha)]^{-1}$.

Now take $\Delta = (1, \alpha, \beta)$ and $D = (1, \gamma, 0)$. The identity $\Delta + D = (\Delta^\perp \cap D^\perp)^\perp$ leads to

$$[1, \; -\gamma^{-1}, \; \gamma^{-1}(\alpha - \gamma)\beta^{-1}] = [1, f(-f(\gamma)^{-1}), \; g(g(\beta)^{-1}(f(\alpha) - f(\gamma))f(\gamma)^{-1})], \quad (187)$$

whose third component leads to

$$g^{-1}(-\delta) = g(\beta)^{-1}(f(\alpha) - f(\gamma))f(\gamma)^{-1}, \text{ with } \delta = \gamma^{-1}(\gamma - \alpha)\beta^{-1}. \quad (188)$$

Using $[-g^{-1}(-\delta)]^{-1} = g(\delta^{-1})$ from the last of Eq. (186), this gives

$$f(\gamma)(f(\gamma) - f(\alpha))^{-1}g(\beta) = g(\beta(\gamma - \alpha)^{-1}\gamma) \quad (189)$$

or

$$f(\gamma)(f(\gamma) - f(\alpha))^{-1} = g(\beta(\gamma - \alpha)^{-1}\gamma)g(\beta)^{-1}. \quad (190)$$

The right-hand side of this equation cannot depend on β since the left-hand side does not. Therefore, $g(\beta\lambda)g(\beta)^{-1} = g(\lambda)g(1)^{-1}$, wherein $\lambda = (\gamma - \alpha)^{-1}\gamma$ can take any non-vanishing value. For the functions

$$F(\alpha) = f(\alpha)k^{-1} \text{ and } G(\alpha) = g(\alpha)l^{-1} \text{ with } k = f(1), \; l = g(1), \quad (191)$$

we have

$$F(\alpha\beta) = F(\beta)F(\alpha) \text{ and } G(\alpha\beta) = G(\beta)G(\alpha) \text{ for any } \alpha \text{ and } \beta. \quad (192)$$

Equations (186) lead to

$$-k\alpha = F(-k)F^2(\alpha), \; -l\alpha = G(-l)G^2(\alpha). \quad (193)$$

Since $F(1) = G(1) = 1$, this gives $F(-k) = -k$ and $G(-l) = -l$; hence

$$F^2(\alpha) = \alpha \text{ and } G^2(\alpha) = \alpha. \quad (194)$$

Equation (190) then leads to

$$F(\gamma) - F(\alpha) = G(\gamma - \alpha)G(\gamma)^{-1}F(\gamma), \quad (195)$$

and, for $\gamma = 1$,

$$F(\alpha) + G(1 - \alpha) = 1. \quad (196)$$

Taking $\alpha = -1, \; 2, \; -1, \; 1/2, \; 4$, and writing F_α for $F(\alpha)$, we get

$$F_{-1} = 1 - G_2, G_{-1} = 1 - F_2, G_3 = 1 - F_{-2}, F_{1/2} + G_{1/2} = 1, F_4 = 1 - G_{-3}. \quad (197)$$

Together with the equations (192), this gives

$$F_2 + G_2 = F_2(F_{1/2} + G_{1/2})G_2 = F_2 G_2 \tag{198}$$

$$F_2{}^2 = F_4 = 1 - G_{-1}G_3 = 1 - (1 - F_2)(1 - F_2 F_{-1}) = F_2 + F_2(1 - F_2)(1 - G_2) = 2F_2, \tag{199}$$

whence $F_2 = 2$, $G_2 = 2$, $G_{-1} = -1$, and $G(-\alpha) = -G(\alpha)$. For any given α' and α'', take $\gamma = (\alpha' + \alpha'')/2$. Then $G(\alpha' - \gamma) = -G(\alpha'' - \gamma)$, and Eq. (195) yields

$$F(\alpha') + F(\alpha'') - 2F(\alpha'/2 + \alpha''/2) = F(\alpha') - F(\gamma) + F(\alpha'') - F(\gamma) = 0, \tag{200}$$

or

$$F(\alpha + \beta) = F(\alpha) + F(\beta), G(\alpha + \beta) = G(\alpha) + G(\beta) \text{ for any } \alpha \text{ and } \beta. \tag{201}$$

From Eq. (195), we now get

$$1 - F(\alpha)F(\gamma)^{-1} = 1 - G(\alpha)G(\gamma)^{-1}, \tag{202}$$

which means that the functions F and G are identical. A star conjugation may now be defined by

$$\alpha^* = F(\alpha) = G(\alpha), \tag{203}$$

and a generalized Hermitian form by

$$<b, a> = X^* x + Y^* ky + Z^* lz \tag{204}$$

for two vectors a and b of coordinates (x, y, z) and (X, Y, Z).

According to relations (185) and (191), the vector a belongs to the plane orthogonal to the vector b of coordinates $(1, \alpha, \beta)$ if and only if

$$x + f(\alpha)y + g(\beta)z = x + \alpha^* ky + \beta^* lz = <a, b> = 0. \tag{205}$$

This remark ends the proof that any projective geometry of dimension 3 is isomorphic to the set of subspaces of a three-dimensional **K***-space. The proof for higher dimension is easily obtained by considering three-dimensional subspaces.

If we combine this result with Birkhoff's earlier result that every irreducible complemented modular lattice defines a projective geometry, we arrive at the conclusion that *every irreducible orthocomplemented modular lattice is isomorphic to the lattice of subspaces of a* **K****-space.* This answers the first question of Birkhoff and von Neumann about their new calculus of propositions. The second question concerns the physical interpretation of the axioms of an orthocomplemented modular lattice. Before examining possible replies to this second question, a few words will be said about the relation between quantum logic and ordinary logic.

Is quantum logic a logic?

Birkhoff and von Neumann clearly did not intend to replace ordinary logic by their new calculus of propositions. By calling this calculus a "quantum logic," they only meant that it was a mathematically natural extension of the Boolean calculus of ordinary logic. In contrast, their physicist and philosopher readers have often taken the expression "quantum logic" literally. They have placed quantum logic and ordinary logic on the same footing, either because they regarded both kinds of logic as empirically founded, or because they regarded ordinary and quantum logic as both justifiable by a priori means. Hilary Putnam famously advocated the first option in his "Is logic empirical?" of 1968:

> I want to begin by considering a case in which "necessary" truths (or rather "truths") turned out to be falsehoods: the case of Euclidean geometry. I then want to raise the question: could some of the "necessary truths" of logic even turn out to be false *for empirical reasons*? I shall argue that the answer to this question is in the affirmative, and that logic is in a certain sense a natural science.

Putnam took von Neumann's quantum logic very seriously and propounded to "just read the logic off from the Hilbert space." In this view, both logic and quantum mechanics would have no a priori necessity, and they would both derive at least in part from experience. Peter Mittelstaedt is a nuanced defender of the second option, which lends some a priori necessity to quantum logic. He indeed believes that both the laws of classical and of quantum logic depend on dialog games or proof trees whose structure depends on pragmatic preconditions of the language of physics:[103]

> [In the operational approach] quantum logic appears as an a-priori structure that is justified more rigorously and under weaker assumptions than the laws of classical logic. This means that first of all the pretended preference of classical logic is no longer justifiable, and secondly that quantum logic contains less empirical contributions—if at all—than classical logic.

In Kantian terms, both Putnam and Mittelstaedt regard the laws of logic as synthetic judgments: a posteriori synthetic for Putnam, a priori synthetic for Mittelstaedt. They thus contradict a long tradition of treating these laws as analytic judgments. Today's philosophers usually do not mind this subversion, because they have assimilated Willard Van Orman Quine's attack on the traditional distinction between synthetic and analytic judgments in his "Two dogma of empiricism" of 1951. If, *pace* Quine, logic is purely analytic, one should not confuse it with a calculus of experimental propositions; and one should not try do derive its laws from experience; conversely, one should not try to derive rules of combined experimental tests from the laws of logic. This is the position defended by Josef Jauch:

> The calculus introduced here has an entirely different meaning from the analogous calculus used in formal logic. Our calculus is the formalization of a set of *empirical* relations which are obtained by making measurements on a physical system. It expresses an objectively given property of the physical world. It is thus the formalization of empirical facts, inductively arrived at and subject to the uncertainty of any such fact. The

---

[103]Putnam 1968, pp. 216, 221; Mittelstaedt 1978; 2004, Chaps. 3, 13; 2011, p. 72. Cf. Gardner 1971; Stachel 1974; Bacciagaluppi 1993, 2009.

calculus of formal logic, on the other hand, is obtained by making an analysis of the meaning of propositions. It is true under all circumstances and even tautologically so. Thus ordinary logic is used even in quantum mechanics of systems with a propositional calculus vastly different from that of formal logic. The two need have nothing in common.

If, contrary to Jauch's opinion, the truths of logic are synthetic truths similar to those of arithmetic or geometry, a strong quantum-logic project becomes conceivable. This thorny issue may be avoided by pragmatically regarding the logic of ordinary reasoning (including that used to defend a new logic!) as unaffected by the physical calculus of propositions. In this view, the question of the necessity of this calculus is decoupled from the question of its interpretation as a logic; its answer should be sought in the naturalness of the implied physical operations, not in pure laws of thought.[104]

The physical interpretation of the logical axioms

Our question is whether the axioms of an orthocomplemented modular lattice can be embodied by natural physical operations. Let us first consider the case of a finite dimension N. The lattice properties are easy to justify, because $a \leq b$ has a well-defined operational interpretation as the statement that the binary test b (Yes-No experiment) always yields a positive result if the test a has just been performed with a positive result.[105] The axioms of partial ordering are evidently satisfied because every ideal test is assumed to be repeatable and because ordinary implication is transitive. Orthocomplementation and its properties are also easy to justify by the duality of the Yes and No answer to binary tests: $\bar{a}$ is the proposition that the test a gives a No answer. Irreducibility holds if and only if there are no non-trivial tests that are compatible (non-interfering) with every other test. This property does not hold in quantum systems obeying superselection rules: for instance, the mass of a particle can be measured without interfering with any other measurement (in other words, the superposition of states of different mass is not allowed). However, generalization to reducible lattices is unproblematic: as was earlier mentioned, a reducible lattice is the direct product of irreducible sublattices, in conformity with the interpretation of superselection rules in terms of non-combining sectors of the Hilbert space of quantum states.[106]

Modularity is somewhat harder to justify. Birkhoff and von Neumann offer a physical argument based on the earlier discussed equivalence of modularity with the existence of a numerical dimension-function such that

- If $a < b$, then $\dim a < \dim b$
- $\dim a + \dim b = \dim(a \vee b) + \dim(a \wedge b)$.

---

[104] Mittelstaedt 1978; Jauch 1968, p. 77.

[105] Here and henceforth I use "binary test" or just "test" instead of Jauch's "Yes–No experiment." Although my tests are most commonly called "measurements" in the quantum logic literature, I prefer to reserve "measurement" to a test involving a concrete operation of addition (in conformity with the Helmholtzian definition of Chapter 6). I loosely use the same letter, say a, to represent a proposition and its test.

[106] Cf. Jauch 1968, pp. 124–6.

If one forgets about the difference between the operations of meet and join and the usual logical operations, these properties make $N^{-1}\dim a$ a probability which Birkhoff and von Neumann interpret as the a priori statistical weight of the associated quantum state (which is a basic notion of quantum statistical mechanics), or as the probability of a positive outcome of test a when nothing is specified as to its preparation.[107] Birkhoff and von Neumann nonetheless judge that "it would be desirable to interpret [modularity] by simpler phenomenological properties of quantum physics."[108]

This can be achieved by replacing modularity with two properties introduced by Constantin Piron in his influential dissertation of 1964: weak modularity and atomicity. A lattice is said to be *weakly modular* if and only if for any two elements a and b, $a \leq b$ implies that a is compatible with b. This condition is evidently met in quantum mechanics, because $a \leq b$ translates into $P_a = P_a P_b = P_b P_a$ and because compatibility corresponds to $P_a P_b = P_b P_a$ for the associate projectors. An orthocomplemented weakly modular lattice is called an *orthomodular lattice*. An *atom* (also called a point because of the geometric interpretation) is a minimal non-zero element of the lattice. An *atomic lattice* is a lattice satisfying the two following axioms:

A_1: Every element contains an atom.

A_2 (*covering law*): If a and b are elements of the lattice and e an atom, one can never have $a < b < a \vee e$ ($a \vee e$ at most covers a).

It will now be proved that *any orthomodular atomic lattice of finite dimension is modular.*[109]

If $a < b$, weak modularity implies that $b = a \vee (b \wedge \bar{a})$. If $b \wedge \bar{a}$ is not an atom, then there exists x such that $0 < x < b \wedge \bar{a}$, which implies that $a \leq a \vee x \leq a \vee (b \wedge \bar{a}) = b$. $a = a \vee x$ is excluded because $x \leq a$ contradicts $x < b \wedge \bar{a} \leq \bar{a}$; $a \vee x = b$ is also excluded because it would imply that $b \wedge \bar{a} = (a \vee x) \wedge \bar{a} = x \wedge \bar{a} = x$ (a, x, and $\bar{a}$ being compatible by weak modularity). Therefore, b does not cover a. By contraposition, if b covers a, then $b \wedge \bar{a}$ is an atom. Consequently, to a maximal chain $0 < a_1 < a_2 < ... < a_n = a$ for the element a of dimension n we can associate the sequence of atoms $e_1 = a_1$, $e_2 = a_2 \wedge \bar{a}_1, ..., e_n = a_n \wedge \bar{a}_{n-1}$ such that $a = e_1 \vee e_2 ... \vee e_n$. The existence of atomic decompositions being now proven, consider a minimal (smallest m) atomic decomposition $a = f_1 \vee f_2 ... \vee f_m$, and define $b_1 = f_1$, $b_2 = f_1 \vee f_2, ..., b_m = a$. The latter sequence is a chain $0 < b_1 < b_2 < ... < b_m = a$ since the former sequence is minimal. As $b_{m-1} \neq a$, among the atoms of the decomposition $a = e_1 \vee e_2 ... \vee e_n$ there must be one, say e_{i_1}, that does not belong to b_{m-1} and for which $b_{m-1} < b_{m-1} \vee e_{i_1} \leq b_{m-1} \vee f_m = a$. By A_2 this implies that $b_{m-1} \vee e_{i_1} = b_{m-1} \vee f_m = a$. The sequence f_i being minimal, $b_{m-2} \vee e_{i_1}$ differs from a, so that there exists an atom e_{i_2} (differing from e_{i_1}) not included in $b_{m-2} \vee e_{i_1}$; whence

---

[107] The latter interpretation is from Jauch 1968, p. 84.

[108] Birkhoff and von Neumann 1936, p. 833.

[109] Piron 1964, pp. 446 (weak modularity), 448 (atomicity), 460 (other proof of the theorem). Also, the theorem indirectly results from Piron's proof that any irreducible complete orthomodular lattice is isomorphic to a projective geometry, since in the finite-dimensional case the lattice associated to a projective geometry is modular.

$$b_{m-2} \vee e_{i_1} < b_{m-2} \vee e_{i_1} \vee e_{i_2} \leq b_{m-2} \vee e_{i_1} \vee f_{m-1} = a. \tag{206}$$

By A_2 this implies that $b_{m-2} \vee e_{i_1} \vee e_{i_2} = a$. Iterating this procedure, we end up with $a = e_{i_1} \vee e_{i_2} \vee ... \vee e_{i_m}$, which requires that $m = n$ because a decomposition into elements derived from a maximal chain cannot be redundant.[110] Therefore, the dimension of an element can be redefined as the minimal number of atoms of which it is the join. For $a \wedge b = 0$, the definition of dimension through chains implies that $\dim(a \vee b) \leq \dim a + \dim b$; whereas the definition through atomic decomposition implies the reverse inequality; so that $\dim(a \vee b) = \dim a + \dim b$. For arbitrary $a \wedge b$, weak modularity implies that

$$a = (a \wedge b) \vee [a \wedge \overline{a \wedge b}] \text{ and } b = (a \wedge b) \vee [b \wedge \overline{a \wedge b}]; \tag{207}$$

hence

$$a \vee b = (a \wedge b) \vee [a \wedge \overline{a \wedge b}] \vee [b \wedge \overline{a \wedge b}]. \tag{208}$$

As $a \wedge b$ and the two expressions in square brackets do not intersect, we have

$$\dim a = \dim(a \wedge b) + \dim[a \wedge \overline{a \wedge b}], \quad \dim b = \dim(a \wedge b) + \dim[b \wedge \overline{a \wedge b}], \tag{209}$$

and

$$\dim a \vee b = \dim(a \wedge b) + \dim[a \wedge \overline{a \wedge b}] + \dim[b \wedge \overline{a \wedge b}], \tag{210}$$

so that relation (174) holds:

$$\dim(a \vee b) = \dim a + \dim b - \dim(a \wedge b).$$

As we already know, this relation implies modularity.

Let us now see whether weak modularity and atomicity have natural justifications. Operationally, weak modularity corresponds to the condition: if a binary test b has a well-defined result whenever a has just been tested, then the testing sequence a, b, a always yields the same result for the two tests of a (and so, too, do the two tests of b in the sequence b, a, b). This seems reasonable, because the premise $a \leq b$ intuitively implies that the test b refines our knowledge of the system without destroying knowledge acquired by the test a. For finite dimension, the atomic axiom A_1 holds necessarily since a chain below any given element of the lattice cannot be indefinitely lengthened by inserting a non-zero element under its least element.

The covering law A_2 is less obvious. In his dissertation, Piron remarked that for propositions x compatible with a given proposition a, the correspondence $x \rightarrow a \vee x$ fills up

---

[110] If it were redundant, there would be an element e_i such that $e_i \leq e_1 \vee ... \vee e_{i-1} \vee e_{i+1} \vee ... \vee e_n = a_{i-1} \vee (a \wedge \bar{a}_i)$. Since $e_i \leq a_i$, this would imply that $e_i \leq a_i \wedge [a_{i-1} \vee (a \wedge \bar{a}_i)] = (a_i \wedge a_{i-1}) \vee (a \wedge \bar{a}_i \wedge a_i) = a_{i-1}$, which is incompatible with $e_i = a_i \wedge \bar{a}_{i-1}$.

the sublattice of propositions containing a, for which a plays the role of a zero. This sub-lattice has a simple physical interpretation: it concerns the tests done on the system when $\bar{a}$ is known to be true, in which case $a \vee x$ is equivalent to x. One should therefore expect the correspondence to turn any atom e of the full lattice (compatible with a but not included in a) into an atom of the sublattice, which means that $a \vee e$ covers a. Piron regarded this remark as a "justification" of the covering law. Alas it cannot be, because weak modularity by itself implies that $a \vee e$ at most covers a if e is compatible with a, and because compatibility is not assumed in the covering law. At best one could try to argue that the validity of the covering law for compatible atoms makes it plausible for incompatible atoms.[111]

It may be noticed, however, that even for incompatible a and e, their join is compatible with a, so that $a \vee e = a \vee d$, with $d = (a \vee e) \wedge \bar{a}$. Since d is compatible with a, the covering law will be justified if we can find a physical reason for d being an atom. This is what Piron managed to do a few years later by assuming the existence of repeatable tests for every proposition. An atom e of the lattice being associated with a maximal repeatable test, it can also represent a state of the system produced by this test (a pure state in quantum-mechanical language). Suppose that the system was originally in this state and that a test a then gives a No answer. After this test, we expect the system to still be in a maximally known state, say f, and we expect this state to be determined by the set of tests that do not interfere with it. This set comprises a second test of a (since such tests are repeatable), and the test of every proposition x containing e and compatible with a. Indeed the latter kind of test may be indifferently performed before or after the (first) test of a (owing to compatibility interpreted as non-interference)[112] and it obviously does not alter the state e in the former case. Compatibility implies that

$$x = (x \vee a) \wedge (x \vee \bar{a}) \geq (e \vee a) \wedge (e \vee \bar{a}) = b. \qquad (211)$$

Therefore, the state f should be fully determined whenever it is known that $f \leq b$ and $f \leq \bar{a}$. Equivalently, there should only be one atom f such that $f \leq \bar{a} \wedge b = (a \vee e) \wedge \bar{a} = d$. This can only be true if d is an atom and $f = d$. Piron thus determined the final state and justified the covering law:

> It is important to remark that without this axiom [the covering law] we cannot determine the final state of the system; and although the measurement may be ideal, [without this axiom] the perturbation results in a loss of information, even if we take the response of the system into account.

This argument implies the notion of conditional tests (depending on the result of a previous test) and of maximally known (pure) states. More recently and within the context of his and Mittelstaedt's schematization of the empirical proof of propositions, Ernst Walther Stachow has shown that the covering law can be derived from the existence of conditional probabilities for tests performed on individual systems. This and other corroborations of

---

[111] Piron 1964, p. 448 (for $b = 1$).

[112] The test is "ideal" in Piron's sense, namely, it does not modify the outcome of a compatible test.

Piron's intuition strongly plead for a certain necessity of the covering law, although not a kind of necessity that jumps to the eyes.[113]

Infinite dimension

In sum, we see that in the case of finite dimension the axioms of an orthocomplemented modular lattice correspond to natural expectations about tests performed on a physical system. Unfortunately, usual quantum mechanics requires a propositional lattice of infinite dimension. The easiest way to deal with this difficulty is to assume that quantum mechanics in finite-dimensional Hilbert space is more basic than its infinite-dimensional counterpart and to derive the latter from the former by simply requiring that finite-dimensional subspaces of the latter should have the structure of the former. This procedure is physically justified inasmuch as quantum processes can concretely be restricted to transitions between a finite number of quantum states, as happens for instance in the two-level approximation of atoms interacting with properly tuned radiation. Operationally, a finite dimension of the propositional lattice corresponds to a maximal length for the chains of refinement of any binary test. This can only happen if the propositions correspond to sets of measurements of quantities that can only take a finite number of (sharply defined) discrete values.

For those who do not wish to assume so much from the start, it is necessary to examine the general case of lattices of infinite dimension. The mathematical component of this generalization is not too problematic. Piron accomplished most of it by adding the axiom of completeness according to which the least upper bound $a \vee b$ of two elements and their greatest lower bound $a \wedge b$ still exist despite the infinite dimension, and by replacing modularity with weak modularity and atomicity. The latter change is necessary because the lattice of subspaces of a Hilbert space of infinite dimension is not modular[114] and because perfectly conceivable quantum measurements, for instance position measurements, violate modularity.[115] Piron managed to prove that *any irreducible complete orthomodular atomic lattice could be represented by the lattice of subspaces of a **K***-space*.[116]

The operational embodiment of Piron's mathematical extension is more troublesome. Even the basic lattice operations, the meet and the join, are hard to justify because their empirical realization requires infinitely many operations. For instance, $a \wedge b$ is true if and only if an infinite sequence of alternate tests of a and b all yield positive results.[117] In 1969 Jauch and Piron tried to circumvent this difficulty by defining propositions as equivalence classes of experimental Yes-No questions and the product of a family of questions

[113]Jauch and Piron 1969, pp. 847–8; Piron 1976, p. 69 (citation); Stachow 1985. On other justifications of the covering law, see Wilce 2009, §5.

[114]Von Neumann hoped to circumvent this difficulty by restricting the permitted projectors and subspaces to a modular subclass. On the failure of this project and on its connection with the von Neumann algebras, cf. Dalla Chiara, Giuntini, and Rédei 2007, pp. 212–17.

[115]See Jauch 1966, pp. 219–21.

[116]Piron 1964. A student of Mackey's, Malcom Donald MacLaren, and two Japanese mathematicians, Ichiro Amemiya and Huzihiro Araki, perfected the proof. Cf. Primas 1981, p. 212.

[117]This has to do with the projector formula $P_{a \wedge b} = \lim_{n \to \infty} (P_a P_b)^n$: cf. Mittelstaedt 1978, p. 20.

as an arbitrary question from the family.[118] The success of this procedure is doubtful, for it involves potentially infinite classes of questions and an unclear notion of arbitrariness. According to Hans Primas, the best one can do is to regard the lattice structure as mathematically convenient. Then, orthocomplementation and weak modularity can be justified in the same manner as in the case of finite dimension. In contrast, atomicity becomes artificial. Why should there be a limit to the refining of a combination of tests? As Primas notes, atomicity does not hold in powerful axiomatic formulations of classical and quantum statistical mechanics. In the lattice of propositions of quantum mechanics, atoms correspond to pure states. This suggests that atomicity only applies to individual systems (the classical case is controversial). A possible justification for the existence of an atom below an element of the lattice would be the finite dimension of this element, despite the infinite dimension of the lattice. But this is not far from assuming the concrete possibility of finite-dimensional lattices of propositions. Then we may as well begin with such lattices and delay infinite-dimensional generalization until the K^*-space representation has been derived.[119]

States, probabilities, and dynamics

Once a calculus of propositions or quantum logic has been defined, it is tempting to define the state of a system through the list of probabilities for the outcomes of all binary tests on the system. The Harvard mathematician George Mackey did so in 1957 in an attempt to base quantum mechanics on a series of plausible axioms. The central notion of his theory was the probability $p(A, S, a)$ for the observable A to take the value a when the system is in the state S. Stated informally, his first four axioms were:

1. p is a probability measure in Kolmogorov's sense.
2. The state S is defined by the function $(A, a) \to p(A, S, a)$, and an observable A by the function $(S, a) \to p(A, S, a)$.
3. For every observable A, one can define an observable $f(A)$ taking the values $f(a)$.
4. Every convex mixture of states is a state.

In the spirit of von Neumann's *Foundations*, Mackey next introduced "questions," that is, observables that take the values zero and one only. In two additional axioms, he required the existence of the sum of mutually exclusive questions and he associated a question to every bivalued probability measure on the space of states. This allowed him to redefine the theory through the lattice of questions (von Neumann's propositions) and through a probability measure on this lattice. In his penultimate axiom, he brutally assumed the lattice of questions to be isomorphic with the lattice of closed subsets of a Hilbert space. In the ultimate axiom, he required the probability of a positive answer to a question to be given by $\mathrm{Tr}\rho P$, wherein P is the orthogonal projector onto the subset associated with the question and ρ is a positive operator of trace *one* (a density matrix).[120]

---

[118] Jauch and Piron 1969; Piron 1976, pp. 20–3.

[119] For a lucid discussion of all these difficulties, cf. Primas 1981, pp. 214–19.

[120] Mackey 1957.

Mackey believed in some "physical plausibility" of all his axioms except the Hilbert-space one. As he did not have in hand the kind of operational justification later given by Piron, he contented himself with showing that this axiom resulted from a most simple and elegant extension of classical logic in which the lattice of propositions would be modular and orthocomplemented in the finite-dimensional case. Mackey of course knew from Birkhoff and von Neumann that the latter properties implied the $\mathbf{K}^*$-space representation, and he and Shizuo Kakutani had made a step toward infinite-dimensional generalization in the modular case. As for the last axiom, he soon heard from his Harvard colleague Andrew Gleason that it was superfluous: *any probability measure on the closed subspaces of a Hilbert space can be represented by a density matrix.*[121]

For adepts of quantum logic, Gleason's theorem is an essential result for it makes the quantum-mechanical representation of states a mere consequence of their definition through probabilities of propositions subjected to the rules of quantum logic. Unfortunately, Gleason's proof of his theorem is difficult, and the many attempts at simplifying it have been moderately successful. Fortunately, the theorem may be replaced by the much simpler variant that the German mathematician Paul Busch stated and demonstrated in 2003, if only the conditions of the latter theorem can be physically justified. Let us begin with a more precise statement of Gleason's theorem in the context of Piron's quantum logic.[122]

In this context, a state S is defined by a probability function $a \to p_S(a)$ over the lattice of propositions a, with $0 \leq p_S(a) \leq 1$, $p_S(0) = 0$, $p_S(1) = 1$. For two compatible propositions a and b, the join and the meet are the ordinary disjunction and conjunction. Piron therefore requires that

$$p_S(a \vee b) = p_S(a) + p_S(b) - p_S(a \wedge b). \tag{212}$$

If in addition the two propositions are mutually exclusive, we have

$$p_S(a \vee b) = p_S(a) + p_S(b). \tag{213}$$

Compatibility and exclusion together require $a \leq \bar{b}$, so that the associated projectors satisfy the orthogonality condition $P_a P_b = P_b P_a = 0$. We may therefore define a probability function w_S on the set of projectors so that

$$w_S(P_a + P_b) = w_S(P_a) + w_S(P_b) \text{ for any two projectors such that } P_a P_b = P_b P_a = 0. \tag{214}$$

By some sort of continuity, Mackey, Gleason, and Piron further require that

$$w_S \left(\sum_{i=1}^{\infty} P_i \right) = \sum_{i=1}^{\infty} w_S(P_i) \tag{215}$$

---

[121] Mackey 1963, pp. 72–3 (Hilbert-space axiom), 74 (Gleason); 1957, pp. 50–1 (Gleason); Gleason 1957.

[122] Busch 2003. On proofs of Gleason's theorem, cf. Dvurečenskij 1993, pp. 130–1.

for any denumerable family of mutually orthogonal projectors. In other words, if $1 = \sum_{i=1}^{\infty} P_i$ is a decomposition of the identity into projectors, then[123] $\sum_{i=1}^{\infty} w_S(P_i) = 1$. According to Gleason's theorem, this condition implies the density-matrix representation of the probability function if and only if the dimension of the Hilbert space is at least three.

Paul Busch replaces the projector decomposition of the identity by the more general decomposition $1 = \sum_{i=1}^{\infty} E_i$, wherein E_i is a positive[124] (therefore self-adjoint) operator, and he assumes the stronger condition $\sum_{i=1}^{\infty} w_S(E_i) = 1$ for any such decomposition. From a physical point of view the operators E_i are called "effects" because they correspond to a broader notion of test obtained by statistically mixing (compatible) binary tests. Indeed, being positive, E_i is also self-adjoint and therefore is (by diagonalization) a linear combination of mutually orthogonal projectors with positive coefficients. Now the move from projectors to effects brings a difficulty: as the various E_i do not necessarily commute, the additivity of the probabilities of the associated test outcomes is not physically obvious.

This difficulty is overcome thanks to a theorem by the Soviet mathematician Mark Aronovich Naimark, according to which for any decomposition of the identity as a sum of positive operators E_i in a Hilbert space $\mathscr{H}$, there is a decomposition of the identity in a larger Hilbert space $\widetilde{\mathscr{H}}$ as a sum of projectors P_i of which the former decomposition is a restriction, namely: $E_i \psi = \Pi P_i \psi$ for every ψ of $\mathscr{H}$ regarded as a subspace of $\widetilde{\mathscr{H}}$, if Π denotes the orthogonal projector onto $\mathscr{H}$. From a physical point of view, this means that an effect on a given system can be realized as an ideal (binary) test on a larger system. The projection Π corresponds to a preliminary test performed on the larger system, so that its resulting states $\widetilde{S}$ can be identified with the states S of a smaller system. The probability function of the latter states is identified with the probability function of the former, leading to $w_S(E_i) = w_{\widetilde{S}}(P_i)$ and

$$w_S(E_i + E_j) = w_{\widetilde{S}}(P_i + P_j) = w_{\widetilde{S}}(P_i) + w_{\widetilde{S}}(P_j) = w_S(E_i) + w_S(E_j). \qquad (216)$$

The converse of Naimark's theorem, according to which the restriction of a projector decomposition of the identity is a positive-operator decomposition of the identity is fairly obvious:

$$\{\forall \psi \in \widetilde{\mathscr{H}}, \langle \psi | P_i | \psi \rangle \geq 0\} \Rightarrow \{\forall \psi \in \widetilde{\mathscr{H}}, \langle \psi | \Pi^+ P_i \Pi | \psi \rangle \geq 0\} \Rightarrow \{\forall \varphi \in \mathscr{H}, \langle \varphi | E_i | \varphi \rangle \geq 0\}$$
$$(217)$$

$$\sum P_i = 1_{\widetilde{\mathscr{H}}} \Rightarrow \forall \varphi \in \mathscr{H}, \sum E_i \varphi = \sum \Pi P_i \varphi = \varphi \Rightarrow \sum E_i = 1_{\mathscr{H}}. \qquad (218)$$

---

[123] The projectors are necessarily mutually orthogonal if they sum up to the identity.

[124] The positivity of E_i means that for every $|\psi\rangle$, $\langle \psi | E_i | \psi \rangle \geq 0$. Its insertion in a decomposition of the identity further requires that $\langle \psi | E_i | \psi \rangle \leq 1$ for any normalized $|\psi\rangle$.

Moreover, for any effect it is easy to construct a projector of which it is an orthogonal restriction. Here is a recipe valid for a Hilbert space of dimension N: first diagonalize the effect, call α_1, α_2, ..., α_N its diagonal elements (which must satisfy $0 \leq \alpha_i \leq 1$ as a consequence of the definition of an effect); introduce $\beta_i = \sqrt{\alpha_i(1-\alpha_i)}$; and form the $2N \times 2N$ matrix that has $1-\alpha_N$, $1-\alpha_{N-1}$, ..., $1-\alpha_1$, α_1, α_2, ..., α_N on its diagonal, $\beta_N, \beta_{N-1}, ..., \beta_1, \beta_1, ..., \beta_{N-1}, \beta_N$ on its antidiagonal, and zeros everywhere else. Unfortunately, this recipe does not translate an effect-decomposition of the identity into a projector-decomposition. For this purpose, Naimark's more ingenious construction is needed. It will be given in the case of a triple decomposition, which is the only one needed here.[125]

Let $1 = E + F + G$ be a decomposition of the identity into three effects and define $E_1 = E$, $E_2 = E + F$, and $E_3 = E + F + G$. For two triplets $\Psi = (\psi_1, \psi_2, \psi_3)$ and $\Phi = (\varphi_1, \varphi_2, \varphi_3)$ of $\mathscr{H}^3$, define the sesquilinear form

$$\langle\langle \Phi \mid \Psi \rangle\rangle = \sum_{i,j} \langle\varphi_i| E_{ij}|\psi_j\rangle \text{ with } E_{ij} = E_{\inf(i,j)} \text{ for } i,j = 1,\ 2,\ 3. \tag{219}$$

This form evidently satisfies $\langle\langle \Phi \mid \Psi \rangle\rangle = \langle\langle \Psi \mid \Phi \rangle\rangle^*$, and it is positive because

$$\langle\langle \Psi \mid \Psi \rangle\rangle = \langle\psi_1 + \psi_2 + \psi_3|E|\psi_1 + \psi_2 + \psi_3\rangle + \langle\psi_2 + \psi_3|F|\psi_2 + \psi_3\rangle + \langle\psi_3|G|\psi_3\rangle. \tag{220}$$

The possible existence of Ψ vectors such that $\langle\langle \Psi \mid \Psi \rangle\rangle = 0$ is remedied by identifying all vectors of $\mathscr{H}^3$ that only differ by such null vectors. We thus reach a new Hilbert space $\widetilde{\mathscr{H}}$ which contains $\mathscr{H}$ as a Hilbert subspace because triplets of the form $\tilde{\psi} = (0,\ 0,\ \psi)$ satisfy $\langle\langle \tilde{\varphi} \mid \tilde{\psi} \rangle\rangle = \langle\varphi \mid \psi\rangle$. In this new space, consider the linear transformations Π, P_1, P_2, P_3 induced by the matrices[126]

$$\mathrm{M} = \begin{pmatrix} 0 & 0 & 0 \\ 0 & 0 & 0 \\ E_1 & E_2 & E_3 \end{pmatrix},\ M_1 = \begin{pmatrix} 1 & 1 & 1 \\ 0 & 0 & 0 \\ 0 & 0 & 0 \end{pmatrix},\ M_2 = \begin{pmatrix} 1 & 0 & 0 \\ 0 & 1 & 1 \\ 0 & 0 & 0 \end{pmatrix},\ M_3 = \begin{pmatrix} 1 & 0 & 0 \\ 0 & 1 & 0 \\ 0 & 0 & 1 \end{pmatrix}. \tag{221}$$

Each of these matrices is equal to its own square and yields a symmetric matrix when multiplied by the metric matrix E_{ij}. Consequently, they generate orthogonal projectors in $\widetilde{\mathscr{H}}$. The projector Π associated with M evidently is the orthogonal projector over the subspace $\mathscr{H}$ of $\widetilde{\mathscr{H}}$. We further have $M_1 M_2 = M_2 M_1 = M_1$, so that the matrices M_1, $M_2 - M_1$, and $M_3 - M_2$ define three orthogonal projectors P, Q, R in $\widetilde{\mathscr{H}}$, with $P + Q + R = 1$. Lastly, the three matrices $\mathrm{M}M_i$ have vanishing first and second lines, and their 33-element is E_i, so that for every $\tilde{\psi}$ vector $\Pi P_i \tilde{\psi} = \widetilde{E_i\psi}$. This means that the effect E_i is the restriction of the projector P_i to the original space $\mathscr{H}$. Consequently, the effect-decomposition $1 = E + F + G$ in $\mathscr{H}$ is the restriction of the projector decomposition $1 = P + Q + R$ in $\widetilde{\mathscr{H}}$, as was to be proved.

[125] For Naimark's construction, cf. Akhiezer and Glazman 1981, pp. 388–93.

[126] M_i is constructed according to the rule $M_i^{jk} = \delta_{jl}$ with $l = \inf(i,k)$.

Naimark's theorem and its physical interpretation being given, in order to prove the density-matrix representation of the probability function w_S we may use the Busch condition that $\sum w_S(E_i) = 1$ for any family of positive operators such that $\sum E_i = 1$. Double and triple decompositions turn out to be sufficient, and there is no restriction on the dimension of the Hilbert space. Firstly (the index S is henceforth omitted),

$$w(E + F) + w(1 - E - F) = 1 \text{ and } w(E) + w(F) + w(1 - E - F) = 1 \qquad (222)$$

imply that

$$w(E + F) = w(E) + w(F) \qquad (223)$$

for any two effects E and F whose sum still is an effect. Consequently, $w(qE) = qw(E)$ for any rational number q and for any effect E such that qE still is an effect. In order to extend this relation to any positive real multiplicand, we need to prove the continuity of w with respect to the norm

$$\|E\| = \operatorname*{Sup}_{|\psi\rangle \in \mathscr{H}, \langle\psi|\psi\rangle = 1} \langle\psi| E |\psi\rangle. \qquad (224)$$

Continuity holds at $E = 0$, because if it did not we would have

$$\exists \varepsilon > 0 \text{ such that } \forall \eta > 0, \ \exists E \text{ with } \|E\| < \eta \text{ and } w(E) \geq \varepsilon; \qquad (225)$$

taking $\eta = 1/N$ with $N\varepsilon > 1$, we would have $\|NE\| < 1$ and $Nw(E) > 1$, hence the absurdity $w(NE) > 1$ (NE is an effect since $\|NE\| < 1$). For $E \neq 0$ and for any sequence E_n converging toward E, the difference $E - E_n$ is a self-adjoint operator that can be regarded as the difference of two positive operators X_n and Y_n who separately converge toward zero (this is easily seen after diagonalization). Therefore, $w(E_n) = w(E + Y_n - X_n) = w(E) - w(X_n) + w(Y_n)$ converges toward $w(E)$ and w is continuous. We may now extend the function w to positive operators A for which $\|A\| > 1$ by taking $w(A) = \|A\| w(A/\|A\|)$. This extension satisfies $w(\lambda A) = \lambda w(A)$ for any positive operator A and for any positive real number λ. Lastly, we extend w to any Hermitian operator A by writing this operator as the difference of two positive operators A' and A'' and setting $w(A) = w(A') - w(A'')$, which is easily seen to be independent of the choice of A' and A''. The resulting w is a linear function on the set of self-adjoint operators regarded as a real vector space:

$$w(\lambda A + \mu B) = \lambda w(A) + \mu w(B) \qquad (226)$$

for any two self-adjoint operators A and B and for any two real numbers λ and μ. A self-adjoint operator A can always be expanded as

$$A = \sum_{m,n} a_{mn} |m\rangle \langle n| \text{ with } a_{mn} = x_{mn} + iy_{mn}, \ x_{mn} = x_{nm}, \ y_{mn} = -y_{nm}; \qquad (227)$$

or

$$A = \sum_{m,n} x_{mn} \frac{|m\rangle\langle n| + |n\rangle\langle m|}{2} - \sum_{m,n} y_{mn} \frac{|m\rangle\langle n| - |n\rangle\langle m|}{2i}, \tag{228}$$

which is a linear combination of self-adjoint operators with real coefficients. Consequently,

$$w(A) = \sum_{m,n} (x_{mn}\alpha_{mn} - y_{mn}\beta_{mn}), \tag{229}$$

with

$$\alpha_{mn} = w\left(\frac{|m\rangle\langle n| + |n\rangle\langle m|}{2}\right), \quad \beta_{mn} = w\left(\frac{|m\rangle\langle n| - |n\rangle\langle m|}{2i}\right). \tag{230}$$

Setting $\rho_{mn} = \alpha_{mn} - i\beta_{mn}$, we reach $w(A) = \sum_{mn} \rho_{nm}a_{mn} = \mathrm{Tr}\rho A$. The operator ρ is self-adjoint by construction; its trace is *one* since $w(1) = 1$; and it is positive since $w(|\psi\rangle\langle\psi|) \geq 0$ for any projector $|\psi\rangle\langle\psi|$. This ends the proof that the probability function w can be represented by a density matrix.[127]

States being defined by a density matrix, it is natural to represent the evolution of a system by a one-to-one correspondence between density matrices ρ and ρ' representing the states of the system at two different times t and t'. In his Harvard lectures, Mackey developed this idea by further requiring this correspondence to depend on the time difference $\tau = t' - t$ only (uniformity of time) and to preserve convex mixtures (conservation of probability). In symbols, $(\alpha\rho + \beta\sigma)' = \alpha\rho' + \beta\sigma'$ for any two density matrices ρ and σ and for any two weights α and β such that $\alpha \geq 0$, $\beta \geq 0$ and $\alpha + \beta = 1$. Obviously, the transformation $\rho' = U\rho U^{-1}$, wherein U is a unitary of anti-unitary operator, meets this condition. With the help of a powerful theorem by Richard Kadison on automorphisms in C*-algebras, Mackey proved that reciprocally *any mixture-preserving one-to-one mapping of the set of density matrices onto itself could be generated by a unitary or anti-unitary operator* (this operator being defined up to a phase factor). The uniformity of time implies that $U(\tau)$ and $U^2(\tau/2)$ only differ by a phase factor. Since the square of an anti-linear operator is linear, the anti-unitary option is excluded for the evolution operator U.[128]

Kadison's proof of his theorem is for experts on C*-algebras. Walter Hunziker has given the following direct proof of Mackey's result. For $N = 2$, a density matrix can be represented as $\rho = \frac{1}{2}(1 + \mathbf{x} \cdot \boldsymbol{\sigma})$, where $\boldsymbol{\sigma}$ denotes the three-vector whose components are the three Pauli matrices and $\mathbf{x}$ is an ordinary three-vector such that $|\mathbf{x}| \leq 1$. Projectors or pure states correspond to $|\mathbf{x}| = 1$. Since the transformation $\rho \to \rho'$ preserves mixtures, it must turn pure states into pure states. The corresponding map of the ball $|\mathbf{x}| \leq 1$ onto itself is the restriction of an isometry S in Euclidean three-space. From the theory of spinors, we have

---

[127] Similar proofs are found in Busch 2003 and in Granström 2006, Chap. 2.

[128] Mackey 1963, pp. 81–2; Kadison 1951; Piron and Jauch favor another approach in which the evolution is assumed to be an automorphism of the lattice of propositions. This leads to the same result by Wigner's theorem, which states that all automorphisms of the lattice of subspaces of a Hilbert spaces are generated by unitary or anti-unitary operators. Cf. Beltrametti and Cassinelli 1981, pp. 252–4.

$$\rho' = \frac{1}{2}[1 + (S\mathbf{x}) \cdot \sigma] = U\rho U^{-1}, \tag{231}$$

where U is a unitary operator if the isometry S is positive ($\det S \geq 0$) and an anti-unitary operator if S is negative ($\det S \leq 0$). The Kadison-Mackey theorem is thus proved for $N = 2$.[129]

For arbitrary N, a pure state is associated to a 1-projector, namely, a projector over a one-dimensional subspace. The image P' of a 1-projector P must be a 1-projector, since the image of a pure state is a pure state. Consider two distinct orthogonal 1-projectors P and Q. Their invariant subspaces span a bidimensional subspace E. If $P_1, P_2, \ldots, P_N$ denotes a set of mutually orthogonal 1-projectors summing to the identity, then the preservation of mixtures implies that the projectors $P_1', P_2', \ldots, P_N'$ also sum to the identity and therefore are mutually orthogonal. Consequently, the transformation $\rho \to \rho'$ turns 1-projectors whose invariant subspace belongs to the 2-space E into projectors whose invariant subspace belongs to another 2-space. Hence there exists a unitary or anti-unitary transformation U_E such that $P' = U_E P U_E^{-1}$ and $Q' = U_E Q U_E^{-1}$. This implies that $\mathrm{Tr}(PQ) = \mathrm{Tr}(P'Q')$.[130] Any linear operator can be regarded as a linear combination of 1-projectors with complex coefficients,[131] so that the transformation $\rho \to \rho'$ admits both a linear and an antilinear extension over the set of linear operators. In this broader context, we have $\mathrm{Tr} AB = \mathrm{Tr} A'B'$ for any two linear (or antilinear) operators A and B.

We are now equipped to construct a unitary or anti-unitary representation of the transformation $\rho \to \rho'$. For a given unit vector e of the Hilbert space $\mathscr{H}$, first define Ue to be any unit vector e' such that $P_{e'} = P_{e'}$ for the associated projectors. Next define $U\lambda e$ to be λUe if the determinant of the earlier defined U_E is positive, and $\lambda^* U_e$ if this determinant is negative.[132] For a vector a not parallel to e, define Ua as $U_E a$, where E is the 2-space spanned by e and a. Hunziker ingeniously proves the linearity and unitarity (or anti-unitarity) of the operator U by considering the linear operator A such that $Ae = a$ and $Au = 0$ when $\langle a|u \rangle = 0$. Since A operates in E and since the transformation $A \to A'$ preserves orthogonality, we also have $A'e' = a'$ and $A'u' = 0$ when $\langle a'|u' \rangle = 0$. Consequently, $\langle a \mid b \rangle = \mathrm{Tr} AB = \mathrm{Tr}(AB)' = \mathrm{Tr} A'B' = \langle a'|b' \rangle$, which, together with $U\lambda a = U_E \lambda a = \lambda Ua$ (or $U\lambda a = \lambda^* a$ if $\det U_E = -1$) implies that U is linear and unitary (or anti-unitary). It remains to be seen that $\rho' = U\rho U^{-1}$ for any density matrix ρ or, sufficiently, that $P' = UPU^{-1}$ for any 1-projector. This results from $P_a' = U_E P_a U_E^{-1}$, $U_E P_a = UP_a$, and $P_a U_E^{-1} = (U_E P_a)^+ = (UP_a)^+ = P_a U^{-1}$ for the projector P_a associated with a. This ends

---

[129] Hunziker 1972.

[130] We are thus reaching the premises of Wigner's theorem: a one-to-one correspondence $P \leftrightarrow P'$ between 1-projectors, satisfying the condition $\mathrm{Tr}(PQ) = \mathrm{Tr}(P'Q')$. The theorem states the existence of a unitary or anti-unitary operator U such that $P' = UPU^{-1}$. The sequel amounts to a proof of this theorem.

[131] This is so because $|m\rangle\langle n|$ is a linear complex combination of the projectors $|m\rangle\langle m|$, $|n\rangle\langle n|$, $(1/2)(|m\rangle + |n\rangle)(\langle m| + \langle n|)$ and $(1/2)(|m\rangle + i|n\rangle)(\langle m| - i \langle n|)$.

[132] This determinant does not depend on the choice of E because it cannot change during a continuous rotation connecting two choices of E (the transformation $A \to A'$ being linear, and the values of the determinant being discrete).

the proof of Kadison's theorem, which implies the unitary representation of quantum evolutions if quantum states are described by density matrices.

Quantum logical necessity

Quantum logic, seen as a calculus of elementary empirical propositions, rests on very broad and fairly natural assumptions about how we may experiment on a physical system. Its only odd feature is the omission of a most natural assumption of classical observation: the possibility of eliminating the mutual interference between successive tests. In other words, quantum logic is an impoverished version of the natural axiomatics of classical observation. Although some axioms, for instance the existence of the least upper bound or the atomicity of the lattice, have frequently been criticized as too restrictive or too artificial, their necessity is evident in the case of finite dimension of the lattice of propositions. Conceptual and mathematical difficulties are mostly confined to the limit of infinite dimension, which may be postponed until the basic structure of quantum mechanics has been obtained for finite dimension.[133]

Birkhoff and von Neumann's axioms in the finite-dimensional case and Piron's axioms in the infinite-dimensional case have far-reaching consequences: the lattice of proposition must be isomorphic to the lattice of subspaces of a $\mathbf{K}^*$-space, namely, a generalized Hilbert space built on a division ring $\mathbf{K}$ equipped with a kind of conjugation. Quantum mechanics corresponds to the case in which $\mathbf{K}$ is the field of complex numbers. For this choice, it is possible to derive the density-matrix representation of states defined through the statistics of binary tests. Furthermore, the evolution of a system is determined by a unitary evolution operator in the manner of quantum mechanics.

The quantum logic approach to quantum mechanics is attractive for its being based on a very simple reduction of experiments to the answer to a series of Yes-No questions. It is also impressive by its ability to (partially) deduce the esoteric Hilbert-space apparatus of quantum mechanics from fairly natural notions. Much of the bad publicity that quantum logic got in some quarters resulted from unnecessary confusion between the calculus of empirical propositions and a genuine logic of thought processes.

The approach is not without defects, however. In its original form, quantum logic is not able to select the division ring $\mathbf{K}$ on which the $\mathbf{K}^*$-space is built. In particular, the field of real numbers or the skew-field of quaternions remain possible. To this day there is no consensus on whether quantum logic can be naturally completed to exclude division rings other than $\mathbf{C}$.[134] Another defect of quantum logic is the level of mathematics required to prove its main theorems, as should be expected when mathematicians of von Neumann's or Mackey's power get involved. The mathematical burden is alleviated by restricting the reasoning to finite lattice dimension and by substituting simpler proofs to the original ones,

---

[133] On later developments of quantum logic and on improvements of its operational grounding, cf. Coecke, Moore, and Wilce 2000; Gabbay, Lehmann, and Engesser 2009; Wilce 2009.

[134] See the discussions in Dalla Chiara, Giuntini, and Rédei 2007, pp. 228–31; Mittelstaedt 2011, p. 64. In 1995 Maria Pia Solèr proved that the choice of $\mathbf{K}$ is restricted to $\mathbf{R}$, $\mathbf{C}$, or $\mathbf{H}$ (quaternions) if and only if there exists an infinite orthonormal sequence of vectors in it. Physical reasons remain to be found for this condition (although it admits a lattice-theoretic expression) and for the exclusion of $\mathbf{R}$ and $\mathbf{H}$.

as I have tried to do in this presentation. Even so, quantum logic requires lattice-theoretical and projective-geometrical notions unfamiliar to most physicists.[135]

8.4 Discreteness, probabilities, and information

Quantum logic was not the only attempt to base quantum mechanics on natural axioms. We already mentioned Mackey's attempt, in which probabilistic axioms play an important role. Among other old axiomatics, most noticeable is Günther Ludwig's, which took off in the mid-1950s and reached its mature form in the mid-1980s. Although Ludwig's declared aim was to base quantum mechanics on "physically interpretable axioms," the demands of mathematical rigor and completeness led him to overabundant formalism. In its final form, his theory has not less than seventy-six axioms, most of which are there only for mathematical reasons. Ludwig starts with formal characterizations of preparation and registration procedures, and defines states (ensembles) and observables (effects) through the statistics of these procedures. After imbedding ensembles and observables in Banach spaces, he ends up deriving a lattice of propositions and using Piron's representation theorem in order to reach the Hilbert-space structure of quantum mechanics.[136]

While Mackey and Ludwig shifted the foundational basis from quantum logic to the structure of a probabilistic state space, they failed to improve on the deductive, rational economy of quantum logic. Indeed they both used quantum logic as an important (though not primitive) bridge between their axioms and quantum mechanics. The real turning point in natural quantum axiomatics was a memoir published in 2001 by the British theoretical physicist Lucien Hardy. As we will now see, Hardy changed the axiomatic game by short-circuiting the representation theorems of quantum logic and by instead deriving quantum mechanics "from five reasonable axioms" about probabilistic state space. In his approach, statistical correlation between discrete measurements is the most basic notion. The states of a system are defined through measurement probability distributions, which may be seen as the expression of information content.[137]

Although the mathematics used by Hardy and his followers tend to be simpler than those of quantum logic, they still involve notions unfamiliar to most physicists. For this reason, I begin with an intuitive introduction to some of the main ideas in the context of the simplest known quantum system: a particle with spin one-half (other degrees of freedom being abstracted away). Then I offer a mathematically light treatment of the general case (with a finite number of discrete outcomes for every measurement), drawing on considerations by Hardy and by Claude Comte. The following sections are devoted to Hardy's theory per se, and to two significant improvements by Borivoje Dakić and Časlav Brukner and by Lluís

---

[135] Still another defect of quantum logic is the difficulty of representing compound systems in a pure lattice-theoretic framework: there is no simple lattice-theoretic counterpart to the tensor product of quantum states. Cf. Wilce 2012, §7, and reference there to the no-go theorems by David Foulis and Charles Randall and by Diederik Aerts.

[136] Ludwig 1983; 1985, pp. 1 (citation), 241 (axioms).

[137] Other more recent attempts include work by Giacomo Mauro D'Ariano and collaborators (see below, p. 329) and Philip Goyal's information-theoretic axioms and his derivation of complex probability amplitudes through the symmetries of a generalized probability theory (see his website).

Masanes and Markus Müller. The last section briefly addresses the axiomatics of Giacomo Mauro D'Ariano, Giulio Chiribella, and Paolo Perinotti, which differs from the former ones by being based exclusively on information-theoretic notions.

The one-half spin system

There is a continuous infinity of possible measurements of this system, giving the angular momentum in any direction of space. In contrast, there are only two possible outcomes for each of these measurements: $+\hbar/2$ and $-\hbar/2$. If the system is found to have the angular momentum $+\hbar/2$ in a given direction, a subsequent measurement performed in a direction making an angle θ with the former direction will give either $+\hbar/2$ and $-\hbar/2$. Let us repeat the same preparation and the same measurement a great number of times. If p_+ and p_- denote the frequencies of the two possible outcomes, we must have

$$p_+ - p_- = \cos\theta \tag{232}$$

in order that the average angular momentum in the direction θ be equal to the projection of the initial angular momentum on this direction. Indeed by a correspondence argument we expect the total angular momentum (or magnetic moment) of a large number of spin-particles to behave as the angular momentum of a macroscopic object. Since $p_+ + p_- = 1$, we have[138]

$$p_+ = \cos^2(\theta/2), \; p_- = \sin^2(\theta/2). \tag{233}$$

To sum up, the double-valuedness of spin, the spatial character of spin measurement, and a correspondence argument together imply the well-known quantum-mechanical expression for the correlations between spin measurements in two different directions. In a different notation, the correlation probability for $+\hbar/2$ spin components in the directions (unit vectors) $\mathbf{u}$ and $\mathbf{u}'$ is

$$p(\mathbf{u}, \mathbf{u}') = \frac{1 + \mathbf{u} \cdot \mathbf{u}'}{2}. \tag{234}$$

In polar coordinates for which $\mathbf{u} = (\cos\theta, \sin\theta \cos\varphi, \sin\theta \sin\varphi)$, we have

$$
\begin{aligned}
p(\mathbf{u}, \mathbf{u}') &= \tfrac{1}{2}[1 + \cos\theta \cos\theta' + \sin\theta \sin\theta' \cos(\varphi - \varphi')] \\
&= \left| \cos(\theta/2)\cos(\theta'/2) + \sin(\theta/2)\sin(\theta'/2)e^{i(\varphi-\varphi')} \right|^2
\end{aligned}
\tag{235}
$$

Introducing a bidimensional Hilbert space with two orthogonal state vectors $|+\rangle$ and $|-\rangle$ corresponding to the spins $+\hbar/2$ and $-\hbar/2$ in the polar direction, the vectors $|+_{\mathbf{u}}\rangle = \cos(\theta/2)|+\rangle + e^{i\varphi}\sin(\theta/2)|-\rangle$ and $|+_{\mathbf{u}'}\rangle = \cos(\theta'/2)|+\rangle + e^{i\varphi'}\sin(\theta'/2)|-\rangle$ are such that

$$p(\mathbf{u}, \mathbf{u}') = |\langle +_{\mathbf{u}} \mid +_{\mathbf{u}'}\rangle|^2. \tag{236}$$

---

[138] A similar reasoning is found in Comte 1996, although Comte uses "homogeneity" (see below p. 312) instead of correspondence.

We thus see that the full quantum-kinematics of a two-level system derives from a very simple combination of discreteness, symmetry, and correspondence.

Now consider a system of two particles with spin one-half, prepared so that its total angular momentum vanishes. Suppose that the spin of the first particle is found to be $+\hbar/2$ in a given direction. Then the spin of the second particle must be $-\hbar/2$ in the same direction by conservation. By a similar correspondence argument, the probabilities $P_{\varepsilon\varepsilon'}(\theta)$ of finding $\varepsilon\,\hbar/2$ and $\varepsilon'\hbar/2$ (with ε, $\varepsilon' = -1$, 1) for the spins of the two particles in two directions making the angle θ must verify

$$P_{++} - P_{+-} = -\frac{1}{2}\cos\theta. \tag{237}$$

Taking into account the normalization

$$P_{++} + P_{+-} = P_{-+} + P_{--} = \frac{1}{2}, \tag{238}$$

we get

$$P_{++}(\theta) = \frac{1}{2}\sin^2(\theta/2) \text{ and } P_{+-}(\theta) = \frac{1}{2}\cos^2(\theta/2). \tag{239}$$

Again, this is the result given by quantum mechanics, with

$$|0\rangle = \frac{1}{\sqrt{2}}\,(|+\rangle\,|-\rangle - |-\rangle\,|+\rangle) \tag{240}$$

for the prepared state, and

$$\langle+|\,\langle+_\theta| = \langle+| \otimes \big(\cos(\theta/2)\,\langle+| + \sin(\theta/2)\,\langle-|\big) \tag{241}$$

for the projecting measurement state in the ++ case. As is well known, these correlations violate the Bell inequalities. Thus, a most astonishing consequence of quantum mechanics, the violation of EPR locality, can be derived from a simple combination of discreteness, conservation, and correspondence.[139]

These two arguments cannot really pass for a rational derivation of quantum mechanical laws, for they involve two empirical facts: the existence of a two-level system for which possible measurements are mapped by unit vectors in geometrical space, and the existence of combined spin states for which the total angular momentum vanishes. They nonetheless suggest some sort of necessity of the quantum mechanics of two-level systems.

N-level systems

Let us now consider an arbitrary N-level system, for which any given maximal measurement performed on the system yields N distinct discrete outcomes. We assume that every such measurement is stable by repetition, so that it can be used as a preparation of the

---

[139] I remember hearing this argument or a similar one from Claude Comte in the early 1990s.

system. A system prepared in this manner is said to be in a pure state. Let the system be continuously transformed by some interaction. The outcome of a second measurement performed on this system is necessarily stochastic, because the system cannot jump from one discrete value to another during a continuous transformation. By a correspondence argument, *we expect the repetition of the measurement on a great number of identically prepared systems to yield a well-defined probability for each possible outcome of the measurement* (axiom A_1). For transformations by internal interaction or by application of an external classical field we should expect the system to be in a pure state before a second interaction, because there is no information loss in this process. For transformations by interaction with another system whose final state is undetermined, there is a loss of information so that the first system can only be in a mixture of pure states.

The most general mixture involves every discrete outcome A_n of every possible measurement A, with the statistical weight $\alpha_n(A)$. The corresponding state S is empirically characterized by the probabilities for the various outcomes B_m of every possible measurement B:

$$P(\mathrm{S}, B_m) = \sum_{\mathrm{A},n} \alpha_n(\mathrm{A}) P(A_n, B_m), \tag{242}$$

in which $P(A_n, B_m)$ denotes the probability of the outcome B_m of the measurement B if the outcome of the earlier measurement A is A_n. The sum over A involves continuous indices. Unless the functions $P(A_n, B_m)$ are chosen in a special way, the number of such combinations is infinite and an infinite number of choices of the measurement B is necessary in order to determine the state of the system. This is excluded by the correspondence condition that a macro-system made of a large number of copies of the system should have a finite number of effective degrees of freedoms: the value of any associated macro-quantity should be a function of the value of a finite number of fixed macro-quantities. Take the example of spin one-half. For a macro-system made of a large number of identically prepared spins, the average spin in the direction **u** should be a function of the average spins along three orthogonal axes:

$$\langle \mathbf{S} \cdot \mathbf{u} \rangle = u_x \langle S_x \rangle + u_y \langle S_y \rangle + u_z \langle S_z \rangle . \tag{243}$$

By correspondence, these ensemble averages should be identical with averages calculated from the probabilities that an individual system be found with the spin $\hbar/2$ in the directions **u**, Ox, Oy, and Oz respectively. Therefore, only three probability measurements are necessary to determine a spin state.

In conformity with this correspondence argument, we will assume that *a finite number of probability measurements is always sufficient to determine the state of the system, despite the infinity of possible measurements* (axiom A_2). In the sequel, the minimal number of required measurements is called K.[140] This means that a state can be represented by the sequence of probabilities $p = (p_1, p_2, ..., p_r, ..., p_K)$ of a fixed set of K determinations, a determination being a measurement performed with a given result (for instance, in the spin case the determinations could be $+\hbar/2$ for S_x, $+\hbar/2$ for S_y, $+\hbar/2$ for S_z). The choice of the

---

[140] This K is Hardy's K minus one.

reference set of determinations is of course not unique. A specific choice will be called a *K-determination*. The state represented by the vector p is mixed if and only if there exist two constants λ and μ and two states p' and p'' such that $0 < \lambda < 1, 0 < \mu < 1, \lambda + \mu = 1$, and $p = \lambda p' + \mu p''$. Since every mixture is allowed, the space of states is a convex, bounded subset of $\mathbf{R}^K$(for a given choice of the set of determinations). Since pure states cannot be mixtures, they necessarily belong to the boundary of this set. Consequently, the subset of pure states is a manifold of dimension inferior or equal to $K - 1$.[141]

We now introduce the additional requirement that *any two pure states can be connected by a continuous reversible transformation within the subset of pure states* (axiom A_3). This requirement is similar to Bohr's old principle of mechanical transformability: it is based on the intuition that a continuously varying action on the system, such as a varying impressed field, cannot induce quantum jumps; and on the intuition that the measurability of a quantity requires its continuous variability. With this complement, the previous assumption has an important consequence: *a mixture of $K + 1$ pure states can always be reduced to a mixture of K pure states (with varying choices of the latter states)*(condition A_2'). This can be proven as follows. By definition, a given state S is a $(K + 1)$-mixture if and only if it belongs to the interior of the polyhedron defined by $K + 1$ pure states $S_1, S_2, ..., S_{K+1}$. By axiom A_3, there is a continuous line of pure states joining S_1 to S_2. Let the state S_1 travel on this line from its original position to the position S_2(the other states being unchanged). At the beginning of this process, the state S remains within the deformed polyhedron (since the dimension of the space of states is K), so that it still is a $(K + 1)$-mixture. Then there are two possibilities: either the state S remains within the deformed polyhedron during the whole process, or there is a stage at which the state S crosses one of the faces of the polyhedron. In the former case, the state S is a K-mixture at the end of the transformation. In the latter case, it is a K-mixture at the crossing stage of the transformation. This ends the proof of the desired result. By repetition of a similar reasoning, every state can be represented as a mixture of K pure states.

A stronger assumption would require *every state to be a mixture of the N eigenstates[142] of a single, properly chosen measurement* (axiom A_2''). For instance, in the spin case the most general mixture yields

$$p_+(\mathbf{u}') = \sum_{\mathbf{u}} \alpha(\mathbf{u})p(\mathbf{u}, \mathbf{u}') = \frac{1}{2} + \frac{1}{2}\mathbf{u}' \cdot \sum_{\mathbf{u}} \alpha(\mathbf{u})\mathbf{u} \qquad (244)$$

for the probability of $+\hbar/2$ being measured in the direction $\mathbf{u}'$. Setting

$$a = \left| \sum_{\mathbf{u}} \alpha(\mathbf{u})\mathbf{u} \right|, \quad \mathbf{u}_0 = a^{-1} \sum_{\mathbf{u}} \alpha(\mathbf{u})\mathbf{u}, \quad \text{and } a_{\pm} = \frac{1 \pm a}{2}, \qquad (245)$$

---

[141] The dimension of the pure-state manifold is not necessarily $K - 1$ because the boundary of the convex state domain may include rectilinear segments within which each state is a mixture of the extremities of the segment. In the case of quantum mechanics, this dimension is $2N - 2$ (rays of a N-dimensional Hilbert space); and $K = N^2 - 1$ (number of real coefficients of a Hermitian matrix of trace 1 in this Hilbert space); so that the value $K - 1$ occurs only in the case $N = 2$.

[142] An *eigenstate* of a (maximal) measurement is a state prepared by this measurement.

we have

$$p_+(\mathbf{u}') = a_+ p_+(\mathbf{u}_0, \mathbf{u}') + a_- p_-(\mathbf{u}_0, \mathbf{u}'), \tag{246}$$

in conformity with the stronger requirement. In the 1990s, Claude Comte and Daniel Fivel introduced this requirement as the defining characteristic of quantum-mechanical states. Comte named it the "principle of homogeneity of statistical ensembles" and used it to derive the form of the quantum-mechanical density matrices in the case of spins of any value. Fivel made it the nerve of an impressive derivation of quantum-mechanical transition probabilities on fairly natural operational-probabilistic grounds. We will see that later quantum axiomatics appeal to this principle or to its weakened form A_2, at least implicitly. In the following I adopt the weakened form, which is easier to justify by a priori means.[143]

In the notation of Eq. (242), a K-mixture S is defined by the existence of a sequence S_r of K pure states (obtained by selecting a given outcome of a given ideal measurement) such that

$$P(S, A_n) = \sum_{r=1}^{K} \alpha_r P(S_r, A_n) \tag{247}$$

for any choice of the measurement A and of the outcome index n. In the two-level case ($N = 2$), the two outcomes of any given measurement are complementary and it is sufficient to consider only one of them. We will now investigate the lowest values of the characteristic number K in this case.[144] The value $K = 1$ is excluded because it would imply the discreteness of the subset of pure states, in contradiction with axiom A_3. It would also imply that the probability of a single determination is sufficient to determine the state of the system, in conformity with the classical probability theory of coin flipping.

For $K = 2$, the dimension of the pure-state manifold must be 1. By axiom A_3, there is a continuous one-parameter group of transformations acting transitively on this manifold. As is well known, any such group is isomorphic to the additive group of real numbers. Call θ an additive parameter, and label the pure states with this parameter. The probability of finding the system in the state θ' when it has been prepared in the state θ is necessarily of the form

$$P(\theta, \theta') = f(\theta - \theta'), \tag{248}$$

because the transformation $T_{\theta'}$ can be combined with the determinations M_θ and $M_{\theta'}$ to yield the determinations $T_{\theta'} M_\theta T_{\theta'}^{-1} = M_{\theta - \theta'}$ and $T_{\theta'} M_{\theta'} T_{\theta'}^{-1} = M_0$. The condition A_2' then requires that for any triplet $\theta_1, \theta_2, \theta_3$ of pure states and for any value of the mixing weights $\alpha_1, \alpha_2, \alpha_3$ there exists a pair θ_1', θ_2' of pure states and mixing weights α_1', α_2' such that for any state θ,

$$\alpha_1 f(\theta - \theta_1) + \alpha_2 f(\theta - \theta_2) + \alpha_3 f(\theta - \theta_3) = \alpha_1' f(\theta - \theta_1') + \alpha_2' f(\theta - \theta_2'). \tag{249}$$

---

[143] Comte 1996; Fivel 1994.

[144] This investigation lacks rigor and generality. Better ones will be given in subsequent sections.

The general solution of this finite-difference equation is a linear combination of solutions of the form $f(\theta) = e^{ik\theta}$. For such solutions, the condition

$$\alpha_1 e^{ik\theta_1} + \alpha_2 e^{ik\theta_2} + \alpha_3 e^{ik\theta_3} = \alpha_1' e^{ik\theta_1'} + \alpha_2' e^{ik\theta_2'} \tag{250}$$

must hold. Since $\alpha_1 + \alpha_2 + \alpha_3 = \alpha_1' + \alpha_2' = 1$, a possible solution obtains by taking $k = 0$ with free choices of θ_1', θ_2', α_1', α_2'. For a non-zero value of k, there are three unknowns (for instance θ_1', θ_2', α_1') and two real equations (the real and imaginary parts of Eq. (250)). The problem is possible, and for the same value of the unknowns the choice $-k$ is also possible since the α coefficients are real. For two distinct and non-opposite values of k, there are four real equations and three unknowns: the problem becomes impossible (except for accidental choices of the parameters for which the equations are not independent). Therefore, the most general solution has the form

$$f(\theta) = a + b e^{ik\theta} + c e^{-ik\theta}. \tag{251}$$

Because f must be real and because the additive parameter of the transformation group can be redefined so that $k = 1$, this formula reduces to

$$f(\theta) = a + 2b\cos\theta, \tag{252}$$

with two real coefficients a and b. The constraints $f(0) = 1$, $0 \le f(\theta) \le 1$ and the existence of a value of θ for which $f(\theta) = 0$ imply that $a = 1/2$ and $b = 1/4$. The end result is

$$f(\theta) = \cos^2\theta/2. \tag{253}$$

Consequently, the pure states may be represented in a two-dimensional Euclidean vector space as

$$|\theta\rangle = \cos(\theta/2)|+\rangle + \sin(\theta/2)|-\rangle, \tag{254}$$

with the transition probability

$$P(\theta, \theta') = |\langle\theta'|\theta\rangle|^2. \tag{255}$$

This is the so-called real-Hilbert-space quantum mechanics of two-level systems.

The next simple choice is $K = 3$. In this case, the pure-state manifold is at most bidimensional since pure states belong to the boundary of a convex bounded domain of $\mathbf{R}^3$. The one-dimensional case is excluded, because a convex domain cannot be formed by patching portions of ruled surfaces (whose interior points cannot be extreme) in such a manner that all extreme points belong to the same connected curve (as required by axiom A_3). In the two-dimensional case, the associated transformation group has one-parameter subgroups that leave a given pure state invariant. This group therefore has the same properties as the group of motions of a rigid body around a fixed point in Helmholtz's theory space.[145] By

---

[145] See Chapter 4, pp. 118–9, 126–9.

the Helmholtz-Lie argument, this group is isomorphic to the SO(3) group of rotations in $\mathbf{R}^3$. With a proper choice of axes (that is, for a proper K-determination), the set of pure states becomes a unit sphere. The subset of states belonging to a disk passing through the origin of the sphere can be treated as in the $K = 2$ case. Since any two pure states $\mathbf{u}$ and $\mathbf{u}'$ on the unit sphere belong to such a disk, the probability $P(\mathbf{u}, \mathbf{u}')$ of finding a system in the state $\mathbf{u}$ when it has been prepared in the state $\mathbf{u}'$ is

$$P(\mathbf{u}, \mathbf{u}') = \frac{1 + \mathbf{u} \cdot \mathbf{u}'}{2}, \tag{256}$$

just as in the quantum mechanics of a two-level system. As we have seen in the case of a one-half spin particle, this probability law implicitly contains the Hilbert-space formalism of quantum mechanics in two (Hilbert-space) dimensions.

In the limited context of two-level systems, the choices $K = 2$ and $K \geq 4$ are permitted. In order to exclude these cases, consider systems of every possible (finite) N and their associated freedom $K(N)$. We may physically combine a N'-level and a N''-level system and investigate the statistical behavior of the combination. It seems reasonable to assume that *this behavior is entirely determined by the statistics of measurements performed on the two components of the combined system* (axiom A_4). In symbols, this means that the state of the combined system is described by the K' probabilities $P(\mathrm{M}'_1), ..., P(\mathrm{M}'_{K'})$ of the determinations $\mathrm{M}_1', ..., \mathrm{M}'_{K'}$ of the first component system, the K'' probabilities $P(\mathrm{M}''_1), ..., P(\mathrm{M}''_{K''})$ of the determinations $\mathrm{M}''_1, ..., \mathrm{M}''_{K''}$ of the second component system, and the $K'K''$ correlation probabilities $P(\mathrm{M}'_r, \mathrm{M}''_s)$ with $1 \leq r \leq K'$ and $1 \leq s \leq K''$. Consequently, the combined system is described by $K' + K'' + K'K''$ measurements. Evidently, the number of levels (distinct measurement outcomes for a given maximal measurement) is the product $N'N''$. Altogether, we have

$$K(N'N'') + 1 = [K(N') + 1][K(N'') + 1] \tag{257}$$

for any two integers N' and N''. As is easily seen, this condition can only be met if

$$K(N) + 1 = N^\kappa, \tag{258}$$

wherein κ is an integer. This implies that for $N = 2$, all even values of K are excluded. Still another argument is needed to exclude odd values of K higher than three.[146]

For the moment, let us assume[147] that *nature has chosen the smallest value of K compatible with the former axioms*, namely $K = N^2 - 1$ (axiom A_5). Evidently, N-level quantum mechanics is compatible with this choice: its most general states are represented by density matrices which are positive (therefore Hermitian) operators of trace *one* in a N-dimensional Hilbert space. Such matrices have $N - 1$ independent real elements on the diagonal, and $N(N - 1)/2$ complex conjugate pairs of elements outside the diagonal, which makes a total of $N - 1 + N(N - 1) = N^2 - 1$ real parameters. We will now examine whether quantum

---

[146] As we will see in a moment, the derivation of Eq. (258) belongs to Hardy.

[147] The assumption is Hardy's.

mechanics is the only theory compatible with the former axioms and with the choice $K = N^2 - 1$.

By axiom A_2, the state S of the system is determined by the sequence of probabilities

$$P(S, M_r) = p_r \tag{259}$$

for the determinations $M_1,...,M_r,...,M_K$. For $K = N^2 - 1$, there exist K unit vectors $|r\rangle$ in a N-dimensional Hilbert space and a Hermitian matrix ρ such that

$$\langle r| \rho |r \rangle = p_r \text{ for } 1 \leq r \leq K. \tag{260}$$

Indeed the $|r\rangle$ vectors can be chosen so that real-coefficient linear combinations of the projectors $|r\rangle \langle r|$ span the space of Hermitian matrices, in which case $\langle r| \rho |r\rangle = \text{Tr}\rho |r\rangle \langle r|$ is the r covariant coordinate of the operator ρ in the basis of these projectors with respect to the scalar product $(\mathbf{A}, \mathbf{B}) \rightarrow \text{Tr}\mathbf{A}\mathbf{B}^+$ for any two Hermitian matrices $\mathbf{A}$ and $\mathbf{B}$.

For an arbitrary measurement M, the probability $P(S, M)$ must be an affine function of the vector p that represents the state S, because the probabilities corresponding to mixtures of two states are the weighted sums of the probabilities corresponding to the individual states. Consequently, there exists a K-vector q and a constant C such that

$$P(S, M) = q \cdot p + C = \sum_r q_r p_r + C. \tag{261}$$

As the constant C can be absorbed in a redefinition of q,[148] this probability can still be expressed by means of the matrix ρ:

$$P(S, M) = \text{Tr}\rho\mathbf{M}, \text{ with } \mathbf{M} = \sum_{r=1}^{K} q_r |r\rangle \langle r|. \tag{262}$$

In plain words, the determination M_r is represented by the projector $|r\rangle \langle r|$ and an arbitrary determination is represented by a linear combination of such projectors.

This does not prove yet that our theory is equivalent to quantum mechanics, because this equivalence further requires the following properties:

1. the matrix ρ is positive and has trace *one*,
2. there is a one-to-one correspondence between ρ matrices and states S,
3. pure states or pure measurements are in a one-to-one correspondence with rays in the N-dimensional Hilbert space.

In order to establish these properties, we will need to know that *any two-level subspace of the space of states of the system behaves as a two-level system* (axiom A_6). This condition may be partially justified as follows.[149]

---

[148] Defining $q_r^{(s)}$ so that $P(S, M_s) = p_s = \sum_r q_r^{(s)} p_r + C$, we have $P(S, M) = \sum_r q_r' p_r$ with $q_r' = q_r + \delta_{rs} - q_r^{(s)}$.

[149] Again, this axiom belongs to Hardy.

Consider a maximal measurement A with the outcomes $A_1, ..., A_N$. The states S for which

$$P(S, A_n) = 0 \text{ for } 3 \le n \le N \qquad (263)$$

define a two-level subspace. Call T a reversible transformation that leaves this subspace invariant. To the determinations M_1 and M_2 corresponding to the outcomes A_1 and A_2 of the measurement A, we may associate the new determinations TM_1T^{-1} and TM_2T^{-1}. The pairs of determinations obtainable in this manner define a new two-level system whose states belong to the defined subspace (since we generally defined a system by its measurements and since every pair of conjugate determinations defines a measurement). Axiom A_1 is trivially satisfied, since the probability $P(S, M)$ for a state S and a determination M belonging to the subspace is the restriction of the original correlation function to the subspace. Axiom A_2 also holds, since the dimension of a subspace of a finite-dimensional space is necessarily finite. By axiom A_3 applied to the original system, we know there is a continuous sequence of transformations connecting any pair of pure states of the subspace. However, the intermediate pure states do not necessarily belong to the two-level subspace. We will nevertheless assume that axiom A_3 is also satisfied, so that the two-level subspace behaves as a genuine two-level system. This system is empirically realizable if the two-level transformations T can be physically realized. Quantum physicists all know that this is the case: when, for instance, an atom in a given stationary state interacts with monochromatic radiation tuned to the frequency of a possible transition to another state, the associated transformation is very nearly a two-level transformation (Rabi cycles).

We are now equipped to prove the desired properties of the ρ matrix representation of the state of a N-level system. We first choose the N first determinations M_r so that they correspond to the different outcomes $A_1, ..., A_N$ of the same maximal measurement A. In addition, we choose the $|r\rangle$ vectors so that the N first projectors $|r\rangle\langle r|$ represent the pure states S_r defined by the determinations M_r. This is possible because the formal expression of this requirement is

$$|\langle r \mid s \rangle|^2 = P(S_r, M_s) \text{ for } 1 \le r \le N \text{ and } 1 \le s \le K; \qquad (264)$$

and because the number NK of such conditions is always inferior to the number $K(2N-1)$ of real parameters that define K rays or $|r\rangle\langle r|$ projectors in the N-dimensional Hilbert space. For any such choice of the N first determinations, the trace of the operator ρ is *one* since the sum of the probabilities of the outcomes $A_1, ..., A_N$ must be one.

Now consider the subspace of states S such that

$$P(S, M_r) = 0 \text{ for } 3 \le r \le N. \qquad (265)$$

By axiom A_6, this subspace should behave as a two-level system and its states should therefore be describable by density matrices in a bidimensional Hilbert space. This implies that any projector $|\varphi\rangle\langle\varphi|$ with

$$|\varphi\rangle = \lambda|1\rangle + \mu|2\rangle \text{ and } |\lambda|^2 + |\mu|^2 = 1 \qquad (266)$$

should represent a possible state S_φ of the system. This state is pure with respect to the subspace. A priori it could still be obtained by mixing states not belonging to the subspace. It is nonetheless pure in the global space because it can be obtained by a reversible transformation (the extension of the reversible subspace transformation that generates it from the state $|1\rangle\langle1|$) and because any state S' related to a pure state S by a reversible transformation T is itself a pure state (since $P(S, M) = 1$ implies that $P(S', M') = 1$ with $S' = TST^{-1}$ and $M' = TMT^{-1}$). We next consider the subspace of states S such that

$$P(S, M_r) = 0 \text{ for } 4 \leq r \leq N \text{ and } P(S, M_{\bar\varphi}) = 0, \tag{267}$$

wherein $M_{\bar\varphi}$ denotes the conjugate of the determination associated with S_φ.

By a similar argument, every projector $|\phi'\rangle\langle\phi'|$ with

$$|\varphi'\rangle = \lambda' |\varphi\rangle + \mu' |3\rangle \text{ and} |\lambda'|^2 + |\mu'|^2 = 1 \tag{268}$$

should represent a possible pure state of the system. Iterating this reasoning until we reach the vector $|N\rangle$, we arrive at the result that every unit vector of the N-dimensional Hilbert space defines a pure state of the system.

Consequently, $\langle\psi| \rho |\psi\rangle$ is positive for any (unit) vector $|\psi\rangle$, which means that the operator ρ is positive. Any positive (therefore Hermitian) operator of trace *one* represents a possible state of the system, since any such operator can be decomposed into a linear combination of its diagonalizing projectors with positive coefficients adding to one, and since this combination corresponds to a mixture of the corresponding pure states. Lastly, any pure state of the system must be represented by a $|\psi\rangle\langle\psi|$ projector, because every other kind of operator would be a mixture of such projectors. This ends the proof that the system is equivalent to an N-level quantum system.[150]

To sum up, we have considered systems defined by a set of maximal measurements that can have N distinct outcomes.[151] We defined pure states as results of maximal measurements, and arbitrary states as statistical mixtures of such states. We then investigated the probability of the various outcomes of every possible measurement for a given state of the system, and proved that this probability has the form given by N-level quantum mechanics if the following axioms hold:

A₁: The repetition of the same measurement on a great number of identically prepared systems yields a well-defined probability for each possible outcome of the measurement.

---

[150]This proof differs from Hardy's by directly introducing the $N \times N$ matrix ρ in Hilbert space instead of the $K \times K$ probability matrix in determination space.

[151]It is not necessary to assume that all maximal measurements have the same number N of outcomes. This follows from the existence of reversible transformations between any two pure states and from the invariance of the state $\hat{S}$ obtained by uniformly mixing all pure states (see below p. 322). Indeed, if M_i is the determination associated with the ith outcome of a maximal measurement (with $1 \leq i \leq N$), for any fixed determination M_0 there is a reversible transformation T_i such that $M_i = T_i M_0 T_i^{-1}$, so that $P(\hat{S}, M_i) = P(\hat{S}, T_i M_0 T_i^{-1}) = P(T_i^{-1} \hat{S} T_i, M_0) = P(\hat{S}, M_0)$ and $1 = \sum_{i=1}^{N} P(\hat{S}, M_i) = NP(\hat{S}, M_0)$. The choice of M_0 being independent of the choice of the maximal measurement, so too must be N. Cf. Chiribella, D'Ariano, and Perinotti 2011, p. 16.

A$_2$: A finite number K of probability measurements is always sufficient to determine the state of the system.

A$_3$: Any two pure states can be connected by a continuous reversible transformation within the subset of pure states.

A$_4$: The behavior of a combined system is entirely determined by the statistics of measurements performed on the components of this system.

A$_5$: The true value of K is the lowest value compatible with the former axioms.

A$_6$: Any two-level subspace of the space of states of the system behaves as a two-level system.

The proof begins with the cases $N = 2$ and $K = 2, 3$, which are treated by means of condition

A$_2'$: A mixture of $K + 1$ pure states can always be reduced to a mixture of K pure states,

which is a consequence of A$_2$ and A$_3$. This condition could have been replaced by Comte's and Fivel's stronger condition,

A$_2''$: Every state is a mixture of the N determinations of a single, properly chosen measurement,

which however does not directly derive from A$_2$ and A$_3$. The proof goes on with the demonstration that axiom A$_4$ implies that $K = N^\kappa - 1$, κ being an integer. The value $\kappa = 1$ being excluded by axiom A$_3$, nature's choice must be $\kappa = 3$ according to A$_5$. In this case and for $N = 2$, the system is equivalent to a two-level quantum system. The equivalence for higher values of N results from the choice $K = N^2 - 1$ and from axiom A$_6$.

The first three axioms can be justified by correspondence arguments. They warrant that a macro-state of a macro-system made of a large number of copies of the system is described by a finite number (A$_2$) of well-defined (A$_1$), continuously modifiable (A$_3$) macro-quantities. The fourth axiom results from the empiricist requirement that the state of a system should always be accessible by measurement and from the correspondence requirement that measurements, being ultimately expressed in terms of classical quantities, should always be analyzable into measurements performed on the components of the system. This axiom is more obvious in the case of very distant components, for which global instantaneous measurement is not conceivable. The sixth axiom is suggested by the concrete possibility of restricting the transitions of a quantum system to two levels. The fifth axiom is the least satisfying. We will see in a moment that it is not needed.

Hardy's axioms

In 2001, Lucien Hardy's posted his "Quantum theory from five reasonable axioms," which is a rationalist attempt to derive quantum mechanics as a natural sort of discrete probability theory:[152]

> Quantum theory is simply a new type of probability theory. Like classical probability theory it can be applied to a wide range of phenomena. However, the rules of classical probability theory can be determined by pure thought alone without any particular appeal to experiment (though, of course, to develop classical probability theory, we

---

[152]Hardy 2001, p. 1.

do employ some basic intuitions about the nature of the world). Is the same true of quantum theory? Put another way, could a 19th century theorist have developed quantum theory without access to the empirical data that later became available to his 20th century descendants? In this paper it will be shown that quantum theory follows from five very reasonable axioms which might well have been posited without any particular access to empirical data.

The basic ingredients of Hardy's approach are devices for preparing, transforming, and measuring a system, and states defined by the probabilities of measurement outcomes. He introduces the "dimension" N and the "number of degrees of freedom" K' of the system, which correspond to my N and my $K + 1$.[153] His axioms read:

H₁ *Probabilities.* Relative frequencies . . . tend to the same value (which we call the probability) for any case where a given measurement is performed on an ensemble of n systems prepared by some given preparation in the limit as n becomes infinite.

H₂ *Simplicity.* K' is determined by a function of N . . . , and for each given N, K' takes the minimum value consistent with the axioms.

H₃ *Subspaces.* A system whose state is constrained to belong to an M dimensional subspace . . . behaves like a system of dimension M.

H₄ *Composite systems.* A composite system consisting of subsystems A and B satisfies $N = N_A N_B$, $K' = K'_A K'_B$.

H₅ *Continuity.* There exists a continuous reversible transformation on a system between any two pure states of that system.

This list of axioms has much in common with my axioms A₁, . . . , A₅. Axiom H₁ is the same as A₁; H₂ is the same as A₅; H₄ can be derived from A₄, as Hardy himself shows;[154] H₅ is the same as A₃. However, Hardy does not regard the finiteness of K (my A₂) as an axiom; he casually introduces this property by arguing that "most physical theories have some structure which relates different measured quantities." The subspace axiom H₃, is a generalization of A₆. Hardy justifies it as follows:[155]

> This axiom is motivated by the intuition that any collection of distinguishable states should be on an equal footing with any other collection of the same number of distinguishable states. In logical terms, we can think of distinguishable states as corresponding to propositions. We expect a probability theory pertaining to M propositions to be independent of whether these propositions are a subset or some larger set or not.

Hardy regards all his axioms as "natural" from the point of view of the theory of probabilities. The axiom H₅ of continuity is the one that excludes classical probability and forces us to adopt quantum probability theory if we accept the simplicity axiom. Hardy's justification of the continuity axiom reads:

> Given the intuition that pure states represent definite states of a system we expect to be able to transform the state of a system from any pure state to any other pure state.

---

[153] Hardy's preference for K' over K comes from his including a probability for the system not to be detected by the measuring device.

[154] Hardy 2001, p. 14.   [155] Hardy 2001, pp. 10, 14.

> It should be possible to do this in a way that does not extract information about the
> state and so we expect this can be done by a reversible transformation.

Implicitly, this is a correspondence argument because the idea of a definite state that can
be transformed continuously is a classical idea. In my opinion, correspondence arguments
are the true justification of most of Hardy's axioms. Indeed the natural character of an
axiom from a probability-theory point of view or from an information-theory point of view
does not imply its natural character from a physical point of view. For instance, definite
probabilities of discrete outcomes may be natural for a probability theorist and continuous
information-preserving transformation may be natural for an information theorist, and
yet their combination leads to the quantum weirdness of superposed and entangled states.
The reason is that the simultaneous, concrete realization of these two axioms involves an
odd mixture of continuity and discontinuity, in a manner imposed by the correspondence
principle.[156]

Hardy's derivation of quantum probability theory from his axioms is more rigorous
but more difficult than the naïve derivation given in the previous section. It begins with
a thorough analysis of the structure of probability relations in the vector space of K-
determinations, which Hardy call "fiducial measurements" (he does not assume pure
measurements and pure states from the start). He exploits this structure to derive the probability relations in the case $N = 2$, $K' = 4$ (the reasoning involves more advanced group
theory than my elementary presentation). He proves the relation $K' = N^\kappa$ from axiom H_4.
Lastly, he sets $\kappa = 2$ and uses the subspace axiom H_3 to construct quantum probability
theory at any N from the $N = 2$ case. He does this in the fiducial vector space, and then
establishes the correspondence of the resulting $K' \times K'$ probability matrices with quantum-
theoretical density matrices. Lastly, Hardy shows that his axioms are compatible with
only two kinds of transformations for closed systems: reversible, probability-conserving
transformations represented by unitary transformations in the quantum-theoretical N-
dimensional Hilbert space, and irreversible transformations associated with measurements
and represented by projectors in Hilbert space.

Dakić and Brukner

The clarity and elegance of Hardy's derivation and the simplicity of his axioms have attracted much legitimate attention. An evident defect of this derivation is the artificial character
of the simplicity axiom H_2. Hardy himself wondered about the possibility of more complicated theories involving higher exponents κ in the relation $K = N^\kappa - 1$.[157] In 2009, the
Vienna-based theorists Borivoje Dakić and Časlav Brukner answered this question negatively by showing that K could not exceed three in the two-level case $N = 2$. They relied on
a different system of axioms:[158]

> D_1 (Information capacity): An elementary system has the information carrying capa-
> city of at most one bit. All systems of the same information carrying capacity are
> equivalent.

---

[156] Hardy 2001, p. 15.   [157] Hardy 2001, p. 13.   [158] Dakić and Brukner 2009.

D_2 (Locality): The state of a composite system is completely determined by local measurements on its subsystems and their correlations.

D_3 (Reversibility): Between any two pure states there exists a reversible transformation.

The axiom D_2 is the same as axiom A_4 and it is directly related to Hardy's axiom H_4. The axiom D_3 is a much weakened form of axioms A_3 or H_5, for it does not assume the existence of a continuous sequence of transformations gradually bringing the first pure state to coincide with the second. This axiom warrants that a compact group acts transitively on the space of pure states. Its weakness is compensated by the strength of axiom D_1, which in fact contains two subaxioms:

D_1': An elementary system has the information carrying capacity of at most one bit.

D_1'': All systems of the same information carrying capacity are equivalent.

Since the information carrying capacity is nothing but the number N of distinct outcomes of a maximal measurement, axiom D_1'' is an information-theoretic rephrasing of Hardy's subspace axiom H_3.[159] Dakić and Brukner translate their information-theoretic axiom D_1' into "any state of a two dimensional system can be prepared by mixing at most two basis (i.e. perfectly distinguishable in a measurement) states." Here we recognize the Comte-Fivel principle A_2'' in the case $N = 2$. Apparently unaware of Comte's and Fivel's works, Dakić and Brukner borrowed axiom D_1' from their Viennese colleague Anton Zeilinger.

In a celebration of Daniel Greenberger's sixty-fifth birthday, Zeilinger suggested to base quantum theory on the principle that *"An elementary system represents the truth value of one proposition"* or, equivalently, that *"An elementary system carries 1 bit of information."* From this principle he derived randomness and entanglement:

> An elementary system can only give a definite result in one specific measurement. The irreducible randomness in other measurements is then a necessary consequence. For composite systems entanglement results if all possible information is exhausted in specifying joint properties of the constituents.

For instance, there can be only one direction of spin measurement for which the spin of a particle can have a definite value because the spin state can only contain the reply to a single Yes–No question. In any other direction, the result of the measurement must be random. For a composite system of two spins, the joint property that the two particles have the same spin in one direction and the other joint property that the two particles have the same spin in another direction exhaust all possible information since the global system has two bits. The answer to other questions, for instance about the spin of one of the particles in a given direction, should be random: this is the signature of an entangled state.[160]

Seduced by this reasoning, Dakić and Brukner turned Zeilinger's information-theoretic principle into the most potent axiom of their theory. Their first remarkable result is that the axioms D_1' and D_3 are sufficient to determine the probability theory for $N = 2$ and for any given value of K. Here is the proof.[161]

[159]The authors credit Grinbaum 2007 for the rephrasing.   [160]Zeilinger 1999, pp. 635, 631.

[161]Dakić and Brukner 2009, pp. 4–6.

By axiom D_3 there exists a group of transformations acting transitively on the states. Consequently, we may construct a totally mixed state $\hat{S}$, such that $P(\hat{S}, M')$ takes the same constant value for any choice of M'. If the group of transformations is a discrete group with ν elements, this is achieved by taking

$$\hat{p} = \frac{1}{\nu} \sum_{i=1}^{\nu} G_i p^{(0)}, \tag{269}$$

where $p^{(0)}$ represents any pure state and G_i represents the action of the ith element of the group (remember that a state S is characterized by the vector $p = (p_1, p_2, ..., p_K)$ defined by the probabilities $p_r = P(S, M_r)$ of the determinations $M_1, M_2, ..., M_K$). If the group of transformations is not discrete, it still is a compact group (because probabilities must be inferior to one), and the discrete sum can be replaced by integration with respect to the Haar measure of this group. For $N = 2$, every determination M has a conjugate determination $\overline{M}$ (corresponding to the other possible value of the associated measurement) such that $P(S', M) + P(S', \overline{M}) = 1$ for any S'. Hence $P(\hat{S}, M) = 1/2$ for any M, and $\hat{p}_r = 1/2$ for any r. Let us define the K-vector x such that $x_r = 2p_r - 1$. Since all transformations preserve $x = 0$ and since they also respect mixtures, they are represented by $K \times K$ matrices.

In the discrete case, we may now form the symmetric matrix

$$Q = \frac{1}{\nu} \sum_{i=1}^{\nu} Z_i^T Z_i, \tag{270}$$

wherein Z_i is the matrix associated with the ith element of the group. The Haar measure allows a similar construct in the continuous compact case. By construction, the matrix Q enjoys

$$Z^T Q Z = Q \text{ for any } Z. \tag{271}$$

The Z matrices being invertible, the matrix Q is strictly positive. Therefore, there exists a symmetric invertible matrix R such that $Q = R^2$. For any transformation Z, the matrix RZR^{-1} is orthogonal because

$$(RZR^{-1})^T(RZR^{-1}) = R^{-1}Z^T R^2 Z R^{-1} = R^{-1}R^2 R^{-1} = \mathbf{1}. \tag{272}$$

We may therefore introduce a new basis in K-space in which the transformations are represented by orthogonal matrices. In this basis, pure states all have the same norm since by axiom D_3 they are all interrelated by transformations. We may further adjust the scale of the basis so that they all belong to the unit sphere S^{K-1}.

For the K-vectors X and $\bar{X}$ of two conjugate pure states, we have $\bar{p}_r = 1 - p_r$ so that $\bar{X} = -X$. According to axiom D_1', the x vector of any state can be put under the form

$$x = \alpha X + \beta \bar{X} = (\alpha - \beta)X. \tag{273}$$

The x vectors can have any possible direction in $\mathbf{R}^K$ because the domain of possible states includes the K-dimensional polyhedron engendered by mixing the S_r states. So too do the associated X vectors. Consequently, every point of the sphere S^{K-1} corresponds to a possible (pure) state of the system.

The probability $P(S, M')$ of the determination M' in the state S must be an affine function of the x vector of the state S since the probability of a mixture of states is the weighted average of the probabilities of these states. In addition, this probability must take the value $1/2$ for the totally mixed state $x = 0$. Therefore, there exists a constant vector a' such that

$$P(S, M') = \frac{1}{2}(1 + a' \cdot x). \tag{274}$$

Call a the vector associated with the determination M that prepares the pure state S of vector X. On the one hand, $P(S, M) = 1$ implies that $a \cdot X = 1$. On the other hand, the probability of the pure state defined by $X' = a/\|a\|$ cannot exceeds *one*, so that $a \cdot X' = \|a\| \leq 1$. These two conditions and $\|X\| = 1$ together imply that $a = X$. For the state S of vector x and for the determination M' that prepares the state S' of vector x', we have

$$P(S, M') = \frac{1}{2}(1 + x \cdot x'). \tag{275}$$

From this formula, it is clear that the basis vector $x^{(r)}$ of components $x_s^{(r)} = \delta_{rs}$ represents both a pure state and a determination. Therefore, in the case $K = 3$ the probability function agrees with the transition probabilities (234) in the quantum mechanics of a two-level system.

The possible transformations of the system must form a transitive subgroup of the orthogonal group $O(K)$. For $K = 2$ and for $K = 3$, it is easy to see that the only such subgroups are the group of rotations or the full orthogonal group. Unfortunately, the cases of higher dimension require more advanced Lie-Group theory.[162] For every odd value of K except 7, it can be proven that any group acting transitively on the sphere of dimension $K - 1$ must contain the group of rotations $SO(K)$. This is not true for even values of K, but those are excluded by axiom D_2. In the rest of this section, we will assume that in any physically relevant case, the group of transformations contains the group of rotations (that is, any orthogonal matrix of determinant *one*).

There remains to be proved that $K = 1$ (classical probability theory) and $K = 3$ (quantum mechanics) are the only two choices compatible with the axioms. By Hardy's argument on product spaces, we already know that even values of K are excluded. Dakić and Brukner further prove that their axioms exclude any value of K higher than 3 by ingenious reasoning of the following kind.

According to axiom D_2, the system obtained by compounding the two-level systems I and II are characterized by the probabilities p_r^I of the determinations for subsystem I, the probabilities p_r^{II} for the determinations of subsystem II, and the joint probabilities p_{rs} for the determinations r and s of the two subsystems. For mathematical convenience, let us introduce the Bloch vectors

---

[162]Cf. Masanes and Müller 2010, p. 8.

$$x_r = 2p_r^{\mathrm{I}} - 1, y_r = 2p_r^{\mathrm{II}} - 1, \tag{276}$$

and the correlation matrix

$$C_{rs} = 4p_{rs} - 2p_r^{\mathrm{I}} - 2p_s^{\mathrm{II}} + 1 \tag{277}$$

that reduces to the product $x_r y_s$ in the uncorrelated case for which $p_{rs} = p_r^{\mathrm{I}} p_s^{\mathrm{II}}$. Note that

$$-1 \leq C_{rs} \leq 1 \tag{278}$$

because C_{rs} may be rewritten as

$$C_{rs} = 1 - 2(p_{rs}^{+-} + p_{rs}^{-+}) \text{ with } 0 \leq p_{rs}^{+-} + p_{rs}^{-+} \leq 1, \tag{279}$$

wherein $p_{rs}^{+-} = p_r^{\mathrm{I}} - p_{rs}$ is the probability for finding the first subsystem in the determination r and the second in the conjugate $\bar{s}$ of the determination s, and $p_{rs}^{-+} = p_s^{\mathrm{II}} - p_{rs}$ is the probability for finding the first subsystem in $\bar{r}$ and the second in s.

A general state of the composite system is then described by a triplet

$$\psi = (x, y, C); \tag{280}$$

and a product state has the form

$$x * y = (x, y, x \otimes y) \tag{281}$$

A similar triplet ψ' can also describe a determination of the system. Call $P(\psi, \psi')$ the probability $P(\psi, \psi')$ of finding the system in the state ψ' when it has been prepared in the state ψ. In the special case for which ψ and ψ' both are product states, we expect this probability to be factorized:

$$P(x * y, x' * y') = \frac{1}{2}(1 + x \cdot x')\frac{1}{2}(1 + y \cdot y') = \frac{1}{4}[1 + x \cdot x' + y \cdot y' + (x \cdot x')(y \cdot y')].$$
$$\tag{282}$$

The latter form and the affine character of $P(\psi, \psi')$ as a function of ψ suggest the general expression[163]

$$P(\psi, \psi') = \frac{1}{4}(1 + \psi \cdot \psi'), \text{ with } \psi \cdot \psi' = x \cdot x' + y \cdot y' + C \cdot C' \text{ and } C \cdot C' = \sum_{rs} C_{rs} C_{rs}'.$$
$$\tag{283}$$

The state ψ is pure if and only if

$$P(\psi, \psi) = \frac{1}{4}(1 + x^2 + y^2 + C^2) = 1, \text{ or } x^2 + y^2 + C^2 = 3. \tag{284}$$

---

[163] For a more rigorous argument, see Dakić and Brukner 2009, p. 7.

Since $x^2 \leq 1$ and $y^2 \leq 1$, we have $C^2 \geq 1$. For a product of two pure states, we have $C^2 = x^2 y^2 = 1$. Reciprocally, any pure state for which $C^2 = 1$ is a product state because the condition

$$P(\psi, \bar{x} * \bar{y}) \geq 0 \quad (\text{with } \bar{x} = -x, \ \bar{y} = -y), \tag{285}$$

implies that

$$(x \otimes y) \cdot C \geq x^2 + y^2 - 1 = 3 - C^2 - 1 = 1, \tag{286}$$

whose compatibility with $(x \otimes y)^2 \leq 1$ and $C^2 = 1$ requires that $C = x \otimes y$.

A maximal measurement in the combined system is obtained by combining two bi-valued measurements in the subsystems. If $+$ and $-$ denote the two measurements outcomes in one of the subsystems and if e denotes the state-vector corresponding to the $+$ outcome, the measurement outcomes in the combined system are $++, --, +-, -+$ and the corresponding state vectors are

$$\psi_{++} = e * e, \psi_{--} = \bar{e} * \bar{e}, \psi_{+-} = e * \bar{e}, \psi_{-+} = \bar{e} * e. \tag{287}$$

According to axiom D_1'', which Dakić and Brukner mean to include Hardy's subspace axiom H_3, the 2-subspace Σ_2 of states ψ such that

$$P(\psi, \psi_{+-}) = 0 \text{ and } P(\psi, \psi_{-+}) = 0 \tag{288}$$

must be equivalent to the state space of a two-level system, which is isomorphic to the spherical ball of radius 1 in $\mathbf{R}^K$. On the one hand, the pure states of this subspace fill the $(K-1)$-dimensional sphere. On the other hand, the only pure product states $x * y$ in this subspace are ψ_{++} and ψ_{--} because they must satisfy

$$(x * y) \cdot (e * \bar{e}) = (x * y) \cdot (\bar{e} * e) = -1, \tag{289}$$

which implies that $x \cdot e = y \cdot e = \pm 1$ (while $x^2 = y^2 = 1$). Consequently, if $K \geq 2$ the subspace Σ_2 must contain entangled states, i.e. states that are not product states.

We will now see that values of K higher than three bring contradiction. For such values, the transformation R_4 that changes the sign of the coordinates x_1, x_2, x_3, x_4 (while keeping the other coordinates unchanged) is permitted because there are at least four independent determinations. It is a physical transformation of the two-level subsystems because it is a rotation (an isometry of determinant *one*) in $\mathbf{R}^K$. We will also use the transformation R_2 that changes the sign of the coordinates x_1, x_2. The K-determination for each of the two-level subsystems can always be chosen so that its first unit vector is the vector e from which the subspace Σ_2 is built. Then any transformation R that changes the sign of x_1 in one of the two-level spaces changes the corresponding e into $\bar{e} = -e$. Call R^I the transformation of the composite system obtained by applying R to the first subsystem. For any vector ψ of the subspace Σ_2, we have

$$P(\psi, R^I \psi) = 0. \tag{290}$$

The definition of this subspace indeed implies that

$$P(\psi, \psi_{+-}) = 0 \text{ and } P(\psi, \psi_{-+}) = 0, \tag{291}$$

with the further consequences

$$P(R^I\psi, R^I\psi_{+-}) = P(R^I\psi, \psi_{--}) = 0 \text{ and } P(R^I\psi, R^I\psi_{-+}) = P(R^I\psi, \psi_{++}) = 0, \tag{292}$$

so that ψ and $R^I\psi$ belongs to two mutually exclusive subspaces. In particular, we have

$$P(\psi, R_2^I\psi) = 0 \text{ and } P(\psi, R_4^I\psi) = 0. \tag{293}$$

For a pure ψ (for which $\psi^2 = 3$), this implies that

$$\psi \cdot (\psi - R_2^I\psi) = \psi \cdot (\psi - R_4^I\psi) = 4, \tag{294}$$

or

$$x_1^2 + x_2^2 + \sum_{r=1}^{K}(C_{1r}^2 + C_{2r}^2) = x_1^2 + x_2^2 + x_3^2 + x_4^2 + \sum_{r=1}^{K}(C_{1r}^2 + C_{2r}^2 + C_{3r}^2 + C_{4r}^2) = 4. \tag{295}$$

This condition requires that

$$x_3 = x_4 = 0 \text{ and } C_{3r} = C_{4r} = 0 \text{ for } 1 \le r \le K. \tag{296}$$

Permuting the coordinates $x_2, x_3, ..., x_K$ and permuting the subsystems I and II, we conclude that x_1, y_1, and C_{11} are the only non-vanishing components of the ψ vector. Since $C^2 \ge 1$ and since $-1 \le C_{rs} \le 1$, we must have $C^2 = C_{11}^2 = 1$. Hence the ψ state must be a product state.

In order to avoid the contradiction with the earlier proven existence of entangled states, we must assume that $K \le 3$. Since the even value $K = 2$ is already excluded, we are left with the options $K = 1$ (classical probabilities) and $K = 3$ (quantum probabilities). By Hardy's consideration of subspaces, we can then show that the states of any N-level system can be represented by a quantum-mechanical density matrix.

Dakić and Brukner's proof of the impossibility of $K \ge 4$ is easily adapted to Hardy's axiomatics, for it only relies on the subspace axiom and on the representation of the pure states of a two-level system by a $(K - 1)$-dimensional sphere. The latter representation can be derived from Hardy's axioms, without appeal to axiom D_1', by exploiting the possibility of representing any compact group by orthogonal matrices.[164] Dakić and Brukner nonetheless prefer their own axiomatics, because it does not require the continuity assumption (for the transformation group) and because it is compatible both with classical and with quantum probabilities. The continuity assumption is only necessary to exclude the classical option.

---

[164] See Hardy 2001, appendix 3.

One may still prefer Hardy's axioms, because they can be justified by correspondence arguments whereas the information capacity axiom seems hard to swallow. Why after all should every two-level system be assimilated to a one-bit information facility? Is it not highly unnatural to assume, when there is a continuum of possibilities of measurement, that every state of the system can be obtained as a mixture of the outcomes of a single measurement? In 2010, Lluís Masanes and Martin Müller replaced this axiom with the "requirement" that in two-level systems "all mathematically well-defined measurements are allowed by the theory" or more precisely: "all tight effects correspond to allowed measurements." Translated in the x-vector language, a "tight effect" is any affine function

$$F(x) = \frac{1}{2}(1 + a \cdot x) \qquad (297)$$

through which the image of the state space is the whole interval [0, 1]. An "allowed measurement" is a F function for which the vector a defines a determination. We will now see that Masanes and Müller's requirement can replace the Comte–Fivel–Zeilinger axiom D_1' in the derivation of the fact that every point of the sphere S^{K-1} defines an allowed (pure) state for a two-level system.[165]

Suppose, *ad absurdum*, that there is a portion of this sphere on which there is no permitted state. Then there cannot be any extreme point in this portion (since such a point would yield a pure state), and the boundary of the convex domain of states must have a facet.[166] Choosing the vector a so that the equation of the facet reads $a \cdot x = 1$, the function $F(x)$ in (297) defines a tight measurement. For the corresponding determination M, the states S' such that $P(S', M) = 1$ define a subspace that has an infinite number of elements (the whole facet). This is absurd because the state S prepared by M should be the only state of the system for which the probability of the determination M is *one*.[167]

Another possible replacement for the axiom D_1' is Hardy's continuity, that is, the possibility of gradually transforming pure states. We will reason by induction, starting with $K = 1$. In this case, the desired property trivially holds, because the only pure states are the two points ($x_1 = \pm 1$) of the sphere S^0. Now suppose that for $K - 1$ degrees of freedom the set of pure states is known to be identical to the sphere S^{K-2}. For K degrees of freedom, the states of the system for which $x_1 = 0$ may be regarded as the states of a system with $K - 1$ degrees of freedom. These states therefore fill the sphere S^{K-2} that is the intersection of the sphere S^{K-1} with the hyperplane $x_1 = 0$, and the subgroup of transformations of the K-system that leave the x_1 axis invariant acts transitively on this sphere (since it evidently includes all transformations of the $(K - 1)$-system). By Hardy's continuity, there must be continuous curves on the sphere S^{K-1} relating any point of the sphere S^{K-2} (which would be the equator in the case $K = 3$) to the poles $x_1 = \pm 1$. Applying transformations of the

---

[165] Masanes and Müller 2010, pp. 1, 5.

[166] For a proof of this intuitive result, see Masanes and Müller 2010, p. 6, ref. [20].

[167] Masanes and Müller do not regard this last point as obvious. They derive it from the subspace axiom, which implies that the subspace of states S' such that $P(S', M) = 1$ should correspond to the unique state of a one-level system.

former subgroup to these curves, we can sweep the whole sphere S^{K-1}. The set of pure states of the K-system is therefore identical with this sphere, as was to be proved.[168]

Chiribella, D'Ariano, and Perinotti

In the wake of Hardy's seminal axiomatics, there has been a growing tendency to formulate and justify the axioms of quantum mechanics by information-theoretical means. This is a natural evolution considering the present importance of researches on quantum-mechanical information processing and quantum computing. In 1990, John Archibald Wheeler famously defined the "It from bit" program for reducing physics to the processing of information. Although this sort of reductionism has often been criticized, it has inspired a few arguments for the information-theoretic necessity of quantum theory. The first of these is found in a memoir of 2003 by three philosophers of physics, Rob Clifton, Jeffrey Bub, and Hans Halvorson (CBH). The gist of their argument is a proof of the three following assertions:

1. The impossibility of supraluminal communication between two systems entails the commutativity of the associated algebras of observables.
2. The impossibility of perfectly broadcasting the information contained in an unknown physical state entails the non-commutativity of the algebra of observables of an individual system.
3. The impossibility of unconditionally secure bit commitment entails the existence of entangled states.

The first impossibility (micro-causality) is a mere consequence of relativity theory; the second and third impossibilities are well-known consequences of quantum mechanics applied to quantum cryptography. CBH regard these three impossibilities as given information-theoretic facts and study their consequences in the powerful language of C*-algebras. This language generalizes the operator algebra on Hilbert spaces and it is meant to encompass every past and future physical theory (see above p. 237).[169]

No matter how interesting the CBH result may be as an information-theoretic characterization of quantum theory, it cannot pass for an argument for the necessity of quantum mechanics. There are three reasons for that. Firstly, the C*-algebraic framework is much too abstractly mathematical to pass for an a priori natural frame in which to formulate

---

[168]This reasoning presupposes that two-level systems exist for any value of K, which is not truly the case. A better reasoning may be obtained by strengthening the continuity axiom so that any two states belonging to the same subsphere of S^{K-1} be continuously related through states belonging to this subsphere. The latter assumption can be justified by a correspondence argument: two states for which a given measurement (for instance, the measurement of a spin-component) yields a zero average should be continuously related through states satisfying the same condition. Hardy (2001, p. 20) treats only the case $K = 3$, for which the group of transformations must be a two-parameter Lie group, through which the orbit of a single point must be the whole sphere.

[169]Wheeler 1990. Clifton, Bub, and Halvorson 2003, pp. 2–3. The no-broadcasting theorem is an extension of the no-cloning theorem of William Wootters, Wojciech Zurek, and Dennis Dieks (1982) to mixed states. Bit commitment is a cryptographic notion defined by Gilles Brassard, David Chaum, and Claude Crépeau in 1988. On the early history of quantum information, cf. Kaiser 2011. For a criticism of information-theoretic reductionism, cf. Deutsch 2003.

physical theories.[170] For a rationalist exploitation of the CBH result, one would first need to derive this framework from simple operational considerations, which does not seem easier than deriving the Hilbert space structure of quantum mechanics. A second shortcoming has to do with the contents of CBH's information-theoretic principles. Even if they could be shown to be natural from an information-theoretic point of view, this would not make their physical realization in elementary systems more natural. Thirdly, CBH do not prove that quantum mechanics results from their principles. What they construct is a generalized quantum theory defined by a C*-algebra satisfying the algebraic constraints that derive from their information-theoretic principles. In fact, they want this generality because they have in mind situations (quantum field theory and quantum gravity) in which it might be needed.

Very recently Giulio Chiribella, Giacomo Mauro D'Ariano, and Paolo Perinotti (CDP) have offered an "informational derivation of quantum theory" that does not have the first and third of the defects of CBH's derivation. CDP arrive at quantum mechanics, and they do so in a purely operational framework based on probability distributions for "circuits" resulting from the connection of physical devices:

> Our principles do not refer to abstract properties of the mathematical structures that we use to represent states, transformations, or measurements, but only to the way in which states, transformations, and measurements combine with each other.

CDP give the following informal statement of their axioms:[171]

C_1 *Causality*: the probability of a measurement outcome at a certain time does not depend on the choice of measurements that will be performed later.

C_2 *Perfect distinguishability*: if a state is not completely mixed (i.e., if it cannot be obtained as a mixture from any other state), then there exists at least one state that can be perfectly distinguished from it.

C_3 *Ideal compression:* every source of information can be encoded in a suitable physical system in a lossless and maximally efficient fashion. Here *lossless* means that the information can be decoded without errors and *maximally efficient* means that every state of the encoding system represents a state in the information source.

C_4 *Local distinguishability*: if two states of a composite system are different, then we can distinguish between them from the statistics of local measurements on the component systems.

C_5 *Pure conditioning:* if a pure state of system AB undergoes an atomic measurement on system A, then each outcome of the measurement induces a pure state on system B.

C_6 *Purification postulate*. Every state has a purification. For fixed purifying system, every two purifications of the same state are connected by a reversible transformation on the purifying system.

---

[170] The same criticism is found in Grinbaum 2007, p. 402.

[171] Chiribella, D'Ariano, and Perinotti 2011, pp. 2, 3.

The causality axiom C_1 is so evident that all other axiomatizers assumed it without stating it. The local distinguishability axiom C_4 is a rewording of A_4 or D_2. The other axioms are more original. The purification postulate means that every state of a system may be regarded as the marginal state of a subsystem of a larger system that is in a pure state. CDP note the affinity with Schrödinger's remark of 1935:

> An optimal knowledge of the whole does not imply an optimal knowledge of its parts—that is the whole mystery. I would not call that *one* but rather *the* characteristic trait of quantum mechanics, the one that enforces its entire departure from classical lines of thought.

Somewhat artificially, CDP include the existence of reversible transformations between any two pure states in the purification postulate. The axioms C_1, C_2, and C_3 serve to derive the duality between pure states and pure determinations as well as the representation of any state as a convex mixture of pure states; whereas in my simplified approach these facts are trivial consequence from the definition of general states as mixtures of states produced by maximal measuring devices. Axioms C_5 and C_6 sustain a proof that $K = N^2 - 1$ and allow the derivation of the density-matrix representation of states. Interestingly, CDP do not need the subspace axiom. They deduce the equivalence (up to a reversible transformation) of all systems with the same dimension from their own axioms.[172]

In the case of two-level systems ($N = 2$), the axiom C_2 of perfect distinguishability results from Masanes and Müller's requirement (MMR) that all tight effects are allowed measurements. Indeed what CDP call an incompletely mixed state is a state that belongs to the boundary of the convex state-space; and the complete distinguishability of two states S and S' translates into $P(S, M') = 0$, where M' is the determination corresponding to the state S'. If an incompletely mixed state is pure, it is perfectly distinguishable from its conjugate state. If it is not pure, it belongs to a facet that defines a tight effect and an allowed measurement (determination) according to MMR, so that it is perfectly distinguishable from the pure state associated with the conjugate determination.

The axiom C_2 may replace MMR in the proof that every point of the sphere S^{K-1} defines a pure state. Indeed if the state S belongs to the boundary of the domain of states, by this axiom there exists a determination M' such that $P(S, M') = 0$. This implies that $P(S, \overline{M}') = 1$, so that S must be identical to the pure state $\overline{S}'$ prepared by $\overline{M}'$. Hence all the states on the boundary are pure and the boundary must coincide with the sphere S^{K-1}.

CDP's proof that axioms C_5 and C_6 imply the relation $K = N^2 - 1$ relies on a difficult and lengthy exploitation of probabilistic teleportation schemes. A much simpler proof can be given by applying the pure-conditioning axiom C_5 directly to the composite system made of the two-level systems I and II. Remember that a state of this composite system is described by the triplet of Eq. (280)

$$\psi = (x, y, C),$$

[172]Chiribella, D'Ariano, and Perinotti 2011, pp. 2 (on Schrödinger), 29 (no subspace axiom); Schrödinger 1935, p. 555. The purification postulate directly implies the no-cloning and the impossibility of bit commitment assumed by CBH.

and that a pure state satisfies Eq. (284)

$$x^2 + y^2 + C^2 = 3.$$

After a measurement of the determination r on system I, using Eqs. (276) and (277), the correlation matrix C_{rs} leads to the state $y^{(r)}$ of system II with coordinates

$$y_s^{(r)} = 2\frac{p_{rs}}{p_r^I} - 1 = \frac{C_{rs} + x_r + y_s + 1}{1 + x_r} - 1 = \frac{C_{rs} + y_s}{1 + x_r}. \tag{298}$$

By axiom C_6, the completely mixed state $x = 0$ of system I can be purified. We will further assume that the two-level system II is sufficient for this purification. It will now be proved that the marginal y of the purifying state ψ must be zero. Suppose, *ad absurdum*, that y does not vanish. Then it is possible to perform a measurement of system II in the direction of y (namely, a determination associated with the pure state $\pm y/\|y\|^2$). The combination of this measurement with the mixing of its two conjugate outcomes does not affect system II and therefore leaves the global state ψ unchanged. According to axiom C_5, to each outcome of this measurement there should correspond a pure state of system I. Therefore, the marginal x of system I for the state ψ of the global system should be a mixture of two pure states with the uneven weights $(1 \pm \|y\|)/2$. Since such a mixture can never be totally mixed, x differs from zero. Therefore, y must vanish if x does.

As a corollary, there exist pure states ψ with totally mixed marginals. For such a state, the state $y^{(r)}$ of system II induced by the determination r of system I has the coordinates

$$y_s^{(r)} = C_{rs}. \tag{299}$$

By axiom C_5 this state is pure, so that

$$(y^{(r)})^2 = \sum_{s=1}^{K} C_{rs}^2 = 1. \tag{300}$$

Consequently, we have

$$C^2 = \sum_{rs} C_{rs}^2 = K. \tag{301}$$

Combined with the condition $C^2 = 3$ for the purity of the state ψ, this implies that $K = 3$, in conformity with $K = N^2 - 1$.

It is possible to obtain the same result with the following weakened form of the purification postulate:

C_6': The combination of two two-level systems admits entangled states; namely, there are pure states of the combined system for which the marginals are not pure.

Using Eq. (298), the purity of the state $y^{(r)}$ of system II induced by the determination r of system I yields

$$\sum_{s=1}^{K} (C_{rs} + y_s)^2 = (1 + x_r)^2.$$ (302)

The similar condition for the permissible state $\psi = (-x, y, -C)$ (derived from ψ by replacing the r determinations with their conjugates) reads:

$$\sum_{s=1}^{K} (-C_{rs} + y_s)^2 = (1 - x_r)^2.$$ (303)

Combining these two conditions and summing over the index r we get

$$C^2 + Ky^2 = K + x^2.$$ (304)

Permuting I and II, we further get

$$C^2 + Kx^2 = K + y^2.$$ (305)

Combining the sum of these two equations with the condition (284)

$$x^2 + y^2 + C^2 = 3$$

for the purity of ψ, we get

$$(K - 3)(2 - x^2 - y^2) = 0.$$ (306)

By axiom C_6', there exists a state ψ for which the second parenthesis does not vanish. Therefore, K must be equal to 3.

Hardy axiomatics and necessity

Let us recapitulate the various deductions of quantum probabilities given in this section. We started with the definition of a system and its pure states by repeatable (maximal) measurement operations that have a fixed number N of discrete outcomes. The selection of a single measurement outcome is what I call a determination, and it is associated to a pure state. The general state S of a system is a mixture of pure states, and it is characterized by the probability $P(S, M')$ for a variable determination M'. A first axiom requires these probabilities to be well defined. A second axiom allows a finite number of such probabilities to determine the state. The minimal value of this number is called K; a possible choice of the K defining determinations is called a K-determination, and the associated probabilities are denoted $p_r = P(S, M_r)$, with $r = 1, ..., K$. The probability $P(S, M')$ may be regarded as a function of the vector $p = (p_1, p_2, ..., p_K)$ characterizing the state S and of the vector p' characterizing the pure state attached to the determination M'. Since states and can be mixed and since probabilities add up by mixing, the probability $P(S, M')$ must be an affine function of the vector p.

By a third axiom, there exist reversible transformations between any two pure states. This implies the existence of a "totally mixed" state $\hat{S}$ for which the probability $P(\hat{S}, M')$ is the same for any choice of M'. In the case $N = 2$, the associated vector is $\hat{p} = (1/2, 1/2, ..., 1/2)$, and it is advantageous to introduce the Bloch vector of coordinates $x_r = 2(p_r - \hat{p}_r) = 2p_r - 1$. The third axiom also implies that for a proper basis in K-space, the Bloch vectors of pure states belong to the $(K - 1)$-dimensional unit sphere.

In order to prove that reciprocally every point of this sphere represents a pure state, we need another axiom. Hardy relies on the continuity axiom for the transformation group, Dakić and Brukner rely on the Comte-Fivel-Zeilinger axiom that two-level systems carry one bit of information, Masanes and Müller on the axiom that every tight effect is a possible measurement, CDP on the axiom of perfect distinguishability of states belonging to the boundary of the convex state domain.

If our purpose is to show the necessity of quantum mechanics, we should pick the most natural of these axioms. Of course, every author believes his axiom to be the most natural: Hardy expects the continuity of quantum transformations by analogy with the continuity of classical evolutions; Dakić and Brukner believe nature to be made of quantum bits; Masanes and Müller regard tight effects as innocent mathematical idealizations of concrete measurements ("mathematically well-defined measurements"); CDP praise their axioms for being phrased in purely information-theoretic and operational terms. Dakić, Brukner, and CDP may be criticized for conflating naturalness with information-theoretic simplicity: Why should physics be reduced to transfers of bits? Is not there a huge gap between the abstract idea of quantum bits and their physical realization? Masanes and Müller similarly confuse mathematical simplicity with physical plausibility: Why should their "tight effects" be plausible idealizations of concrete measurements? The main advantage of their axiom is that it leads to the desired result (the Bloch sphere of pure states) in the most direct manner. We are left with Hardy's continuity, whose necessity derives from a more compelling correspondence argument. Some liberty in the choice of axioms is nonetheless welcome as a further indication of the necessity of the consequences.

Once the domain of pure states is known to be the unit sphere, it is easy to see that the probability function takes the form (275)

$$P(S, M') = \frac{1}{2}(1 + x \cdot x'),$$

where x' denote the Bloch vector of the state S' that is prepared by M'. In order to arrive at the Bloch-sphere representation of two-level quantum mechanics, there remains to prove that $K = 3$. The axiom that the state of a combined system is characterized by the statistics of the determinations of its components leads to the relation (258),

$$K = N^{\kappa} - 1,$$

which excludes all even values of K. A further restriction is reached by considering the combination of two two-level systems. On the one hand, a restriction of Hardy's subspace axiom, according to which certain subspaces of states can be regarded as the states of a two-level systems, implies the existence of entangled states for the combined system if $K \geq 2$. On the other hand, by Dakić and Brukner's argument or by the simpler argument given at the end of the previous paragraph, entangled states are impossible for K larger than three. The leftover possibilities are $K = 1$, which corresponds to classical probability theory, and $K = 3$, which corresponds to quantum probability theory. Hardy's continuity then excludes the first option. For an N-level system, we must have $K = N^2 - 1$, which is the number of degrees of freedom compatible with the representation of states by a density matrix in N-dimensional Hilbert space. As we saw, the adequacy of this representation can be established by repeated application of the subspace axiom. This ends the proof that Hardy's axioms or the variants by Dakić, Brukner, Masanes, and Müller lead to quantum probabilities.

A strikingly simple derivation of $K = 3$ for two-level systems is obtained by replacing the subspace axiom with two of CDP's axioms, the existence of entangled states (or the stronger postulate of purification) and the pure conditioning axiom. The latter assumption, according to which for a combined system in a pure state a determination of one subsystem implies a pure state of the other subsystem, is fairly natural. It is a particular case of a broader axiom that would require any incomplete measurement of a system originally in a pure state to leave the system in a (generally different) pure state.[173] In other words, if we have maximal information on a system, we still have maximal information on this system after performing an ideal (yet incomplete) measurement. Unfortunately, the purification postulate or the resulting existence of entangled states is harder to swallow.[174] To assume this postulate is to admit in the very basis of the theory the quantum oddities deplored by Schrödinger. As explained by CDP, the true advantage of their approach is its providing direct illuminating links between the now information-theoretic axioms of quantum mechanics and the various quantum-information theorems to which physicists have lately devoted much attention.[175]

CDP criticize earlier axiomatics for involving uninterpreted mathematical assumptions.[176] This charge certainly applies to Ludwig's old axiomatics, despite Ludwig's intention to provide physically justified axioms; it also applies to Masanes and Müller's "tight effect" axiom; but it does not truly apply to Hardy's approach because his only uninterpreted axiom, the simplicity axiom, is now known to be superfluous; and the charge has no grip on Dakić and Brukner's axiomatics. Altogether, an improved version of Hardy's axiomatics provides a convincing demonstration that the consistent melding of the discontinuity of measurement results with the continuity of measurement possibilities necessarily leads to the density-matrix representation of physical states.

There are three limitations to this class of necessity arguments. The first is the assumption of a finite value for the maximal number N of distinct measurement outcomes. This is not a very serious limitation, because in the laboratory quantum processes involve only finite-dimensional Hilbert subspaces (as a consequence of effective infrared and ultraviolet

---

[173] As we saw earlier (p. 297), this supposition is made to justify the covering law in quantum logic.

[174] CDP (2011, p. 2) try to justify the purification principle by having it express the possibility of reducing thermodynamic irreversibility to reversible interaction with an uncontrolled environment.

[175] Chiribella, D'Ariano, and Perinotti 2011, p. 38.   [176] Chiribella, D'Ariano, and Perinotti 2011, p. 2.

cutoffs). The second limitation concerns the evolution of systems. So to say, the axiomatics of Hardy and his followers provides only the kinematics of quantum mechanics, that is, its representation of physical states. It does not tell us how the states evolve, except that probability should be conserved. In the continuous case in which no measurement is performed, the latter property implies the existence of a Hamiltonian operator from which the evolution derives. Hardy regards the precise expression of the Hamiltonian as a contingent fact to be drawn from experience. However, since his axiomatics implicitly involves correspondence arguments, it would seem natural to extend kinematic correspondence to dynamic correspondence in order to arrive at the usual quantization rules in Heisenberg's or in Schrödinger's form. This would bring in Planck's constant, which otherwise does not belong to Hardy's axiomatics.

The third and most fundamental limitation to a rationalist exploitation of Hardy's axiomatics is inherent in the assumption of strictly discrete (ideal) measurement outcomes. There is no direct empirical difference between an isolated discrete value and a very narrow continuous spread of values around a central value. Yet the two options lead to very different intuitions of the possible correlations between successive measurements: only in the first option does one expect well-defined probabilities for these correlations; in the second option, the fine structure of the spectrum of possible values should naturally affect the correlations. In sum, Hardy's axiomatics shows the necessity of quantum mechanics to the extent that the strict discontinuity of measurement results is judged necessary. Historically, the latter necessity has sometimes been regarded as empirical, for instance as a consequence of the discreteness and universality of atomic spectra; and sometimes as intertheoretical, as the only escape from paradoxes regarding the interaction between radiation and a large assembly of atoms (ultraviolet catastrophe). These arguments in favor of discontinuous measurement outcomes are not as compelling as the deduction of quantum mechanics from this discontinuity in axiomatics à la Hardy. In the present state of this approach, we should probably content ourselves with the insight that quantum discontinuity, if it is admitted as a fundamental feature of the microworld and if it is complemented with natural axioms concerning the relation between micro- and macro-world, necessarily leads to quantum mechanics as we know it.

We are now in a position to compare the Hardy kind of axiomatics with quantum logic. A first difference is the manner in which non-classicality is introduced. In quantum logic, the classical reference is the Boolean logic of binary measurement; what causes departure from that logic is the admission of incompatible measurements. In Hardy axiomatics, the classical reference is the classical theory of probabilities of discrete events; what causes departure from this theory is the continuity of measurement possibilities. The classical reference being different, it would not make much sense to say that one approach better justifies departure from classicality than the other. Rather, we should compare the manners in which the two approaches purport to derive quantum mechanics.

Let us first compare the crucial ingredients of these derivations. In the quantum logic approach, the most evident quantum-like ingredient is the assumption of incompatible measurements; in the Hardy approach, it is the discreteness of measurements outcomes combined with the continuity of measurement possibilities. At first glance, quantum logic seems more economic, since incompatible measurements are easier to conceive than an intrinsic discontinuity of physical quantities. This difference is tenuous, however. If we believe in Bohr's intuition of quantum discontinuity, the discreteness of physical quantities and the incompatibility of measurements both result from the existence of the quantum action: a measurement generally implies a finite and uncontrollable perturbation of its object, perturbation that randomly affects the result of a subsequent measurement on a correlated object. Moreover, it is doubtful that quantum logic can truly dispense with an assumption of discreteness. As we saw, it is only in the discrete, finite-dimensional case that its axioms are natural enough.

Another difference in the two kinds of derivations of quantum mechanics is the nature of the employed mathematics. For the average physicist, the mathematics of quantum logic is exotic as it involves deep interconnections between lattice theory, projective geometry, and generalized Hilbert spaces. The mathematics of Hardy axiomatics is globally simpler. Whenever it gets more difficult, it is through its reliance on the theory of Lie groups, which is abundantly used by modern theoretical physicists. By restricting itself to purely logical axioms, quantum logic raises the mathematical stakes much higher than Hardy axiomatics does.

A last and most decisive difference is the degree to which the two approaches succeed in deriving quantum mechanics. In this respect, quantum logic is losing because it allows for generalizations of quantum mechanics in which the field of complex numbers is replaced by other fields or division rings. The fuller success of Hardy axiomatics seems to result from its reliance on axioms regarding composite systems and subsystems. Quantum logic, in its original version, lacks any such axiom in its foundations.[177] This deficiency is not a sufficient reason to condemn quantum logic: it may well continue to be a productive enterprise in the golden triangle of mathematics, physics, and philosophy. Yet, globally judging from the number and naturalness of the axioms, from the accessibility of the math, and from the fullness of the deductions, Hardy axiomatics offers more convincing arguments for the rational necessity of quantum mechanics.

Conclusions

The history of quantum theory is a first remedy for the mathematical abruptness of standard quantum mechanics. It provides an understandable genesis of both the matrix and the wave form of this theory. However, the historical development is too complex and two impregnated with empirical arguments to be regarded as a rational justification. It only gives hints at such justifications. A first hint is Bohr's correspondence principle, whose historical success suggests that the asymptotic agreement of classical and quantum theory should be a good guide in constructing the formalism of quantum mechanics. Another hint is the general feeling that a melding of continuity and discontinuity, properly orchestrated by the correspondence principle, should lead to quantum mechanics.

The first hint found spectacular confirmation in Lichnerowicz's and Gutt's proofs that the phase-space formulation of quantum mechanics is the unique (up to an isomorphism) one-parameter deformation of the Poisson algebra of classical mechanics. This result is purely mathematical. It does not imply that the mathematically generated deformation

---

[177] As was mentioned in n. 135 above (p. 307), the representation of a compound system in a pure lattice-theoretic framework turns out to be impossible.

should be a physical theory. However, this deformation is constructed so as to possess a Lie algebra of infinitesimal evolutions. If we share Poincaré's belief in the synthetic a priori character of transformation groups, this makes the deformation a good candidate for being some sort of dynamics. Compared to other arguments for the necessity of quantum mechanics, the deformation approach has an important advantage: in addition to the general description of states and their evolution, it gives the expression of the Hamiltonian and other observables.

The other hint from history, that quantum mechanics should result from a correspondence-guided melding of continuity and discontinuity, is confirmed by Hardy's axiomatics, which generates the density-matrix representation of physical states and their unitary evolution by postulating discrete measurement outcomes and continuous variations of the choice of the measured quantity. The latter principle of continuity, and most of Hardy's other axioms are implicit consequences of some correspondence between classical and quantum theory. The least convincing of Hardy's axioms, the one requiring the lowest value for the number of degrees of freedom associated to a given dimension, is now known to derive from his other axioms. The theorists who corrected this defect moved toward a fuller information-theoretic expression of the axioms. Although this tendency has the advantage of bringing quantum mechanics closer to its applications to the processing of information, it diminishes the necessity of the axioms by severing them from the correspondence arguments that were available in Hardy's original formulation.

In the popular perception of quantum mechanics, its main peculiarities are quantum discontinuity (discrete character of physical quantities that used to be continuous), the existence of incompatible measurements (uncertainty relations), and the existence of entangled states (in Schrödinger's sense). Hardy's axiomatics postulates quantum discontinuity; and the information-theoretic axiomatics of Chiribella, D'Ariano, and Perinotti postulates entangled states through their principle of purification. A much earlier kind of axiomatics, the quantum logic initiated by von Neumann and Birkhoff, begins with incompatible measurements. Its focus on experimental Yes-No questions makes it an impoverished logic based on a non-distributive lattice. Its elegance lies in the economy of its presuppositions and in the power of the mathematics deployed to derive a generalized Hilbert-space representation of the lattice of propositions. Its main defects, compared to Hardy's approach, are the required level of mathematical competence and the failure to single out the Hilbert-space representation among all representations compatible with the basic lattice structure. The axioms of quantum logic are most natural in the finite-dimensional case, and a bit contrived in the infinite-dimensional case developed by Piron and others. Once completed with a definition of states through the statistics of binary tests, they lead to the density-matrix representation of states and to their unitary evolution if the Hilbert-space representation of the logical lattice is selected.

In order to appreciate the kind of necessity of the various axiomatics encountered in this chapter, one must examine the nature of the primitive notions needed to formulate the axioms. In the case of quantum logic, the basic notion is that of a repeatable binary test (Piron's "measurements of the first kind"). The theory is constructed from this highly idealized notion, without any information regarding the concrete realization of the tests. But there is little doubt that the existence of such tests is a minimal requirement about the possibility of experimentation: We must somehow be able to determine the properties of a system through reliable tests, and every test is evidently traceable to a set of binary tests; a clear-cut answer to the latter kind of test is a plausible idealization. As long as the axioms regarding the combination of binary tests are physically reasonable, their consequences seem quite necessary. This extreme generality is the main attraction of quantum logic. It also is its main deficiency. Even after being completed by a statistical definition of states in Mackey's manner, the theory does not give any concrete instructions about how to perform tests and measurements. It only suggests that single measurements should correspond to mutually compatible tests that generate a decomposition of the identity as a sum of orthogonal projectors. In order to associate a measurement value to a given projector and in order to bring in Planck's constant, further considerations are needed, perhaps symmetry considerations (in particular, the Hamiltonian is an observable related to the uniformity of time) or correspondence arguments. So to say, quantum logic is an empty shell waiting for an imbedded theory of measurement.[178]

This state of affairs brings out a novel aspect of the relation between mathematics and physics. In Chapter 6, I have insisted on Helmholtz's and Poincaré's doctrine that the possibility of measurement (in a strong sense including the concrete addition of quantities) is responsible for much of the mathematical structure found to be necessary in formulating physical theories. In quantum logic, the basic notions do not include measurement; they only include binary tests. Interestingly, the specific lattice structure of these tests implies much mathematics, including generalized Hilbert spaces, which are much more advanced mathematical constructs than the real numbers associated with ideal measurement. This mathematics does not imply numbers determinable by experiment. Such numbers only occur at the probabilistic stage at which the statistical correlation between successive tests is defined. Measurement *stricto sensu* remains irrelevant until the quantum logic shell is filled with appropriate metric notions. In a neo-Kantian reading, quantum logic may be seen as a very basic precondition of experiment, prior to the measurability conditions expressed in Helmholtz's doctrine.

At first glance, Hardy's axiomatics does not seem to share this pre-metric quality of quantum logic, since it presupposes "single shot measurements" with definite (discrete) numerical outcomes. In fact, the basic notion of this theory is that of states defined through the probabilities of the various measurement outcomes. This difference with quantum logic is not so great, however, because Hardy's measurements are only defined *in abstracto*, without any prescription for their concrete realization. A more significant difference stems from Hardy's introducing the statistical concept of state at the very beginning of his theory. This difference implies a different sort of necessity for the axioms in the two approaches. Whereas in quantum logic necessity is inherent in the logic of tests, in Hardy's theory it derives from the correspondence arguments that implicitly sustain the axioms. The latter kind of necessity seems vaguer than the former. Yet it leads to more determinate conclusions.

In conformity with the general theme of this book, I have given much attention to the degree in which various developments can pass for rational derivations of quantum mechanics. This does not mean, however, that the actors of these developments truly had

---

[178] Jauch was well aware of this shortcoming and did his best to remedy it in Jauch 1968.

rationalist ambitions. Birkhoff and von Neumann define the aim of their foundational paper as "to discover what logical structure one may hope to find in physical theories which, like quantum mechanics, do not conform to classical logic." Although they occasionally ask for "a plausible physical motivation" for their axioms, they seem to be more concerned with mathematical fertility. Mackey's main purpose clearly is clean axiomatization in Hilbert's sense. Other authors profess an operational approach in which the deployed mathematics should be mostly dictated by idealized operations. Yet they do not place the necessity of the operational axioms at the top of their agenda. Piron and Jauch purport to define the most adequate language of quantum theory. They spend relatively little time justifying their axioms. Hardy and his followers insist on the "natural" or "reasonable" character of their axioms in the context of probability theory or information theory. Although they probably mean this naturalness to imply a kind of necessity, it should be distinguished from applicability to physical systems.[179]

To summarize, we have three sorts of arguments for the necessity of quantum mechanics: historical, mathematical, and operational. The historical ones are too impregnated with empirical knowledge to be regarded as rational; the mathematical ones are purely rational but they leave the physical significance of the deduced theory open; the operational ones (quantum logic, or probabilistic states) come closest to rational deductions of quantum mechanics, although their inventors usually avoided to claim so much, and although some of the basic assumptions, especially the discreteness of measurement outcomes and the existence of incompatible measurements, remain largely empirical. At any rate, one cannot help being impressed by the fact that these assumptions, together with basic preconditions of experience and correspondence arguments, lead to the Hilbert-space formalism of quantum mechanics. We thus understand why microphysics needs a kind of mathematics earlier believed to belong to the pure mathematician. We also become convinced that quantum mechanics is the only plausible generalization of classical mechanics that takes into account the basic atomicity of physical phenomena. This of course does not preclude that quantum mechanics would someday be seen as a limiting case of a more fundamental theory.

[179] Birkhoff and von Neumann 1936, pp. 823, 837; Hardy 2001.

9

NECESSITY, THEORIES, AND MODULES

The most successful of the necessity arguments given in the previous chapters involve the manner in which theory is connected to experiments and also the way different theories are related to each other. As long as a theory is considered from a purely mathematical point of view, as long as its mode of application is left in the dark, as long as it is treated as an independent whole, effective necessity arguments remain unconceivable. Are they allowed in a more refined conception of physical theory? The present chapter is devoted to this question.

The first section briefly addresses the nature of necessity arguments and their compatibility with a few influential conceptions of physical theory. I first argue that the best necessity arguments all have to do with the comprehensibility of nature in a Helmholtzian sense, and I catalog various comprehensibility principles including measurability, causality, homogeneity, correspondence, and reduction. Then I compare these principles with attempts to redefine or relativize Immanuel Kant's a priori, from Hermann Helmholtz to Michael Friedman. Despite evident similarities, I argue that the latter attempts have a different, broadly epistemological purpose and that they are not meant to lead to necessity arguments for existing theories. Next, I show that comprehensibility requires idealized definitions of quantities, idealized operations and measurements, as well as a variety of intertheoretic relations. The rest of this first section is a discussion of the compatibility of these notions with the linguistic, semantic, and structuralist views of physical theory.

In the second section, I propose a view somewhat similar to the structuralist view but better adapted to my purpose. I reached this view inductively, by examination of a large sample of existing theories. A first central feature is the inclusion of mathematical blueprints of experimental devices, which I call *interpretive schemes*. Another is the importance of *modules* defined as essential theoretical components that are themselves theories with different domains of application. In the third section, I argue that modularity is necessary to the effective application, comparison, construction, and communication of theories. In the fourth and last section I show that the necessity arguments defended in this book are conceivable within the modular conception of physical theory.[1]

---

[1] For a more detailed exposition of my views on theories and modules, see Darrigol 2008.

Physics and Necessity. First Edition. Olivier Darrigol.
© Olivier Darrigol 2014. Published in 2014 by Oxford University Press.

9.1   The preconditions of necessity

Necessity and comprehensibility

In the previous chapters we encountered various types of necessity arguments: theological, metaphysical, transcendental, Helmholtzian, and empirical.[2] Theological or metaphysical arguments are bound to specific religious or ontological beliefs and are therefore likely to have limited credence and lifespan. Kantian providers of transcendental arguments purport to avoid deeper ontological questions; they focus on the intellectual preconditions of our experience of the world; and they derive necessary features of our physical theories from these conditions. Unfortunately, they tend to exaggerate our ability to determine these conditions by purely rational means, so that the features they deem necessary tend to disappear in the next generation of physical theories. The Helmholtzian type of necessity shares the Kantian focus on the comprehensibility of nature. However, for a Helmholtzian the various conditions of comprehensibility, no matter how natural they may seem, are not to be regarded as a priori certain. Only experience can tell us to what extent these conditions are met. Thus for a Helmholtzian, when a certain theory or certain laws are argued to be necessary, it is only in the limited sense that they derive from natural but fallible preconditions for the comprehensibility of nature. Lastly, necessity arguments may be based on the extreme broadness of some empirical fact, for instance the impossibility that heat goes spontaneously from a cold body to a hot body. The Helmholtzian type may be seen as the subtype of this broad-empirical type for which the empirical fact is some aspect of the comprehensibility of nature. Or the converse may be argued: as will be seen in a moment, broad empirical truths can often be seen as forms of causality, which is a kind of comprehensibility.

The Helmholtzian type of necessity arguments is the only one to combine sufficient robustness with high structuring power. These arguments are not purely rational, since their premises are ultimately empirical. However, the premises are few and they have an appealing simplicity and naturalness. Let us try to classify the various kinds of comprehensibility.

Firstly, we may require the measurability of some basic quantities. A broad consequence of this requirement is the relevance of mathematical analysis in the formulation of physical theories. More detailed consequences depend on the type of quantity and on the way in which the measurement process is idealized. Space measurement by rigid bodies leads to the locally Euclidean character of space; time measurement by inertial motion, together with space measurement, to the locally Minkowskian structure of spacetime; space and time measurement by light signals and free-falling particles leads to a Weyl spacetime. Field measurement by point-like particles leads to the accepted classical field theories. Considerations of measurability often go hand in hand with a requirement of objectivity: measurements performed by different observers or with different conventions should be interrelated in a consistent manner. The principle of relativity or general covariance expresses this sort of objectivity.

---

[2]In the preface (p. viii) and in Chapter 1 (p. 45), I had the "conceptual" type instead of the "Helmholtzian" type: the latter is what is left of the former after proper filtration.

A second kind of comprehensibility rests on the applicability of correspondence principles. In this case, the necessity of some features of a theory T is derived from its agreement with a theory T′ known to be (approximately) true in a restricted domain of experience. This agreement implies the existence of sub-theories that are approximations of the theory T, as well as the identity or the equivalence of one of these sub-theories with the theory T′. In combination with other arguments, a correspondence argument can be used to show the necessity of a theory. The strength of the demonstration of necessity depends on the quality of the other arguments and on the necessity of the restricted theory. The latter may be established empirically, or it may itself be derived by necessity arguments. Both circumstances are met in the case of non-relativistic mechanics as a correspondence-basis of relativistic mechanics, or in the case of classical mechanics as a correspondence-basis for quantum mechanics.

The third kind of comprehensibility rests on varieties of causality. The broadest variety is the stability of statistical correlations between measurements performed on the same system. This is the one admitted for quantum systems, and exploited together with "natural" assumptions on the type of correlations. In the classical case, we require a stricter kind of causality according to which the same cause creates exactly the same effect in similar circumstances. In addition, we may require causal relations at a given scale to be stable when the causes fluctuate at a finer scale: this is the secular principle, which can be used together with the previous principle in a derivation of Newton's second law and d'Alembert's principle. We may also assume the impossibility of perpetual motion, which is a kind of causality since it boils down to the impossibility of elevating a weight without a compensating alteration of the rest of the world. This principle not only contributes to a proof of the necessity of classical mechanics, but it also helps justifying the energy principle and the first principle of thermodynamics without appealing to mechanical reduction. The uniqueness of thermodynamic equilibrium, which we used in arguing some necessity of Gibbsian statistical mechanics, may also be seen as a kind of causality since it requires the uniqueness of the macrostate of a system under given macroscopic circumstance. The second principle of thermodynamics, expressed as the impossibility of spontaneous heat flow from a cold body to a hot body, derives from the uniqueness of equilibrium if we regard the state of equal temperatures as an equilibrium state.

Fourthly, the comprehensibility of nature may imply the possibility of obtaining large or complex systems and their evolutions by combining simple, homogenous units. This is the sort of argument used by Poincaré to establish the synthetic a priori character of group theory, to base the doctrine of space on group theory, and to explain the pervasiveness of (partial) differential equations in physics. Although this form of comprehensibility comes closest to Kantian transcendentalism, it is not as useful in deriving specific theories as the three other kinds of comprehensibility.

A last and most questionable kind of comprehensibility is the reduction of various theories to a single, more fundamental theory. For example, the energy principle, the principle of least action, and the laws of thermodynamics have all been justified by appeal to a reducing mechanics or dynamics. In these derivations the necessity of the result depends on the necessity of the reducing basis. Unfortunately, this basis is established for a limited range of observable phenomena and its extension to the full range of the derived principles is unwarranted. Historically, the general principles tend to be more stable than their reducing basis. This is not to say that nothing is learned from the reductionist attempts, or that their history should come to an end.

Comprehensibility and relativized transcendentalism

The comprehensibility of nature, as it is understood in this book, has evident similarities with attempts to relativize the a priori of Kant's transcendental philosophy. In both cases, the mind imposes its own conditions on our experience of the world, but there is no unique, forever determined choice of these conditions. Two strong motivations for modifying Kant's system were the nineteenth-century revolution in the foundations of geometry and the advent of relativity theory in the early twentieth century. An outstanding example of the first motivation is Helmholtz's attempt to rescue some of the Kantian doctrine after proposing his own foundations of geometry. According to Helmholtz, Kant had erred in regarding the axioms of Euclidean geometry as a necessary component of our understanding of space; but the general idea of space as a necessary form of external intuition was still valid. In addition, Helmholtz's assumption of the measurability of space through rigid bodies, which leads to spaces of constant curvature, can be seen as some categorical demand of our understanding, if one forgets—as many of Helmholtz's commentators have done to this day—that for Helmholtz this measurability is an empirical fact (*eine Thatsache*) whose validity may not be unlimited. Poincaré, in his own philosophy of geometry, departed even further from Kant by evacuating the whole idea of space as an external form of intuition and making the choice of the geometrical axioms a matter of convention. He nonetheless maintained a synthetic a priori core of geometry: the concept of continuous group that enables us to conceive arbitrary combinations of homogenous operations or displacements.[3]

In their philosophies of geometry, Helmholtz and Poincaré replaced Kant's static notion of the a priori with other static notions compatible with the removal of Euclidean geometry from its apodictic throne. They did not envision a more temporary conception of the a priori. This came in 1920 from two philosophers, Ernst Cassirer and Hans Reichenbach, as a reaction to Albert Einstein's relativity theory. Reichenbach clearly distinguished between two aspects of Kant's a priori: its apodictic certainty and its constitutive function. While he preserved the latter aspect as the faculty of the mind to constitute objects of experience, he argued that the constitutive principles should be modifiable when difficulties accumulate in the integration of new empirical results. He defined these principles as "principles of coordination" (*Prinzipien der Zuordnung*)[4] that gave empirical meaning to theoretical laws, for example the admission of arbitrary systems of coordinates in general relativity. He was not too clear on what gave constitutive value to such principles. In his later empiricist philosophy, he replaced them with "coordinative definitions" that were highly conventional and lacked any constitutive value.[5]

---

[3] On Helmholtz relation to Kant, cf. Hatfield 1990, DiSalle 2006b; Hyder 2009. On Poincaré's relation to Kant, cf. Ly 2007, Principe 2012.

[4] He borrowed this set-theoretical expression from Moritz Schlick, who used it in a narrowly empiricist philosophy in which theoretical concepts were mapped (*zugeordnet*) to empirical reality.

[5] Reichenbach 1920, pp. 46–7. Cf. Ryckman 2005, pp. 28–39.

Cassirer's own attempt to relativize Kant's a priori had broader scope and longer life. Cassirer saw Einstein's relativity theory as an outstanding illustration of the need to adapt our transcendental apparatus to new domains of experience, leaving aside every intellectual prejudice except the most basic demands of synthetic unity and coherence. Thus, Euclidean geometry and Newtonian mechanics had to yield to new rules of the understanding (*Regeln des Verstandes*) such as general covariance and the principle of equivalence. For Cassirer, no such rule is final or universal. It is only a "principle that the understanding uses hypothetically, as a norm of investigation, in the interpretation of experience." In his *Philosophy of symbolic forms* (1923–9), Cassirer broadly defended a multidimensional theory of knowledge in which different modalities of sense-giving were tentatively applied to different domains of experience. He advocated a productive interaction between various transcendental demands of determinateness, univocity, stability, simplicity, and unity. The precise expression and the relative weight of these demands depended on the explored domain, in a flexible manner solely conditioned by epistemic efficiency. His transcendentalism was open and dynamic, and it even offered suggestions for how to improve the frames in which we try to make sense of experience.[6]

Most recently, Michael Friedman has defended a notion of the "relativized a priori" which shares features of Reichenbach's and Cassirer's varieties of neo-Kantianism. Friedman admits the overall importance of constitutive principles that provide empirical meaning to the abstract formalism of a theory. These principles are necessary, but they are subject to revision. At a given moment of history, they are a response to a contingent intellectual situation. In his times Kant identified them with Euclidean geometry and Newton's laws of motion. In the times of special relativity, the light principle and the principle of relativity became the new constitutive principles. In the case of general relativity, the equivalence principle plays this role. Typically, Friedman's constitutive principles are former empirical laws elevated to the rank of presuppositions. They do not emanate from purely intellectual considerations, and it is impossible to predict what the next constitutive principles will be.[7]

This quick survey of a few attempts to relativize Kant's a priori should be sufficient to detect differences with the notion of the comprehensibility of nature that is promoted in this book. Firstly, the general endeavor is different. Whereas Reichenbach, Cassirer, and Friedman want to construct a new theory of knowledge, my purpose is to show that various physical theories result from ideas of the comprehensibility of nature. Reichenbach's or Friedman's constitutive principles, or Cassirer's rules of the understanding do not purport to determine the mathematical formalism of a theory. At best, they constrain the choice of this formalism. For Reichenbach and Friedman, the main purpose of the constitutive principles is to give physical meaning to the formalism. Friedman's principles further differ from comprehensibility principles in the way they relate to experience: whereas Friedman's principles originate in experimental laws whose validity came as a surprise (for instance the universality of free fall or the undetectability of motion with respect to the ether), comprehensibility principles correspond to natural demands of post-Aristotelian physics such

---

[6] Cassirer 1921, p. 82; 1923–9. Cf. Ryckman 2005, 29–46; Schmitz-Rigal 2002, 2009.

[7] Friedman 2001, 2009.

as measurability or causality (which makes them more similar to Cassirer's rules of the understanding). Typically, the next generation of theory corresponds to a weakening of older comprehensibility principles, that is, to a renunciation of formerly natural demands.[8] In contrast, Friedman's principles of relativistic physics are new principles which jeopardize former constitutive principles.[9] These principles match some basic presuppositions of Einstein's theories in their original formulation, whereas the comprehensibility principles used in the EPS or in the Helmholtzian derivation of relativistic spacetime express broad considerations of measurability. A last difference lies in the refutability of the principles. By definition, Friedman's principles are beyond refutation: they are preconditions for expressing any experimental result in theoretical language, and therefore would have to be assumed to be true in any attempted refutation. In contrast, most comprehensibility principles can be contradicted by experience, either separately or in combination with other principles.

In brief, comprehensibility principles have a theory-generating power and an empirical immediacy or "naturalness" that the principles of the relativized a priori do not have. The conjunction of these two qualities may seem paradoxical: how could principles so easily found in common experience have so far-reaching theoretical implications? As we are about to see, the answer lies in the circumstance that comprehensibility principles are not applied in an epistemological vacuum. They presuppose a general conception of physical theory in which comprehensibility can take a mathematically and empirically precise form.

Comprehensibility and theory

Comprehensibility arguments cannot be deployed without some prior theoretical notions. As is already clear in the mechanical case treated in the two first chapters, they require a mathematical definition of idealized systems, idealized experiments, and idealized measurements. These definitions usually involve partial theories that have a more limited domain of application than the theory under investigation. In addition, the comprehensibility arguments may directly imply intertheoretic relations, as is evidently the case for correspondence or reduction arguments. Consequently, necessity arguments are conceivable only to the extent that our general concept of physical theory meets the two following requirements:

1. A physical theory must be developed in a well-defined mathematical frame.
2. This mathematical frame must be structured so as to allow for applications to idealized physical systems and for a variety of intertheoretic relations.

The first requirement may be regarded as a consequence of basic arguments for the necessity of mathematics in physics, as argued in Chapter 6. The second requirement has to do with the fact that the comprehensibility of phenomena is not a constraint on the mathematical apparatus of a physical theory per se; it is a constraint on the way this apparatus conceivably relates to experiment.

---

[8] A similar remark is found in Mittelstaedt 2011.

[9] I suppose that Friedman's equivalence principle is stronger than the mere equality of inertial and gravitational mass, which already held in the Galileo–Newtonian theory of gravitation.

Is there a received view of physical theory that meets these requirements? Considering the important role of measurability as a comprehensibility requirement, it would seem that the operationalist view is a plausible candidate. In that view, however, quantities are directly defined by concrete measurement, whereas the deployment of necessity arguments requires mathematical models of idealized measurement procedures. As we saw for instance in the case of mechanics, the concrete realization of these procedures cannot be rigid. It typically evolves and diversifies as the domain of application of the theory widens. What matter for the necessity arguments is not the specific, variable concrete realizations of the measurement procedures but a shared mathematical model of them.[10]

A second candidate for meeting the aforementioned requirements is the "linguistic" or "syntactic" view of physical theory inherited from logical positivism. In this view, a theory is regarded as a set of propositions derivable from a small number of axioms and harboring empirical terms and propositions that have a stable concrete referent. Although this view acknowledges the necessity of purely theoretical terms in any theory that goes beyond a mere description of previously observed facts, it shares with naive operationalism the defect of rigidifying the connection between theory and experiment. Its distinction between theoretical and empirical terms is too sharp, and it generates pseudo-questions of the kind: "Is $\mathbf{f} = m\mathbf{a}$ a law or a definition of force?" At any rate, this view does not reflect the fact that physical theories are usually characterized in mathematical rather than linguistic terms.[11]

For these and other reasons, philosophers of physics nowadays favor the "semantic view" initiated by Patrick Suppes in the 1960s and championed by Frederick Suppe, Bas van Fraassen, and Ronald Giere among others. In this view, a theory is defined as a family of models, model being here understood in the sense of the mathematical "theory of models," namely: as a set-theoretical, real-number construct that may interpret a system of axioms if the theory is axiomatized. Thus, the basic definition of a physical theory is purely mathematical. The interpretation of the mathematical formalism is obtained by establishing a partial isomorphism between the models of the theory and concrete experimental devices or procedures. The linguistic description of the theory becomes secondary, because the same model may have different linguistic descriptions and because the models, not the terms of the description, link the theory to the real world.[12]

Unfortunately, in its most common forms the semantic view remains very vague on the way the correspondence between models and real systems is established. Patrick Suppes imagines a hierarchy of models of decreasing abstraction between the models of the theory and models of experimental data. Giere (in the case of mechanics) selects concretely realizable models among the theoretical models. Van Fraassen similarly evokes "empirical substructures." There is, however, a variety of the semantic view that goes deeper into the relation between models and experiment, through model-theoretic characterizations of "intended applications" and through intertheoretic relations that allow for piecewise relations between aspects of the model and experimental devices. This is the "structuralist" view developed by Joseph Sneed, Wolfgang Balzer, Ulises Moulines, and others. This

---

[10]On operationalism, cf. Chang 2009.

[11]On linguistic vs. semantic views, cf. Suppe 1989, Chap. 2; Vorms 2011, Chap. 2.

[12]Cf. Suppes 1960, 1967; Suppe 1974; Giere 1988; van Fraassen 1980; Vorms 2011, Chap. 3.

view improves on the bare semantic view by capturing some of the ways physicists actually connect theory to experiment. And it seems rich enough to be compatible with the deployment of comprehensibility arguments. Yet I will not follow it, because I find it safer to obtain a general definition of physical theory by inspection of existing theories. This definition will be found to partially overlap with the structuralist definition, with alterations that bring it closer to the physicists' practice. [13]

9.2 Theories and modules

Theories defined

The theories of modern physics share the following characteristics:

(a) The definition of a *symbolic universe* in which systems, states, transformations, and evolutions are defined by means of various magnitudes based on powers of $\mathbf{R}$ (or $\mathbf{C}$) and on derived functional spaces and algebras.

(b) The postulation of *theoretical laws* that restrict the behavior of systems in the symbolic universe.

(c) The description of *interpretive schemes* that relate the symbolic universe to idealized experiments.

(d) Methods of *approximation* that enable us to derive the consequences that the theoretical laws have on the interpretive schemes.

The strictness with which these characteristics are distinguished from each other and the order in which they appear in the exposition of a theory are matters of taste. In the actual practice of theoretical physics, they are almost never formulated in a rigorous, systematic manner. They are usually impregnated with intuitive, semi-concrete considerations. Yet it is hard to think of a physical theory that does not share them.

Roughly, (a) and (b) correspond to what physicists usually call the formalism of a theory, and (c) to its interpretation. The distinction between (a) and (b) is somewhat conventional, because some laws may be integrated in the definition of the components of the symbolic universe. Theoretical laws should not be confused with empirical laws. Whereas the former belong to the theory and exactly hold in its symbolic universe, the latter are meant to have an empirical meaning that does not depend on their insertion in any particular theory. The interpretive schemes vary during the life of a theory. Some of them are invented during the construction of the theory. Others appear in the course of subsequent applications. For the theory to have any empirical relevance, some of the interpretive schemes and some of the idealized experiments based on them must admit concrete realizations. Other interpretive schemes are purely imaginary and may support only thought-experiments. Even in concretely intended cases, the schemes do not include any predetermined correspondence rules between theoretical objects and concrete objects. As will later be clear, they belong to the symbolic universes of the theory and of its modules. With few exceptions, the application of the theoretical laws to interpretive schemes leads to equations that

---

[13]Suppes 1962; van Fraassen 1980, p. 64; Giere 1988, p. 49; Sneed 1971; Balzer, Moulines, and Sneed 1997. On the structuralist approach, cf. Torretti 1980, Chap. 3.

cannot be solved exactly. Approximations therefore play an essential role in preparing the confrontation with experiments. In some theories, such as quantum electrodynamics, the symbolic universe itself is given only through successive approximations.

As was announced, this characterization of physical theories has antecedents in the semantic and structuralist views of theories. The symbolic universe corresponds to the "family of models" of the semantic view, or to Sneed's "potential models." Sneed's "models of the theoretical core" are those of the potential models that satisfy the laws of the theory. His "intended applications" resemble my interpretive schemes for they refer to the concrete situations one wishes to grasp with the models. The resemblance is only vague: as will be clear in the following, the two notions differ in important respects. My own notions of a symbolic universe and its interpretive schemes will be illustrated by a quick survey of a few theories of physics.[14]

As a first simple example, take the Newtonian mechanics of a finite number of mass points. The first component of the symbolic universe is the definition of a generic system by the number N of mass points, the list of their masses, the $3N$-dimensional configuration space (the Nth power of a three-dimensional Euclidean space), the one-dimensional affine space of time, the $3N$-dimensional space of forces (the Nth power of another three-dimensional Euclidean space), and an unspecified force function that determines the forces for a given configuration or for a given time. The symbolic universe also contains evolutions of the configuration as (differentiable) functions of time. The basic law of the theory is Newton's law relating force, mass, and acceleration.

The interpretive schemes can be of various kinds. One possibility is that the scheme consists in the selection of a system together with the description of ideal procedures for measuring spatial configuration, time, force, and mass. Then an idealized experiment may consist in the verification of the motion predicted by the theory for given initial positions and velocities of the mass points and for a properly selected frame of reference. Another possibility is that the observation of the motion of a particle is used to determine the force field on which the forces impressed on the mass points depend. Subsequent experiments can verify that the motion proceeds according to Newton's law of acceleration for given initial conditions and for an adequate choice of the masses. The predictive power of the theory derives from the assumption that the forces are the same functions of the configuration of the system in every conceivable experiment.[15] In well-known domains of applications, the force field is determined by adopting a specific force law, for instance Newton's law of gravitation, in the definition of the system; the interpretive schemes then involve time and space measurements only.

In practice, there are two ways of determining the adequate force field. Firstly, the measured trajectory of one of the point masses can be used to compute its acceleration and a functional relation with the measured configuration of the other masses is sought. This procedure can only succeed if the motion is simple enough. In the case of direct action at a distance between the various mass points, this implies that only two masses are present. If the number of masses is beyond experimental control (as is the case in celestial

---

[14] Balzer, Moulines, and Sneed 1987, chap. 2; pp. 38,86–9 for the intended applications.

[15] Balzer, Moulines, and Sneed (1987, pp. 40–7) take this into account.

mechanics), a strategy of approximation is necessary in which the two-body problem is taken as a starting point. Alternatively, physicists may compute the motion in interpretive schemes in which the force field is any simple given function of the configuration. With some luck, they may find out that for one of these simple choices of the force function the derived motion fits experimental data.[16]

From this brief analysis of the case of the mechanics of mass points, we see that interpretive schemes contain parameters (the masses and the force field) that are sometimes directly measured, sometimes predetermined in the symbolic universe, sometimes induced from experiments, sometimes obtained by trial and error. This is a general subtlety in the function of interpretive schemes to which I will not return, although it concerns a basic presupposition, the generic character of physical systems, without which physics would not exist at all. The only point I wish to emphasize is the inclusion in the interpretive schemes of ideal measuring procedures, whose number and variety depends on the intended ways of applying the theory. In most cases, only a fraction of the quantities defined in the symbolic universe are intended to be measured. How the measuring procedures are defined will be the subject of later considerations.

As a second example of a physical theory, take the thermodynamics of a gas system. The symbolic universe entails a two-dimensional manifold on which pressure, volume, and temperature fields are defined, as well as differential forms of heat and work. Basic laws are the statements that the integral of the sum of the heat and work forms vanishes over a cycle (energy conservation) and the statement that the integral of the heat form divided by the temperature vanishes over a reversible cycle (existence of the entropy function). Interpretive schemes may involve measurements of various thermodynamic coefficients and serve to verify the relations that the laws of thermodynamics imply between those coefficients. Alternatively, thermodynamic engines can be described with the gas as the working substance. The theory has well-known verifiable consequences on the efficiency of these engines. Again the interpretive schemes involve ideal measuring procedures for pressure, volume, temperature, and ideal processes for conveying heat and work to the system (in the second case).

A third simple case is that of general relativity. Its symbolic universe is made of various fields, including the metric field, defined over a 4-dimensional, differentiable, and locally Minkowskian manifold. Fundamental laws are Einstein's equations and generally covariant equations for the non-metric fields. Interpretive schemes imply procedures of geodesic surveying mostly based on light signals, and measurements of quantities associated with non-metric fields. Typical verifiable predictions are deviations from motions predicted by Newton's theory of gravitation, the gravitational red-shift, the gravitational deviation of light, gravitational waves, black holes, etc.

The symbolic universe of statistical mechanics can be defined through probability distributions in the phase space of mechanical systems with many degrees of freedom. One basic law is the microcanonical distribution for representing the equilibrium properties of an isolated system. Interpretive schemes imply the identification of measurable macroscopic quantities with ensemble averages of some functions of the phase of the system.

---

[16]This is the situation emphasized in Giere's (1988) and Cartwright's (1983, pp. 135–9) conceptions of models.

Idealized experiments are meant to test relations between those quantities, and sometimes to measure their statistical dispersion.

Quantum mechanics has a fairly complex symbolic universe, namely, a (rigged) Hilbert space of infinitely many dimensions in addition to other relevant magnitudes such as time, mass, and external fields. The basic law is given by Schrödinger's equation. Interpretive schemes involve correlations between the measures of classically defined attributes of the particles (such as position and momentum) and of the fields (such as intensity and frequency in the case of electromagnetic radiation). Although most of these correlations are irreducibly statistical, they may also include non-statistical results such as the energy spectrum of the stationary bound states of a system of charged particles. The complexity of the symbolic universe and of the interpretive schemes varies with the type of system considered (single particle in external fields, several interacting particles, quantum fields).

Judging from this sample of theories, an interpretive scheme consists of a given system of the symbolic universe together with a list of characteristic quantities that satisfy the following three properties:

1. They are selected among or derived from the (symbolic) quantities that define the state of this system.
2. At least for some of them, ideal measuring procedures are known.
3. The laws of the symbolic universe imply relations of a functional or statistical nature among them.

In general, the schematic quantities represent only a tiny fraction of what could be conceived in the symbolic universe. Some filtering or extracting process is needed to move from symbols to schemes. This is especially true in the case of statistical and quantum mechanics, for which no direct correspondence exists between states of systems of the symbolic universe and measured quantities.

The characteristic quantities of a given interpretive scheme are divided into measured quantities and more theoretical quantities. Only the former quantities are measured in experiments based on the scheme. The theoretical quantities either are the unknowns that the experiments aim to determine, or their value is taken from empirical laws established by preliminary experiments. I do not mean the measured quantities to be non-theoretical. In the case of indirect measurement, the measuring procedure is dictated by the theory itself. As noted by Roberto Torretti, even in supposedly direct measurements (that is, based on purely empirical recipes or on external theories) the laws of the theory may play a role in filtering out background noise or in identifying components of the measuring device.[17] It remains true, however, that the interpretive schemes usually involve more quantities than are meant to be measured and that the laws of the theory do not dictate all the measuring procedures included in the interpretive schemes.[18]

The interpretive schemes do not automatically follow from the symbolic universe. The laws of the symbolic universe in themselves do not determine the filtering process

----

[17] Torretti 1980, pp. 147–8 for Pickering's example of background noise in the discovery of the neutral current, pp. 136–7 for reference systems in mechanics.

[18] For this reason, my interpretive schemes differ both from Sneed's "partial potential models" and from his "intended applications."

that yields the schemes. Nor does the symbolic universe contain the distinction between measured and theoretical quantities. In general, the class of interpretive schemes increases in time together with the variety of applications of the theory. This evolution sometimes involves the promotion of some symbolic quantities to the schematic level. For instance, late-nineteenth-century studies of gas discharge and cathode rays provided experimental access to the invisible motions assumed in the electron theories of Hendrik Lorentz, Joseph Larmor, and Emil Wiechert.

Modules defined

The necessity and the variability of interpretive schemes raise a few fundamental questions. How do physicists arrive at them? How do they relate to concrete devices? How should we conceive the articulation between symbols, schemes, and experiments? I believe the modular structure of theories is the key to a reasonable answer to these questions. Namely, every theory contains or is related to other theories with different (usually more restricted) domains of application, which I call *modules*.[19]

Modules already occur in the symbolic universe. For instance, the definition of mechanical systems and their states involves a geometric module. In thermodynamics, the definition of pressure and work implies a mechanical module; other thermodynamic parameters may involve other modules (electrodynamical, chemical, etc.); temperature and heat involve a calorimetric module. Mechanical ether theories imply a mechanical module in the construction of their symbolic universe. Statistical mechanics implies a mechanical module for the definition of phase space.[20]

The occurrence of modules in the symbolic universe may or may not reflect the needs of schematic interpretation. It does so in the case of geometry as a module of mechanics and in the case of mechanics as a module of thermodynamics; it does not in the case of mechanics as a module of statistical mechanics or of mechanical ether theories. In the latter case, the mechanical module serves to construct the entire symbolic universe of the theory, and most of the mechanical attributes of symbolic systems are not found in interpretive schemes. For instance, James Clerk Maxwell's rotating-wheel mechanism of the ether is not meant to be accessible to any conceivable experiment. Such theories are to be contrasted with phenomenological theories in which the defining modules do not define more of the symbolic universe than is needed for schematic interpretation.

Interpretive schemes involve modules that may or may not be part of the symbolic universe. In phenomenological theories such as mechanics, thermodynamics, and macroscopic electrodynamics, all modules belong to the symbolic universe. In microphysical theories such as statistical mechanics or quantum mechanics, the schemes bring additional modules.

---

[19] Modularity has often been evoked in biology, as a condition for the evolvability of living organisms. It is also a central feature of Jerry Fodor's cognitive psychology, which ascribes different modules to lower-level tasks and a central module that coordinates them (Fodor 1983). Whereas Fodor had in mind a static modularity of the brain, other psychologists have argued for a gradual and adaptable modularization (e.g. Karmiloff-Smith 1992). Despite evident analogies, I did not base my reflections on these considerations.

[20] There seems to be a partial overlap between my notion of defining modules and Friedman's above-mentioned constitutive principles (Friedman 2001, pp. 45–6).

This happens because at least some of the schematic quantities are not primitive quantities of the symbolic universe. They are derived from combinations and averages of such quantities, and they may, in some useful approximation, be related by a module. The knowledge of this module is important both in conceiving the schemes and in using them as a basis for experimentation. The schemes of statistical mechanics thus imply a thermodynamic module; those of quantum mechanics imply a classical module.

The latter kind of schematic modules may be judged inessential if they are in some sense derivable from the laws of the symbolic universe. In practice, however, we would not know how to apply the theory without being aware of them. Modular structure seems indispensable in defining the quantities that are actually measured in experiments, even when the investigated relation between these quantities contradicts their modular affiliation. For example, the schematic expression $\lambda \propto \sqrt{Tt}$ of the average distance λ that a Brownian particle travels in the time t involves the thermodynamic temperature T, even though the thermodynamic module does not admit Brownian fluctuation.

At any rate, modules play an important role in imagining tests of a theory. The conception of quantum experiments usually appeals to much classical or semi-classical reasoning. Microphysical experiments always suppose a classically described environment. In the case of early cathode-ray experiments, macro-modules directly apply to the electrons. This is so because the trajectories of the electrons can be visualized, despite their microphysical character. In the case of quantum mechanics, macro-modules are used only in the description of measurements. For example, classical electromagnetism and classical mechanics are used in velocity measurement by magnetic deflection; classical optics is used in the spectral analysis of the light emitted by atoms.

According to Carl Friedrich Gauss and Niels Bohr, all observations in physics are ultimately macro-mechanical. Hence the schemes must involve macro-mechanical quantities belonging to modules or sub-modules of the theory. This does not mean, however, that interpretive schemes involve only such quantities. This sort of instrumentalism would miss the importance of theoretical quantities and of higher-level modules in the schemes belonging to any non-trivial theory. The generality and unity of theory implies a departure from the world of immediate experience, even at the level of schemes.

For a given class of schemes and when a given parameter of the theory becomes very large (or very small) the relations between the schematic quantities may reach a limiting form that is deducible from a simpler theory. I regard this approximating theory as a module, because once it has been identified physicists and engineers use it instead of the more general theory whenever their concern is limited to phenomena in which the approximation is valid. Such modules may or may not be known before the theory of which they are an approximation. When they are, the older theory is usually said to have been reduced to the newer one.[21]

For properly selected classes of interpretive schemes and under a given condition on the characteristic quantities, the theory may be exactly equivalent to a more specialized theory.

---

[21] As will be discussed in a moment, this common way of expression is philosophically dangerous, for at least three reasons: it wrongly suggests the dispensability of the older theory; it does not take into account the contextual character of the reduction (reference to a specific class of interpretive schemes); and it confuses the two distinct concepts of reduction involved in reducing modules and in approximating modules.

The symbolic universe of the specialized theory is obtained by extracting a substructure of the symbolic universe of the more general theory, and its laws must be such that their implications for this substructure are equivalent to those of the original laws. For instance, the electrodynamics of bodies at rest applies to interpretive schemes in which all material bodies are at rest; electrostatics applies to such interpretive schemes under the additional condition that every current vanishes and other electric quantities (charge, potential) do not change in time. This sort of specialization pervades the teaching and application of theories.

A last way to relate a theory to another is to simplify its symbolic universe while preserving the interpretive schemes. The resulting theory may no longer be a decent theory of the investigated domain; some of its predictions may be evidently false. Yet theoretical physicists often explore such theories when the more realistic theory resists mathematical analysis. They thus hope to explain features of the investigated domain that are assumed or known to resist the simplification. The Ising model of ferromagnetism or Maxwell's use of a $1/r^5$ force law in his kinetic theory of gases, are good examples of such simplified theories. Their usefulness is contingent to the mathematical workability of the original theory, which may change in time. They are nonetheless an essential part of our understanding of theories, as they reveal generic structures that do not depend on details of the systems under consideration.

To summarize, modules occur in the symbolic universe, in the schemes, and in limits of the schemes. Since by definition they are themselves theories, they also contain modules, submodules, and so forth until the most elementary modules are reached. There are (at least) five sorts of modules. In reductionist theories such as the mechanical ether theories of the nineteenth century, there is a *reducing module* diverted from its original domain to build the symbolic universe of another domain. In many theories, the symbolic universe also appeals to *defining modules* that define some of the basic quantities. For instance, mechanics is a defining module of thermodynamics. When the symbolic universe is too complex to be directly understood, it can be explored through *outlining modules* obtained by simplifying the symbolic universe and keeping the same interpretive schemes. There are *schematic modules* that occur at the level of interpretive schemes. There are *specializing modules* that are exact substitutes of a theory for subclasses of schemes under certain conditions. For instance, electrostatics is a specializing module of electrodynamics. Lastly, there are *approximating modules* that can be obtained by taking the limit of the theory for a given subclass of schemes. For instance, geometrical optics is an approximating module of wave optics. These categories are not mutually exclusive: for example, a schematic module can also be a defining module or an approximating module.

Thus we see that there are diverse ways in which the full exposition of a given theory calls for other theories. My choice of the word "module" is intended to convey metaphorically this diversity as well as the fact that the same theory can be a module of a number of different theories. For instance, classical mechanics is a module of electrodynamics, thermodynamics, quantum mechanics, general relativity, etc.; and it can be so in different ways. At any given time, any non-trivial theory has a modular structure, namely: it includes a number of modules of the above-defined kinds.

The modular structure of a theory is not unique and invariable. It depends on a number of factors: the conception we have of this theory, the type of experience that is conceivable at a given period of time, the degree of elaboration of the theory, etc. As an example of the first factor, for some nineteenth-century physicists mechanics was a reducing module of electrodynamics; for phenomenologists it was only a defining module; for believers in the electromagnetic worldview, it was a schematic module. As an example of the second factor, approximating modules for the description of stochastic processes appeared in statistical mechanics only after the development of relevant experiments. As an example of the third factor, the boundary-layer approximating module of hydrodynamics appeared only at a late stage of its evolution, even though it concerned an old domain of experience.

This ambiguity and variability of modular structure may explain why philosophers of physics have paid little attention to it. It seems to elude any formal, rigorous epistemology. It seems too fleeting and too vague to embody the epistemic virtues that philosophers wish to find in physical theories. Against these appearances, I will now argue that modular structure is essential to the application of theories, to their comparison, to their construction, and to their communication. These four aspects of theorizing activity will thus appear to be intimately related to each other. Moreover, modular structure will acquire some sort of necessity: without it physical theories would remain paper theories.

9.3   The necessity of modularity

Theories applied

The symbolic universe of a theory never applies directly to a concrete situation. The application is mediated through interpretive schemes that describe ideal devices and quantitative properties of these devices. In order to build a concrete counterpart of a scheme, we must know the correspondence between ideal device and real device, as well as concrete operations that yield the measured quantities. In any advanced theory this correspondence obtains in a piecewise manner, through the modules involved in the scheme. So to say, the schemes are blueprints, and the modules help us select the materials for realizing them. The most superficial observer of a modern test of a theory cannot fail noticing the contrast between the simplicity of the theoretical statement to be tested and the complexity of the experimental setting. What enables physicists to make sense of this complexity is, for a good part, the modular structure of schemes.[22]

The modules enable us to exploit the competence we have already acquired in applying the modular theories. This application may involve submodules and their schemes, and so forth until the concrete operations become so basic that their description can be expressed in ordinary language. Take the relatively simple case of mechanics. The schemes involve a geometric module, which one already knows how to realize by means of surveying with rigid rods (for example). This knowledge is essential in building the apparatus and realizing the relevant measurements. Other useful modules may be kinematics and statics.

Although modular structure is a necessary guide in the application of theories, it is not a sufficient one. Much of the necessary knowledge has usually been acquired during the formation and earlier applications of the theory and its modules. For instance, in the applications of the mechanics of connected systems, we have learned how to realize

---

[22] This is somewhat similar to the hierarchy of mediating models in Suppes 1962; 1967, pp. 62–4.

(approximately) frictionless gliding, or rolling without gliding. This implies some know-how for the choice of materials, the preparation of surfaces, lubrication, etc. Much of this knowledge may be tacit, learned by manipulating rather than reading.

Even from a purely theoretical point of view, schematic modules may not completely determine measured quantities. For instance, the module of Euclidean geometry does not suffice to the determination of spatial relations in a mechanical system. The reference system must be chosen in conformity with the law of inertia, which is a basic law of the symbolic universe.

Moreover, not every measured quantity is defined through a module. This could happen only at the earliest stages of the application of a theory, before any measuring scheme specific to this theory has been developed. The application of mature theories usually involves indirect measurements that imply the laws of these theories themselves. For example, the measurements of time, mass, and force in mechanics are not determined by pre-mechanical modules. Their connection to modular space measurement appeals to conventions, to empirical laws, and to the laws of mechanics. In the case of electrodynamics, the initial stages of theory formation involved only mechanical measurements. The mature theory relied on the laws discovered trough such measurements (Coulomb's law, Biot and Savart's law) to define measurements of charge and current. Note that these definitions were done through electrostatic and quasi-stationary modules that existed before full electrodynamics existed. In more advanced physics, measuring devices or detectors are too complex to allow for a complete theoretical analysis of their functioning, no matter how detailed a modular structure is available; theory tells us only that there should (approximately) be a correspondence between input and output. More precise knowledge of this correspondence requires calibration and simulation.[23]

Lastly, it is not even true that the applied theory and its modules are the only theories used in the applications. Very frequently, the relevant measurements involve techniques that rely on external theories. For instance, the determination of the spatial configuration of a mechanical system often relies on optical methods based on the laws of optics. This is possible because geometry is a module of optics. In general, the measurement of a modular quantity may involve any theory of which the relevant modular theory is a module.

To sum up, the modular structure of a theory greatly helps in the realization of its schemes, although non-theoretical knowledge and external theories are also involved. This complication indirectly points to two additional virtues of modules. Firstly, the non-theoretical knowledge implied in the application of a given theory can be exploited in the application of any other theory that contains this theory as a module. Secondly, when two theories share the same module, the applications of one theory may benefit from the other theory in the measurement of modular quantities. For instance, electronics can be used in building galvanometers, relativistic mechanics in building oscilloscopes, and optics in measuring distances; aerodynamics can be used for correcting pendulum measurements, elasticity theory for correcting weighing scales, etc.

The modular structure of theories affects the discussion of their refutability. The freedom in defining the schemes and the tacit knowledge involved in their concrete realization seem

---

[23]On the latter point, cf. Franklin 1990; Galison 1997, Chap. 9.

to leave plenty of room for protecting theories from refutation. In reality, the modular structure severely limits protective strategies because it restricts the form of the schemes and because it tends to confine tacit knowledge in the application of well-understood modules. To the extent that experimental error and reasoning lapses can be avoided, the accommodation of contradictory experiments is made difficult. Surely there still is some sort of Lakatosian protective belt: as long as no better alternative theory is available, physicists prefer to modify the symbolic universe or the non-modular components of the schemes. But the modules themselves usually remain untouched. Duhemian holism, or unrestricted "open-endedness" do not occur in the actual practice of physics. The modular structure of theories conveys to them much more rigidity in their adaptation to the empirical world than some historians would have it.[24]

So far I have done as if the experiments that test theories were always designed for that purpose. This is obviously not the case. In Ian Hacking words, "experimentation has a life of its own," and its relevance to higher-level theory often comes a posteriori.[25] In this case, the interpretive schemes of a theory cannot be used to describe the experimental protocol. In an autonomous experimental world, experiments are defined through *descriptive schemes* that characterize the system, relevant physical quantities, and procedures in a formal, ideal manner without regard to the nitty-gritty of concrete realization. For example, the determination of the emission spectrum of a vapor involves the description of a partially evacuated tube with electrodes subjected to a generator of high electric tension, as well as a spectrometer that may include a diffraction grating and a system of lenses. Clearly, the description of the system involves a combination of the interpretive schemes of various partial theories. The definition of relevant physical quantities is also done by partial theories, even if the measuring devices are used as black boxes (a most frequent occurrence). By analogy with the modules of theories, we may conveniently speak of a modular use of theories in experiments. Theories are modules for a given experiment when they serve only to describe the experiment, not to derive its outcome.[26]

This use of modules permits an autonomous experimental physics, yielding experimental laws defined as relations between the quantities occurring in the descriptive schemes of experiments. A higher-level theory is said to explain an experimental law if a correspondence can be established between a system in the symbolic universe of this theory and an experimental system, and if the laws of the symbolic universe imply the law under consideration.

Theories compared

The comparison of two theories obviously requires a non-vanishing intersection of their domains of application. In my terminology, this means that the two theories should share the same subset of schemes. More exactly, the characteristic quantities for a subset of schemes in one theory should be the same for a subset of schemes of the other theory.

---

[24] On constraints in theory construction, cf. Galison 1995 (in favor) and Pickering 1995 (against).

[25] Hacking 1983, p. xiii.

[26] My descriptive schemes are somewhat similar to the "theory of experimental design" discussed in Suppe 1989, pp. 134–41.

This can only be the case if the characteristic quantities are defined through modules that belong to both theories. Once this condition is met, the predictions of the two theories are said to agree if and only if the laws of the two theories imply the same relations between the schematic quantities in the compared subsets of schemes. The physicists' practice of comparison always involves schematic quantities defined by shared modules. Radical incommensurability is only a philosophical fiction.

In particular, there are crucial experiments that enable physicists to decide between two competing theories. The crucial character of an experiment requires the sharing of the modules involved in its scheme, as well as the exclusion of ad hoc modifications of the symbolic universe. A famous example is that of the experiment that François Arago performed at the Paris Academy of Sciences in answer to an objection to Augustin Fresnel's diffraction theory. A supporter of Newton's older theory, Siméon Denis Poisson, had noted that according to Fresnel's theory there should be a bright point of light in the middle of the shadow cast by a disc. The corpuscular theory, even in a version including deflections of the rays by the rims of the disc, could not possibly yield this bright point. Arago's experiment confirmed the prediction of the wave theory. The experimental setup only involved geometric and primitive photometric modules that both theories shared. Their predictions were clear-cut, with no tolerable tampering on their symbolic universe.[27]

Interestingly, this allegedly crucial experiment did not immediately persuade Poisson to give up the corpuscular theory. The reason for this apparent stubbornness was that the wave theory of light, which worked so well in the explanation of interference and diffraction, had not yet been proven to contain geometrical optics as an approximation. It is only after Poisson himself provided such a proof that he gave up the corpuscular theory. More generally, the lack of decision between two incompatible theories may come from insufficient development of their modular structure.

Another, slightly more refined example of a crucial experiment is the one Helmholtz performed to decide between his potential-based electrodynamics of 1870 and Maxwell's theory. In this experiment, a bowtie-shaped blade rotates within a perpendicular, homogenous magnetic field. The extremities of the blade periodically face two fixed plates that are connected to an electric condenser whenever the same blade faces the same plate. According to Maxwell's theory, the electric current induced in the rotating blade causes a potential difference between its extremities so that the condenser gradually acquires a detectable charge. According to Helmholtz's theory, there is no induced current and no charge at all in the condenser. To Helmholtz's surprise, the outcome refuted his own theory! It could do so because the scheme of the experiment only involved shared modules of magnetostatics, electrokinetics, and electrostatics (and a geometric sub-module, of course).[28]

In rare cases, the two compared theories do not share basic defining modules such as Euclidean geometry or mechanics. This happens for instance when the predictions of classical and relativistic electron dynamics are compared, or when the predictions of Newton's theory of gravitation are compared with those of general relativity. It would seem that in such cases the shared interpretive schemes could only involve pre-spatial

---

[27] Cf. Buchwald 1989, appendix.   [28] Cf. Darrigol 2000, pp. 232–3.

and pre-mechanical observations about the coincidence of two small material objects or the emission and reception of light flashes. This very limited conception of interpretive schemes may in principle allow the comparison of the two theories, for it permits an idealized coordination between theory and simple concrete procedures.[29] In practice, however, physicists never work on a tabula rasa devoid of Euclidean theory, Newtonian mechanics, and other pre-relativistic theories. Comparative schemes involve approximate, local use of these older theories in a complex manner that would deserve systematic study. At any rate, the astronomical tests of general relativity all involve earth-based or satellite-based instruments whose internal design requires earlier accepted geometry and optics, even though the tested spacetime relations are essentially non-Euclidean and non-Minkowskian.

The comparison between two theories may lead to the approximate inclusion of one theory into the other, also called reduction. In this case, the schemes of the reduced theory must correspond to a subset of those of the more general theory. The sharing of schematic modules is trivial, since by our definition of modules, the reduced theory is itself a module of the general theory. What is less trivial is the necessity of defining the schemes of the reduced theory. In a common misconception, the reduction of a theory to another is regarded as a mere limiting process involving a characteristic parameter of the more general theory (for instance c in relativistic mechanics, h in quantum mechanics) and some correspondence between the theoretical quantities of the two theories. In reality, one must introduce the schemes that define the domain of the reduced theory. Limits performed in the symbolic universe alone are ambiguous and lack definite empirical applicability.[30]

In many cases, the limiting or approximation process that leads to the allegedly reduced theory is non-rigorous and partly guided by intuition or experimental data. For instance, there exists no rigorous generic proof of the validity of the boundary-layer approximation for resisted flow at high Reynolds number. As often happens with modular dependence, the relation between hydrodynamics and its boundary-layer module is a deep and useful relation and yet cannot be regarded as a straightforward deduction.[31]

Another frequent feature of approximating modules is that they exhibit qualitatively novel properties. This happens when the limiting process leading to the module is singular. For instance, discontinuous phase transitions appear in the infinite-volume limit of statistical mechanics.[32] Moreover, the alleged reduction may fail for some of the shared interpretive schemes. This happens at critical points in the case of the reduction of thermodynamics to statistical mechanics. It also happens for the so-called catastrophic optics that occurs in the vicinity of caustics in the case of the reduction of geometrical optics to wave optics. This does not mean that no simple regularity exists in these singular situations. They may exhibit special symmetries such as scale invariance that imply a "universal" behavior independent of the details of the system under consideration. Loosely speaking, these behaviors may be regarded as "asymptotic" modules that replace the approximating modules

[29] Friedman (2001, pp. 76–8) suggests so much.

[30] A striking example of this ambiguity is that of Galilean electrodynamics as an approximation to relativistic electrodynamics: cf. Le Bellac and Lévy-Leblond 1973. For criticisms of naïve reductionism, cf. Sklar 1967; Schaffner 1967; Torretti 1980, pp. 155–60; Barberousse 2000, pp. 88–94.

[31] Cf. Heidelberger 2006; Darrigol 2005, pp. 283–302.   [32] Cf. Liu 1999.

wherever these fail. As Robert Batterman recently argued, the physics that occurs in these singular situations is so rich and so novel to deserve the epithet "emergent."[33]

Although the sharing of schemes and their modules is necessary and sufficient for the empirical comparison of two theories, this comparison can be supplemented with strategies that affect the symbolic universes of the two theories. For example, the comparison between geometrical optics and wave optics is eased by the Hamiltonian, eikonal reformulation of geometrical optics, in which the surfaces orthogonal to bundles of rays become the counterpart of wave surfaces. Another example is Helmholtz's comparison between Maxwell's electrodynamics and German electrodynamics, which relied on a reformulation of the former theory as a module of a generalized potential theory (with polarizable ether) that also included German theories as particular cases. Although such strategies do not avoid the necessary definition of comparative schemes, they ease the finding of schemes in which the differences between the predictions of the two theories best come out.[34]

Theories constructed

From history we learn that theory construction is a very complex process, depending on diverse resources both internal and external to the investigated domain. This complexity of what Reichenbach called the "context of discovery" has often discouraged philosophers from finding any rationality in it. Yet a closer analysis of the practice of modern theoretical physics shows that the construction of theories is highly constrained and that at some stages it may proceed almost automatically, as if the plan were known in advance. This is the sort of historical necessity we have seen at work in the history of thermodynamics or in the history of quantum mechanics. Well-known constraints in theory construction are experimental laws and general principles such as the conservation of energy or the principle of least action. Less appreciated is the fact that the construction of a new theory always relies on earlier theories in specifiable ways. In other words, some anticipation of the modular structure of a theory efficiently guides its construction.

Most generally, theory construction depends on defining modules whose validity is assumed from the start. For example, the construction of Newtonian mechanics presupposed the module of Euclidean geometry; and the construction of electrodynamics presupposed the module of mechanics (at least in the definition of forces). Such defining modules occur both in the symbolic universe and in the interpretive schemes. They sustain our theoretical imagination in a concrete manner, in direct connection with measurement possibilities.[35]

In a less universal and less concrete mode, theory construction may rely on reducing modules, as was for instance the case in Maxwell's first derivation of his electromagnetic field equations. The analogy between magnetic phenomena and rotational motion inspired Thomson's and Maxwell's idea that the electromagnetic ether could be a connected system with internal rotations to be identified with the magnetic field. And the consistent

---

[33] Batterman 2002.

[34] This is only one case of the importance of formulations and reformulations of a theory, as argued in Vorms 2011.

[35] The defining modules thus share some virtues of Friedman's constitutive principles, as already pointed out in n. 20.

development of this idea led to Maxwell's equations. Although Maxwell suppressed the mechanical model in the final version of his theory, he retained a broader principle of Lagrangian structure. This is only one example of a historical process in which a reducing module evolves into a general principle of a more abstract nature. Our theories are full of such vestiges of past modular reductions.[36]

In the development of his mechanical model of the ether, Maxwell was also guided by his desire to integrate electrodynamic, electrostatic, and optical modules in the same theory. In this case modular structure played a double role: in founding a reductionist strategy, and in bringing together different partial theories as modules of a new theory. To sum up, reduction and unification are modes of theory construction that explicitly depend on modular structure. There are two kinds of reduction of a theory to another: one in which the second theory is a reducing module of the second, and another in which the first theory becomes an approximating module of the second. The unification of two or more theories is a process following which these theories end up being approximating or specializing modules of the same theory.

When a reduction begins to fail, this failure may have an interesting feedback on the reducing theory. For instance, Maxwell's reduction of optics to electromagnetism failed to account for optical dispersion and for basic facts of the optics of moving bodies. This failure motivated the electron theories of Lorentz, Larmor, and Wiechert. The classically-based kinetic theory of gases gave wrong values for their specific heats, unless some mysterious freezing of internal degrees of freedom was assumed. Although this failure did not by itself trigger the quantum revolution, its later elucidation in a quantum approach gave a boost to the nascent quantum theory.[37]

Failed unifications may also be strong incentives for conceptual innovation. For instance, Max Planck's failure at a unified kinetic theory of gases and thermal radiation led to the quantum. In Einstein's view, the proper framework for unifying gas and radiation theory was statistical mechanics regarded as a reciprocal connection between a largely unknown microdynamic module of and a thermodynamic macro-module. In particular, Ludwig Boltzmann's relation between entropy and probability served to infer some of the quantum properties of the micro-world. Modular structure was therefore essential to Einstein's method, in a peculiar way in which a basic module of the future theory was regarded as an unknown.[38]

There is another interesting way in which modular structure mattered for the emergence of early quantum ideas. In Jürgen Renn's terms, Planck and Einstein reasoned on "mental models" that combined different theories of physics. For instance, Planck mixed electromagnetism and thermodynamics through the model of resonators immersed in cavity radiation. Einstein mixed electromagnetism, kinetic gas theory, and thermodynamics through the model or resonators interacting both with an ambient gas and radiation. In my terminology, these mental models correspond to combinations of interpretive schemes of different theories. They do not need to have concrete counterparts. Their hybrid character

---

[36] Cf. Buchwald 1985; Siegel 1991.

[37] Cf. Buchwald 1985 on electron theories; Kuhn 1978, Chap. 9, on specific heats.

[38] Cf. Kuhn 1978; Klein 1963.

calls for a combination of the theories themselves, either by making them modules of an overarching theory or by making them modules of each other. Renn calls this process "integration of knowledge" and sees in it an incentive to revise and mutually adapt the partial theories. In some cases, such as Einstein's resonator-radiation-gas model, the consequences were truly revolutionary.[39]

Similarly, thought-experiments are modularity-dependent devices that play an important role in theory construction. In some cases, they can be used to test the compatibility of various theories, because they rely on the sort of combined interpretive schemes that has just been mentioned. In other cases, such as the computation of the entropy of a given state of a substance by an imaginary reversible transformation from a reference state to that state, the thought-experiments are used to develop the consequences of theoretical laws in a semi-concrete manner. This is possible when these laws are expressible directly as constraints on interpretive schemes, as happens when the second law of thermodynamics is expressed as the vanishing of the Clausius integral over a reversible cycle. In other cases, such as Poincaré's electromagnetic radiators of 1900, the thought-experiment serves to judge the compatibility of the theory with a general principle of physics, here the equality of action and reaction.[40] In still other cases, such as Einstein's free-falling elevator experiment, it serves to suggest a new principle (the equivalence principle). What all these cases have in common is the necessary recourse to schematic modules in the conception of the thought experiments. It is precisely because these modules are better known than the investigated theory that the thought-experiments can play their exploratory role.

Theory construction also depends on the important modular constraint that the new theory should contain earlier successful theories as approximations. In our terminology, the earlier theories should be approximating modules of the newer theory. This constraint is usually called the *correspondence principle*, in reverence to Bohr's endeavor to construct quantum theory in a way that ensured the asymptotic validity of classical electrodynamics. Together with some assumed symmetries or some general postulates, this principle may completely define the sought-after theory. This happened in the case of relativistic dynamics and in the case of quantum mechanics. In Bohr's conception of the latter theory, the classical module is important not only in the construction of its symbolic universe but also in the definition of the interpretive schemes. Indeed for Bohr any measurement ultimately relies on classical modules.[41]

The pursuit of modularity does not always bring progress. In some cases, theories that had long been used as defining or reducing modules must be thrown away or relegated to the humbler modular role of approximation. For instance, most nineteenth-century physicists regarded mechanical reduction as a legitimate and accessible aim for the whole of physics. They were blinded by the success of early reductions of this kind. Toward the end of the century, the pragmatist or positivist convictions of a few physicists confined mechanics to the more modest function of a defining module. The downgrading of classical

[39] Büttner, Renn, and Schemmel 2003.

[40] Poincaré 1900, in which Poincaré discusses the theoretical recoil of unidirectional radiator (Hertzian resonator at the focus of a parabolic mirror).

[41] Friedman (2001, pp 59–63) similarly insists on correspondence principles. On Bohr's views, cf. Chevalley 1985, 1991.

mechanics went on at the beginning of the twentieth century, when it appeared to be an approximating module of a more fundamental relativistic or quantum mechanics. In this process, even the defining modules of Euclidean geometry and Galilean kinematics came under attack. They were ultimately replaced by and became approximating modules of the pseudo-Riemannian geometry of general relativity theory.

The lesson to be drawn from this evolution is that the modular structure of a theory should never be regarded as definitive. The most we can say is that any theory that has been successful in a given domain of physics is likely to remain, after adequate purification or reformulation, a module of future, more general theories. But its modular function may evolve in time. As we saw, classical mechanics once played the role of a defining or a reducing module. It remains a defining module in useful macroscopic theories. But it is only an approximating module for the most fundamental theories such as general relativity or quantum field theory. Theories all have modular structure. But the modular structures of successive theories may only bear a partial resemblance. Any stiffening of our modular habits could become an obstacle to further progress.

In order to describe the evolution of science, Jed Buchwald and Silvan Schweber use the metaphor of a scaffold that periodically undergoes partial restructuring and extensions, and they compare the growing pragmatic efficiency of science with the stability of the scaffold. "The connections and linkages inherent in a scaffolding," they write, "are precisely the sources of its stability, on the one hand, and they also form the resources that permit the extensions in new directions; for our scaffolding indeed involves over time." Although Buchwald and Schweber have in mind the many elements of practice involved in any science, their metaphor neatly applies to the more limited case of physical theories if the elements of the scaffolding are identified with the modules of a theory.[42]

Theories communicated

The modular structure of theories is essential to successful communication among practitioners belonging to different social groups. Physicists who belong to different local subcultures may adhere to different theories of the same domain. As Poincaré and Boltzmann forcefully argued, this cultural diversity is usually beneficial to science, because it favors the exploration of a greater variety of symbolic universes and thus increases chances to find the one that best fits the widest domain. It can be so only if communication is possible between the different subcultures. Maximal communication, in which the physicists of one subculture perfectly understand the theories of the other, almost never occurs. It is not even to be wished, because it would interfere with the creative energy of each group.[43] More commonly, the two groups communicate through interpretive or descriptive schemes that only involve shared modules. The shared modules are in part given, or they are constructed by a few "bilingual" individuals who labor for the easy communicability of science.

An instructive example is that of electrodynamics in the nineteenth century. British physicists favored a field-based approach; German physicists favored direct action at a distance.

---

[42] Buchwald and Schweber 1995, p. 347.   [43] Cf. Biagoli 1993, Chap. 4.

Yet these two communities were able to benefit from each other's results and to compare the predictions of their theories. In part, this was possible because of their inheriting mechanical, electrostatic, magnetostatic, and electrodynamic modules from the same French sources (Coulomb, Poisson, Ampère). For the rest, William Thomson played a crucial role in designing modular concepts that could be used equally well by physicists and engineers of any country. For example, he defined the electric potential through the mechanical concept of energy, independently of any deeper interpretation in the competing symbolic universes. As a consequence or as a motivation, electrometers and other electrical apparatus could be traded between the two cultures, because the modules necessary to their use were made available on both sides. [44]

As was just mentioned in the Thomson case, modules are also essential in the communication between physicists and engineers. Engineers almost never have a detailed knowledge of the deeper theories through which physicists would understand some aspects of their practice. Yet they are constantly benefiting from these theories because they master the modules that are sufficient for their own purposes. More generally, division of work necessitates modular communication between groups who have unequal access to deeper theory. In some domains of modern physics such as particle physics, there are separate subcultures of theorists, experimenters, and instrument makers. As Peter Galison has argued, the necessary communication between these various groups leads to the formation of "trading zones," that is, virtual places of exchange in which the various protagonists can benefit from each other's competences without ever acquiring all of them. Theoretical modules play a crucial role in this sort of trade. In some cases, physicists forge the modules just for this purpose. This does not mean, however, that modules only are an arbitrary product of a social consensus formed in the trading zone. The structure they reflect is an inherent structure of the embedding theories, and it becomes part of our ultimate understanding of these theories. [45]

More broadly, Terry Shinn has pleaded for a "transversalist sociology of science" in which social interactions between well differentiated disciplinary fields play a significant role. Besides Galison's trading zones, he identifies another transversal mechanism involving actors who design multi-purpose instruments in the interstices between disciplinary fields. In his view, this transversal regime permits a sort of intellectual convergence and contributes to the unity of science. Typically, the generic instruments carry with them a "decontextualized fundamental theory" that circulates and becomes "recontextualized" whenever the interstitial actors temporarily penetrates established fields. This process clearly resembles the insertion of a theoretical module within various theories, although Terry Shinn's notion of a theory and its contextual embeddings seems much broader than envisioned in this chapter. [46]

Modularity is also important in the communication and understanding of theories within the same subculture of physicists. Physics courses and textbooks are divided

[44] Cf. Smith and Wise 1989; Darrigol 2000, pp. 113–25.

[45] Galison 1997, Chap. 9. My modules should not be confused with those of Galison 1998, which refer to the elementary processes combined in Feynman's diagrams, not to component-theories.

[46] Shinn and Ragouet 2005, pp. 179–82.

into chapters which often correspond to approximating or specializing modules of the theory to be taught. For instance, a textbook of electrodynamics typically has chapters of electrostatics, electrokinetics, quasi-stationary electrodynamics, and electromagnetic radiation. Within each chapter, exemplars are given of interpretive schemes for which the consequences of the laws of the relevant modules can be fully worked out. Eventually, approximation methods are taught for dealing with systems that somewhat depart from these exemplars. Thomas Kuhn, Ronald Giere, and Nancy Cartwright all insist on the importance of exemplars in teaching theories, although they do so for different purposes. Kuhn has in mind the communication of tacit knowledge that is not conveyed in general statements of the theory; Giere and Cartwright want to drive the semantic view of theories according to which interpreting models are prior to a linguistic statement of laws. What I want to emphasize is that exemplars usually concern approximating or specializing modules of a theory rather than the whole theory, and that their treatment almost always involves defining and schematic modules that the students have already learned in other contexts.[47]

Modularity may also be illustrative. Its purpose may be to feed the intuition. Many British physicists of the nineteenth century believed that a theory could not be properly understood without illustrating some of its parts by other well-understood theories. They relied on *illustrating modules*, namely: reducing mechanical modules that worked for limited classes of interpretive schemes of the global theory. For instance, in "On Faraday's lines of forces," Maxwell illustrated the electrostatic, magnetostatic, and electrokinetic modules of electrodynamics by means of the mechanics of resisted flow in a porous medium. He thus insufflated some life in the dry symbols of potential theory. From these parallel illustrations he extracted a classification of quantities in the symbolic universe of the theory of electrodynamics. As Norton Wise and Peter Harman have shown, this classification played an important role in the subsequent construction of Maxwell's own theory of electrodynamics. More generally, Maxwell's British contemporaries liked to flesh out the equations of their theories by attaching them to partial reducing modules. As Jordi Cat has argued, this sort of fictitious concreteness is part of any understanding of theories. Besides its pedagogical virtue, it eases the mental associations through which a theory can evolve and fuse with other theories.[48]

A logic of modularity?

So far, I have argued for the necessity of modular structure in the verification, comparison, construction, and communication of theories. I have not formally proven the possibility of modular structure. The illustrations I have given could be illusions; they could result from superficial distinctions that vanish in a more rigorous analysis. Can theories truly contain the sort of self-consistent, fairly autonomous subsystems of relations that modules require? Is not there a conflict between the limited scope of the modules and the wider scope of the theory in which they are used?

[47] Kuhn 1962; Giere 1988; Cartwright 1983, essay 7: "Fitting fact to equations."
[48] Wise 1979; Harman 1987; Cat 2001.

In the broadly semantic view adopted in this chapter, these questions are in essence mathematical. They concern the structure of the symbolic world and its interpretive schemes both conceived in an abstract, mathematical manner. They have nothing to do with the empirical meaning and success of the theory. In order to answer them, one could try to mathematically construct a generic theory that exhibits modular structure. To some extent this has already been done by Sneedian structuralists, as I realized after arriving at modularity by historical induction. My defining modules roughly corresponds to their notion of *theoretization* of the non-theoretical concepts of a theory by another theory; my specializing modules to their notion of the *specialization* of a theory by another; my approximating modules to their notion of *reduction* of a theory by another.[49]

These structuralist constructs offer a convincing proof of the possibility of some basic modular functions. However, the confrontation of these constructs with actual scientific practice leads to some difficulties. As Torretti has shown, some structuralist distinctions are stricter than in real life, for instance the distinction between the models and the potential models (models without the laws) of a theory, or the distinction between theoretical and non-theoretical terms. This is why I have stuck to my idiom of symbols, schemes, and modules even after I began to appreciate the structuralist accomplishment. Another reason is the diversity and variability of modular relations.[50]

For those who dislike or distrust the icy abstraction of structuralist constructs, it is sufficient to point to theories of physics in which the existence of modular structure can be demonstrated with reasonable rigor. In the case of reducing modules, the symbolic universe of the theory usually is the module itself, applied to a selected class of imaginary systems. Therefore, the module trivially is a consistent substructure of the theory. If the module only applies to fragments of the symbolic universe, as is for instance the case in of the linear-elastic module in the crack-theory of rupture, it does so by construction and its existence therefore cannot be questioned. In the case of approximating modules, we have the example of optics, in which it can rigorously be shown that the short-wavelength solutions of the wave equation in a medium of variable optical index are approximately compatible with the predictions of geometrical optics, except for well-understood singularities. In the case of defining modules, the relative autonomy of the module comes from the existence of hybrid schemes that contain subsystems obeying the laws of the module. Again, the modularity is there by construction. It may or may not reflect deeper properties of the symbolic universe. The geometric module of mechanics does so, for it involves a global symmetry through the Euclidean group; so does too the mechanical module of electrodynamics, for it involves a splitting of the Lagrangian into matter, field, and interaction terms. In contrast, the mechanical module of thermodynamics implies no more than the existence of hybrid schemes, as long as the theory remains phenomenological.

Laws and principles

Granted that the modular structure of theory is possible and necessary, it may be used to clarify a variety of epistemological questions, for instance the relation between models

---

[49]Balzer, Moulines, and Sneed 1997, Chap. 6.   [50]Torretti 1980, pp. 109–30, 158–62.

and theories, the unity or disunity of physics, and the nature of physical laws. Here I will only address this last question, for it is related to the possibility of some necessity arguments.[51]

Laws can be classified according to their relation to the modular structure of theories. Firstly, there are fundamental laws that hold strictly for every system in the symbolic universe. A good example is Newton's law of acceleration in the mechanics of systems of mass points. Some of the fundamental laws belong to the defining modules, and for this reason they tend to be excluded from the traditional list of laws of a theory. For instance, Pythagoras' theorem is usually not regarded as a law of mechanics.

Secondly, there are phenomenological laws that result from the application of the fundamental laws to classes of interpretive schemes. For example, the law of the conservation of angular momentum is a consequence of the fundamental laws of mechanics applied to a mass point subjected to a central force. The selection of a class of interpretive schemes may involve a restriction on the range of some of the characteristic quantities, in which case the phenomenological law is said to be approximate. For instance, the law $T = 2\pi\sqrt{l/g}$ for the period of oscillation of a pendulum of length l in the gravity g is an approximate law, valid only for small oscillations. Such laws may be the fundamental laws of an approximating module, as is for instance the case for the laws of refraction in wave optics.

Thirdly, there are empirical laws that are obtained by generalizing the numerical outcome of the concrete realization of a class of interpretive schemes. An empirical law sometimes gets upgraded to the status of fundamental law if the class of interpretive schemes to which this law (or a natural generalization of it) applies is found to be large enough. For instance, Newton's law of gravitation promptly became a fundamental law of celestial mechanics after Newton understood its ability to explain the Kepler motion. Even in the absence of any such upgrading, these empirical laws are relative to the theories in which the interpretive schemes are defined. Newton's law of gravitation cannot even be expressed outside its mechanical context, even though the observations from which it is inferred only rely on pre-mechanical modules. As was earlier mentioned, there also are purely experimental laws that are reached through the concrete realization of *descriptive* schemes without reference to a higher-level theory.

A phenomenological theory can be defined as a theory for which every fundamental law is also a phenomenological law (as noted above, this implies that the same modules occur in the symbolic universe and in the interpretive schemes). In other words, every law of the theory can be expressed as a relation between schematic quantities. Mechanics clearly is a phenomenological theory. The molecular theory of elasticity is not, because its basic laws apply at a level that is not accessible to macroscopic experiments. A priori, the laws of fundamental theories have less factual content than phenomenological laws, for at least three reasons: the symbolic universe of a fundamental theory may involve surplus structure that does not affect the interpretive schemes; the various laws of the symbolic universe are not tested separately, it is their joint application to interpretive schemes that is tested; the ideal systems of the symbolic universe often imply a deliberate caricature of natural systems. For

---

[51] On models/theories and on unity/disunity, see Darrigol 2008, pp. 195–6, 215–16, 219–22. See also Morrison 2000 on the many faces of unification in physics.

these and other reasons, Nancy Cartwright speaks of the "lying" of fundamental laws, and ascribes more factual content to phenomenological laws.[52]

Although this view nicely captures some often neglected features of fundamental theories, it makes sense only if the symbolic universe and the interpretive schemes of a fundamental theory have been given once forever. History teaches us the contrary. Usually, the symbolic universe evolves in such a way that surplus structure is gradually eliminated. For instance, Maxwell eliminated the rotating-cell mechanism of his earlier theory of the electromagnetic field in order to reach a "dynamical theory" in which all fundamental variables became empirically controllable. Similarly, at the turn of the nineteenth and twentieth century a few physicists, including Einstein, eliminated the ether from Lorentz's electrodynamics. The domain of interpretive schemes also evolves. For instance, in his study on Brownian motion Einstein imagined an interpretive scheme in which the theoretical fluctuations of statistical mechanics became manifest; similarly, Lorentz and Larmor soon understood that their electron theory of matter could be provided with interpretive schemes that related them to recent experiments on electric discharge in rarefied gases. Thanks to this parallel purification of the symbolic universe and extension of the interpretive schemes, the laws of the symbolic universe may ultimately be all phenomenological. Theories that once raised the phenomenologists' eyebrows have now become commonplace phenomenology. Such theories are rightly regarded as superior to the earlier phenomenological theories with which they initially competed, for they enjoy a larger domain of validity and a simpler symbolic universe.

It remains true that all the laws of a theory, be they fundamental or phenomenological, are only true in the symbolic universe of the theory. Since testing experiments involve schemes, modules, sub-modules, and much practical know-how, they can never establish the truth of any particular law. As van Fraassen puts it, experiments at best establish the "empirical adequacy" of the theory.[53]

Lastly, some laws are shared by several theories, or even by every theory of physics. Such laws are usually called *principles*. Typical examples are the energy principle, the principle of least action, and the two principles of thermodynamics. The formulation and application of these principles depends on the sharing of modules by a wide class of theories. Thus, the energy principle involves a common measure of energy given by a process described in a mechanical module (e.g. the hoisting of a weight). The least action principle involves Lagrangian dynamics abstractly used as a defining module. The principles of thermodynamics rely on shared mechanical and calorimetric modules, which are defining modules in phenomenological theories, or approximating modules in statistico-mechanical theories. As we saw, principles may originate in the generalization of an empirical law, or in abstraction from a reductionist attempt. Modules are involved in both cases: defining modules in the former case, and reducing modules in the latter.

Conclusion: The possibility of necessity

Although the above-given definition of physical theory and its modular structure was reached independently of any consideration of the necessity of our theories, it provides the

---

[52] Cartwright 1983, essay 6: "For phenomenological laws."   [53] Van Fraassen 1980, pp. 45–6, 64.

basic notions needed to express the comprehensibility arguments discussed in this book. Firstly, the measurability and causality arguments rely on idealized operations, for instance rigid-rod surveys in the chapter on space, the pulley-thread mechanisms in the chapters on mechanics, or the elevated weights and heat transfers in the chapter on thermodynamics. These operations belong to what I called the interpretive schemes of the theory. They may or may not be susceptible of a concrete realization, or this realization may be limited and regional, as is the case for interpretive schemes in general. They define the form of a class of interpretive schemes of the theory, and thus indirectly determine aspects of the symbolic universe inhabited by the interpretive schemes. Of course, this determination can only be partial, although the underdetermination of the symbolic universe and its laws can be reduced by minimizing its surplus content.

Interpretive schemes involve defining or approximating modules, and so too do the idealized operations in measurability and causality arguments. Most frequently, a space module is needed to define the ideal systems manipulated in the operations; or a mechanical module is needed to define some measurable attributes of this system. In addition, correspondence arguments require approximating modules, which in turn require the sharing of interpretive schemes by the older and newer theories. Reductionist arguments evidently rely on reducing modules to construct the symbolic universe of the new theory.

To sum up, we see that the general notions of interpretive schemes and theoretical modules, whose own necessity derives from the applicability, comparability, constructability, and communicability of theories, are just what we need to express the best necessity arguments, those based on varieties of Helmholtzian comprehensibility. This remark only shows the possibility of such arguments. Any evaluation of their cogency requires attention to the details of the deductions, including aspects that are not expressed in the basic requirements of comprehensibility. There is always a risk that some details be overlooked, and others wrongly judged to be inevitable. With sufficient care, however, we may reasonably judge that some theory is the only possible one in a given domain of experience if this domain is to be understood in a certain manner.

History does not contradict judgments of this moderately rationalist kind. The dethroning of older fundamental theories only shows that what we take to be a natural, even necessary manner of understanding in a given region of experience may not extend to a broader field of experience. A scientific revolution is a previous necessity denied, more exactly: the confining of a necessity to a limited domain of experience. In this process, the necessity arguments are still useful in telling us which basic presuppositions we must give up when facing experimental contradiction. At the end of a revolution, a new sort of comprehensibility must take place and the older kind remains valid in important regions of experience.[54]

By definition, a physical theory must help us comprehend the world. The conditions of this comprehensibility, be they contingent or not, severely restrict the spectrum of imaginable theories. In satisfying these conditions, nature permitted a few triumphs of impure reason.

---

[54]This is what Mittelstaedt (2011, p. x) calls the "stepwise reduction of prejudices."

ABBREVIATIONS

AHES *Archive for the history of exact sciences*
AP *Annalen der Physik und der Chemie*
BB Akademie der Wissenschaften zu Berlin, mathematisch-physikalische Klasse, *Sitzungsberichte*
BJPS *British journal for the philosophy of science*
CD Jammer 1966
CQ Darrigol 1992
CR Académie des Sciences, *Comptes-rendus hebdomadaires des séances*
FP *Foundations of physics*
GN Königliche Gesellschaft der Wissenschaften und der Georg August Universität zu Göttingen, *Nachrichten*
HD Mehra & Rechenberg 1982–1987
HPA *Helvetica physica acta*
HSPS *Historical studies in the physical sciences*
JRAM *Journal für die reine und angewandte Mathematik*
MAB Académie Royale des sciences et des belles-Lettres de Berlin, *Mémoires*
MAS Académie Royale des Sciences, *Mémoires (de physique et de mathématiques)*
PM *Philosophical magazine*
PRS Royal Society of London, *Proceedings* A
PR *Physical Review*
PRSE Royal Society of Edinburgh, *Proceedings*
PT Royal Society of London, *Philosophical transactions*
PZ *Physikalische Zeitschrift*
RHS *Revue d'histoire des sciences et des techniques*
SHPMP *Studies in history and philosophy of modern physics*
SHPS *Studies in history and philosophy of science*
TCPS Cambridge Philosophical Society, *Transactions*
TRSE Royal Society of Edinburgh, *Transactions*
WB Kaiserliche Akademie der Wissenschaften in Wien, Mathematisch-Naturwissenschaftlische Klasse, *Sitzungsberichte*
ZP *Zeitschrift für Physik*

BIBLIOGRAPHY

Afriat, Alexander. 2009 How Weyl stumbled across electricity while pursuing mathematical justice. *SHPMP*, 40: 20–5.

Akhiezer, Naum Ilyich, and Izrail Markovič Glazman. 1981 *Theory of linear operators in Hilbert space*, 2 vols. London: Pitman.

Álvarez, Enrique. 1989 Quantum gravity: An introduction to some recent results. *Reviews of modern physics*, 61: 561–604.

Bacciagaluppi, Guido. 1993 Critique of Putnam's quantum logic. *International journal of theoretical physics*, 32: 1835–46.

——. 2009 Is Logic Empirical? In Gabbay, Lehmann, and Engesser 2009, 49–78.

Balzer, Wolfgang, Ulises Moulines, and Joseph Sneed. 1997 *An architectonic for science: The structuralist program*. Dordrecht: North Holland.

Barberousse, Anouk. 2000 *La physique face à la probabilité*. Paris: Vrin.

Barbour, Julian. 1999 *The end of time: The next revolution in physics*. Oxford: Oxford University Press.

—— and Bruno Bertotti. 1982 Mach's principle and the structure of dynamical theories. *PR*, A 382: 295–306.

Barkan, Diana. 1993 The witches' Sabbath: The first international Solvay congress in physics. *Science in context*, 6: 59–82.

Batterman, Robert. 2002 *The devil in the details: Asymptotic reasoning in explanation, reduction, and emergence*. Oxford: Oxford University Press.

Bayen, François, Moshé Flato, Christian Fronsdal, André Lichnerowicz, and Daniel Sternheimer. 1978a Deformation theory and quantization. I. Deformations of symplectic structures. *Annals of physics*, 111: 61–110.

——. 1978b Deformation theory and quantization. II. Physical applications. *Annals of physics*, 111: 111–51

Bélidor, Bernard Forest de. 1819 *Architecture hydraulique, ou l'art de construire, d'élever, et de ménager les eaux pour les besoins de la vie*, critical edition by Claude Louis Navier. Paris: Didot.

Bell, John L., and Herbert Korté. 2011 Hermann Weyl. In Edward N. Zalta (ed.), *The Stanford encyclopedia of philosophy* (Spring 2011 edn). Available at http://plato.stanford.edu/archives/spr2011/entries/weyl/ (last accessed Aug. 2013).

Beller, Mara. 1990 Born's probabilistic interpretation: A case study of "concepts in flux". *SHPS*, 21: 563–88.

——. 1999 *Quantum dialogue: The making of a revolution*. Chicago: University of Chicago Press.

Belna, Jean-Pierre. 1996 *La notion de nombre chez Dedekind, Cantor, Frege: Théories, conceptions, philosophie*. Paris: Vrin.

Beltrametti, Enrico, and Gianni Cassinelli.1981 *The logic of quantum mechanics*. Reading, MA: Addison-Westley.

Beltrami, Eugenio. 1868 Saggio di interpretazione della geometria non-euclidea. *Giornale di matematiche*, 6: 284–312.

Bernoulli, Daniel. 1726 Examen principiorum mechanicae, et demonstrationes geometricae de compositione et resolutione virum. *Commentarii Academiae scientiarum imperialis Petropolitanae*, 1: 126–42. Also in D. Bernoulli, *Werke*, vol. 3 (Basel: Birkhäuser, 1987), 119–35.

——. 1738 *Hydrodynamica, sive de viribus et motibus fluidorum commentarii*. Strasbourg: Decker.

Bernoulli, Jacob. 1703 Démonstration générale du centre de balancement ou d'oscillation, tiré de la nature du levier. *MAS* (for 1703, pub. 1720), 78–84.

Bernoulli, Johann. 1724 Discours sur les lois de communication du mouvement. In *Opera ommia*, vol. 3 (Lausanne: Bousquet, 1742), 7–107.

——. 1735 De vera notione virium virarum earumque usu in dynamicis. In *Opera ommia*,vol. 3 (Lausanne: Bousquet, 1742), 239–60.

Berzi, Vottorio, and Vittorio Gorini. 1969 Reciprocity principle and the Lorentz transformations. *Journal of mathematical physics*, 10: 1518–24.

Bevilacqua, Fabio. 1993 Helmholtz's *Über die Erhaltung der Kraft*: The emergence of a theoretical physicist. In David Cahan, ed., *Hermann von Helmholtz and the foundations of nineteenth-century science* (Berkeley: University of California Press, 1993), 291–333.

Biagoli, Mario. 1993 *Galileo courtier*. Chicago: University of Chicago Press.

Birkhoff, Garrett. 1935 Combinatorial relations in projective geometries. *Annals of mathematics*, 36: 743–48.

——. 1940 *Lattice theory*. New York: American Mathematical Society.

—— and John von Neumann. 1936 The logic of quantum mechanics. *Annals of mathematics*, 37: 823–43.

Bitbol, Michel, Pierre Kerszberg, and Jean Petitot (eds). 2009 *Constituting objectivity: Transcendental perspectives on modern physics*. Berlin: Springer.

Bohr, Niels. 1913 On the constitution of atoms and molecules. *PM*, 26: 1–25, 476–502, 857–75.

——. 1914 On the effect of electric and magnetic fields on spectral lines. *PM*, 27: 506–24.

——. 1918 On the quantum theory of line spectra, Part I: On the general theory; Part II: On the hydrogen spectrum. Det Kongelige Danske Videnskabernes Selskab, *Matematisk-fysiske Meddelser*, 4: 1–36, 36–100.

——, Hendrik Kramers, and John Slater. 1924 The quantum theory of radiation. *PM*, 47: 785–822.

Boltzmann, Ludwig. 1868 Studien über das Gleichgewicht der lebendigen Kraft zwischen bewegten materiellen Punkten, *WB*, 58: 517–60. Also in Boltzmann 1909, vol. 1, 49–96.

——. 1871a Über das Wärmegleichgewicht zwischen mehratomigen Gasmolekülen. *WB*, 63: 397–418. Also in Boltzmann 1909, vol. 1, 237–58.

——. 1871b Einige allgemeine Sätze über Wärmegleichgewicht. *WB*, 63: 679–711. Also in Boltzmann 1909, vol. 1, 259–87.

——. 1871c Analytischer Beweis des zweiten Hauptsatzes der mechanischen Wärmetheorie aus den Sätzen über das Gleichgewicht der lebendigen Kraft. *WB*, 63: 712–32. Also in Boltzmann 1909, vol. 1, 288–308.

——. 1872 Weitere Studien über das Wärmegleichgewicht unter Gasmolekülen. *WB*, 66: 275–370. Also in Boltzmann 1909, vol. 1, 316–402.

——. 1877a Bemerkungen über einige Probleme der mechanischen Wärmetheorie, *WB*, 75: 62–100. Also in Boltzmann1909, vol. 2, 112–48.

——. 1877b Über die Beziehung zwischen dem zweiten Hauptsatze der mechanischen Wärmetheorie und der Wahrscheinlichkeitsrechnung respektive den Sätzen über das Wärmegleichgewicht. *WB*, 76: 373–435. Also in Boltzmann 1909, vol. 2, 164–223.

——. 1881a Über einige das Wärmegleichgewicht betreffende Sätze. *WB*, 84: 136–45. Also in Boltzmann 1909, vol. 2, 572–81.

——. 1881b [Review of Maxwell 1879]. *Beiblätter zu Annalen der Physik*, 5: 403–17. Also in Boltzmann 1909, vol. 2, 582–95.

——. 1885 Über die Eigenschaften monozyklischer und anderer damit verwandter Systeme. *JRAM*, 98: 68–94. Also in Boltzmann 1909, vol. 3, 122–52.

——. 1891–1893 *Vorlesungen über die Maxwell'sche Theorie der Electricität und des Lichtes*, 2 vols. Leipzig: Barth.

——. 1909 *Wissenschaftliche Abhandlungen*, 3 vols. Leipzig: Barth.

Borda, Jean-Charles de. 1766 Sur l'écoulement des fluides par les orifices des vases. *MAS*, 579–607.

——. 1767 Mémoire sur les roues hydrauliques. *MAS*, 270–87.

Born, Max. 1924 Über Quantenmechanik. *ZP*, 26: 379–95.

——. 1926a Zur Quantentheorie der Stossvorgänge. *ZP*, 37: 863–67.

——. 1926b Quantenmechanik der Stossvorgänge. *ZP*, 38: 803–27.

—— and Pascual Jordan. 1925 Zur Quantenmechanik. *ZP*, 35: 557–615.

——, Werner Heisenberg, and Pascual Jordan. 1926 Zur Quantenmechanik II. *ZP*, 35: 557–615.

Bošković, Ruđer Josip. 1758 *Philosophiæ naturalis theoria: reducta ad unicam legem virium in natura existentium*. Vienna: Libreria Kaliwodiana.

Bracco, Christian, and Jean–Pierre Provost. 2006 La relativité de Poincaré de 1905 et les transformations actives. *AHES*, 60: 337–51.

Briginshaw, Anthony. 1979 The axiomatic geometry of space-time: An assessment of the work of A. A. Robb. *Centaurus*, 22: 315–23.

Briot, Charles. 1861 *Leçons de mécanique conformes aux programmes officiels*. Paris: Dunod.

Broglie, Louis de. 1923a Ondes et quanta. *CR*, 177: 507–10.

——. 1923b Quanta de lumière, diffraction et interférences. *CR*, 177: 548–50.

——. 1923c Les quanta, la théorie cinétique des gaz et le principe de Fermat. *CR*, 177: 630–2.

——. 1924 *Recherches sur la théorie des quanta*. Paris: Masson.

Brown, Harvey. 2005 *Physical relativity: Space-time structure from a dynamical perspective*. Oxford: Oxford University Press.

—— and Oliver Pooley. 2006 Minkowki space-time: A glorious non-entity. In Dennis Dieks (ed.), *The ontology of spacetime* (Amsterdam: Elsevier), 67–89.

—— and Jos Uffink. 2001 The origins of time-asymmetry in thermodynamics: The Minus First Law. *SHMP*, 32: 525–38.

——, ——, and Wayne Myrvold. 2009 Boltzmann's H-theorem, its discontents, and the birth of statistical mechanics. *SHMP*, 40: 174–91.

Brush, Stephen. 1976 *The kind of motion we call heat: A history of the kinetic theory of gases in the 19th century*, 2 vols. Amsterdam: North-Holland.

Buchwald, Jed. 1985 *From Maxwell to microphysics: Aspects of electromagnetic theory in the last quarter of the nineteenth century*. Chicago: University of Chicago Press.

——. 1989 *The rise of the wave theory of light: Optical theory and experiment in the early nineteenth century*. Chicago: University of Chicago Press.

—— and Silvan Schweber. 1995 Conclusion. In Jed Buchwald (ed.), *Scientific practice: Theories and stories of doing physics* (Chicago: University of Chicago Press), 345–51.

Burtt, Edwin Arthur. 1925 *The metaphysical foundations of modern physical science: A historical and critical essay*. London: Ketan, Trench, Trubner.

Busch, Paul. 2003 Quantum states and generalized observables: A simple proof of Gleason's theorem. *Physical reviews letters*, 91: 1–4.

——, Marian Grabowski, and Pekka J. Lahti. 1995 *Operational quantum physics*. Heidelberg: 1995.

Büttner, Jochen, Jürgen Renn, and Matthias Schemmel. 2003 Exploring the limits of classical physics: Planck, Einstein, and the structure of a scientific revolution. *SHPMP*, 34: 35–59.

Caneva, Kenneth. 1993 *Robert Mayer and the conservation of energy*. Princeton: Princeton University Press.

Cao, Tian Yu, and Silvan Schweber. 1993 The conceptual foundations and the philosophical aspects of renormalization theory. *Synthese*, 97: 33–108.

Capecchi, Danilo. 2012 *History of virtual work laws: A history of mechanics prospective*. Berlin: Birkhäuser.

Cardwell, Donald. 1971 *From Watt to Clausius*. Ithaca, NY: Cornell University Press.

——. 1989 *James Joule: A biography*. Manchester: Manchester University Press.

Carnot, Lazare. 1786 *Essai sur les machines en général*. Dijon: Defay.

——. 1803 *Principes fondamentaux de l'équilibre et du mouvement*. Paris: Deterville.

Carnot, Sadi. 1824 *Réflexions sur la puissance motrice du feu*. Paris: Bachelier.

Carrier, Martin. 1994 Geometric facts and geometric theory: Helmholtz and 20th-century philosophy of physical geometry. In Lorenz Krüger (ed.), *Universalgenie Helmholtz. Rückblick nach 100 Jahren* (Berlin: Akademie Verlag), 277–91.

Cartwright, Nancy. 1983 *How the laws of physics lie*. Oxford: Oxford University Press.

Cassirer, Ernst. 1921 *Zur Einstein'schen Relativitätstheorie. Erkenntnistheoretische Betrachtungen*. Berlin: B. Cassirer.

——. 1923–9 *Philosophie der symbolischen Formen*, 3 vols. Berlin: B. Cassirer.

Castagnino, Mario. 1971 The Riemannian structure of space-time as a consequence of a measurement method. *Journal of mathematical physics*, 12: 2203–11.

Casullo, Albert. 2003 *A priori justification*. Oxford: Oxford University Press.

Cat, Jordi. 2001 On Understanding: Maxwell on the methods of illustration and scientific metaphor. *SHPM*, 32: 395–441.

Chabot, Hugues. 2004 Nombre et approximations dans la théorie de la gravitation de Lesage. *Sciences et Techniques en Perspective*, 8: 179–98.

Chang, Hasok. 2004 *Inventing temperature: Measurement and scientific progress*. Oxford: Oxford University Press.

——. 2009 Operationalism. In Edward N. Zalta (ed.), *The Stanford encyclopedia of philosophy* (Fall 2009 edn). Available at http://plato.stanford.edu/archives/fall2009/entries/operationalism/ (last accessed July 2013).

Chevalley, Catherine. 1985 Complémentarité et langage dans l'interprétation de Copenhague. *RHS*, 38: 251–92.

Chevalley, Catherine. 1991 Le dessin et la couleur. Introduction to Niels Bohr, *Physique atomique et connaissance humaine* (Paris: Gallimard), 19–140.

Chiribella, Giulio, Giacomo Mauro D'Ariano, and Paolo Perinotti. 2011 Informational derivation of quantum theory. *PR*, A84, 012311.

Clausius, Rudolf. 1850 Über die bewegende Kraft der Wärme und die Gesetze, welche sich daraus für die Wärmelehre selbst ableiten lassen. *AP*, 79: 368–97, 500–24.

——. 1854 Über eine veränderte Form des zweiten Hauptsatzes der mechanischen Wärmetheorie. *AP*, 93: 481–506.

——. 1857 Über die Acht der Bewegung, welche wir die Wärme nennen. *AP*, 100: 353–80.

——. 1858 Über die mittlere Länge der Wege, welche bei der Molecularbewegung gasförmiger Körper von den einzelnen Molecülen zurückgelegt werden; nebst einigen anderen Bemerkungen über die mechanische Wärmetheorie. *AP*, 105: 239–58.

——. 1865 Über verschiedene für die Anwendung bequeme Formen der Hauptgleichungen der mechanischen Wärmetheorie. *AP*, 125: 353–400.

Clifton, Rob, Jeffrey Bub, and Hans Halvorson. 2003 Characterizing quantum theory in terms of information-theoretic constraints. *FP*, 33: 1561–91.

Coecke, Bob, David Moore, and Alexander Wilce. 2000 Operational quantum logic: An overview. In *Current research in operational quantum logic* (Dordrecht: Kluwer), 1–36.

Comte, Claude. 1986 Leibniz aurait-il pu découvrir la relativité? *European journal of physics*, 7: 225–35.

——. 1996 Symmetry, relativity and quantum mechanics. *Il nuovo cimento*, 111B: 937–956.

Coope, Ursula. 2004 *Time for Aristotle.* Oxford: Oxford University Press.

Coriolis, Gaspard. 1829 *Du calcul de l'effet des machines.* Paris: Carilian-Goeury.

——. 1835 Mémoire sur la manière d'étendre les différents principes de la mécanique pour des systèmes de corps, en les considérants comme des assemblages de molécules. *Journal de l'Ecole Polytechnique*, vol. 15, cahier 24: 93–125.

Cornu, Alfred. 1875 *Cours de Physique, première division, 1874–1875*, cours autographié. Paris: Ecole Polytechnique.

Corry, Leo. 1996 *Modern algebra and the rise of mathematical structures.* Basel: Birkhäuser.

Costabel, Pierre. 1960 *Leibniz et la dynamique.* Paris: Hermann.

Crosland, Michael. 1967 *The Society of Arcueil.* Cambridge, MA: Harvard University Press.

Curtwright, Thomas, and Cosmas Zachos. 2011 *Quantum mechanics in phase space.* arXiv: 1104.5269v2 [physics.hist-ph] Dec. 26, 2011.

Dakić, Borivoje, and Časlav Brukner. 2009 Quantum theory and beyond: Is entanglement special? arXiv: 0911.0695v1 [quant-ph] 3 Nov 2009.

D'Alembert, Jean le Rond. 1743 *Traité de dynamique, dans lequel les loix de l'équilibre et du mouvement des corps sont réduites au plus petit nombre possible, et démontrées d'une manière nouvelle, et où l'on donne un principe général pour trouver le mouvement de plusieurs corps qui agissent les uns sur les autres, d'une manière quelconque.* Paris: David.

——. 1751 Discours préliminaire. In *Encyclopédie, ou dictionnaire raisonné des sciences, des arts et des métiers* (Paris: Briasson, David, Le Breton, Durand), vol. 1, pp. i–lv.

——. 1758 *Traité de dynamique*, 2nd edn. Paris: David.

Dalla Chiara, Maria Luisa, Roberto Giuntini, and Miklos Rédei. 2007 The history of quantum logic. In Dov M. Gabbay and John Woods (eds), *The many valued and nonmonotonic turn in logic*, vol. 8 of *Handbook of the history of logic* (11 vols., Amsterdam: Elsevier, 2004–11), 205–83.

Darrigol, Olivier. 1992 *From c-numbers to q-numbers: The classical analogy in the history of quantum theory.* Berkeley: University of California Press.

——. 1993 Strangeness and soundness in Louis de Broglie's early work. *Physis*, 30: 303–72.

——. 1994 Helmholtz's electrodynamics and the comprehensibility of nature. In Lorenz Krüger (ed.), *Universalgenie Helmholtz. Rückblick nach 100 Jahren* (Berlin: Akademie Verlag), 216–42.

——. 1995 Henri Poincaré's criticism of *fin de siècle* electrodynamics. *SHPMP*, 26: 1–44.

——. 1997 Classical concepts in Bohr's atomic theory (1913–1925). *Physis*, 34: 547–67.

——. 2000 *Electrodynamics from Ampère to Einstein.* Oxford: Oxford University Press.

——. 2001 God, waterwheels, and molecules: Saint-Venant's anticipation of energy conservation. *HSPS*, 31: 285–353.

——. 2002 Between hydrodynamics and elasticity theory: The first five births of the Navier-Stokes equation. *AHES*, 56: 95–150.

——. 2003a Number and measure: Hermann von Helmholtz at the crossroads of mathematics, physics, and psychology. *SHPS*, 34: 515–73.

——. 2003b La termodinamica. In *Storia della scienza*, vol. 7: *L'ottocento* (Rome: Istituto della Enciclopedia Italiana), 470–81.

——. 2003c Quantum theory and atomic structure, 1900–1927. In Nye, Mary Jo (ed.), *The Cambridge history of science*, vol. 5: *The modern physical and mathematical sciences* (Cambridge: Cambridge University Press), 331–49.

——. 2005 *Worlds of flow: A history of hydrodynamics from the Bernoullis to Prandtl*. Oxford: Oxford University Press.

——. 2006 The genesis of the theory of relativity. In T. Damour, O. Darrigol, B. Duplantier, V. Rivasseau (eds), *Einstein 1905–2005: Poincaré seminar 2005* (Basel: Birkhäuser, 2006), 1–31.

——. 2007a A Faradayan principle for selecting classical field theories. *International studies in the philosophy of science*, 21: 35–56.

——. 2007b On the necessary truth of the laws of classical mechanics. *SHPMP*, 38: 757–800.

——. 2007c A Helmholtzian approach to space and time. *SHPS*, 38: 528–42.

——. 2008 The modular structure of physical theories. *Synthese*, 162: 195–223.

——. 2009 A simplified genesis of quantum mechanics. *SHPMP*, 40: 151–66.

——. 2010 James MacCullagh's ether: An optical route to Maxwell's equations? *European physics journal H*, 35: 133–72.

—— and Jürgen Renn. 2000 Statistical mechanics. *Preprint* 139 of Max Planck Institut für Wissenschaftsgeschichte, Berlin. Published as "La nascita della meccanica statistica," in *Storia della scienza*, vol. 7: *L'ottocento* (Rome: Istituto della Enciclopedia Italiana, 2003), 496–507.

Daub, Edward. 1971 Clausius, Rudolf. In C. C. Gillispie (ed.), *Dictionary of scientific biography*, vol. 3 (New York: Scribner), 303–11.

de Courtenay, Nadine. 1999 *Science et philosophie de Ludwig Boltzmann. La liberté des images par les signes*. Thèse de doctorat. Université de Paris 4.

——. 2008 Mesure et formation des concepts physiques: Rudolph Carnap et Norman Campbell. In Jacques Bouveresse and PierreWagner (eds), *Mathématiques et expérience* (Paris: Odile Jacob), 211–51.

Dedekind, Richard. 1872 *Stetigkeit und irrationale Zahlen*. Braunschweig: Vieweg.

Delaunay, Charles. 1856 *Traité de mécanique rationnelle*. Paris: Masson.

del Monte, Guidobaldo. 1577 *Mechanicorum liber*. Pesaro: Hieronymus Concordia. *Facsimile* and commentary in Jürgen Renn and Peter Damorow (eds), *Guidobaldo del Monte's* Mechanicorum liber (Berlin: Max Planck research library for the history and development of knowledge, 2010).

de Montagu, Sybil. 2010 *L'émergence de l'analyse dimensionnelle: L'analyse dimensionnelle en France et en Grande-Bretagne au XIXe siècle*. Thèse de doctorat: Université Denis Diderot.

Descartes, René. [1637] *Explication des engins par l'ayde desquels on peut, avec l'aide d'une petite force, lever un fardeau fort pesant*. Annex to a letter to Constantyn Huygens of 5 Oct. 1637. In Descartes, *Oeuvres*, ed. Charles Adam and Paul Tannery, vol. 1 (Paris: Cerf, 1897), 435–47.

——. 1644 *Principia philosophiae*. Amsterdam: Ludovicus Elzevirius.

Deser, Stanley. 1970. Self-interaction and gauge invariance. *General relativity and gravitation*, 1: 9–18.

Deutsch, David. 2003 It from qubit. In J. Barrow, P. Davies, and C. Harper (eds), *Science and ultimate reality* (Cambridge: Cambridge University Press), 90–102.

Dhombres, Jean. 1987 Un style axiomatique dans l'écriture de la physique mathématique au XVIIIe siècle: Daniel Bernoulli et la composition des forces. *Rivista di storia della scienza*, 4: 265–318.

—— and Patricia Radelet de Grave. 1991 Contingence et nécessité en mécanique. Etude de deux textes inédits de Jean d'Alembert. *Physis*, 28: 35–114.

Díez, José. 1997 A hundred years of numbers, an historical introduction to measurement theory 1887–1990. *SHPS*, 28: 167–85, 237–65.

Dijksterhuis, Eduard Jan. 1961 *The mechanization of the world picture: Pythagoras to Newton*. Oxford: Oxford University Press.

Dirac, Paul. 1925 The fundamental equations of quantum mechanics. *PRS*, 109: 642–53.

——. 1926 On the theory of quantum mechanics. *PRS*, 112: 661–77.

——. 1927 The physical interpretation of quantum dynamics. *PRS*, 113: 621–41.

——. 1930 Note on exchange phenomena in the Thomas atom. *Mathematical proceedings of the Cambridge Philosophical Society*, 26: 376–85.

Dirac, Paul. 1939 The relation between mathematics and physics (lecture delivered on presentation of the James Scott prize, February 6, 1939). Royal Society of Edinburgh, *Proceedings*, vol. 59, part 2: 122–9.

DiSalle, Robert. 1993 Helmholtz's empiricist philosophy of mathematics: Between laws of perception and laws of nature. In David Cahan (ed.), *Hermann von Helmholtz and the foundations of nineteenth-century science* (Berkeley: University of California Press), 498–521.

——. 2006a *Understanding space-time: The philosophical development of physics from Newton to Einstein.* Cambridge: Cambridge University Press.

——. 2006b Kant, Helmholtz, and the meaning of empiricism. In Michael Friedman and Alfred Nordmann (eds), *The Kantian legacy in nineteenth-century science* (Cambridge, MA: MIT Press), 123–40.

Drago, Antonino, Salvatore Domenico Manno, and Giuseppe Mauriello. 2001 Una presentazione concettuale della meccanica di Lazare Carnot. *Giornale di fisica della Società italiana di fisica*, 42: 131–56.

Dresden, Max. 1987 *H. A. Kramers: Between tradition and revolution.* Berlin: Springer.

Drude, Paul. 1894 *Physik des Aethers auf elektromagnetischer Grundlage.* Stuttgart: Enke.

Du Bois-Reymond, Paul. 1882 *Die allgemeine Functionentheorie. Erster Theil. Metaphysik und Theorie der mathematischen Grundbegriffe: Grösse, Grenze, Argument und Function.* Tübingen: Laupp.

Duchesneau, François. 1994 *La dynamique de Leibniz.* Paris: Vrin.

Dugac, Pierre. 1976 *Richard Dedekind et les fondements des mathématiques.* Paris: Vrin.

Dugas, René. 1950 *Histoire de la mécanique.* Paris: Dunod.

Duhamel, Jean-Marie Constant. 1845–1846 *Cours de mécanique de l'Ecole Polytechnique*, 2 vols. Paris: Bachelier.

Duhem, Pierre. 1905–1906 *Les origines de la statique*, 2 vols. Paris: Hermann.

Duncan, Anthony, and Michel Janssen. 2007 On the verge of *Umdeutung* in Minnesota: Van Vleck and the correspondence principle. *AHES*, 61: 553–671.

Dvurečenskij, Anatolij. 1993 *Gleason's theorem and its applications.* Dordrecht: Kluwer.

Earman, John. 1989 *World enough and spacetime: Absolute versus relational theories of space and time.* Cambridge, MA: MIT Press.

——, Clark Glymour, and John Stachel (eds). 1977 *Foundations of space-time theories. Minnesota studies in the philosophy of science*, vol. 8. Minneapolis: University of Minnesota Press.

Eckert, Michael. 1993 *Die Atomphysiker: Eine Geschichte der theoretischen Physik am Beispiel der Sommerfeldschen Schule.* Brauchschweig: Vieweg.

——. 2013 *Arnold Sommerfeld. Science, life and turbulent times 1868–1951.* New York: Springer.

Eddington, Arthur Stanley. 1920 *Space, time and gravitation.* Cambridge: Cambridge University Press.

——. 1923 *The mathematical theory of relativity.* Cambridge: Cambridge University Press.

——. 1939 *The philosophy of physical science.* Cambridge: Cambridge University Press.

Ehlers, Jürgen. 1973 Survey of general relativity theory. In Werner Israel (ed.), *Relativity, astrophysics and cosmology* (Dordrecht: Reidel, 1973), 1–125.

——, Felix Pirani, and Alfred Schild. 1972 The geometry of free fall and light propagation. In Lochlainn O'Raifeartaigh (ed.), *General relativity papers in honour of J. L. Synge* (Oxford: Clarendon Press, 1972), 63–84.

Einstein, Albert. 1903 Eine Theorie der Grundlagen der Thermodynamik. *AP*, 11: 170–187.

——. 1905a Über einen die Erzeugung und Verwandlung des Lichtes betreffenden heuristischen Gesichtspunkt. *AP*, 17: 132–48.

——. 1905b Zur Elektrodynamik bewegter Körper. *AP*, 17: 891–921.

——. 1906 Zur Theorie der Lichterzeugung und Lichtabsorption. *AP*, 20: 199–206.

——. 1907a Über das Relativitätsprinzip und die aus demselben gezogenen Folgerungen. *Jahrbuch der Radioaktivität und Elektronik*, 4: 411–62.

——. 1907b Die Plancksche Theorie der Strahlung und die Theorie der spezifischen Wärme. *AP*, 22: 569–72.

——. 1909a Zum gegenwärtigen Stand des Strahlungsproblems. *PZ*, 10: 185–93.

——. 1909b Über die Entwickelung unserer Anschauungen über das Wesen und Konstitution der Strahlung. *PZ*, *10*: 817–25.

——. 1911 L'état actuel du problème des chaleurs spécifiques. In Paul Langevin and Maurice de Broglie (eds), *La théorie du rayonnement et les quanta. Rapports et discussions de la réunion tenue à Bruxelles, du 30 octobre au 3 novembre 1911. Sous les auspices de M. E. Solvay* (Paris: Gauthier-Villars), 407–35.

——. 1916a Die Grundlage der allgemeinen Relativitätstheorie. *AP*, 49: 769–822.

——. 1916b Strahlungs-Emission und-Absorption nach der Quantentheorie. Deutsche Physikalische Gesellschaft, *Verhandlungen*, 18: 318–23.

——. 1917 Zur Quantentheorie der Strahlung. *PZ*, 18: 121–8.

——. 1919 What is the theory of relativity? *The London times*, Nov. 28, 1819.

——. [1920] Grundgedanken und Methoden der Relativitätstheorie, in ihrer Entwicklung dargestellt. Unpub. draft of a paper for *Nature*, in M. Janssen, R. Schulmann et al. (eds), *The collected papers of Albert Einstein*, vol. 7 (Princeton: Princeton University Press, 2002), 245–78.

——. 1924 Quantentheorie des einatomigen idealen Gases. *BB* (1924), 261–67.

——. 1925a Quantentheorie des einatomigen idealen Gases. Zweite Abhandlung. *BB* (1925), 3–14.

——. 1925b Zur Quantentheorie des idealen Gases. *BB* (1925), 18–25.

——. 1949 Autobiographical notes. In Paul Arthur Schilpp (ed.), *Albert Einstein: Philosopher-Scientist* (Evanston, IL: Library of Living Philosophers, vol. 7), 1–94.

——and Adriaan Fokker. 1914 Die Nordströmsche Gravitationstheorie vom Standpunkt des absoluten Differentialkalküls. *AP*, 44: 321–8.

——and Marcel Grossmann. 1913 Entwurf einer verallgemeinerten Relativitätstheorie und einer Theorie der Gravitation. I. Physikalischer Teil von A. Einstein II. Mathematischer Teil von M. Grossmann. *Zeitschrift für Mathematik und Physik*, 62: 225–44, 245–61.

Euclid. [*c.* 300 BC] *The elements*. See Heath and Heiberg 1908.

Euler, Leonhard. 1736 *Mechanica, sive motus scientia analytice exposita*, 2 vols. Petersburg: Typographia Academiae Scientiarum.

——. 1744 *Methodus inveniendi lineas curvas maxime minimive proprietate gaudentes, sive solutio problematis isoperimetrici latissimo sensu accepti*. Lausanne and Geneva: Bousquet.

——. 1748a Recherches sur les plus grands et les plus petits qui se trouvent dans les actions des forces. *MAB*, 4: 149–88.

——. 1748b Réflexions sur quelques loix générales de la nature qui s'observent dans les effets des forces quelconques. *MAB*, 4: 189–218.

——. 1750a Découverte d'un nouveau principe de mécanique. *MAB*, 6: 185–217.

——. 1750b Recherches sur l'origine des forces. *MAB*, 6: 419–47.

——. 1751a Harmonie entre les principes généraux de repos et de mouvement de M. de Maupertuis. *MAB*, 7: 169–98.

——. 1751b Essay d'une démonstration métaphysique du principe général de l'équilibre. *MAB*, 7: 246–54.

Faraday, Michael. 1839. *Experimental researches*, vol. 1. London: Taylor & Francis.

Feynman, Richard. 1965 *The character of physical law*. Cambridge, MA: MIT Press.

——. 1995 *Feynman lectures on gravitation* [Caltech, 1962–3], ed. by Brian Hatfield. Reading, MA: Addison-Wesley.

Fichant, Michel (ed.). 1994 *Gottfried Wilhelm Leibniz. La réforme de la dynamique. De corporum concursu (1678) et autres textes inédits*. Paris: Vrin.

Firode, Alain. 2001 *La dynamique de d'Alembert*. Paris: Vrin.

FitzGerald, George Francis. 1880 On the electromagnetic theory of the reflection and refraction of light. *PT*, 171: 691–711.

Fivel, Daniel. 1994 How interference effects in mixtures determine the rules of quantum mechanics. *PR*, A50: 2108–19.

Flato, Moshé, André Lichnerowicz, and Daniel Sternheimer. 1975 Déformations 1-différentiables des algèbres de Lie attachées à une variété symplectique ou de contact. *Compositio mathematica*, 31: 47–82.

Fodor, Jerry. 1983 *Modularity of mind: An essay on faculty psychology*. Cambridge: Cambridge University Press.

Forman, Paul. 1969 Why was it Schrödinger who developed de Broglie's ideas. *HSPS*, 1: 291–314.

Fourier, Joseph. 1822 *Théorie analytique de la chaleur*. Paris: Didot.

Fox, Robert. 1974 The rise and fall of Laplacian physics. *HSPS*, 4: 89–136.

Fox Keller, Evelyn. *Making sense of life: Explaining biological development with models, metaphors, and machines*. Cambridge, MA: Harvard University Press.

Franklin, Alan. 1990 *Experiment, right or wrong*. Cambridge: Cambridge University Press.

Fraser, Craig. 1983 Lagrange's early contributions to the principles and methods of mechanics. *AHES*, 28: 197–241.

Friedman, Michael. 1983 *Foundations of spacetime theories: Relativistic physics and philosophy of science.* Princeton: Princeton University Press.

——. 2001 *Dynamics of reason: The 1999 Kant lectures at Stanford University.* Stanford, CA: Stanford University Press.

——. 2009 Einstein, Kant, and the relativized *a priori.* In Bitbol, Kerszberg, and Petitot 2009, 253–267.

——. 2013 *Kant's construction of nature: A reading of the Metaphysical foundations of natural science.* Cambridge: Cambridge University Press.

Frigg, Roman, Joseph Berkovitz, and Fred Kronz. 2011 The ergodic hierarchy. In Edward Zalta (ed.), *The Stanford encyclopedia of philosophy* (Summer 2011 edn). Available at http://plato.stanford.edu/archives/sum2011/entries/ergodic-hierarchy/ (last accessed Jan. 2013).

Gabbay, Dov, Daniel Lehmann, and Kurt Engesser (eds). 2009 *Handbook of quantum logic and quantum structures: Quantum logic.* Amsterdam: Elsevier.

Galileo, Galilei. 1623 *Il saggiatore.* Rome: Mascardi.

Galison, Peter. 1995 Context and constraints. In Jed Buchwald (ed.), *Scientific practice: Theories and stories of doing physics* (Chicago: University of Chicago Press), 13–41.

——. 1997 *Image and logic: A material culture of microphysics.* Chicago: University of Chicago Press.

——. 1998 Feynman's war: Modeling weapons, modeling nature. *SHPMP*, 29: 391–434.

——. 2003 *Einstein's clocks, Poincaré's maps: Empires of time.* New York: Norton.

Gallavotti, Giovanni. 1994 Ergodicity, ensembles, irreversibility in Boltzmann's and beyond. *Journal of statistical physics*, 78: 1571–89.

Garber, Daniel. 1992 *Descartes' metaphysical physics.* Chicago: University of Chicago Press.

——. 2009 *Leibniz: Body, substance, monad.* New York: Oxford University Press.

Gardner, Michael. 1971 Is quantum logic really logic? *Journal of the Philosophy of Science Association*, 38: 508–29.

Garner, Lynn. 1981 *An outline of projective geometry.* New York: North Holland.

Gaukroger, Stephen, John Shuster, and John Sutton. 2000 *Descartes' natural philosophy.* London: Routledge.

Gauss, Carl Friedrich. 1828 *Disquisiones generales circa superficies curvas.* Göttingen: Dieterich.

Gearhart, Clayton. 2002 Planck, the quantum, and the historians. *Physics in perspective*, 4: 170–215.

Geroch, Robert. 1978 *General relativity from A to B.* Chicago: University of Chicago Press.

Giere, Ronald. 1988 *Explaining science: A cognitive approach.* Chicago: University of Chicago Press.

Gleason, Andrew. 1957 Measures on the closed subspaces of a Hilbert space. *Journal of mathematics and mechanics*, 6: 885–93.

Goenner, Hubert. 2004 On the history of unified field theories. Available at http://relativity.livingreviews.org/Articles/lrr-2004-2/index.html (last accessed Jan. 2013).

Gooding, David. 1978 Conceptual and experimental bases of Faraday's denial of electrostatic action at a distance. *SHPS*, 9: 117–49.

Gosson, Maurice de, and Basil Hiley. 2009 The symplectic camel and the uncertainty principle: The tip of an iceberg? *FP*, 99: 194–204.

——. 2011 Imprints of the quantum world in classical physics. *FP*, 11: 1415–36.

Granström, Helena. 2006 *Gleason's theorem.* Master thesis, Stockholm University.

Grattan-Guinness, Ivor. 1984 Work of the workers: Advances in engineering mechanics and instruction in France, 1800–1830. *Annals of science*, 41: 1–33.

Grätzer, George. 2011 *Lattice theory: Foundations.* Basel: Springer.

Gray, Jeremy. 2007 *Worlds out of nothing: A course in the history of geometry in the 19th century.* London: Springer.

Greenberg, Marvin Jay. 2007 *Euclidean and non-Euclidean geometries: Development and history*, 4th edn. New York: Freeman.

Grinbaum, Alexei. 2007 Reconstruction of quantum theory. *BJPS*, 58: 387–408.

Groenewold, Hilbrand. 1946 On the principles of elementary quantum mechanics. *Physica*, 12: 405–60.

Gupta, Suraj. 1954 Gravitation and electromagnetism. *PR*, 96: 1683–5.

Gutt, Simone. 1979 Equivalence of deformations and associated *-products. *Letters in mathematical physics*, 3: 297–309.

—— and John Rawnsly. 1998 Equivalence of star products on a symplectic manifold: An introduction to Deligne's Čech cohomology classes. *Journal of geometry and physics*, 29: 347–92.

Gyeong Soon, Im. 1996 Experimental constraints on formal quantum mechanics: The emergence of Born's quantum theory of collision processes in Göttingen, 1924–1927. *AHES*, 90: 73–101.

Haas, Arthur Erich. 1909 *Die Entwicklungsgeschichte des Satzes von der Erhaltung der Kraft*. Vienna: Hölder.

Hacking, Ian 1983 *Representing and intervening: Introductory topics in the philosophy of natural science*. Cambridge: Cambridge University Press.

Hall, Alfred Rupert. 1966 Mechanics and the Royal Society, 1668–70. *BJHS*, 3: 24–38.

Hankins, Thomas. 1967 The influence of Malebranche on the science of mechanics during the eighteenth century. *Journal for the history of ideas*, 28: 193–210.

——. 1970 *Jean d'Alembert: Science and the enlightenment*. Oxford: Clarendon Press.

——. 1980 *Sir William Rowan Hamilton*. Baltimore: Johns Hopkins University Press.

Hanle, Paul. 1977 The coming of age of Erwin Schrödinger: His quantum statistics of ideal gases. *AHES*, 17: 165–92.

Hardy, Lucien. 2001 Quantum theory from five reasonable axioms. arXiv:quant-ph/0101012v4 25 Sep 2001.

Harman, Peter. 1982 *Energy, force, and matter: The conceptual development of nineteenth-century physics*. Cambridge: Cambridge University Press.

——. 1987 Mathematics and reality in Maxwell's dynamical physics: The natural philosophy of James Clerk Maxwell. In R. Kargon and P. Achinstein (eds), *Kelvin's Baltimore lectures and modern theoretical physics: Historical and philosophical perspectives* (Cambridge, MA: MIT Press), 267–97.

Haroche, Serge, and Jean-Michel Raimond. 2006 *Exploring the quantum: Atoms, cavities, and photons*. Oxford: Oxford University Press.

Hatfield, Gary. 1990 *The natural and the normative: Theories of spatial perception from Kant to Helmholtz*. Cambridge, MA: MIT Press.

Heath, Thoma Little. 1908 Introduction and commentary in Heath and Heiberg 1908, vol. 1.

—— and Johan Ludvig Heiberg. 1908 *The thirteen books of Euclid's elements translated from the text of Heiberg with introduction and commentary by T. L. Heath*, 3 vols. Cambridge: Cambridge University Press.

Heidelberger, Michael. 1986 Zur Philosophie der Messung im 19. Jahrhundert. In *Die historische Metrologie in den Wissenschaften* (St. Katharinen: Scripta mercaturae), 159–68.

——. 2006 Applying models in fluid dynamics. *International studies in philosophy of science* 20: 49–67.

Heilbron, John. 1964 *A History of the problem of atomic structure from the discovery of the electron to the beginning of quantum mechanics*. PhD diss., University of California at Berkeley.

——. 1974 *H. G. J. Moseley: The life and letters of an English physicist, 1887–1915*. Berkeley: University of California Press.

——. 1977 Lectures on the history of atomic physics. In Charles Weiner (ed.), *History of twentieth century physics* (New York: Academic Press), 40–108.

——. 1983 The origins of the exclusion principle. *HSPS*, 13: 261–310.

——. 1993 *Weighing imponderables and other quantitative science around 1800*. Berkeley: University of California Press.

—— and Thomas Kuhn. 1969 The genesis of the Bohr Atom. *HSPS*, 1: 211–90.

Heimann, Peter (Peter Harman). 1974 Helmholtz and Kant: The metaphysical foundations of *Über die Erhaltung der Kraft*. *SHPS*, 5: 205–38.

Heinzmann, Gerhard. 2001a The foundations of geometry and the concept of motion: Helmholtz and Poincaré. *Science in context*, 14: 457–70.

——. 2001b L'occasionalisme de Poincaré: l'élément unificateur de sa philosophie des sciences. Colloque "Poincaré et la théorie de la connaissance," CEPERC, Aix-en-Provence. Available at http://poincare.univ-nancy2.fr/digitalAssets/153508_occasionnalisme_poincare.pdf (last accessed Jan. 2013).

Heisenberg, Werner. 1925 Über die quantentheoretische Umdeutung kinematischer und mechanischer Beziehungen. *ZP*, 33: 879–93.

Helmholtz, Hermann. 1845 Bericht über die Theorie der physiologischen Wärmeerscheinungen für 1845. In *Die Fortschritte der Physik im Jahre 1845* (Berlin: Reimer, 1947), 346–55. Also in Helmholtz 1882–95, vol. 1, 4–11.

——. 1847 *Über die Erhaltung der Kraft, eine physikalische Abhandlung*. Berlin: Reimer. Also in Helmholtz 1882–95, vol. 1, 12–68.

——. 1868a Über die Thatsächlichen Grundlagen der Geometrie. In Helmholtz 1882–95, vol. 2, 610–17.

Helmholtz, Hermann. 1868b Über die Thatsachen, die der Geometrie zum Grunde legen. In Helmholtz
    1882–95, vol. 2, 618–39.
——. 1870 Über den Ursprung und die Bedeutung der geometrischen Axiome. In Helmholtz 1884, vol. 2,
    1–31.
——. 1878a Über die Thatsachen in der Wahrnehmung. In Helmholtz 1884, vol. 2, 213–48.
——. 1878b Über den Ursprung und Sinn der geometrischen Sätze; Antwort gegen Herrn Professor Lang.
    In Helmholtz 1882–95, vol. 2, 640–60.
——. 1882–95 Wissenschaftliche Abhandlungen, 3 vols. Leipzig: Barth.
——. 1884 Vorträge und Reden, 2 vols. Braunschweig: Vieweg.
——. 1886 Über die physikalische Bedeutung des Princips der kleinsten Wirkung. JRAM. Also in
    Helmholtz 1882–95, vol. 3, 203–48.
——. 1887 Zählen und Messen, erkenntnistheoretisch betrachtet. In Helmholtz 1882–95, vol. 3, 356–91.
——. 1921 Schriften zur Erkenntnistheorie, ed. Paul Hertz and Moritz Schlick. Berlin: Springer.
Hendry, John. 1984 The creation of quantum mechanics and the Bohr-Pauli dialogue. Dordrecht: Reidel.
Hertz, Paul. 1921 Notes to Helmholtz 1921.
Hilbert, David. 1899 Grundlagen der Geometrie. Leipzig: Teubner.
Hölder, Otto. 1900 Anschauung und Denken in der Geometrie (inaugural lecture for the Leipzig Academy,
    July 22, 1899). Leipzig: Teubner.
——. 1901 Die Axiome der Quantität und die Lehre vom Mass. Königlich Sächsische Gesellschaft der
    Wissenschaften zu Leipzig, Mathematisch-physische Classe, Berichte über die Verhandlungen, 53:
    1–46.
Howard, Don, and John Stachel (eds). 1989 Einstein and the history of general relativity, Einstein studies,
    vol. 1. Boston: Birkhäuser.
Hund, Friedrich. 1967 Geschichte der Quantentheorie. Manheim: Biographisches Institut.
——. 1974 The history of quantum theory. London: Harrap.
Hunziker, Walter. 1972 A note on symmetry operations in quantum mechanics. HPA, 45: 233–36.
Huygens, Christiaan. 1669 Extrait d'une lettre de M. Hugens à l'auteur du journal. Journal des sçavans
    (18 March 1669), 22–4.
——. 1673 Horologium oscillatorium, sive de motu pendulorum ad horologia aptato demonstrationes geomet-
    ricae. Paris: Muguet.
——. 1703 De motu corporum ex percussione. In Christiani Hugenii Zelemii, dum viveret, toparchae opuscula
    postuma, quae continent dioptricam. Commentarios de vitris figurandis. Dissertationem de corona &
    parheliis. Tractatum de motu. Tractatum de vi centrifuga. Descriptionem automati planetarii (Leiden:
    Boutesteyn, 1703), 369–400.
Hyder, David. 2001 Physiological optics and physical geometry. Science in context, 14: 419–56.
——. 2009 The determinate world: Kant and Helmholtz on the physical meaning of geometry. Berlin: de
    Gruyter.
Ignatowski, Woldemar von (Vladimir Sergeyevitch). 1910 Einige allgemeine Bemerkungen zum
    Relativitätsprinzip. Deutsche Physikalische Gesellschaft, Verhandlungen, 12: 788–96.
Jammer, Max. 1957 Concepts of force: A study in the foundation of dynamics. Cambridge, MA: Harvard
    University Press.
——. 1961 Concepts of mass, in classical and modern physics. Cambridge, MA: Harvard University Press.
    Revised edition (Princeton: Princeton University Press, 2001).
——. 1966 The conceptual development of quantum mechanics. New York: McGraw-Hill.
——. 1974 The philosophy of quantum mechanics: The interpretations of quantum mechanics in historical
    perspective. New York: Wiley.
Janiak, Andrew. 2004 (ed.) Newton: Philosophical writings. Cambridge: Cambridge University Press.
——. 2009 Newton's Philosophy. In Edward N. Zalta (ed.), The Stanford encyclopedia of philosophy (Winter
    2009 edn). Available at http://plato.stanford.edu/archives/win2009/entries/newton-philosophy/ (last
    accessed Jan. 2013).
Janis, Allen. 2002 Conventionality of simultaneity. In Edward N. Zalta (ed.), The Stanford encyclopedia of
    philosophy (Fall 2002 edn). Available at http://plato.stanford.edu/archives/fall2002/entries/spacetime-
    convensimul/ (last accessed Jan. 2013).
Janssen, Michel. 1995 A comparison between Lorentz's theory and special relativity in the light of the
    experiments of Trouton and Noble. PhD. diss., University of Pittsburgh.
——. 2005 Of pots and holes: Einstein's bumpy road to general relativity. AP, 14, supplement: 58–85.

Jarzinsky, Christopher. 2010 Equalities and inequalities: Irreversibility and the second law of thermodynamics at the nanoscale (paper given at Poincaré séminaire XV: *Le temps*). Available at http://www.bourbaphy.fr/jarzynskitemps.pdf (last accessed Jan. 2013)

Jauch, Josef. 1968 *Foundations of quantum mechanics*. Reading, MA: Adisson-Westley.

—— and Constantin Piron. 1963 Can hidden variables be excluded in quantum mechanics? *HPA*, 36: 827–837.

——. 1969 On the structure of quantal proposition systems. *HPA*, 42: 842–48.

Joas, Christian, and Christoff Lehner. 2009 The classical roots of wave mechanics: Schrödinger's transformations of the optical-mechanical analogy. *SHPMP*, 40: 338–51.

Jordan, Pascual. 1927 Über eine neue Begründung der Quantenmechanik. *ZP*, 40: 809–38 and 44: 1–25.

Jouguet, Emile. 1908 *Lectures de mécanique*, 2 vols. Paris: Gauthier-Villars.

Joule, James. 1847 Matter, living force, and heat [May 1847]. In *Scientific papers*, 2 vols. (London: Taylor & Francis, 1884, 1887), vol. 1, 265–76.

Jurkowitz, Edward. 2010 Helmholtz's early empiricism and the *Erhaltung der Kraft*. *Annals of science*, 67: 39–78.

Kadison, Richard. 1951 Isometries of operator algebras. *Annals of mathematics*, 54: 325–38.

Kaiser, David. 2011 *How the hippies saved physics: Science, counterculture, and the quantum revival*. New York: Norton.

Kant, Immanuel. 1770 *De mundi sensibilis atque intelligibilis forma et principiis*. Königsberg: Kanter.

——. 1781 *Critik der reinen Vernunft*. Riga: Hartknoch.

——. 1783 *Prolegomena zu einer jeden künftigen Metaphysik die als Wissenschaft wird auftreten können*. Riga: Hartknoch.

——. 1786 *Metaphysische Anfangsgründe der Naturwissenschaft*. Riga: Hartknoch.

——. 1787 *Critik der reinen Vernunft*, 2nd edn. Riga: Hartknoch.

Karmiloff-Smith, Annette. 1992 *Beyond Modularity: A developmental perspective on cognitive science*. Cambridge, MA: MIT Press.

Klein, Felix. 1890 Zur nicht-Euklidischen Geometrie. *Mathematische Annalen*, 37: 544–72. Also in *Gesammelte mathematische Abhandlungen*, 3 vols (Berlin: Springer, 1921–3), vol. 1, 352–83.

Klein, Martin. 1963 Einstein's first paper on quanta. *The natural philosopher*, 2: 59–86.

——. 1964 Einstein and the wave-particle duality. *The natural philosopher*, 3: 1–49.

——. 1965 Einstein, specific heats, and the early quantum theory. *Science*, 148: 173–80.

——. 1969 Gibbs on Clausius. *HSPS*, 1: 127–49.

——. 1970a *Paul Ehrenfest*, vol. 1: *The making of a theoretical physicist*. Amsterdam: North-Holland.

——. 1970b Maxwell, his demon, and the second law of thermodynamics. *American scientist*, 58: 84–97.

——. 1970c The first phase of the Bohr–Einstein dialogue. *HSPS*, 2: 1–39.

——. 1972 Mechanical explanation at the end of the nineteenth century. *Centaurus*, 17: 58–82.

——. 1973 The development of Boltzmann's statistical ideas. In E. G. D. Cohen and Walther Thirring (eds), *The Boltzmann equation: Theory and applications* (Vienna: Springer), 53–106.

——. 1977 The beginnings of quantum theory. In Charles Weiner (ed.), *History of twentieth century physics* (New York: Academic Press), 1–39.

——. 1979 Einstein and the development of quantum physics. In Anthony Philip French (ed.), *Einstein: A centenary volume* (Cambridge, MA: Harvard University Press), 133–51.

——. 1982 Fluctuations and statistical physics in Einstein's early work. In Gerald Holton and Yehuda Elkana (eds), *Albert Einstein: Historical and cultural perspectives. The centennial symposium in Jerusalem* (Princeton: Princeton University Press), 39–58.

Koenigsberger, Leo. 1902–3 *Hermann von Helmholtz*, 3 vols. Braunschweig: Vieweg.

Konno, Hiroyuki. 1978 The historical roots of Born's probability interpretation, *Japanese studies in history of science*, 17: 129–45.

Koyré, Alexandre. 1943 Galileo and Plato. *Journal of the history of ideas*, 4: 400–28.

Kragh, Helge. 1979 Niels Bohr's second atomic theory. *HSPS*, 10: 123–86.

——. 1982 Erwin Schrödinger and the wave equation: The crucial phase. *Centaurus*, 26: 154–97.

——. 1990 *Dirac: A scientific biography*. Cambridge: Cambridge University Press.

——. 1999 *Quantum generations: A history of physics in the twentieth century*. Princeton: Princeton University Press.

——. 2011 *Higher Speculations: Grand theories and failed revolutions in physics and cosmology*. Oxford: Oxford University Press.

Kragh, Helge. 2012 *Niels Bohr and the quantum atom: The Bohr model of atomic structure 1913–1925*. Oxford: Oxford University Press.

Kramers, Hendrik. 1924 The quantum theory of dispersion. *Nature*, 114: 310–11.

—— and Werner Heisenberg. 1925 Über die Streuung von Strahlung durch Atome. *ZP*, 31: 681–708.

Krantz, David, Patrick Suppes, Robert Duncan Luce, and Amos Tversky. 1971 *Foundations of measurement*, vol. 1. New York: Academic Press.

Krüger, Lorenz. 1994 Helmholtz über die Begreiflichkeit der Natur. In Krüger (ed.), *Universalgenie Helmholtz. Rückblick nach 100 Jahren* (Berlin: Akademie Verlag), 201–15.

Kuhn, Thomas. 1959 Energy conservation as an example of simultaneous discovery. In *The essential tension: Selected studies in scientific tradition and change* (Chicago: University of Chicago Press, 1977), 66–104.

——. 1962. *The structure of scientific revolutions*. Chicago: University of Chicago Press.

——. 1978 *Black-body theory and the quantum discontinuity, 1894–1912*. Chicago: University of Chicago Press.

Lacki, Jan. 2000 The early axiomatizations of quantum mechanics: Jordan, von Neumann and the continuation of Hilbert's program. *AHES*, 54: 279–318.

Lagrange, Joseph Louis. 1788 *Méchanique analitique*. Paris: Desaint

——. 1798. Sur le principe des vitesses virtuelles. *Journal de l'École Polytechnique*, vol. 2, *cahier* 5: 115–18.

——. 1811–15 *Mécanique analytique*, 2nd edn, 2 vols. Paris: Courcier.

——. 1867–92 *Œuvres de Lagrange*, 14 vols. Paris: Gauthier-Villars.

Lanczos, Cornelius. 1966 *The variational principles of mechanics*. Toronto: University of Toronto Press.

——. 1970 *Space through the ages: The evolution of geometrical ideas from Pythagoras to Hilbert and Einstein*. Academic Press: London and New York.

Landau, Lev Davidovich, and Evgeny Mikhailovich Lifshitz. 1951 *The classical theory of fields*. Reading, MA: Addison-Westley. French transl. (Paris: Mir, 1970).

——. 1960 *Mechanics*. Oxford: Pergamon Press.

Laplace, Pierre Simon de. 1796 Pierre Simon de Laplace, *Exposition du système du monde*, vol. 1. Paris: Cercle Social.

Le Bellac, Michel, and Jean-Marc Lévy-Leblond. 1973 Galilean electromagnetism. *Nuovo Cimento*, B14: 217–33.

Leibniz, Gottfried Wilhelm. 1686 Brevis demonstratio erroris memorabilis Cartesii et aliorum circa legem naturae, secundum quam volunt a Deo eandem semper quantitatem motus conservari; qua et in re mechanica abutuntur. *Acta eruditorum* (1686), 161–2.

——. [*c.*1686] On contingency. In *Philosophical essays*, ed. by Roger Ariew and Daniel Garber (Indianapolis: Hacket, 1989), 28–30.

——. [*c.*1688] Specimen inventorum de admirandis naturae generalis arcanis. In Leibniz 1875–90, vol. 7, 309–18.

——. [1692] Essay de dynamique. In Costabel 1960, 97–106.

——. 1695 Specimen dynamicum, pro admirandis naturae legibus circa corporum vires et mutuas actiones detegendis, et ad suas causas revocandis [part 1]. *Acta eruditorum* (1695), 145–57. Also in Leibniz 1849–63, vol. 6, 234–46 (part 1), 246–54 (part 2).

——. [1698] Essay de dynamique sur les loix du mouvement, où il est monstré, qu'il ne se conserve pas la même quantité de mouvement, mais la même force absolue, ou bien même quantité de l'action motrice. In Leibniz 1849–60, vol. 6, 215–31.

——. 1849–63 *Mathematische Schriften*, ed. C. I., Gerhardt, 7 vols. Berlin: Asher.

——. 1875–90 *Die philosophischen Schriften*, ed. C. I. Gerhardt, 7 vols. Berlin: Weidmann.

Lenoir, Timothy. 1982 *The strategy of life: Theology and mechanics in 19th-century biology*. Dordrecht: Reidel.

Levi-Civita, Tullio. 1917 Nozione di parallelismo in una varietà qualunque e consequente specificazione geometrica della curvatura Riemanniana. *Rendiconti del Circolo Matematico di Palermo*, 42: 73–205.

Lévy-Leblond, Jean-Marc. 1976 One more derivation of the Lorentz transformation. *American journal of physics*, 44: 1–13.

Lichnerowicz, André. 1979 Equivalence et existence des $*_\nu$-produits sur une variété symplectique. *CR*, A 289: 349–53.

——. 1982 Déformation d'algèbres associées à une variété symplectique (les $*_\nu$-produits). *Annales de l'institut Fourier*, 32: 157–209.

Lie, Sophus. 1886 Bemerkungen zu v. Helmholtzs Arbeit: Über die Thatsachen, die der Geometrie zugrunde legen. In *Gesammelte Abhandlungen*, vol. 2 (Leipzig: Teubner, 1935), 374–9.

——. 1890. Über die Grundlagen der Geometrie. In *Gesammelte Abhandlungen*, vol. 2 (Leipzig: Teubner, 1935), 380–413 (Abh. I), 414–468 (Abh. II).

Liu, Chuang. 1999 Explaining the emergence of cooperative phenomena. *Philosophy of science*, 66: 92–106.

Lorentz, Hendrik Antoon. 1923 De Bepaling van het g-veld in de algemeene relativiteitstheorie met behulp van de wereldlijnen van lichtsignalen en stoffelijke punten, met eenige opmerkingen over de lengte van staven en den duur van tijdsintervallen en over de theorieën van Weyl en Eddington. Koninklijke Akademie van Wetenschappen te Amsterdam, *Verslagen van de Gewone Vergaderingen der Wis- en Natuurkundige Afdeeling,* 32: 383–402. Translated as "The determination of the potentials in the general theory of relativity, with some remarks about the measurement of lengths and intervals of time and about the theories of Weyl and Eddington" in Royal Academy of Amsterdam, *Proceedings*, 29: 383–99. Also in Lorentz, *Collected papers*, 9 vols (The Haghe: Nijhoff, 1934–9), vol. 5, 363–82.

Lovelock, David. 1971 The Einstein tensor and its generalizations. *Journal of mathematical physics*, 12: 498–501.

Luce, Robert Duncan, and Patrick Suppes. 1981 Measurement, theory of. *Encyclopedia Britannica*, 15th edn. *Macropaedia*, 11: 739–45.

Ludwig, Günther. 1983 *Foundations of quantum mechanics*, 2 vols. Berlin: Springer.

——. 1985 *An axiomatic basis for quantum mechanics*, 2 vols. Berlin: Springer.

Lützen, Jesper. 2005 *Mechanistic images in geometric form: Heinrich Hertz's principles of mechanics*. Oxford: Oxford University Press.

Ly, Igor. 2007 *Géométrie et physique dans l'oeuvre de Henri Poincaré*. Thèse, Université Nancy 2.

MacCullagh, James. 1848 An essay towards the dynamical theory of crystalline reflexion and refraction.

——. [read Dec. 9, 1839]. *Transactions of the Royal Irish Academy*, 21: 17–50.

McGuire, John. 1970 Atoms and the "Analogy of Nature": Newton's third rule of philosophizing. *SHPS*, 1: 3–58.

Mach, Ernst. 1872 *Die Geschichte und die Wurzel des Satzes der Erhaltung der Arbeit*. Prague: Calve.

——. 1883 *Die Mechanik in ihrer Entwicklung. Historisch-kritisch dargestellt*. Leipzig: Brockhaus.

——. 1896 *Die Prinzipien der Wärmelehre. Historisch-kritisch dargestellt*. Leipzig: Barth.

Mackey, George. 1957 Quantum mechanics and Hilbert space. *American mathematical monthly, supplement*, 64(8): 45–57.

——. 1963 *The mathematical foundations of quantum mechanics*. Reading, MA: Benjamin/Cummings.

McMullin, Ernan. 1985 Galilean idealization. *SHPS*, 16: 247–73.

Mariotte, Edme. 1679 *Traité de la percussion ou choc des corps*. Paris: Michallet.

Martínez, Alberto. 2009 *Kinematics: The lost origins of Einstein's relativity*. Baltimore: Johns Hopkins University Press.

Marzke, Robert. 1959 *The theory of measurement in general relativity*. PhD diss., Princeton University.

—— and John Archibald Wheeler. 1964 Gravitation as geometry—I: The geometry of space-time and the geometrodynamical standard meter. In Hong-Yee Chiu and Willam F. Hoffmann (eds), *Gravitation and relativity* (New York: Benjamin), 40–64.

Masanes, Lluís, and Markus Müller. 2010 A derivation of quantum theory from physical requirements. arXiv:1004.1483v2 [quant-ph] 5 Oct 2010.

Maupertuis, Pierre Louis Moreau de. 1740 Loi du repos des corps. *MAS* (1740), 170–6.

——. 1744 Accord de différentes loix de la nature qui avoient jusqu'ici paru incompatibles. *MAS* (1744), 417–26.

——. 1746 Les loix du mouvement et du repos déduites d'un principe métaphysique. *MAB*, 2: 267–94.

——. 1756 Examen philosophique de la preuve de l'existence de Dieu employée dans l'essai de cosmologie. *MAB*, 14: 389–424.

Maxwell, James Clerk. 1860 Illustrations of the dynamical theory of gases. *PM*, 19: 19–32 and 20: 21–37.

——. 1861 On physical lines of force. Parts I and II. *PM*, 21: 161–75, 281–91, 338–48. Also in Maxwell 1890, vol. 1, 451–88.

——. 1865 A dynamical theory of the electromagnetic field. *PT*, 155: 459–512. Also in Maxwell 1890, vol. 1, 586–97.

——. 1867 On the dynamical theory of gases. *PT*, 157: 49–88.

——. 1873 *A treatise on electricity and magnetism*, 2 vols. Oxford: Clarendon Press.

——. 1878 Diffusion. *Encyclopedia Britannica*. Also in Maxwell 1890, vol. 2, 625–46.

Maxwell, James Clerk. 1879 On Boltzmann's theorem on the average distribution of energy in a system of material points. *TCPS*, 12: 547–70.

——. 1890 *The scientific papers of James Clerk Maxwell*, 2 vols. Cambridge: Cambridge University Press.

——. 1990–2002 *The scientific letters and papers of James Clerk Maxwell*, ed. by Peter Harman, 3 vols. Cambridge: Cambridge University Press.

Mayer, Julius Robert. 1842 Bemerkungen über die Kräfte der unbelebten Natur. *Annalen der Chemie und Pharmacie*, 42: 233–40.

——. 1845 *Die Organische Bewegung in ihrem Zusammenhange mit dem Stoffwechsel. Ein Beitrag zur Naturkunde*. Heilbronn: Drechsler.

——. 1851 *Bemerkungen über das mechanische Äquivalent der Wärme*. Heilbronn: Landherr.

Mazzoni, Jacobi. 1587 *Universam Platonis et Aristotelis philosophiam praeludia, sive de comparatione Platonis et Aristotelis*. Venice: Guerilio.

Mehra, Jagdish, and Helmut Rechenberg. 1982–1987 *The historical development of quantum theory*, 5 vols. New York: Springer.

Merker, Joël. 2010 *Sophus Lie, Friedrich Engel et le problème de Riemann-Helmholtz*. Paris: Hermann.

Meyer-Abich, Karl. 1965 *Korrespondenz, Individualität und Komplementarität: Eine Studie zur Geistgeschichte der Quantentheorie in den Beiträgen Niels Bohrs*. Wiesbaden: Franz Steiner.

Michell, John. 1993 The origins of the representational theory of measurement: Helmholtz, Hölder, and Russell. *SHPS*, 24: 185–206.

——. 1999 *Measurement in psychology. Critical history of a methodological concept*. Cambridge: Cambridge University Press.

Minkowski, Hermann. 1909 [Sept. 21, 1908] *Raum und Zeit*. Leipzig: Teubner.

Mittelstaedt, Peter. 1978 *Quantum logic*. Dordrecht: Reidel.

——. 2003 Interpreting quantum mechanics—in the light of quantum logic. In Lutz Castell and Otfried Ischebeck (eds), *Time, quantum and information* (Berlin: Springer), 281–90.

——. 2011 *Rational reconstructions of modern physics*. Heidelberg: Springer.

—— and Paul Weingartner. 2005 *Laws of nature*. Berlin: Springer.

Montucla, Jean Etienne. 1799–1802 *Histoire des mathématiques dans laquelle on rend compte de leurs progrès depuis leur origine jusqu'à nos jours; où l'on expose le tableau et le développement des principales découvertes dans toutes les parties des mathématiques, les contestations qui se sont élevées entre les mathématiciens, et les principaux traits de la vie des plus célèbres*, 4 vols. Paris: Agasse.

Morrison, Margaret. 1989 Hypotheses and certainty in Cartesian science. In James Rorbert Brown and Jürgen Mittelstrass (eds), *An intimate relation: Studies in the history and philosophy of science presented to Robert E. Butts on his 60th birthday* (Dordrecht: Kluwer Academic), 43–64.

——. 2000 *Unifying scientific theories: Physical concepts and mathematical structure*. Cambridge: Cambridge University Press.

Moyal, Ana. 2006 *Maverick mathematician: The life and science of J. E. Moyal*. Canberra: Australian National University E Press. Also at http://epress.anu.edu.au (last accessed Jan. 2013).

Moyal, José Enriques. 1949 Quantum mechanics as a statistical theory. *Proceedings of the Cambridge Philosophical Society*, 45: 99–124.

Nagel, Ernest. 1931 Measurement. *Erkenntnis*, 2: 313–33.

Needell, Allan. 1980 *Irreversibility and the failure of classical dynamics: Max Planck's work on the quantum theory, 1900–1915*. PhD diss., University of Michigan, Ann Arbor.

Neumann, Johann von [János Neumann Margittai]. 1930 Allgemeine Eigenwerttheorie Hermitescher Funktionaloperatoren. *Mathematische Annalen*, 102: 49–131.

——. 1931 Die Eindeutigkeit der Schrödingerschen Operatoren. *Mathematische Annalen*, 104: 570–8.

——. 1932 *Mathematische Grundlagen der Quantenmechanik*. Berlin: Springer.

Newton, Isaac [MS page numbers are those of *The Newton project*, http://www.newtonproject.sussex.ac.uk/]. [*c*.1664] Questiones quaedam philosophicae. MS, Cambridge University library, Add. 3996.

——. [*c*.1668?] De gravitatione et aequipondio fluidorum et solidorum in fluidis. MS, Cambridge University library, Add. 4003.

——. 1687 *Philosophiae naturalis principia mathematica*. London: Streater & Smith.

——. 1707 *Arithmetica universalis; sive de compositione et resolutione arithmetica liber*. Cambridge: Typis Academicis.

——. 1713 *Philosophiae naturalis principia mathematica*, 2nd edn, ed. Roger Cotes. Cambridge: Crownfield.

——. 1720 *Universal arithmetick: or, a treatise of arithmetical composition and resolution*, transl. Raphson and Cunn. London: Senex, Taylor, Warner, and Osborn.

——. 1729 *The mathematical principles of natural philosophy*, transl. Andrew Motte, 2 vols. London: Motte.

Nordström, Dunnar. 1913a Träge und Schwere Masse in der Relativitätsmechanik. *AP*, 40: 856–78.

——. 1913b Zur Theorie der Gravitation vom Standpunkt des Relativitätsprinzips. *AP*, 42: 533–54.

Norton, John D. 1989a What was Einstein's principle of equivalence? In Howard and Stachel 1989, 5–47.

——. 1989b How Einstein found his field equations, 1912–1915. In Howard and Stachel 1989, 101–59.

——. 1993 General covariance and the foundations of general relativity: eight decades of dispute. *Reports on progress in physics*, 56: 591–858.

——. 2004 The hole argument. In Edward N. Zalta (ed.), *The Stanford encyclopedia of philosophy* (Spring 2004 edn). ed. Available at http://plato.stanford.edu/archives/spr2004/entries/spacetime-holearg/ (last accessed Jan. 2013)

Pais, Abraham. 1982 *Subtle is the Lord: The science and the life of Albert Einstein*. Oxford: Oxford University Press.

——. 1991 *Niels Bohr's times, in physics, philosophy, and polity*. Oxford: Clarendon Press.

Pal, Palash. 2003 Nothing but relativity. *European journal of physics*, 24: 315–19.

Panza, Marco. 1995 De la nature épargnante aux forces généreuses: le principe de moindre action entre mathématiques et métaphysique. Maupertuis et Euler, 1740–1751. *RHS*, 48: 435–520.

Paty, Michel. 1993 *Einstein philosophe: La physique comme pratique philosophique*. Paris: Presses Universitaires de France.

Pauli, Wolfgang. 1929 Theorie der schwarzen Strahlung. In *Müller-Pouillets Lehrbuch*, 11th edn, vol. 2, part 2, chap. 27 (Braunschweig: Vieweg), 1483–1553.

Perlick, Volker. 1987 Characterization of standard clocks by means of light rays and freely falling particles. *General relativity and gravitation*, 19: 1059–73.

——. 2007 On the radar method in general-relativistic spacetimes. arXiv:0708.0170v1 [gr-qc] 1 Aug 2007.

Petruccioli, Sandro. 1993 *Atoms, metaphors and paradoxes: Niels Bohr and the construction of a new physics*. Cambridge: Cambridge University Press.

Piron, Constantin. 1964 Axiomatique quantique. *HPA*, 37: 439–68.

——. 1976 *Foundations of quantum mechanics*. Reading, MA: Benjamin.

Planck, Max. 1887 *Das Prinzip der Erhaltung der Energie*. Leipzig: Teubner.

Poincaré, Henri. 1887 Sur les hypothèses fondamentales de la géométrie. *Bulletin de la Société Mathématique de France*, 15: 203–16.

——. 1890 *Electricité et optique. I. Les théories de Maxwell et la théorie électromagnétique de la lumière*. [Sorbonne lectures, second semester 1887–88]. Paris: Carré.

——. 1891 Les géométries non euclidiennes. *Revue générale des sciences pures et appliquées*, 2: 761–74.

——. 1892a *Thermodynamique* [Sorbonne lectures, first semester 1888–9], ed. J. Blondin. Paris: Gauthier-Villars.

——. 1892b Sur les géométries non euclidiennes. *Revue générale des sciences pures et appliquées*, 3: 74–5.

——. 1893 Le continu mathématique. *Revue de métaphysique et de morale*, 1: 26–34.

——. 1898 La mesure du temps. *Revue de métaphysique et de morale*, 6: 371–84.

——. 1900a La théorie de Lorentz et le principe de la réaction. In *Recueil de travaux offerts par les auteurs à H.A. Lorentz à l'occasion du 25ème anniversaire de son doctorat le 11 décembre 1900; Archives néerlandaises*, 5: 252–78.

——. 1900b Sur les rapports de la physique expérimentale et de la physique mathématique. In *Rapports présentés au congrès international de 1900*, vol. 1 (Paris), 1–29. Also in Poincaré 1902, Chaps. 9–10.

——. 1902 *La science et l'hypothèse*. Paris: Flammarion.

——. 1904 L'état actuel de la physique mathématique. *Bulletin des sciences mathématiques*, 28: 302–24.

——. 1905 Sur la dynamique de l'électron. *CR*, 140: 1504–8.

——. 1906 Sur la dynamique de l'électron. *Rendiconti del Circolo Matematico di Palermo*. Also in *Œuvres*, vol. 9 (Paris: Gautier-Villars, 1954), 494–550.

——. 1908 La dynamique de l'électron. *Revue générale des sciences pures et appliquées*, 19: 386–402.

Poisson, Siméon Denis. 1811 *Traité de mécanique*, 2 vols. Paris: Courcier.

——. 1828 Mémoire sur l'équilibre et le mouvement des corps élastiques. *MAS*, 8: 357–570, 623–7.

Primas, Hans. 1981 *Chemistry, quantum mechanics and reductionism*. Berlin: Springer.

Principe da Silva, João. 2012 Sources et nature de la philosophie de la physique d'Henri Poincaré. *Philosophia scientiae*, 16: 197–222.

Pulte, Helmut. 1989 *Das Prinzip der kleinsten Wirkung und die Kraftkonzeptionen der rationalen Mechanik: Eine Untersuchung zur Grundproblematik bei Leonhard Euler, Pierre Louis Moreau de Maupertuis und Joseph Louis Lagrange*. Stuttgart: Steiner.

Putnam, Hilary. 1968 Is logic empirical? In Mark Wartowsky and Robert Cohen (eds), *Boston studies in the philosophy of science* (Dordrecht: Reidel), vol. 5, 216–41.

Rabouin, David. 2009 *Mathesis universalis. L'idée de « mathématique universelle » d'Aristote à Descartes*. Paris: Presses Universitaires de France.

Radelet de Grave, Patricia. 1987 Daniel Bernoulli et le parallélogramme des forces. *Sciences et techniques en perspective*, 11: 69–90.

Rédei, Miklós. 2007 The birth of quantum logic. *History and philosophy of logic*, 28: 107–22.

Reichenbach, Hans. 1920 *Relativitätstheorie und Erkenntnis apriori*. Berlin: Springer.

——. 1928 *Philosophie der Raum-Zeit-Lehre*. Berlin and Leipzig: de Gruyter.

Renn, Jürgen. 1997 Einstein's controversy with Drude and the origin of statistical mechanics: A new glimpse from the "love letters," *AHES*, 51: 315–54.

——. 2007 (editor and contributor) *The genesis of general relativity*, 4 vols. Berlin: Springer.

——. 2013 Schrödinger and the genesis of wave mechanics. In W. L. Reiter and J. Yngvason (eds), *Erwin Schrödinger—50 years after* (Zürich: European Mathematical Society), 9–36.

—— and Tilman Sauer. 2003. Errors and insights: Reconstructing the genesis of general relativity from Einstein's Zurich notebook. In F. L. Holmes, J. Renn, and H. J. Rheinberger, *Reworking the bench: Research notebooks in the history of science* (Dordrecht and Boston: Kluwer), 253–68.

Reynolds, Osborne. 1903 *Papers on mechanical and physical subjects*, vol. 3: *The sub-mechanics of the universe*. Cambridge: Cambridge University Press.

Richards, Joan. 1977 The evolution of empiricism: Hermann von Helmholtz and the foundations of geometry. *BJPS*, 28: 235–53.

Riemann, Bernhard. [1861] Commentatio mathematica, qua respondere tentatur quaestioni ab Illma Academia Parisiensis propositae: "Trouver quel doit être l'état calorifique d'un corps solide homogène indéfini pour qu'un système de courbes isothermes, à un instant donné, restent isothermes après un temps quelconque, de telle sorte que la température d'un point puisse s'exprimer en fonction du temps et de deux autres variables indépendantes." In Riemann 1990, 433–36, with commentary by Heinrich Weber and Richard Dedekind, 437–55.

——. 1867 [*Habilitationsvortrag* of 1854] Über die Hypothesen, welche der Geometrie zu Grunde liegen. Königliche Gesellschaft der Wissenschaften zu Göttingen, *Abhandlungen*, 13: 132–52. Translated by William Kingdon Clifford in *Nature*, 8 (1873), 14–17, 36–7.

——. 1990 *Gesammelte mathematische Werke, wissenschaftlicher Nachlass und Nachträge. Collected papers*, ed. Heinrich Weber, Richard Dedekind, and Rhahavan Narasimhan. Berlin: Springer.

Robb, Alfred Arthur. 1914 *A theory of time and space*. Cambridge: Cambridge University Press.

——. 1921 *The absolute relations of time and space*. Cambridge: Cambridge University Press.

——. 1936 *Geometry of time and space*. Cambridge: Cambridge University Press.

Roberval, Gilles de. 1636 *Traité de méchanique des poids soustenus par des puissances sur les plans inclinez à l'horizon. Des puissance qui soustiennent un poids suspendu à deux cordes*. Paris: Charlemagne.

Roche, John. 1998 *The mathematics of measurement: A critical history*. London: Athlone Press.

Rüdinger, Erik. 1985 The correspondence principle as a guiding principle. *Rivista di storia della scienzia*, 2: 357–67.

Russo, Arturo. 1981 Fundamental research at Bell laboratories: The discovery of electron diffraction. *HSPS*, 12: 117–60.

Ryckman, Thomas. 1994 Weyl, Reichenbach and the epistemology of geometries. *SHPS*, 25: 831–70.

——. 2003 The philosophical roots of the gauge principle: Weyl and transcendental phenomenological idealism. In Katherine Brading and Elena Castellani (eds), *Symmetries in physics: Philosophical reflections* (Cambridge: Cambridge University Press), 61–88.

——. 2005 *The reign of relativity: Philosophy in physics 1915–1925*. Oxford: Oxford University Press.

Saint-Venant, Adhémar Barré de. [1834] Sur les théorèmes de la mécanique générale. In Archives de l'Académie des Sciences, *pochette de séance* for 14 Apr. 1834. Also in Alfred Flamant, *Mécanique générale* (Paris: Tignol, 1888), IX–XXXII.

——. [1835] Commencement d'un traité de mécanique. Manuscript B5 in Fond Saint-Venant (Ecole Polytechnique), *carton* 19.

——. 1851 *Principes de mécanique fondés sur la cinématique*. Paris: Bachelier.

Schaffner, Kenneth. 1967 Approaches to reduction. *Philosophy of science*, 34: 137–47.

Scheibe, Erhard. 1994 Zwischen Rationalismus und Empirismus: Der Weg der Physik. In F. Fulda and R. P. Horstmann (eds), *Vernunftbegriffe der Moderne. Stuttgarter Hegel-Kongress 1993* (Stuttgart: Klett-Cotta), 73–95. Transl. in Scheibe 2001, Chap. 1.

——. 2001 *Between rationalism and empiricism: Selected papers in the philosophy of physics*, ed. Brigitte Falkenburg. New York: Springer.

Schiemann, Gregor. 1997 *Wahrheitsgewissheitsverlust: Hermann von Helmholtz' Mechanismus im Anbruch der Moderne: eine Studie zum Übergang von klassischer zu moderner Naturphilosophie*. Darmstadt: Wissenschaftliche Buchgesellschaft.

——. 2009 *Hermann von Helmholtz's mechanism: The loss of certainty* [translation of Schiemann 1997]. New York: Springer.

Schmit, Christophe. 2011 Les articles de mécanique dans l'Encyclopédie, ou D'Alembert lecteur de Varignon. *Recherches sur Diderot et sur l'Encyclopédie*, 46: 169–200.

Schmitz-Rigal, Christiane. 2002 *Die Kunst offenen Wissens: Ernst Cassirers Epistemologie und Deutung der modernen Physik*. Hamburg: Meiner.

——. 2009 Ernst Cassirer: Open constitution by functional a priori and symbolical structuring. In Bitbol, Kerszberg, and Petitot 2009, 75–93.

Scholz, Erhard. 1980 *Geschichte des Mannigfaltigkeitsbegriffs von Riemann bis Poincaré*. Basel: Birkhäuser.

——. 2001 Weyl's infinitesimalgeometrie, 1917–1925. In E. Scholz (ed.), *Hermann Weyl's Raum – Zeit – Materie and a general introduction to his scientific work* (Basel: Birkäuser), 48–104.

Schrödinger, Erwin. 1926a Quantisierung als Eigenwertproblem (Erste Mitteilung). *AP*, 79: 361–76.

——. 1926b Quantisierung als Eigenwertproblem (Zweite Mitteilung). *AP*, 79: 489–527.

——. 1926c Über das Verhältnis der Heisenberg-Born-Jordanschen Quantenmechanik zu der meinen. *AP*, 79: 734–56.

——. 1926d Quantisierung als Eigenwertproblem (Dritte Mitteilung). *AP*, 80: 437–90.

——. 1926e Quantisierung als Eigenwertproblem (Vierte Mitteilung). *AP*, 81: 109–39.

——. 1935 Discussion of probability relations between separated systems. *Proceedings of the Cambridge Philosophical Society*, 31: 555–63; 32 (1936): 446–51.

Schüller, Volkmar. 1994 Das Helmholtz–Liesche Raumproblem und seine ersten Lösungen. In Lorenz Krüger (ed.), *Universalgenie Helmholtz. Rückblick nach 100 Jahren* (Berlin: Akademie Verlag), 260–75.

Schwarzschild, Karl. 1903 Zur Elektrodynamik. I. Zwei Formen des Prinzips der kleinsten Wirkung in der Elektronentheorie. *GN* (1903), 126–31.

——. 1916 Zur Quantenhypothese, *BB* (1916), 548–68.

Séris, Jean-Pierre. 1987 *Machine et communication: du théâtre des machines à la mécanique industrielle*. Paris: Vrin.

Seth, Suman. 2010 *Crafting the quantum: Arnold Sommerfeld and the practice of theory, 1890–1926*. Cambridge, MA: MIT Press.

Shinn, Terry, and Pascal Ragouet. 2005 *Controverses sur la science: Pour une sociologie transversaliste de l'activité scientifique*. Paris: Raisons d'agir.

Siegel, Daniel. 1991 *Innovation in Maxwell's electromagnetic theory: Molecular vortices, displacement current, and light*. Cambridge: Cambridge University Press.

Sklar, Laurence. 1967 Types of intertheoretic reduction. *BJPS*, 18: 109–24.

——. 1977 Facts, conventions, and assumptions in the theory of space-time. In Earman, Glymour, and Stachel 1977, 206–74.

——. 1993 *Physics and chance: Philosophical issues in the foundations of statistical mechanics*. Cambridge: Cambridge University Press.

Smadja, Ivahn. 2010 Tuning up mind's pattern to nature's own idea: Eddington's early twenties case for variational derivatives. *SHPMP*, 41: 128–45.

Smeaton, John. 1794 Memoirs for the Royal Society of London, *Philosophical Transactions*, reprinted as *Experimental enquiry concerning the natural powers of wind and water to turn mills and other machines depending on a circular motion* [1759]. *And an experimental examination of the quantity and proportion of mechanic power necessary to be employed in giving different degrees of velocity to heavy bodies from a state of rest* [1776] *and also new fundamental experiments upon collisions of bodies* [1782]. London: Taylor.

Smith, Crosbie. 1998 *The science of energy: A cultural history of energy physics in Victorian Britain*. Chicago: Chicago University Press.

Smith, Crosbie and Norton Wise. 1989 *Energy and empire: A biographical study of Lord Kelvin*. Cambridge: Cambridge University Press.

Sneed, Joseph. 1971 *The logical structure of mathematical physics*. Dordrecht: Reidel.

Sommerfeld, Arnold. 1916 Zur Quantentheorie der Spektrallinien. *AP*, 51: 1–94, 125–67.

Stachel, John. 1974 The "Logic" of "Quantum Logic". *Proceedings of the biennial meeting of the Philosophy of Science Association* (Berlin: Springer, 1974), 515–26.

——. 1977 Introduction to Earman, Glymour, and Stachel 1977.

——. 1986 Einstein and the quantum: Fifty years of struggle. In Robert Colodny (ed.), *From quarks to quasars: Philosophical problems of modern physics*. Pittsburgh: Pittsburgh University Press.

——. 1989a The rigidly rotating disc as the "missing link" in the history of general relativity. In Howard and Stachel 1989, 48–62.

——. 1989b Einstein's search for general covariance, 1912–1915. In Howard and Stachel 1989, 63–100.

——. 1995 History of relativity. In L. Brown, A. Pais, B. Pippard (eds), *Twentieth century physics* (New York: American Institute of Physics Press), 249–356.

Stachow, Ernst-Walther. 1975 Structures of quantum languages for individual systems. In P. Mittelstaedt and W. Stachow (eds), *Recent developments in quantum logic* (Zürich: Bibliographisches Institut), 129–46.

Stein, Howard. 1977 Some philosophical prehistory of general relativity. In Earman, Glymour, and Stachel 1977, 3–49.

——. 1981 "Subtler forms of matter" in the period following Maxwell. In G. Cantor and M. J. S. Hodge (eds), *Conceptions of ether: Studies in the history of ether theories, 1740–1900* (Cambridge: Cambridge University Press, 1981), 309–40.

Sternheimer, Daniel. 1998 Deformation quantization: Ten years after. arXiv:math/9809056v1 [math.QA] 10 Sep 1998.

Stevens, Stanley Smith. 1946 On the theory of scales of measurement. *Science*, 103: 667–80.

Stevin, Simon. 1586 *De beghinselen der weeghconst*. Leiden: Plantijn.

Straumann, Norbert. 2000 Gauge theory and gravitation. In Dirk Graudenz and Valeri Markushin (eds), *Proceedings of the summer school on phenomenology of gauge interactions: August 13–19, 2000, Zuoz (Engadin), Switzerland*. Villigen: Paul Scherrer Institut. Also at http://ltpth. web.psi.ch/zuoz_school/previous_summerschools/zuoz2000/zuoz2000proc.htm (last accessed Jan. 2013).

Straus, Ernst Gabor. 1956 Assistant bei Albert Einstein. In Carl Seelig (ed.), *Helle Zeit, dunkle Zeit. In memoriam Albert Einsteins* (Zürich: Europa), 65–73.

Stuewer, Roger. 1975 *The Compton effect: Turning point in physics*. New York: Science History Publications.

Suppe, Frederick. 1974 *The structure of scientific theories*. Urbana: University of Illinois Press.

——. 1989 *The semantic conception of theories and scientific realism*. Urbana: University of Illinois Press.

Suppes, Patrick. 1951 A set of independent axioms for extensive quantities. *Portugaliae mathematica*, 10: 163–72.

——. 1960 A Comparison of the meaning and uses of models in mathematics and the empirical sciences. *Synthese*, 12: 287–301.

——. 1962 Models of data. In E. Nagel, P. Suppes, and A. Tarski (eds), *Methodology and philosophy of science: Proceedings of the 1960 International Congress* (Palo Alto, CA: Stanford University Press), 252–61.

——. 1967 What is a scientific theory? In S. Morgenbesser (ed.), *Philosophy of science today* (New York: Basic Books), 55–67.

Synge, John Lighton. 1960 *Relativity: The general theory*. Amsterdam: North Holland.

Szabó, István. 1979 *Geschichte der mechanischen Prinzipien und ihrer wichtigsten Anwendungen*. Basel: Birkhäuser.

Thomson, William (Lord Kelvin). 1849 An account of Carnot's theory of the motive force of heat; with numerical results from Regnault's experiments on steam. *TRSE*, 16: 541–74. Also in Thomson 1882–1911, 113–55.

——. 1851 On the dynamical theory of heat, with numerical results deduced from Mr. Joule's equivalent of a thermal unit, and M. Regnault's observations of steam. Parts I–III. *TRSE*, 20: 261–88, 475–82. Also in Thomson 1882–1911, vol. 1, 174–210.

——. 1852 On a universal tendency in nature to the dissipation of energy. *PRSE*, 3: 139–42. Also in Thomson 1882–1911, vol. 1, 194–7.

——. 1857 [read May 1, 1854] On the dynamical theory of heat. Part V: Thermo-electric currents. *TRSE*, 21: 123–71. Also in Thomson 1882–1911, vol. 1, 232–323.

——. 1882–1911 *Mathematical and physical papers*, 6 vols. Cambridge: Cambridge University Press.

Torretti, Roberto. 1978 *Philosophy of geometry from Riemann to Poincaré*. Dordrecht: Reidel.

——. 1980 *Creative understanding: Philosophical reflections on physics*. Chicago: University of Chicago Press.

——. 1984 *Philosophy of geometry from Riemann to Poincaré*, 2nd edn. Dordrecht: Reidel.

Torricelli, Evangelista. 1644 *De motu gravium naturaliter descendentium et projectorum libri duo*. In *Opera geometrica* (Florence: Masse and Landis), 95–243.

Truesdell, Clifford. 1954 Rational fluid mechanics, 1687–1765. In Leonhard Euler, *Opera omnia*, series 2, vol. 12 (Lausanne, 1954), I–CXXV.

——. 1968a Whence the law of moment of momentum? In *Essays in the history of mechanics* (Berlin: Springer), 239–71.

——. 1968b Early kinetic theory of gases. In *Essays in the history of mechanics* (Berlin: Springer), 272–304.

Uffink, Jos. 2007 Compendium to the foundations of classical statistical physics. In Jeremy Butterfield and John Earman (eds), *Handbook for the philosophy of physics* (Amsterdam Elsevier), 924–1074.

van Fraassen, Bas. 1980 *The scientific image*. Oxford: Clarendon Press.

——. 1991 *Quantum mechanics: An empiricist view*. Oxford: Clarendon Press.

Varignon, Pierre. 1687 *Projet d'une nouvelle méchanique avec un examen de l'opinion de M. Borelli, sur les propriétez des poids suspendus par des cordes*. Paris: Veuve Martin.

——. 1725 *Nouvelle mécanique ou statique, dont le projet fut donné en 1687*, 2 vols. Paris: Jombert.

Vey, Jacques. 1975 Déformation du crochet de Poisson sur une variété symplectique. *Commentarii mathematici Helvetici*, 50: 421–54.

Vilain, Christiane. 1996 *La mécanique de Christiaan Huygens: La relativité du mouvement au XVIIe siècle*. Paris: Blanchard.

——. 2000 La question du "centre d'oscillation" de 1660 à 1690; de 1703 à 1743. *Physis* 37: 21–51, 439–66.

Violle, Jules. 1883–92 *Cours de physique*, 2 vols. Paris: Masson.

Vitrac, Bernard. 1994 Introduction to Euclide, *Les éléments*, vol. 2, Livres V–VI. Paris: Presses Universitaires de France.

Vorms, Marion. 2011 *Qu'est-ce qu'une théorie scientifique?* Paris: Vuibert.

Waerden, Bartold van der. 1960 Exclusion principle and spin. In Markus Fierz and Viktor Weisskopf (eds), *Theoretical physics in the twentieth century* (New York: Interscience), 199–244.

Wald, Robert. 1986 Spin-two fields and general covariance. *PR*, D33: 3613–25.

Wallis, John. 1668 A summary account given by Dr. John Wallis of the general laws of motion. *PT*, 3: 864–6.

Walter, Scott. 1999 The non-Euclidean style of Minkowskian relativity. In Jeremy Gray (ed.), *The symbolic universe: Geometry and physics* (Oxford: Oxford University Press), 91–127.

——. 2008 Hermann Minkowski's approach to physics. *Mathematische Semesterberichte*, 55: 213–35.

——. 2009 Hypothesis and convention in Poincaré's defense of Galilei spacetime. In Michael Heidelberger and Gregor Schiemann (eds), *The significance of the hypothetical in the natural sciences* (Berlin: de Gruyter), 193–219.

Weinberg, Steven. 1972 *Gravitation and cosmology: Principles and applications of the general theory of relativity*. New York: John Wiley.

Wessels, Linda. 1979 Schrödinger's route to wave-mechanics. *SHPS*, 10: 311–40.

Weyl, Hermann. 1917 Zur Gravitationstheorie. *AP*, 54: 117–45.

——. 1918a *Raum · Zeit · Materie: Vorlesungen über allgemeine Relativitätstheorie*. Berlin: Springer.

——. 1918b Reine Infinitesimalgeometrie. *Mathematische Zeitschrift*, 2: 384–411. Also in Weyl 1968, vol. 2, 1–28.

——. 1918c Gravitation und Elektrizität. *BB* (1918), 465–80. Also in Weyl 1968, vol. 2, 29–42.

——. 1919a *Raum · Zeit · Materie: Vorlesungen über allgemeine Relativitätstheorie*, 3rd edn. Berlin: Springer.

——. 1919b Introduction and commentary to Bernhardt Riemann, *Über die Hypothesen, welche der Geometrie zu Grunde liegen. Neu hrsg. und erläutert von H. Weyl*, iii–iv, 23–48. Berlin: Springer. Also in Riemann 1990, 739–68.

——. 1921a *Raum · Zeit · Materie: Vorlesungen über allgemeine Relativitätstheorie*, 4th edn. Berlin: Springer.

Weyl, Hermann. 1921b Zur Infinitesimalgeometrie: Einordnung der projektiven und konformen Auff-
    assung. *GN* (1921), 99–112. Also in Weyl 1968, vol. 2, 195–207.
——. 1922a Das Raumproblem. *Jahresbericht der Deutschen Mathematikervereinigung*, 31: 205–21. Also in
    Weyl 1968, vol. 2, 328–44.
——. 1922b Die Einzigartigkeit der Pythagoreischen Maßbestimmung. *Mathematische Zeitschrift*, 12: 114–
    46. Also in Weyl 1968, vol. 2, 263–95.
——. 1923 *Raum · Zeit · Materie: Vorlesungen über allgemeine Relativitätstheorie*, 5th edn. Berlin: Springer.
——. 1927 Quantenmechanik und Gruppentheorie. *ZP*, 46: 1–46.
——. 1931 Geometrie und Physik. *Die Naturwissenschaften*, 19: 49–58. Also in Weyl 1968, vol. 3, 336–45.
——. 1968 *Gesammelte Abhandlungen*, 4 vols. Ed. by K. Chandrasekharan. Berlin: Springer.
Wheaton, Bruce. 1983 *The tiger and the shark: Empirical roots of wave-particle dualism*. Cambridge:
    Cambridge University Press.
Wheeler, John Archibald. 1990 Information, physics, quantum: The search for links. In Wojciech Zurek
    (ed.), *Complexity, entropy, and the physics of information* (Redwood City, CA: Addison-Wesley), 3–28.
Wigner, Eugen. 1932 On the quantum correction for thermodynamic equilibrium. *PR*, 40: 749–59.
——. 1960 The unreasonable effectiveness of mathematics in the natural sciences (Richard Courant lecture
    in mathematical sciences delivered at New York University, May 11, 1959). *Communications on pure
    and applied mathematics*, 13: 1–14.
Wilce, Alexander. 2009 Quantum logic and probability theory. In Edward N. Zalta (ed.), *The
    Stanford encyclopedia of philosophy* (Spring 2009 edn). Available at http://plato.stanford.edu/
    archives/spr2009/entries/qt-quantlog/ (last accessed Jan. 2013).
Wilczek, Frank. 2006 Reasonably effective I: Deconstructing a miracle. *Physics today*, 59N11: 8–9.
——. 2007 Reasonably effective II: Devil's advocate. *Physics today*, 60N5: 8–9.
Winnie, John. 1977 The causal theory of space-time. In Earman, Glymour, and Stachel 1977, 134–205.
Winter, Eduard and Maria Winter. 1957 *Die Register der Berliner Akademie der Wissenschaften,1746–1766:
    Dokumente für das Wirken Leonhard Eulers in Berlin zum 250. Geburtstag*. Berlin: Akademie-Verlag.
Wise, Norton. 1979 The mutual embrace of electricity and magnetism. *Science*, 203: 1310–18.
——. 1981 German concepts of force, energy, and electromagnetic ether. In G. Cantor and M. J. S. Hodge
    (eds), *Conceptions of ether: Studies in the history of ether theories, 1740–1900* (Cambridge: Cambridge
    University Press), 269–307.
Wren, Christoffer. 1668 Lex naturae de collisione corporum. *PT*, 3: 867–8.
Zachos, Cosmas, David Fairlie, and Thomas Curtright. 2005 *Quantum mechanics in phase space: An
    overview with selected papers*. Singapore: World Scientific.
Zeilinger, Anton. 1999 A foundational principle for quantum theory. *FP*, 29: 631–43.

INDEX